## PHYSICAL CONSTANTS AND CONVERSION FACTORS

| | | |
|---|---|---|
| Avogadro's number | $N_A$ | $6.022 \times 10^{23}$ particles/mole |
| Fundamental charge | $e$ | $1.602 \times 10^{-19}$ C |
| Speed of light | $c$ | $2.998 \times 10^{8}$ m/s |
| Permittivity of empty space | $\varepsilon_0$ | $8.854 \times 10^{-12}$ $C^2/N{\cdot}m^2$ |
| Permeability of empty space | $\mu_0$ | $4\pi \times 10^{-7}$ $N/A^2$ |
| Planck's constant | $h$ | $6.626 \times 10^{-34}$ J·s |
| | | $4.136 \times 10^{-21}$ MeV·s |
| Electron mass | $m_e$ | $9.1095 \times 10^{-31}$ kg |
| | | 0.5110 $MeV/c^2$ |
| Proton mass | $m_p$ | $1.673 \times 10^{-27}$ kg |
| | | 938.29 $MeV/c^2$ |
| Neutron mass | $m_n$ | $1.675 \times 10^{-27}$ kg |
| | | 939.57 $MeV/c^2$ |
| Bohr magneton | $\mu_B$ | $9.268 \times 10^{-24}$ $A{\cdot}m^2$ |
| 1 eV = $1.602 \times 10^{-19}$ J | | 1 tesla (T) = $10^4$ gauss (G) |

## SPECIAL FUNCTIONS AND SERIES

Delta Function: $\delta(x-a) = 0 \text{ for } x \neq a \qquad \delta(x-a) = \delta(a-x)$

$$\left.\begin{aligned} &\int \delta(x-a)\,dx = 1 \\ &\int f(x)\,\delta(x-a)\,dx = f(a) \end{aligned}\right\} \text{for } x \text{ integration limits that encompass } a$$

Theta Function:

$$\theta(x-a) = \begin{cases} 1, & \text{for } x > a \\ 0, & \text{for } x < a \end{cases} \qquad \frac{d}{dx}\theta(x-a) = \delta(x-a)$$

Fourier Series:

$$f(x) = \sum_{n=0}^{\infty}\left(A_n \cos\frac{n\pi x}{a} + B_n \sin\frac{n\pi x}{a}\right) \qquad A_0 = \frac{1}{2a}\int_{-a}^{+a} f(x)\,dx$$

$$A_n = \frac{1}{a}\int_{-a}^{+a} f(x)\cos\frac{n\pi x}{a}\,dx \quad (n \neq 0) \qquad B_n = \frac{1}{a}\int_{-a}^{+a} f(x)\sin\frac{n\pi x}{a}\,dx$$

Legendre Series:

$$\Phi(r, \theta) = \sum_n (A_n r^n + B_n r^{-n-1})P_n \text{ cc}$$

# Classical Electricity and Magnetism

## A Contemporary Perspective

Vernon D. Barger

Martin G. Olsson

University of Wisconsin—Madison

Allyn and Bacon, Inc. Boston London Sydney Toronto

*To our families:*
Annetta, Victor, Amy, and Andrew
Sallie, Marybeth, Nina, and Anne

---

**Library of Congress Cataloging-in-Publication Data**

Barger, V. (Vernon), 1938–
Classical electricity and magnetism.

Includes index.
1. Electromagnetism. I. Olsson, Martin, 1938–
II. Title.
QC760.B27 1987 537 86-7960
ISBN 0-205-08758-2

Printed in the United States of America.
10 9 8 7 6 5 4 3 2 1 90 89 88 87 86

**Acknowledgments:** Figure 2.16 reprinted by permission: 25 MV tandem electrostatic accelerator, Holifield Heavy Ion Research Facility, Oak Ridge National Laboratory. Figure 10.11 reprinted by permission: Lick Observatory Photo.

# Contents

# Preface

This book is a comprehensive junior–senior-level college text. A textbook on electromagnetism at the intermediate level should present the experimental foundations of the theory in a straightforward manner, with emphasis on the fundamental ideas. At the same time, a variety of practical applications should illustrate the concepts in an interesting and understandable way. These were our standards in writing this book.

A conventional development of the subject motivated by experimental discoveries is pursued. Our approach is to first establish the theory in empty space and then make the generalization to matter by including induced charges and currents. Chapters 1, 2, and 3 deal with electric fields in empty space, and Chapter 4 is devoted to electric polarization. Chapters 5 and 6 deal with magnetic fields in empty space, and Chapter 7 is devoted to magnetism in matter. Maxwell's equations are developed and wave solutions obtained in Chapter 8. Chapter 9 gives a description of how matter affects wave propagation. In the chapters on matter, the macroscopic descriptions are presented first, then microscopic models are introduced to explain the molecular origins of the macroscopic phenomena. In Chapter 11 the implications of electromagnetic theory for the nature of space and time are explored. The basic concepts of relativity and covariant methods are introduced.

The textbook is designed to permit maximum flexibility in a junior–senior-level course. The contents of most chapter sections are self-contained so that the material can be included, omitted, or used in interchangeable order, as desired. The application sections are interspersed in the text. The textbook is suitable for either a one- or a two-semester course.

Mathematical methods involving vector calculus and the delta function are integrated into the text as necessitated by the physics. For students who are unfamiliar with the mathematics, this approach is a more motivated approach than one that sets the mathematics apart as a separate chapter at the beginning of the course. For a class in which the students have a strong mathematical background, the proofs of the vector calculus integral theorems can be omitted.

Numerous natural phenomena and practical applications are discussed. Typical examples are the electric lens, electrostatic energy release in nuclear fission, electrostatic precipitation, atmospheric electricity, flip coil, magnetic mirrors, rainbows, pulsars, and antenna design. Computer graphics are utilized to give precise renditions of field and potential configurations. A carefully chosen range of problems is given at the end of each chapter to sharpen the student's understanding of the material covered.

It is a pleasure to acknowledge numerous helpful interactions with our colleagues. Professor C. Goebel generously acted as a continuous advisor and substantially influenced the presentation. W. F. Long and J. Madsen provided valuable assistance in problem solving, graphics, and editing. Several colleagues provided useful feedback after using preliminary versions of the manuscript in courses. Many thanks go to L. Dolan and to S. Quatrini for typing the numerous drafts of the manuscript.

# 1 Electric Field and Potential from Static Charges

CHAPTER CONTENTS

Matter in the universe interacts through only four basic forces, according to present understanding. The strong force binds nucleons together to form nuclei of atoms; without this force there would be no atomic nuclei. The electromagnetic forces bind electrons to nuclei to form atoms and molecules. The weak force causes radioactive decay and makes nuclear energy generation in stars possible. The gravitational force governs the motion of planets and stars and causes the accretion of cosmic matter to form these bodies. The four forces have vastly different strengths and ranges of interaction. The strengths of the strong to electromagnetic to weak to gravitational forces are roughly in the ratios $1:10^{-2}:10^{-10}:10^{-38}$ at low energy. The strong and weak forces fall exponentially with distance; their ranges are effectively of order $10^{-15}$ m and $10^{-18}$ m, respectively. Hence the strong and weak forces are irrelevant for interactions at distances of the atomic scale, $10^{-10}$ m, or larger. The electromagnetic and gravitational forces are said to be of infinite range, although their strength declines as the square of the distance between two interacting particles. Since the strength of the electromagnetic force between elementary particles is overwhelmingly larger than that of gravity, it is the most important force for understanding the composition and behavior of matter. The normal state of matter is electrically neutral, consisting of an equal number of positively and negatively charged particles (protons and electrons), and the universe itself is electrically neutral. The gravity force is significant for massive objects only because of the cancellation of electric forces in neutral matter.

The evolution of classical electromagnetic theory required over a century of experimental and theoretical effort. The development of the subject was highly phenomenological, with empirical laws for electric and magnetic phenomena deduced from experimental observations. The culmination of the classical theory took place about 1865 when James Clerk Maxwell succeeded in describing the apparently different phenomena of electricity, magnetism, and optics in terms of a single unified theory. The set of differential equations for electric and magnetic fields postulated by Maxwell encompassed all of the known experimental laws of electromagnetism. The Maxwell theory also correctly predicted the existence of electromagnetic waves and provided the framework for Einstein's theory of special relativity. The nineteenth-century discoveries of the physical laws of electromagnetism made possible the industrial technology of the twentieth century.

For macroscopic phenomena the Maxwell equations are exact, as far as we know. Space probes confirm the classical Maxwell theory out to the order of 100 earth radii by measurements of the earth's magnetic field. At molecular and smaller distances electromagnetism is correctly described by Maxwell's equations in a quantum mechanical framework. This theory of quantum electrodynamics developed from 1925 to 1950 has survived all experimental challenges down to the smallest distances probed, $\approx 10^{-18}$ m. Thus a truly universal electromagnetic theory exists that is valid over a scale range of at least 26 orders of magnitude. The electromagnetic force is the best understood force in all of physics and serves as an invaluable probe in the study of all natural phenomena.

In this book we introduce the fundamentals of electromagnetic theory and demonstrate its application to a variety of natural and practical phenomena. This first chapter introduces the Lorentz electromagnetic force, electric fields from static sources, and the scalar potential from which the electric field can be calculated.

## 1-1 LORENTZ FORCE

The electromagnetic force on a particle with charge $q$ and with valocity $\mathbf{v}$ at position $\mathbf{r}$ and time $t$ is found experimentally to be proportional to $q$ and to have two parts, one independent of $\mathbf{v}$ and the other proportional to $v$, as follows:

**Lorentz Force**

$$\mathbf{F}_q(\mathbf{r}, t) = q[\mathbf{E}(\mathbf{r}, t) + \mathbf{v} \times \mathbf{B}(\mathbf{r}, t)] \tag{1-1}$$

This equation is the Lorentz force law. Here **E** is called the electric field and **B** is called the magnetic field. The **E** and **B** fields are determined by charges and currents in a way that we will look into later. In Equation (1-1) $\mathbf{v} \times \mathbf{B}$ is the vector cross product of $\mathbf{v}$ and **B**, given in Cartesian coordinates by

$$\mathbf{v} \times \mathbf{B} = (v_y B_z - v_z B_y)\hat{\mathbf{x}} + (v_z B_x - v_x B_z)\hat{\mathbf{y}} + (v_x B_y - v_y B_x)\hat{\mathbf{z}} \tag{1-2}$$

The electric part of the Lorentz force, independent of the particle's velocity, can be separated by measuring the force when the charge is at rest. The magnetic part of the force is proportional to the magnitude of the velocity and perpendicular to both the velocity direction and the magnetic field direction. The Lorentz force law can be tested by observing the behavior of a beam of electrons in an oscilloscope tube in the presence of electromagnetic fields. A quantitative measure of electric charge will have to await a discussion and definition of units in Section 5-4.

In general, $\mathbf{E}(\mathbf{r}, t)$ and $\mathbf{B}(\mathbf{r}, t)$ are defined at all $\mathbf{r}$. Any physical quantity that can be specified at all points in space is called a field—hence the terminology *electromagnetic fields* for **E** and **B**.

For a charge $q$ that is moving uniformly, the **E** and **B** fields that appear in the Lorentz force law are due entirely to charges other than $q$. However, if the charge $q$ is accelerating, the field of $q$ itself also contributes in the electromagnetic force. This additional contribution is usually negligible in practice.

The Lorentz force law, Equation (1-1), provides a way to measure the fields **E** and **B** due to charges other than $q$. However, the charge $q$ at the point **r** will exert forces on the charges responsible for the fields and hence move them from where they would have been if the charge $q$ had been elsewhere, which in turn changes the fields **E** and **B** at the point **r** (this phenomenon is known as polarization). A simple way to avoid polarization is to make the magnitude of the charge $q$ very small; it is then known as a test charge.

## 1-2 APPLICATION: ELECTRON MOTION IN CROSSED E AND B FIELDS

As an application of the Lorentz force law, we discuss the motion of an electron in constant electric and magnetic fields that are perpendicular to each other. This configuration, called crossed fields, is used in a number of devices.

When $q$ is the electron charge $-e$ in the Lorentz force of Equation (1–1), the Newton equation of motion is

$$m\ddot{\mathbf{r}} = (-e)(\mathbf{E} + \dot{\mathbf{r}} \times \mathbf{B}) \tag{1-3}$$

where

$$\dot{\mathbf{r}} = \frac{d\mathbf{r}}{dt} \quad \text{and} \quad \ddot{\mathbf{r}} = \frac{d^2\mathbf{r}}{dt^2}$$

The Cartesian components of this equation are

$$\begin{aligned} \ddot{x} &= -\frac{e}{m}(E_x + \dot{y}B_z - \dot{z}B_y) \\ \ddot{y} &= -\frac{e}{m}(E_y + \dot{z}B_x - \dot{x}B_z) \\ \ddot{z} &= -\frac{e}{m}(E_z + \dot{x}B_y - \dot{y}B_x) \end{aligned} \tag{1-4}$$

where $(E_x, E_y, E_z)$ are the components of **E**. We choose our coordinate system such that the crossed **E** and **B** fields are

$$\mathbf{E} = E\hat{\mathbf{y}} \qquad \mathbf{B} = B\hat{\mathbf{z}} \tag{1-5}$$

as illustrated in Figure 1–1. The equations of motion simplify to

$$\ddot{x} = -\omega_c \dot{y} \qquad \ddot{y} = -\omega_c\left(\frac{E}{B} - \dot{x}\right) \qquad \ddot{z} = 0 \tag{1-6}$$

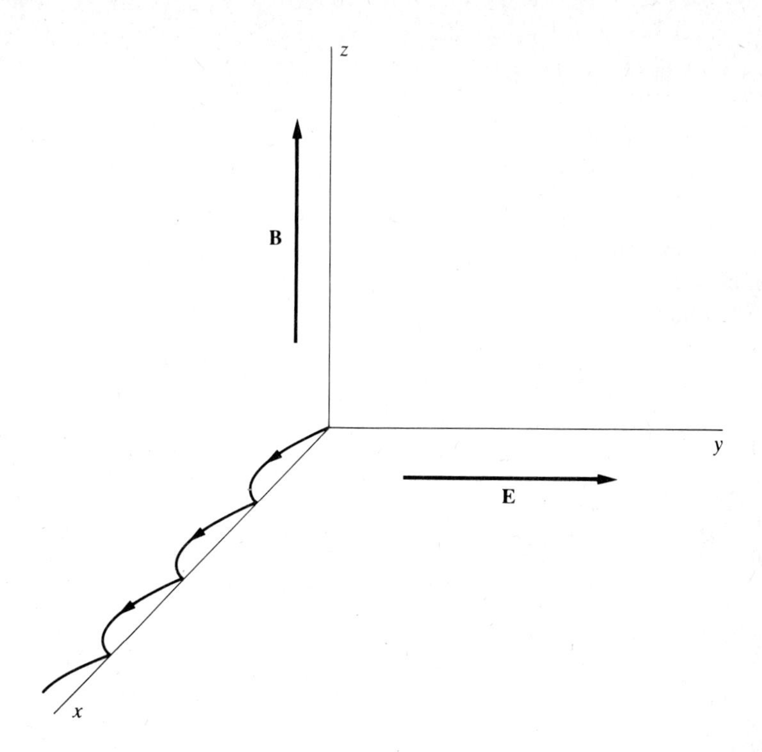


**FIGURE 1-1** Cycloid trajectory of a charged particle moving in crossed electric and magnetic fields

where we have introduced the definition

$$\omega_c \equiv \frac{e}{m} B \tag{1-7}$$

The motion along the $z$ direction is uniform, namely,

$$z = v_z^0 t + z^0 \tag{1-8}$$

We use the zero superscripts to denote initial values at time $t = 0$. For simplicity we take the initial position of the electron to be at $x = 0$, $y = 0$, $z = 0$. A first integral over time can immediately be carried out in Equations (1–6), giving

$$\dot{x} = -\omega_c y + v_x^0 \qquad \dot{y} = -\omega_c \frac{E}{B} t + \omega_c x + v_y^0 \tag{1-9}$$

When Equations (1–9) are substituted into Equations (1–6), we arrive at the uncoupled differential equations

$$\begin{aligned} \ddot{x} + \omega_c^2 x &= \omega_c^2 \frac{E}{B} t - \omega_c v_y^0 \\ \ddot{y} + \omega_c^2 y &= -\omega_c \frac{E}{B} + \omega_c v_x^0 \end{aligned} \tag{1-10}$$

These equations can be solved by standard methods. The homogeneous solutions are simple harmonic in $\sin \omega_c t$ and $\cos \omega_c t$. The particular solutions can be obtained by inspection. The complete solutions that satisfy the initial conditions are

$$\begin{aligned} x &= \frac{1}{\omega_c}\left(v_x^0 - \frac{E}{B}\right) \sin \omega_c t + \frac{v_y^0}{\omega_c}(\cos \omega_c t - 1) + \frac{E}{B} t \\ y &= \frac{v_y^0}{\omega_c} \sin \omega_c t - \frac{1}{\omega_c}\left(v_x^0 - \frac{E}{B}\right)(\cos \omega_c t - 1) \end{aligned} \tag{1-11}$$

This result can be verified by substitution into Equations (1–6) and (1–9).

We notice that the $xy$ motion satisfies the equation

$$\left(x + \frac{v_y^0}{\omega_c} - \frac{E}{B} t\right)^2 + \left(y - \frac{v_x^0}{\omega_c} + \frac{E}{\omega_c B}\right)^2 = \left[\left(v_x^0 - \frac{E}{B}\right)^2 + (v_y^0)^2\right]\left(\frac{1}{\omega_c}\right)^2 \tag{1-12}$$

which is the locus of a circle of radius

$$R = \frac{1}{\omega_c}\left[\left(v_x^0 - \frac{E}{B}\right)^2 + (v_y^0)^2\right]^{1/2} \tag{1-13}$$

with center at $xy$ coordinates

$$x_c = -\frac{v_y^0}{\omega_c} + \frac{E}{B} t \qquad y_c = \frac{1}{\omega_c}\left(v_x^0 - \frac{E}{B}\right) \tag{1-14}$$

For a magnetic field alone ($\mathbf{E} = 0$) the center of the circle is stationary and has the radius

$$R = \frac{v_\perp}{\omega_c} = \frac{m v_\perp}{eB} = \frac{p_\perp}{eB} \tag{1-15}$$

where $v_\perp$ is the velocity component perpendicular to $\mathbf{B}$,

$$v_\perp = [(v_x^0)^2 + (v_y^0)^2]^{1/2} \tag{1-16}$$

and $p_\perp$ is the corresponding momentum. Hence the angular frequency

$$\omega_c = \frac{v_\perp}{R} = \frac{e}{m} B \tag{1-17}$$

is just the familiar *cyclotron frequency* for circular motion in a constant $B$ field.

For a simpler derivation of Equation (1–15) we note that for circular motion the magnetic force is constant in magnitude and always perpendicular to the velocity. Thus it must balance the centrifugal force

$$\frac{mv_{\perp}^{2}}{R} = ev_{\perp}B \tag{1-18}$$

This result is the same as Equation (1–15). When the particle has the $z$ component of motion in Equation (1–8), the general trajectory in a magnetic field is the composition of the circular $xy$ motion and the uniform translational $z$ motion. The trajectory is thus a cylindrical helix with axis oriented along **B**, as illustrated in Figure 1–2.

Bending magnets used at nuclear and particle accelerators select particles of given momentum by passing the beam through a uniform magnetic field, as illustrated in Figure 1–3. The magnetic field is perpendicular to the beam direction and to the plane of desired deflection. Particles in the beam are moving nearly parallel but have different momenta. The radius of curvature in the magnetic field region is proportional to $p$, so a detector

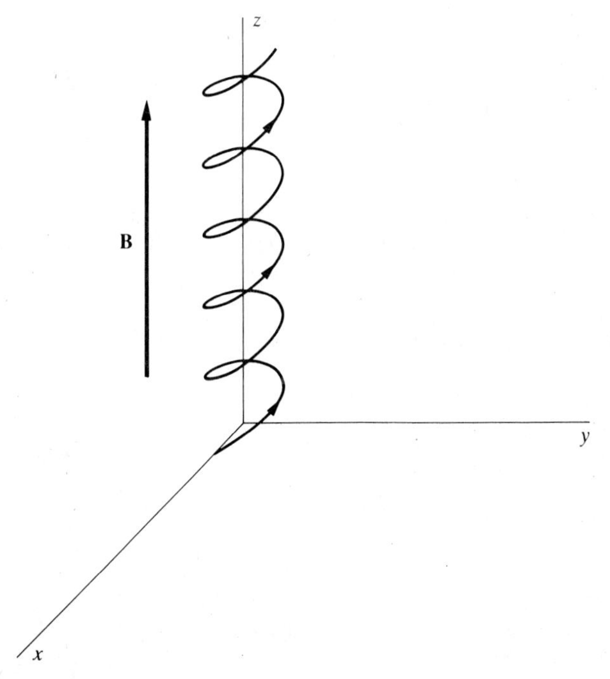


**FIGURE 1–2** Helical trajectory of a charged particle with an initial velocity component along the magnetic field

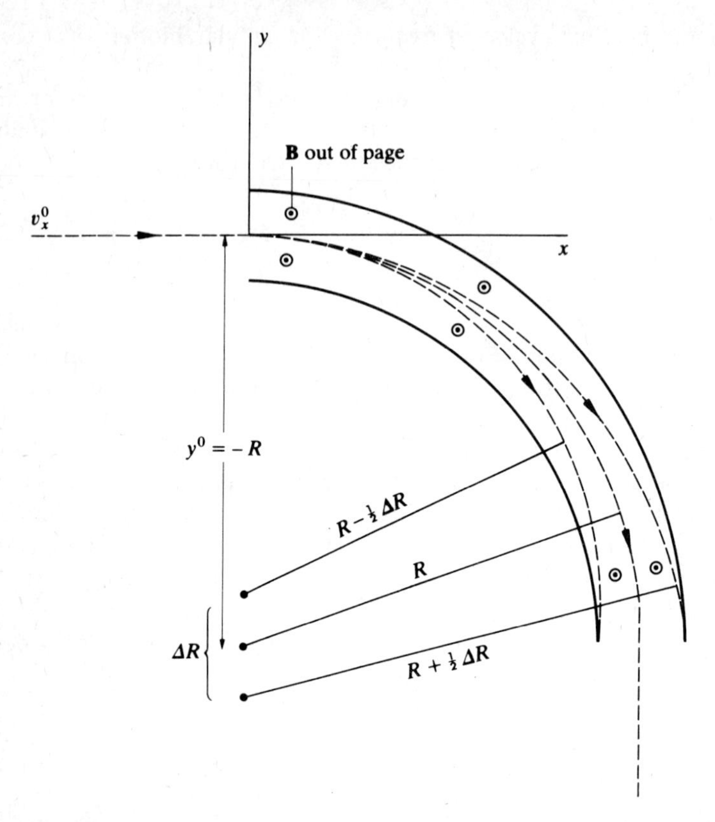


**FIGURE 1-3** Momentum selector used in nuclear and particle accelerator experiments

placed at $R$ with aperature $\Delta R$ will accept only particles with momentum between $p_1 = eB(R - \frac{1}{2}\Delta R)$ and $p_2 = eB(R + \frac{1}{2}\Delta R)$. Such momentum selection instruments are sometimes called momentum spectrometers or momentum analyzers.

For crossed electric and magnetic fields the general motion described by Equation (1–12) is a circular orbit that moves in the $x$ direction with *drift velocity*

$$v_D = \frac{E}{B} \tag{1-19}$$

Figure 1–4 illustrates the orbit in the $xy$ plane for various $v_x^0$ with $v_y^0 = 0$. For $v_x^0 = E/B$ the electron is undeflected from the $x$ axis. If $v_y^0$ is not zero, the motion is similar to the $v_y^0 = 0$ case illustrated in Figure 1–4 but shifted in the $xy$ plane.

We can understand the representative motion in Figure 1–5 qualitatively as follows. As the electron moves against **E**, it is accelerated. The

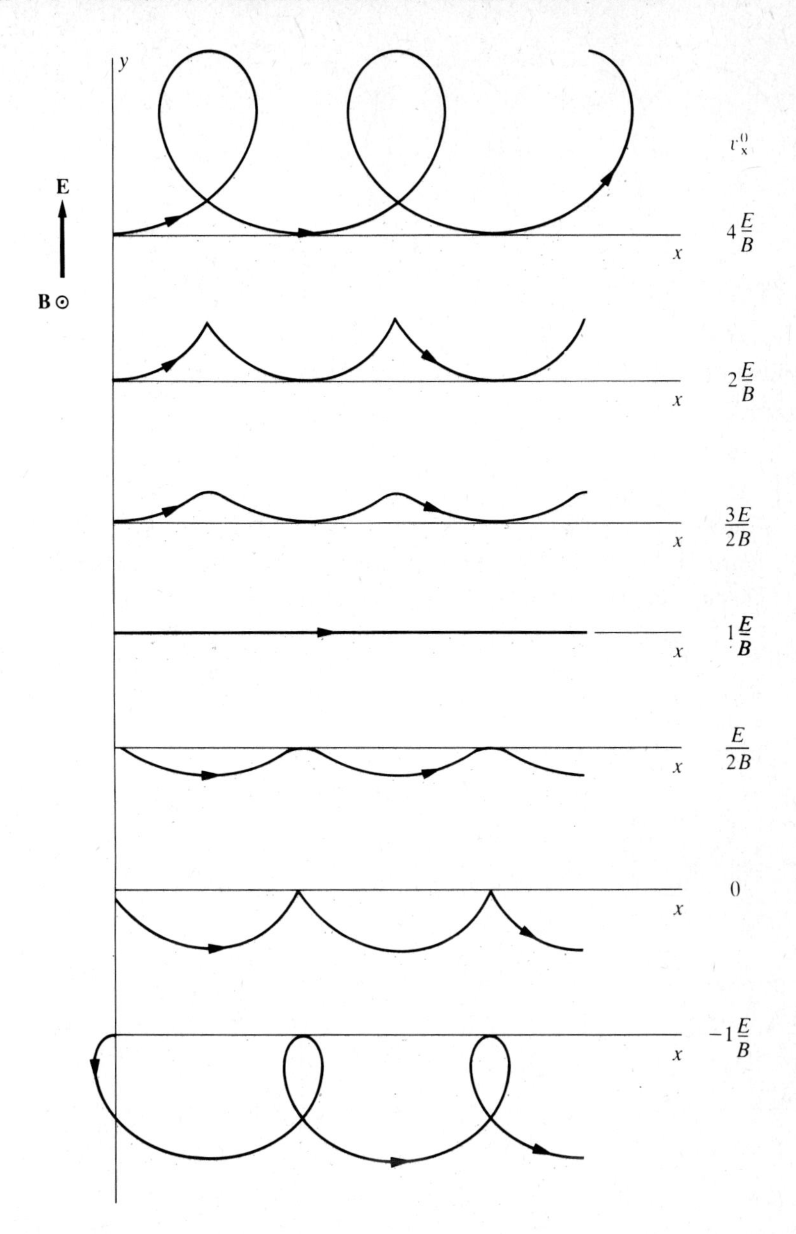


**FIGURE 1-4** Trajectories in the $xy$ plane of a charged particle in crossed $E_y$ and $B_z$ fields for various initial velocities along the $x$ axis

resulting higher velocity makes the orbit less curved by the magnetic field. As the electron moves along **E**, it is slowed down and curved more by **B**. On the average, the net result is a drift in the direction of **E** × **B**.

A cathode ray tube with crossed **E** and **B** fields was used by the

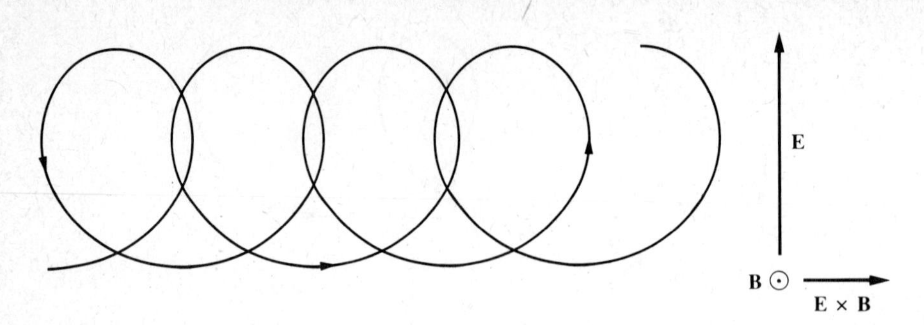


**FIGURE 1-5** Drift velocity of a charged particle in crossed electric and magnetic fields, in the **E** × **B** direction

English physicist J. J. Thomson in 1897 in his experimental discovery of the electron and the measurement of its charge-to-mass ratio, as illustrated in Figure 1–6. The speed of the particles emitted by the cathode was determined by balancing the magnetic and electric forces,

$$qvB = qE \tag{1-20}$$

so that the particles were undeflected. From the known field strengths this equation gives

$$v = \frac{E}{B} \tag{1-21}$$

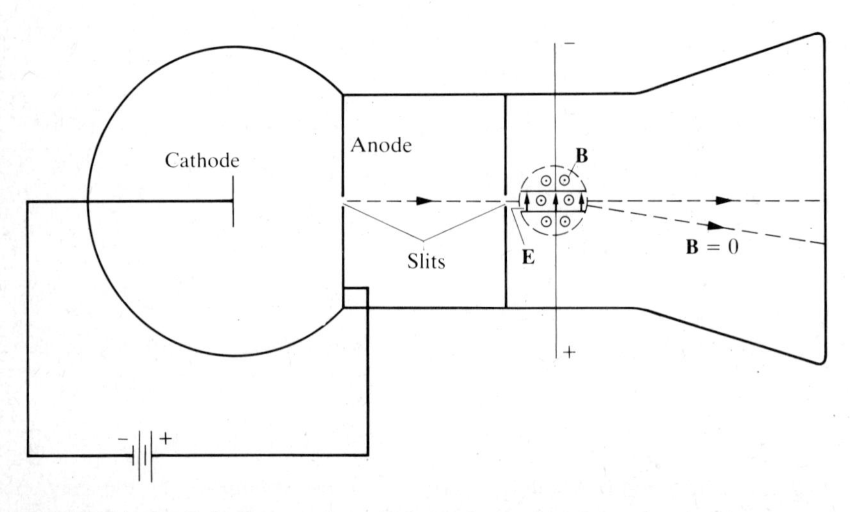


**FIGURE 1-6** Cathode ray tube, which can be used to determine the charge-to-mass ratio of electrons. The path of the electrons emitted from the cathode is determined by the relative strength of the electric and magnetic fields

which corresponds to Equation (1–11) with $v_y^0 = 0$, $v_x^0 = E/B$, $y = 0$, and $x = (E/B)t$. Knowing the speed of the particles, Thomson deduced the sign of the charge and the magnitude of the $q/m$ ratio from their deflection in an electric field or in a magnetic field. The experiment established that cathode rays consisted of particles (electrons) with a definite charge-to-mass ratio. Later, in 1909, Robert A. Millikan demonstrated in his classic oil drop experiment that electric charge is quantized in multiples of the electron charge. In SI units the electron charge is $-e = -(1.60210 \pm 0.00007) \times 10^{-19}$ C.

## 1-3 COULOMB'S LAW FOR THE ELECTRIC FIELD OF STATIC CHARGES

Up to this point we have been concerned with the force on a charged particle due to given electric and magnetic fields. We next turn our attention to the electric field arising from a system of static electric charges.

The static force between two charges $q$ and $q_1$ at positions $\mathbf{r}$ and $\mathbf{r}_1$, respectively, in empty space is empirically given by the Coulomb force law

**Coulomb Force Law**

$$\mathbf{F}_q(\mathbf{r}) = -\mathbf{F}_{q_1}(\mathbf{r}_1) = \frac{1}{4\pi\varepsilon_0}\frac{qq_1}{R_1^3}\mathbf{R}_1 = \frac{1}{4\pi\varepsilon_0}\frac{qq_1}{R_1^2}\hat{\mathbf{R}}_1 \qquad \textbf{(1-22)}$$

where $\mathbf{F}_q$ is the force on charge $q$, $\mathbf{F}_{q_1}$ is the force on charge $q_1$, and

$$\mathbf{R}_1 = \mathbf{r} - \mathbf{r}_1 \qquad R_1 = |\mathbf{r} - \mathbf{r}_1| \qquad \hat{\mathbf{R}}_1 = \frac{\mathbf{R}_1}{R_1} \qquad \textbf{(1-23)}$$

In a Cartesian coordinate system with unit vectors $\hat{\mathbf{x}}$, $\hat{\mathbf{y}}$, $\hat{\mathbf{z}}$ along the coordinate axes, we have

$$\begin{aligned} \mathbf{R}_1 &= (x - x_1)\hat{\mathbf{x}} + (y - y_1)\hat{\mathbf{y}} + (z - z_1)\hat{\mathbf{z}} \\ R_1 &= [(x - x_1)^2 + (y - y_1)^2 + (z - z_1)^2]^{1/2} \end{aligned} \qquad \textbf{(1-24)}$$

In Equation (1–22), $(4\pi\varepsilon_0)^{-1}$ is a proportionality constant whose value depends on the system of units for charge, distance, and force. We will show in Section 4–3 that in dielectric matter $\varepsilon_0$ is replaced by $\varepsilon$, an empirically determined parameter that depends on the nature of the dielectric. The charges $q$ and $q_1$ can be of either algebraic sign. Figure 1–7 illustrates the force between two like-sign charges.

The Coulomb force determines the basic atomic and molecular properties of matter. A hydrogen atom consists of a negatively charged electron bound to the positively charged proton by the attractive Coulomb

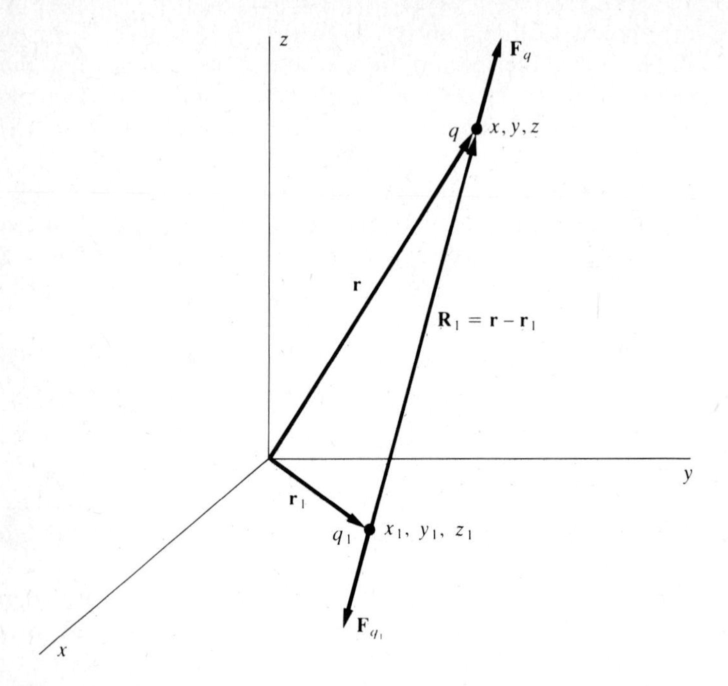


**FIGURE 1-7** Coulomb forces between two charges $q_1$ and $q$ of the same sign

force. Complex atoms, molecular structure, and chemical forces are also determined largely by the electrostatic force, though quantitative calculations using quantum mechanics can tax even the most sophisticated electronic computers.

The electrostatic force is involved in many industrial processes. For example, coatings are applied electrostatically to almost every manufactured object. The pages of this book have been coated to give a surface quality suitable for printing reproduction. Automobiles receive several successive coatings to protect the metal from corrosion. Even the clear plastic used for food wrappings in supermarkets may have been coated several times to enhance physical strength, moisture resistance, and heat-sealing properties. In electrostatic coating a fine mist of the coating material is given an electric charge, and the surface is oppositely charged. The mist is attracted to the surface in a controllable way to form the coating.

From Equations (1–1) and (1–22) the electric field at the static charge $q$ due to the charge $q_1$ is

$$\mathbf{E}(\mathbf{r}) = \frac{\mathbf{F}_q(\mathbf{r})}{q} = \frac{1}{4\pi\varepsilon_0}\frac{q_1}{R_1^3}\mathbf{R}_1 = \frac{1}{4\pi\varepsilon_0}\frac{q_1}{R_1^2}\hat{\mathbf{R}}_1 \qquad \textbf{(1-25)}$$

Since this result is a vector equation, it holds in any coordinate system. In a Cartesian system the electric field components are

$$E_x = \frac{q_1}{4\pi\varepsilon_0}\frac{(x - x_1)}{R_1^3} \qquad E_y = \frac{q_1}{4\pi\varepsilon_0}\frac{(y - y_1)}{R_1^3} \qquad E_z = \frac{q_1}{4\pi\varepsilon_0}\frac{(z - z_1)}{R_1^3} \tag{1-26}$$

For a static distribution of discrete charges at positions $\mathbf{r}_1, \mathbf{r}_2, \ldots, \mathbf{r}_n$, the electric field at a position $\mathbf{r}$ is given by a superposition of the electric fields from all charges:

**Coulomb's Law**

$$\mathbf{E}(\mathbf{r}) = \frac{1}{4\pi\varepsilon_0}\left(\frac{q_1\hat{\mathbf{R}}_1}{R_1^2} + \frac{q_2\hat{\mathbf{R}}_2}{R_2^2} + \cdots + \frac{q_n\hat{\mathbf{R}}_n}{R_n^2}\right) = \frac{1}{4\pi\varepsilon_0}\sum_{i=1}^{n}\frac{q_i\hat{\mathbf{R}}_i}{R_i^2} = \frac{1}{4\pi\varepsilon_0}\sum_{i=1}^{n}\frac{q_i\mathbf{R}_i}{R_i^3} \tag{1-27}$$

where a vector $\mathbf{R}_i = \mathbf{r} - \mathbf{r}_i$ points from the source position $\mathbf{r}_i$ to the field position $\mathbf{r}$; see Figure 1–8. An $\mathbf{E}$ field of the form of Equation (1–27) is called

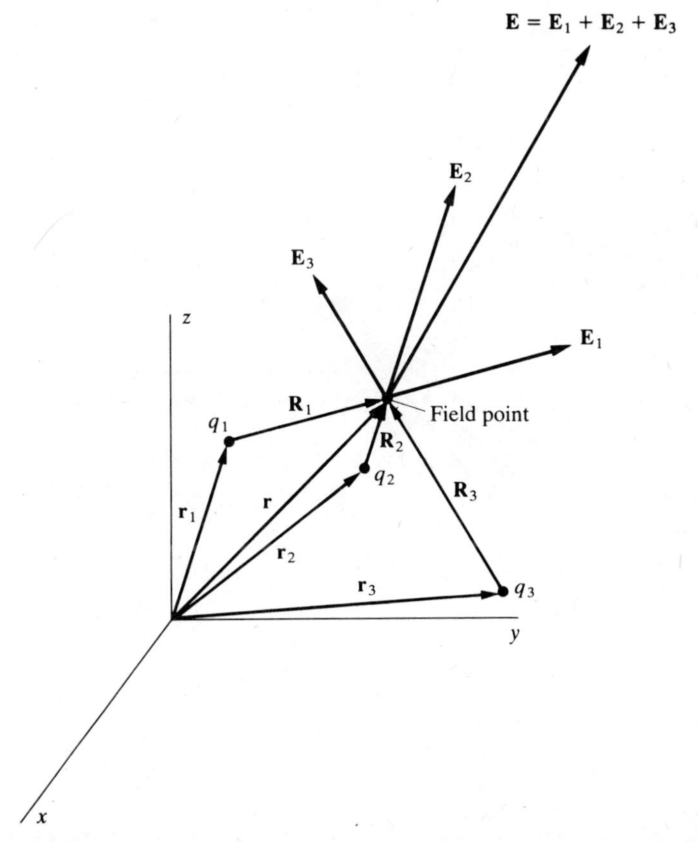


**FIGURE 1–8** Resultant electric field at a position **r** from a collection of charges; given by a superposition of the electric fields from all charges

an electrostatic field. From Equation (1–27), when two field sources $a$ and $b$ are simultaneously present, the resultant electric field is given by the vector sum of the fields produced separately by sources $a$ and $b$:

$$\mathbf{E} = \mathbf{E}_a + \mathbf{E}_b \tag{1-28a}$$

This *superposition principle* is, in fact, a general property of the complete electromagnetic theory described by Maxwell's equations. A similar property holds for magnetic fields:

$$\mathbf{B} = \mathbf{B}_a + \mathbf{B}_b \tag{1-28b}$$

Sometimes, a pictorial representation of **E** or **B** is used in which *lines* representing the *field* are drawn tangent everywhere to the direction of

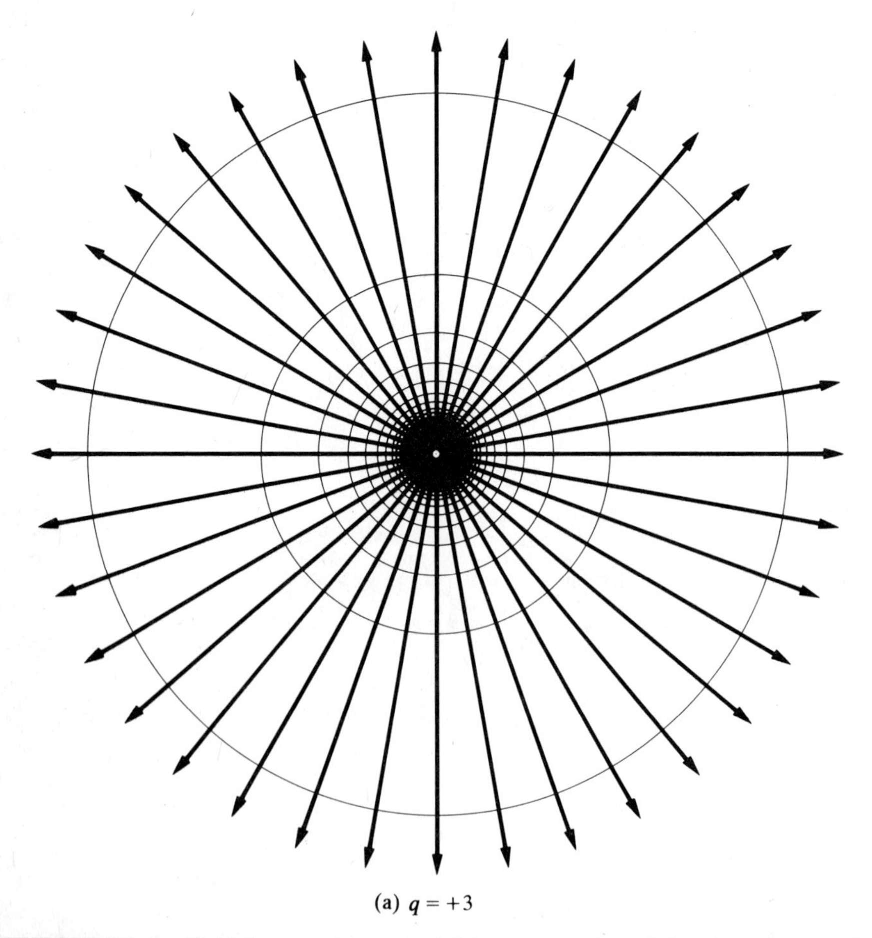

(a) $q = +3$

**FIGURE 1-9** Field lines used as a pictorial representation of electric and magnetic fields. The density of the lines is proportional to the strength of the field, as can be seen by comparing (a), where $q = +3$, with (b), where $q = -2$.

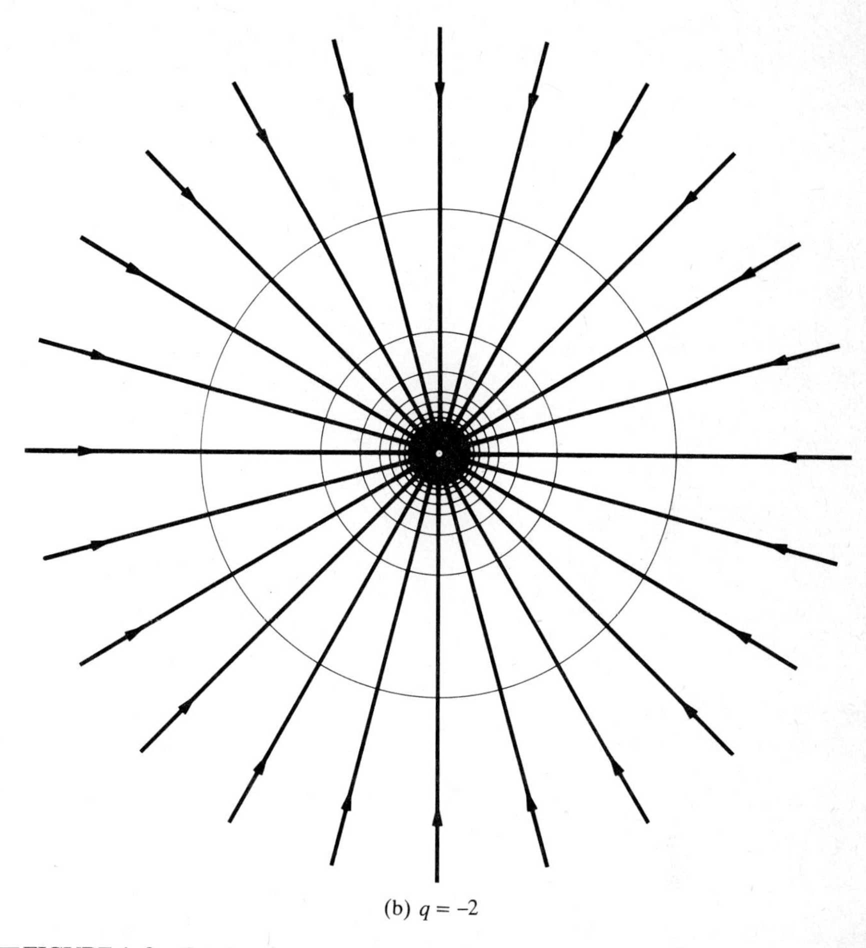

(b) $q = -2$

**FIGURE 1-9** *Continued*

the field vectors. The density of the lines is proportional to the magnitude of the field [see Equation (2–9)]. The electric field lines for a positive charge of $+3$ and a negative charge of $-2$ are separately illustrated in Figure 1–9. The field lines for the combined system of these charges are shown in Figure 1–10. The field line concept is often useful; its main weakness is that the superposition principle cannot be expressed in this picture. This weakness is evident from Figures 1–9 and 1–10, where the field lines shown have no obvious relation to a superposition of field lines from each of the individual charges. Field lines do not end except on charges. Since there are no magnetic charges, so far as is known, the lines of **B** must have no ends, although they are not necessarily closed.

In classical mechanics a conservative vector force can be expressed in terms of a scalar potential energy. A similar relation holds in electrostatics

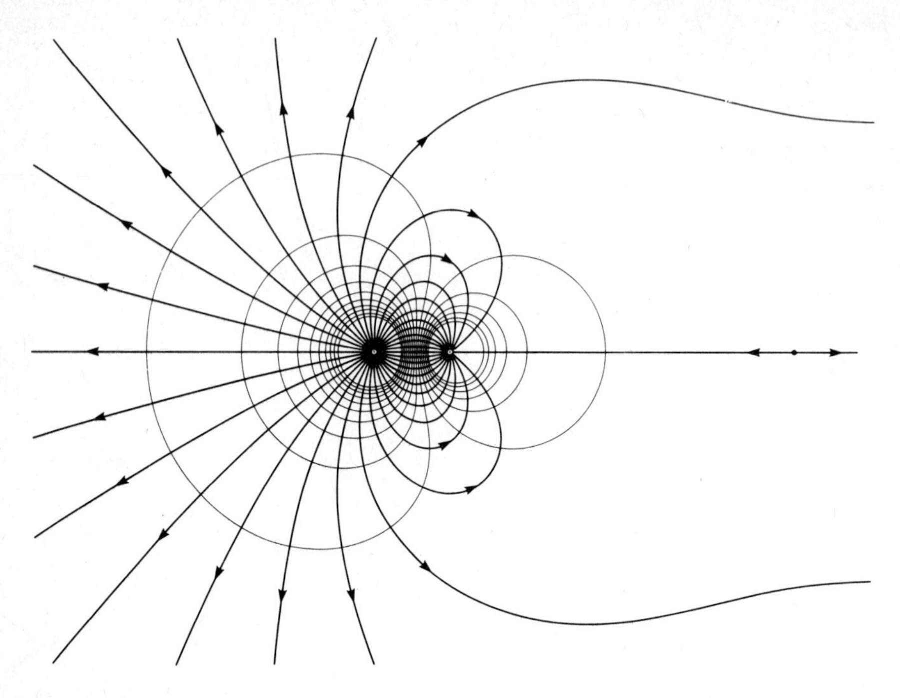

**FIGURE 1-10** Electric field lines (with arrows) and equipotentials for a system consisting of two charges of values $+3q$ and $-2q$. The electric field lines are shown only in a plane that contains the charges. The lines without arrows are the intersections of the equipotential surfaces with this plane.

between the vector electric field and a scalar potential. This relation is derived in the next section. Many electrostatic problems can be more conveniently solved in terms of the electric potential.

## 1-4 ELECTRIC POTENTIAL

According to Equation (1–1), the electromagnetic force on a charge $q$ is

$$\mathbf{F} = q(\mathbf{E} + \mathbf{v} \times \mathbf{B})$$

Consequently, the work done by the electromagnetic field on the charge as it moves from $\mathbf{r}_a$ at time $t_a$ to $\mathbf{r}_b$ at time $t_b$ along a path $\mathbf{r}(t)$ is

$$W = \int_{\mathbf{r}_a}^{\mathbf{r}_b} \mathbf{F} \cdot d\mathbf{r} = q \int_{\mathbf{r}_a}^{\mathbf{r}_b} \mathbf{E} \cdot d\mathbf{r} = q \int_{t_a}^{t_b} \mathbf{E}(\mathbf{r}(t), t) \cdot \frac{d\mathbf{r}}{dt}\, dt \qquad \textbf{(1-29)}$$

where in the last form the notation is more explicit. Note that the magnetic term makes no contribution because $(\mathbf{v} \times \mathbf{B}) \cdot \mathbf{v} = 0$. The line integral of Equation (1–29) is illustrated in Figure 1–11.

If $\mathbf{E}$ is an electrostatic field, $\mathbf{E} \cdot d\mathbf{r}$ is a perfect differential, and so the work equation, (1–29), depends only on the endpoints (i.e., $q\mathbf{E}$ is a conservative force). To show this result, we use the Coulomb law for $\mathbf{E}$, Equation (1–27),

$$\mathbf{E} \cdot d\mathbf{r} = \sum_i \frac{q_i}{4\pi\varepsilon_0} \frac{\mathbf{R}_i \cdot d\mathbf{r}}{R_i^3}, \qquad \mathbf{R}_i = \mathbf{r} - \mathbf{r}_i$$

Because the $\mathbf{r}_i$ are fixed, we can write $d\mathbf{r} = d\mathbf{R}_i$. Since

$$\mathbf{R}_i \cdot d\mathbf{R}_i = \tfrac{1}{2} d(\mathbf{R}_i \cdot \mathbf{R}_i) = \tfrac{1}{2} d(R_i^2) = R_i \, dR_i$$

we have, then,

$$\mathbf{E} \cdot d\mathbf{r} = \sum_i \frac{q_i}{4\pi\varepsilon_0} \frac{\mathbf{R}_i \cdot d\mathbf{R}_i}{R_i^3} = -\sum_i \frac{q_i}{4\pi\varepsilon_0} d\left(\frac{1}{R_i}\right)$$

which is a perfect differential. We define *electric potential*

$$\Phi(\mathbf{r}) \equiv \frac{1}{4\pi\varepsilon_0} \sum_i \frac{q_i}{R_i} = \frac{1}{4\pi\varepsilon_0} \sum_i \frac{q_i}{|\mathbf{r} - \mathbf{r}_i|} \tag{1-30}$$

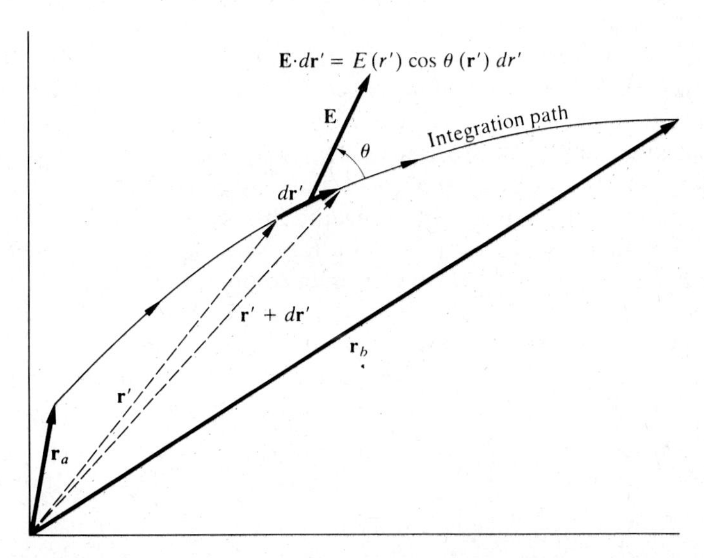


**FIGURE 1–11** Line integral of a vector field, given by the integral over the path of the vector field dotted into the differential path length. In Cartesian coordinates $\mathbf{E} \cdot d\mathbf{r}' = E_x dx' + E_y dy' + E_z dz'$ and the line integral is a sum of three rectilinear integrations.

so that

$$\mathbf{E} \cdot d\mathbf{r} = -d\Phi \tag{1-31}$$

Hence in an electrostatic field the work done on a charge as it moves from $\mathbf{r}_a$ to $\mathbf{r}_b$ is

$$W = q \int_{\mathbf{r}_a}^{\mathbf{r}_b} \mathbf{E} \cdot d\mathbf{r}' = -q[\Phi(\mathbf{r}_b) - \Phi(\mathbf{r}_a)] \tag{1-32}$$

In the absence of nonelectric forces the work is the difference in kinetic energies,

$$W = -q[\Phi(\mathbf{r}_b) - \Phi(\mathbf{r}_a)] = K(\mathbf{r}_b) - K(\mathbf{r}_a)$$

Since $\mathbf{r}_a$ and $\mathbf{r}_b$ are arbitrary points, this equation is equivalent to

$$K(\mathbf{r}) + q\Phi(\mathbf{r}) = \text{constant} \tag{1-33}$$

which has the interpretation of energy conservation for a charged particle moving in an electrostatic field, with the electrostatic potential energy of the charge given by

$$U = q\Phi(\mathbf{r}) \tag{1-34}$$

The term $\Phi(\mathbf{r})$ itself, the electrostatic potential energy per unit charge, is called the electrostatic potential.

From Equation (1–32) we see that only differences in potential are determined by $\mathbf{E}$. Hence we could have added an arbitrary constant to the definition of $\Phi(r)$ in Equation (1–30). With the choice in Equation (1–30) the potential is zero at $r = \infty$. More generally, by adding a constant term, we can arbitrarily specify the potential at some chosen reference point; the potential at other points is then given with respect to this reference value.

We can obtain an expression for $\mathbf{E}$ in terms of the scalar potential $\Phi$ by expressing Equation (1–31) in Cartesian components, using

$$d\mathbf{r} = dx\,\hat{\mathbf{x}} + dy\,\hat{\mathbf{y}} + dz\,\hat{\mathbf{z}} \tag{1-35a}$$

$$\mathbf{E} = E_x\hat{\mathbf{x}} + E_y\hat{\mathbf{y}} + E_z\hat{\mathbf{z}} \tag{1-35b}$$

$$d\Phi = \frac{\partial\Phi}{\partial x}dx + \frac{\partial\Phi}{\partial y}y + \frac{\partial\Phi}{\partial z}dz \equiv \nabla\Phi \cdot d\mathbf{r} \tag{1-35c}$$

The resulting equation,

$$E_x\,dx + Ey\,dy + E_z\,dz = -\frac{\partial\Phi}{\partial x}dx - \frac{\partial\Phi}{\partial y}dy - \frac{\partial\Phi}{\partial z}dz \tag{1-36a}$$

implies

$$E_x = -\frac{\partial \Phi}{\partial x} \qquad E_y = -\frac{\partial \Phi}{\partial y} \qquad E_z = -\frac{\partial \Phi}{\partial z} \tag{1-36b}$$

In vector form this equation can be written as

$$\mathbf{E} = -\left(\hat{\mathbf{x}}\frac{\partial}{\partial x} + \hat{\mathbf{y}}\frac{\partial}{\partial y} + \hat{\mathbf{z}}\frac{\partial}{\partial z}\right)\Phi(\mathbf{r}) \tag{1-37}$$

The operator in parentheses is a vector operator (see Problem 1–12) commonly called the *gradient* or *del* operator:

$$\nabla \equiv \hat{\mathbf{x}}\frac{\partial}{\partial x} + \hat{\mathbf{y}}\frac{\partial}{\partial y} + \hat{\mathbf{z}}\frac{\partial}{\partial z} \tag{1-38}$$

Thus **E** is the negative of the gradient of the potential:

$$\mathbf{E} = -\nabla\Phi(\mathbf{r}) \tag{1-39}$$

Since **E** and $\Phi$ are related by a linear operation [e.g., doubling $\Phi(\mathbf{r})$ implies $\mathbf{E}(r)$ also is doubled, etc.], the principle of superposition is valid for sources of $\Phi$; see Equation (1–30). In calculations of **E** it is often simpler to first calculate $\Phi$ and then take the gradient, because $\Phi$ is a single quantity, whereas **E** has three components.

A spatial surface on which $\Phi(\mathbf{r})$ = constant is known as an *equipotential surface*. According to Equation (1–39), the electric field vanishes in all directions tangential to an equipotential surface going through that point. Hence the direction of **E** is normal to the equipotential surface. A family of equipotential surfaces provides a pictorial description of the field. Once the equipotential surfaces are drawn, the electric field lines can be easily sketched by their orthogonality to equipotentials. Equipotentials and electric field lines for particular examples are illustrated in Figures 1–9 and 1–10.

The potential due to a point charge $q_1$ at $\mathbf{r}_1$ is

$$\Phi(\mathbf{r}) = \frac{q_1}{4\pi\varepsilon_0 R_1} \tag{1-40}$$

and the electric field expression is

$$\mathbf{E}(\mathbf{r}) = -\frac{q_1}{4\pi\varepsilon_0}\nabla_{\mathbf{r}}\left(\frac{1}{R_1}\right) \tag{1-41}$$

where

$$R_1 = |\mathbf{r} - \mathbf{r}_1| = [(x - x_1)^2 + (y - y_1)^2 + (z - z_1)^2]^{1/2}$$

The subscript **r** on $\boldsymbol{\nabla}$ denotes the variable of the differential del operator. The Cartesian components of Equation (1–41) are

$$E_x = \frac{q_1}{4\pi\varepsilon_0} \frac{1}{R_1^2} \frac{\partial R_1}{\partial x} = \frac{q_1}{4\pi\varepsilon_0} \frac{(x - x_1)}{R_1^3}$$

$$E_y = \frac{q_1}{4\pi\varepsilon_0} \frac{1}{R_1^2} \frac{\partial R_1}{\partial y} = \frac{q_1}{4\pi\varepsilon_0} \frac{(y - y_1)}{R_1^3} \tag{1-42}$$

$$E_z = \frac{q_1}{4\pi\varepsilon_0} \frac{1}{R_1^2} \frac{\partial R_1}{\partial z} = \frac{q_1}{4\pi\varepsilon_0} \frac{(z - z_1)}{R_1^3}$$

Noting that

$$\mathbf{R} = (x - x_1)\hat{\mathbf{x}} + (y - y_1)\hat{\mathbf{y}} + (z - z_1)\hat{\mathbf{z}}$$

we can write Equations (1–42) in vector form as

$$\mathbf{E} = \frac{q_1}{4\pi\varepsilon_0} \frac{\mathbf{R}_1}{R_1^3} = \frac{q_1}{4\pi\varepsilon_0} \frac{\hat{\mathbf{R}}_1}{R_1^2}$$

which corresponds to Equation (1–25).

The *curl* $\boldsymbol{\nabla} \times \mathbf{E}$ of a vector field **E** is the cross product of the del operator of Equation (1–38) with **E**:

$$\boldsymbol{\nabla} \times \mathbf{E}(\mathbf{r}) \equiv \begin{vmatrix} \hat{\mathbf{x}} & \hat{\mathbf{y}} & \hat{\mathbf{z}} \\ \dfrac{\partial}{\partial x} & \dfrac{\partial}{\partial y} & \dfrac{\partial}{\partial z} \\ E_x & E_y & E_z \end{vmatrix} \tag{1-43}$$

The curl of an electrostatic field vanishes:

$$\boldsymbol{\nabla} \times \mathbf{E} = -\boldsymbol{\nabla} \times \boldsymbol{\nabla}\Phi = 0 \tag{1-44}$$

Since the electric potential of an electrostatic field is single-valued [e.g., given by Equation (1–40) for a point charge], the line integral of $\mathbf{E} \cdot d\mathbf{r}$ around a closed path vanishes, by Equation (1–32):

$$\oint \mathbf{E} \cdot d\mathbf{r} = 0 \tag{1-45}$$

Here the circle on the integral sign denotes a closed path. Fields for which Equation (1–45) holds or equivalently for which the curl is zero are called *conservative*. The **E** field of moving charges does not satisfy $\boldsymbol{\nabla} \times \mathbf{E} = 0$, and $\mathbf{E} = -\boldsymbol{\nabla}\Phi$ does not hold. We will consider the general nonstatic case later.

## 1-5 DEL OPERATOR IN SPHERICAL AND CYLINDRICAL COORDINATES

To determine **E** from $\Phi$ for a charge distribution with spherical symmetry, we will want an expression for the gradient in spherical coordinates. To derive this expression, we first use Equations (1–35a) and (1–35c) in Cartesian coordinates to write

$$d\Phi = d\mathbf{r} \cdot \nabla\Phi \tag{1-46}$$

The transformation from Cartesian to spherical coordinates is

$$x = r \sin\theta \cos\phi \qquad y = r \sin\theta \sin\phi \qquad z = r\cos\theta \tag{1-47}$$

From Figure 1–12 the differential of the coordinate vector in spherical coordinates is

$$d\mathbf{r} = \hat{\mathbf{r}}\, dr + \hat{\boldsymbol{\theta}} r\, d\theta + \hat{\boldsymbol{\phi}} r \sin\theta\, d\phi \tag{1-48}$$

where ($\hat{\mathbf{r}}$, $\hat{\boldsymbol{\theta}}$, $\hat{\boldsymbol{\phi}}$) are orthogonal unit vectors in the directions of increasing ($r$, $\theta$, $\phi$). From Equations (1–46) and (1–48) we have

$$d\Phi = dr\, \hat{\mathbf{r}} \cdot \nabla\Phi + r\, d\theta\, \hat{\boldsymbol{\theta}} \cdot \nabla\Phi + r \sin\theta\, d\phi\, \hat{\boldsymbol{\phi}} \cdot \nabla\Phi \tag{1-49}$$

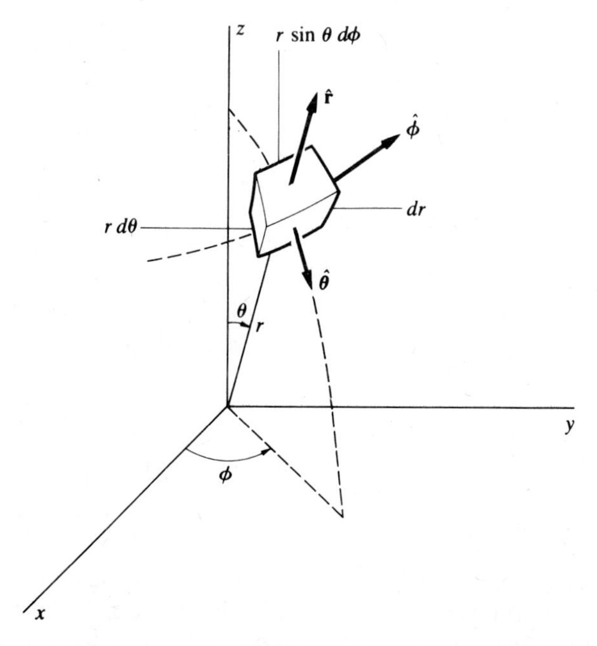


**FIGURE 1–12** Differential volume element in spherical coordinates ($r$, $\theta$, $\phi$)

The expression for the total differential $d\Phi$ in terms of partial differentials is

$$d\Phi = \frac{\partial \Phi}{\partial r}\,dr + \frac{\partial \Phi}{\partial \theta}\,d\theta + \frac{\partial \Phi}{\partial \phi}\,d\phi \tag{1-50}$$

Equality of Equations (1–49) and (1–50) requires

$$\hat{\mathbf{r}} \cdot \nabla = \frac{\partial}{\partial r} \qquad \hat{\boldsymbol{\theta}} \cdot \nabla = \frac{1}{r}\frac{\partial}{\partial \theta} \qquad \hat{\boldsymbol{\phi}} \cdot \nabla = \frac{1}{r \sin \theta}\frac{\partial}{\partial \phi} \tag{1-51}$$

Hence from Equation (1–51) the gradient operator in spherical coordinates is

$$\nabla = \hat{\mathbf{r}}\frac{\partial}{\partial r} + \hat{\boldsymbol{\theta}}\frac{1}{r}\frac{\partial}{\partial \theta} + \hat{\boldsymbol{\phi}}\frac{1}{r \sin \theta}\frac{\partial}{\partial \phi} \tag{1-52}$$

As an alternative evaluation of the $\nabla$ operation in Equation (1–41), we take the differential of the equation

$$\mathbf{R}_1 = \mathbf{r} - \mathbf{r}_1 \tag{1-53}$$

for a fixed source point $\mathbf{r}_1$ to get

$$d\mathbf{R}_1 = d\mathbf{r} \tag{1-54}$$

This result implies that

$$\nabla_{\mathbf{R}_1} = \nabla_{\mathbf{r}} \tag{1-55}$$

Then Equation (1–41) becomes

$$\mathbf{E}(\mathbf{r}) = -\frac{q_1}{4\pi\varepsilon_0}\nabla_{\mathbf{R}_1}\left(\frac{1}{R_1}\right) \tag{1-56}$$

Since $\mathbf{r}_1$ is constant, we can evaluate the gradient in a coordinate system centered at $\mathbf{r}_1$ (or equivalently, $\mathbf{R}_1 = 0$). Using the spherical coordinate form of del in Equation (1–52) with origin at $\mathbf{R}_1 = 0$, we obtain

$$\begin{aligned} \mathbf{E}(\mathbf{r}) &= -\frac{q_1}{4\pi\varepsilon_0}\hat{\mathbf{R}}_1\frac{\partial}{\partial R_1}\left(\frac{1}{R_1}\right) \\ &= \frac{q_1}{4\pi\varepsilon_0}\frac{\hat{\mathbf{R}}_1}{R_1^2} \end{aligned} \tag{1-57}$$

An analogous calculation to that above for spherical coordinates can be carried out to obtain $\nabla$ in cylindrical coordinates $(\imath, \phi, z)$. The relation between Cartesian and cylindrical coordinates is

$$x = \imath \cos \phi \qquad y = \imath \sin \phi \qquad z = z \tag{1-58}$$

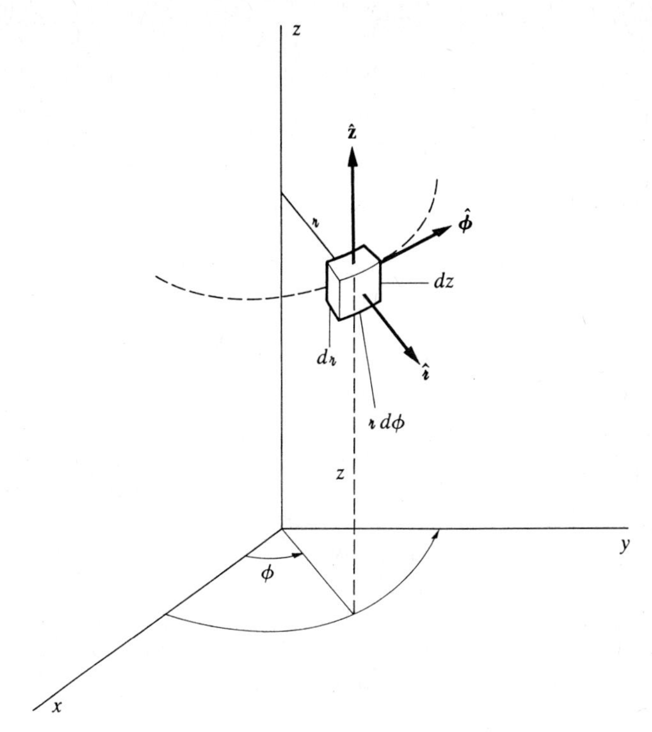


**FIGURE 1-13** Differential volume element in cylindrical coordinates ($r$, $\phi$, $z$)

The form of the $\boldsymbol{\nabla}$ operator in cylindrical coordinates is

$$\boldsymbol{\nabla} = \hat{\mathbf{r}}\frac{\partial}{\partial r} + \hat{\boldsymbol{\phi}}\frac{1}{r}\frac{\partial}{\partial \phi} + \hat{\mathbf{z}}\frac{\partial}{\partial z} \tag{1-59}$$

where ($\hat{\mathbf{r}}$, $\hat{\boldsymbol{\phi}}$, $\hat{\mathbf{z}}$) are the cylindrical coordinate unit vectors, as illustrated in Figure 1–13. In the following section we discuss an electrostatic problem with cylindrical symmetry.

## 1-6 APPLICATION: ELECTRIC LENS

An electric field with axial symmetry can be used as an electric lens to focus a beam of electrons, just as an optical lens can be used to focus light. Particle focusing by electric or magnetic fields has a wide range of applications. Electrostatic lenses are used in the cathode ray tubes of oscilloscopes and in electron microscopes. These lenses make electrons that are emitted from a point source to converge at another point (e.g., on the screen of an oscilloscope).

The focusing property of a converging-light lens is illustrated in Figure 1–14 for light rays originating from a point $P$ on the lens axis. For

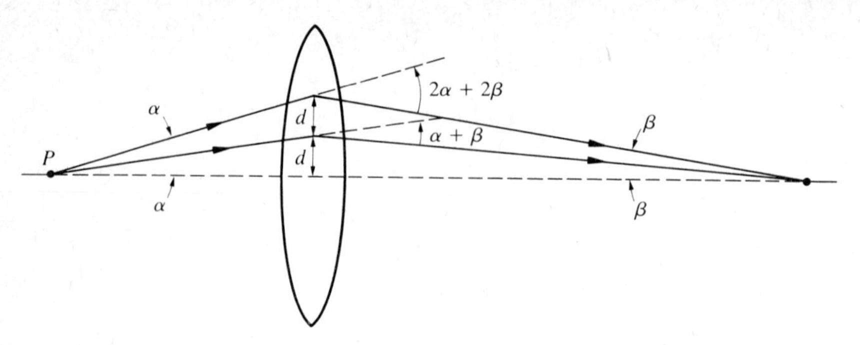


**FIGURE 1-14** Light rays originating from a point $P$ on the lens axis focused by a converging optical lens

rays making a small angle with the axis, the angular deflection $\alpha + \beta$ is proportional to the distance $d$ of the rays from the axis. The trajectories of electrons can be deflected in a similar way by an axially symmetric electric field. With the electric lens the deflection occurs in a continuous manner over the entire trajectory. The focal length of the electric lens can be changed merely by changing the electric field intensity. Unfortunately, aberrations are generally greater and more difficult to correct in electric lenses than in optical lenses.

An example of an electrostatic lens is sketched in Figure 1–15. Three successive conducting cylinders of equal radii are charged to potentials $V_0$, $V_1$, $V_0$, respectively. Figure 1–15 depicts typical electric field lines within the lens for the case $V_1 > V_0$. The field lines are directed from regions of higher potential to lower potential. The electric field is strongest at the junctions between the conductors, where the potential changes most rapidly. Each gap acts as a lens. There are two gaps for the practical reason that both ends of the tube can then be grounded.

To understand the operation of the electric lens, we consider electrons incident from the left of the first gap in Figure 1–15. A blowup of this gap is shown in Figure 1–16. The electric field can be resolved into radial and axial components, $E_{\imath}$ and $E_z$. By symmetry, $E_{\imath} = 0$ at $\imath = 0$. Close to the axis $E_{\imath}$ can be approximated by the first term in a power series expansion in $\imath$ as

$$E_{\imath}(\imath, z) = C(z)\imath \qquad \textbf{(1-60)}$$

We denote the region to the left of the gap by $A$ and the region to the right of the gap by $B$. From the direction of the **E** field lines in Figure 1–16, we conclude that the electric force $\mathbf{F} = -e\mathbf{E}$ on the electrons satisfies

$$\begin{aligned} F_{\imath}(A) &< 0 \qquad F_z(A) > 0 \\ F_{\imath}(B) &> 0 \qquad F_z(B) > 0 \end{aligned} \qquad \textbf{(1-61)}$$

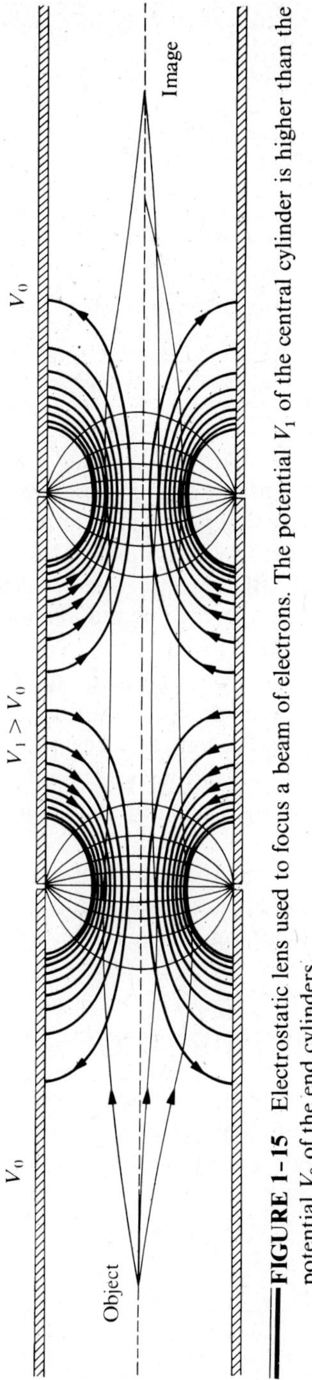


**FIGURE 1–15** Electrostatic lens used to focus a beam of electrons. The potential $V_1$ of the central cylinder is higher than the potential $V_0$ of the end cylinders.

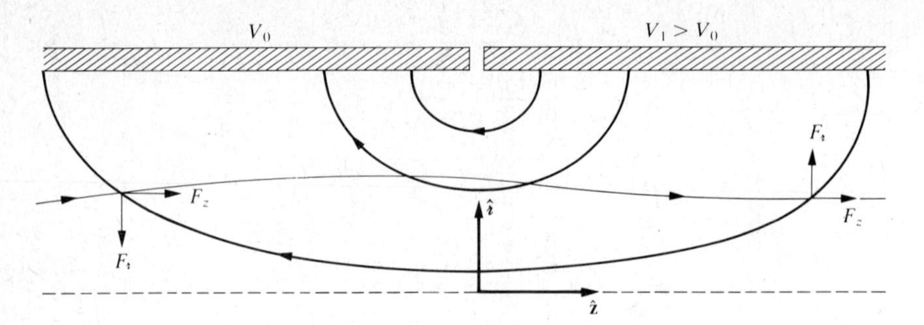


**FIGURE 1-16** Detail of the first gap in the electrostatic lens where the potential changes

Thus region $A$ gives convergence and region $B$ divergence of the electron trajectory. For electrons that are initially moving nearly parallel to the axis and that undergo small deflections, the focusing property is simple to establish. The electrons are further from the axis in $A$ than in $B$; and since $F_t$ is proportional to $t$, $|F_t(A)|$ is greater than $|F_t(B)|$. The electrons also spend less time in $B$ than in $A$ because of the acceleration by $F_z$. The radial impulse from region $A$ is therefore greater, and the net effect is a deflection of the electrons toward the axis. The reader can make a similar analysis to show that the second gap in Figure 1–15 also leads to convergence toward the axis. As long as the first-order approximation in Equation (1–60) is valid and the electrons have the same initial momentum, all trajectories from the incident focal point are imaged to the same point on the axis by the geometric lens property of Figure 1–14.

The lens in Figure 1–15 will focus an electron beam for either $V_1 > V_0$ or $V_1 < V_0$. If $V_1$ is varied while $V_0$ is held constant, the focal length can be varied without changing the electron velocity on either side of the lens.

## 1-7 CONTINUOUS CHARGE DISTRIBUTIONS

Frequently, the number of charges in even a small region of space is so large that to carry out a summation over the point charges in calculating the electric field is impractical. Then a convenient procedure is to introduce a continuous-charge distribution. If the charge in a small region of volume $\Delta V_i$ at position $\mathbf{r}_i$ is $\Delta q_i$, the resultant electric potential is

$$\Phi(\mathbf{r}) = \frac{1}{4\pi\varepsilon_0}\sum_i \frac{\Delta q_i}{|\mathbf{r} - \mathbf{r}_i|} = \frac{1}{4\pi\varepsilon_0}\sum_i \frac{\rho_i\,\Delta V_i}{|\mathbf{r} - \mathbf{r}_i|} \tag{1-62}$$

where $\rho_i \equiv \Delta q_i/\Delta V_i$ is the average charge density in region $i$. This equation is valid if we observe the average of the field over regions of the order of the

$\Delta V_i$. If we make the regions infinitesimally small, the summation over the discrete points $\mathbf{r}_i$ is replaced by an integral over a continuous source variable $\mathbf{r}'$:

$$\Phi(\mathbf{r}) = \frac{1}{4\pi\varepsilon_0}\int \frac{dq(\mathbf{r}')}{|\mathbf{r} - \mathbf{r}'|} = \frac{1}{4\pi\varepsilon_0}\int \frac{\rho(\mathbf{r}')\,dV'}{|\mathbf{r} - \mathbf{r}'|} \tag{1-63}$$

where the integration extends over all space. Of course, regions where $\rho = 0$ contribute nothing and can be omitted.

The charge density $\rho(\mathbf{r}')$ has been defined by

$$dq(\mathbf{r}') \equiv \rho(\mathbf{r}')\,dV' \tag{1-64}$$

where $dV'$ is the differential volume element at position $\mathbf{r}'$. If the charges lie on a surface, a convenient technique is to define a surface charge density $\sigma(\mathbf{r}')$ by

$$dq(\mathbf{r}') \equiv \sigma(\mathbf{r}')\,dS' \tag{1-65}$$

where $dS'$ is the differential surface area element at $\mathbf{r}'$. Likewise, for charges distributed on a line, we define a line charge density $\lambda(\mathbf{r}')$ by

$$dq(\mathbf{r}') \equiv \lambda(\mathbf{r}')\,dr' \tag{1-66}$$

where $dr'$ is a differential line element. Equation (1–63) for the potential takes the various forms

$$\Phi(\mathbf{r}) = \frac{1}{4\pi\varepsilon_0}\int \frac{\rho(\mathbf{r}')}{R}\,dV' \tag{1-67a}$$

$$\Phi(\mathbf{r}) = \frac{1}{4\pi\varepsilon_0}\int \frac{\sigma(\mathbf{r}')}{R}\,dS' \tag{1-67b}$$

$$\Phi(\mathbf{r}) = \frac{1}{4\pi\varepsilon_0}\int \frac{\lambda(\mathbf{r}')}{R}\,dr' \tag{1-67c}$$

where $R = |\mathbf{r} - \mathbf{r}'|$.

From the expression for $\Phi(\mathbf{r})$ the corresponding result for $\mathbf{E}(\mathbf{r})$ can be obtained by taking the gradient of $\Phi(\mathbf{r})$. For example, we consider the potential of Equation (1–67a) for a three-dimensional charge distribution:

$$\mathbf{E}(\mathbf{r}) = -\nabla_{\mathbf{r}}\Phi(\mathbf{r}) = -\frac{1}{4\pi\varepsilon_0}\int \rho(\mathbf{r}')\,\nabla_{\mathbf{r}}\left(\frac{1}{R}\right)dV' \tag{1-68}$$

An $\mathbf{r}$ subscript on $\nabla$ is used here to label the differentiation variable. The $\nabla$ operator could be moved inside the integral because the variable $\mathbf{r}'$ and the limits of the integration are independent of $\mathbf{r}$. To evaluate $\nabla_{\mathbf{r}}(1/R)$, we express $R$ in Cartesian coordinates:

$$\mathbf{R} = (x - x')\hat{\mathbf{x}} + (y - y')\hat{\mathbf{y}} + (z - z')\hat{\mathbf{z}}$$
$$R = [(x - x')^2 + (y - y')^2 + (z - z)^2]^{1/2} \tag{1-69}$$

For fixed $x'$, $y'$, and $z'$, the partial derivatives of $R$ are

$$\frac{\partial R}{\partial x} = \frac{x - x'}{R} \qquad \frac{\partial R}{\partial y} = \frac{y - y'}{R} \qquad \frac{\partial R}{\partial z} = \frac{z - z'}{R} \tag{1-70}$$

Thus for fixed $\mathbf{r}'$ the gradient of $1/R^n$ with respect to $\mathbf{r}$ is

$$\begin{aligned}\nabla_{\mathbf{r}}\left(\frac{1}{R^n}\right) &= \left(\hat{\mathbf{x}}\frac{\partial}{\partial x} + \hat{\mathbf{y}}\frac{\partial}{\partial y} + \hat{\mathbf{z}}\frac{\partial}{\partial z}\right)\left(\frac{1}{R^n}\right) \\ &= -\frac{n}{R^{n+1}}\left(\hat{\mathbf{x}}\frac{\partial R}{\partial x} + \hat{\mathbf{y}}\frac{\partial R}{\partial y} + \hat{\mathbf{z}}\frac{\partial R}{\partial z}\right) \\ &= -\frac{n}{R^{n+2}}[(x - x')\hat{\mathbf{x}} + (y - y')\hat{\mathbf{y}} + (z - z')\hat{\mathbf{z}}] \\ &= -\frac{n\mathbf{R}}{R^{n+2}} = -\frac{n\hat{\mathbf{R}}}{R^{n+1}}\end{aligned} \tag{1-71}$$

These derivatives are well defined only for $R \neq 0$.

A simpler method for deriving Equation (1–71) comes from the realization that the differential of the equation $R = r - r'$ at fixed $r'$ yields $dR = dr$, from which

$$\nabla_{\mathbf{r}}\left(\frac{1}{R^n}\right) = \nabla_{\mathbf{R}}\left(\frac{1}{R^n}\right) = -\frac{n\hat{\mathbf{R}}}{R^{n+1}}$$

Inserting Equation (1–71) for $n = 1$ in Equation (1–68), we obtain expressions for the electric fields resulting from continuous distributions:

$$\mathbf{E}(\mathbf{r}) = \frac{1}{4\pi\varepsilon_0}\int \frac{\hat{\mathbf{R}}}{R^2}\rho(\mathbf{r}')\,dV' \tag{1-72a}$$

$$\mathbf{E}(\mathbf{r}) = \frac{1}{4\pi\varepsilon_0}\int \frac{\hat{\mathbf{R}}}{R^2}\sigma(\mathbf{r}')\,dS' \tag{1-72b}$$

$$\mathbf{E}(\mathbf{r}) = \frac{1}{4\pi\varepsilon_0}\int \frac{\hat{\mathbf{R}}}{R_2}\lambda(\mathbf{r}')\,dr' \tag{1-72c}$$

where $\mathbf{R} = \mathbf{r} - \mathbf{r}'$ and $\hat{\mathbf{R}} = \mathbf{R}/R$. We could have directly obtained these electric field expressions by generalizing Equation (1–27) to continuous charge distributions.

## 1-8 APPLICATION: ELECTRIC FIELD OF AN INFINITE CHARGE SHEET

As an example application with a continuous charge distribution, we calculate the electric field due to a flat sheet of infinite extent that has a uniform surface charge density $\sigma$. We will use cylindrical coordinates with the location of the origin in the charge sheet directly below the observation point, as illustrated in Figure 1–17. From Equation (1–72b) the electric field is

$$\mathbf{E}(z) = \frac{\sigma}{4\pi\varepsilon_0} \int \frac{\hat{\mathbf{R}}}{R^2}\, dS' \tag{1-73}$$

To do the integration, we consider a differential area element $dS'$ at a distance $r'$ from the origin:

$$dS' = (r'\, d\phi')\, dr' \tag{1-74}$$

The direction of **E** is along a line from $dS'$ to the field point and can be resolved into components perpendicular and parallel to the plane. When we calculate

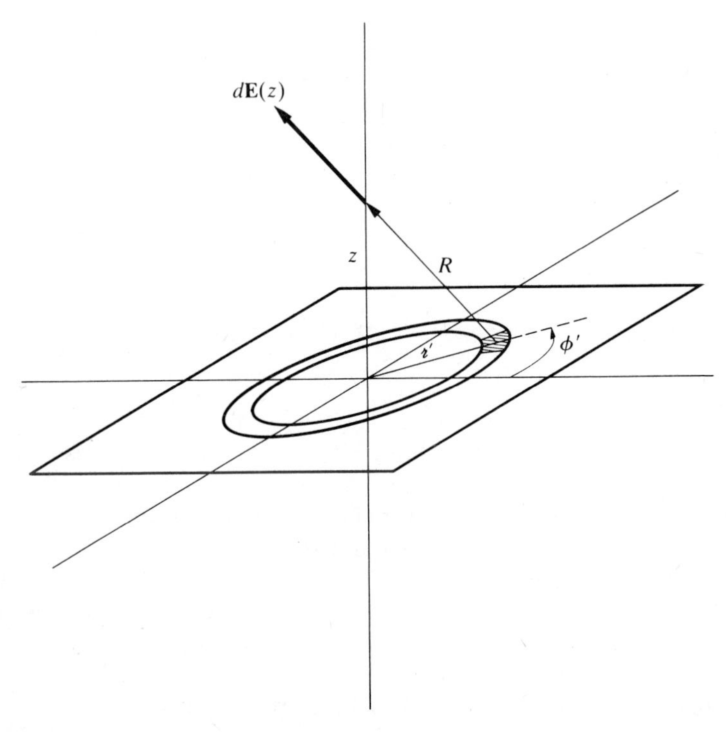


**FIGURE 1-17** Differential electric field due to a differential area element on an infinite sheet of charge

the field due to a circular ring of charge (i.e., integrating over $\phi'$ from 0 to $2\pi$ at fixed $\imath'$), the component of **E** parallel to the plane integrates to zero by symmetry. The nonvanishing component of **E** is

$$E_z = \frac{\sigma}{4\pi\varepsilon_0} \int \frac{\hat{\mathbf{R}} \cdot \hat{\mathbf{z}}}{R^2} \imath' \, d\imath' \, d\phi' \tag{1-75}$$

From the geometry of Figure 1–17 we see that

$$R^2 = \imath'^2 + z^2 \qquad \hat{\mathbf{R}} \cdot \hat{\mathbf{z}} = \frac{z}{R}$$

Since the integrand is $\phi'$-independent, the $\phi'$ integral gives $2\pi$; and the $\imath'$ integral is

$$E_z = \frac{\sigma z}{2\varepsilon_0} \int_0^\infty \frac{\imath' \, d\imath'}{(\imath'^2 + z^2)^{3/2}} \tag{1-76}$$

We find

$$E_z = -\frac{\sigma z}{2\varepsilon_0} \frac{1}{\sqrt{z^2 + \imath'^2}} \bigg|_0^\infty = \frac{\sigma}{2\varepsilon_0} \frac{z}{|z|} \tag{1-77}$$

Thus for $z > 0$

$$E_z = \frac{\sigma}{2\varepsilon_0} \tag{1-78}$$

and for $z < 0$

$$E_z = \frac{-\sigma}{2\varepsilon_0} \tag{1-79}$$

For a positive charge density the electric field points away from the sheet and is independent of the distance from the sheet. Because the sheet is uniform and infinite in extent, symmetry requires that **E** not vary with horizontal position, and hence **E** is a function only of $z$.

The choice of the origin of the coordinate system used above simplified the calculation because of the symmetry, but the same result for **E** would be obtained for any choice of coordinate system.

We could equally well determine the electrostatic potential and use it to calculate **E**. From Equation (1–67b) the potential is

$$\Phi(z) = \frac{\sigma}{4\pi\varepsilon_0} \int \frac{dS'}{R} = \frac{\sigma}{4\pi\varepsilon_0} \int \frac{\imath' \, d\imath' \, d\phi'}{(\imath'^2 + z^2)^{1/2}} \tag{1-80}$$

The $\phi'$ integral again gives $2\pi$, but the $\imath'$ integral is divergent. To avoid the divergence problem, we integrate $\imath'$ up to a radius $\Lambda$, which is much larger than $z$, and later take the limit $\Lambda \to \infty$:

$$\Phi(z) = \frac{\sigma}{2\varepsilon_0} \sqrt{\imath'^2 + z^2}\,\Big|_0^{\Lambda} = \frac{\sigma}{2\varepsilon_0} (\sqrt{\Lambda^2 + z^2} - |z|) \tag{1-81}$$

We can make an expansion of $\Phi$ in powers of $|z|/\Lambda$:

$$\Phi(z) = \frac{\sigma}{2\varepsilon_0} \left( \Lambda - |z| + \frac{z^2}{2\Lambda} + \cdots \right)$$

where the other terms involve higher powers of $\Lambda^{-1}$. Since we can always change $\Phi$ by an arbitrary constant, we can drop the first term. In the limit $\Lambda \to \infty$ only the second term survives:

$$\Phi(z) = -\frac{\sigma}{2\varepsilon_0} |z| \tag{1-82}$$

The corresponding electric field,

$$E_z = -\frac{\partial}{\partial z} \Phi(z)$$

reproduces the result obtained previously in Equations (1–77) to (1–79).

## 1-9 DELTA FUNCTION FOR REPRESENTING POINT CHARGES

In Section 1–7 we generalized the expressions for the electric potential and field due to point charges to the case of continuous charge distributions. Inversely, the results for continuous charge distributions can be applied to point charges. We consider first the case of a discrete charge $q$ at $x = a$; the line charge density $\lambda(x)$ on the $x$ axis then has the properties

$$\lambda(x) = 0 \qquad \text{for} \qquad x \neq a \tag{1-83a}$$

$$\int_{x_1}^{x_2} \lambda(x)\, dx = q \tag{1-83b}$$

when the limits on the $x$ integration include the point $x = a$; the integral is zero, otherwise. We write for this charge density

$$\lambda(x) = q\delta(x - a) \tag{1-83c}$$

where $\delta(x - a)$ is known as the *delta function*. The delta function was introduced in physics by P. A. M. Dirac. The properties of the delta function are

$$\delta(x - a) = 0 \qquad \text{for} \qquad x \neq a \tag{1-84a}$$

$$\int \delta(x - a)\, dx = 1 \tag{1-84b}$$

$$\int f(x)\delta(x - a)\, dx = f(a) \tag{1-84c}$$

*for $x$ integration limits that encompass $a$.* In Equation (1–84c), $f(x)$ is an arbitrary continuous function. The delta function is an even function of its argument:

$$\delta(x - a) = \delta(a - x) \tag{1-85}$$

as can be shown from the defining properties. The delta function is frequently used as a calculation tool for a continuous description of discrete quantities and for the approximate description of nearly discontinuous charge distributions.

The delta function $\delta(x - a)$ can be defined as the limit of an ordinary function $g(x - a)$ whose definite integral is

$$\int_{-\infty}^{+\infty} g(x - a)\, dx = 1$$

but whose values at all points except $x = a$ tend toward 0, $\lim g(x - a) \to 0$ if $x \neq a$. As an example, consider the function

$$g(x - a) = \frac{1}{\pi} \frac{\Delta}{(x - a)^2 + \Delta^2} \tag{1-86}$$

which is plotted in Figure 1–18 for several values of the parameter $\Delta$. We see that in the limit of small $\Delta$, $g(x - a)$ becomes sharply peaked around $x = a$, approaching $\infty$ as $\Delta \to 0$. The integral of $g(x - a)$ from $x = -\infty$ to $x = +\infty$ is

$$\int_{-\infty}^{+\infty} g(x - a)\, dx = \frac{1}{\pi} \arctan \frac{x - a}{\Delta}\Bigg|_{-\infty}^{+\infty} = 1 \tag{1-87}$$

The value of the integral is independent of the value of $\Delta$. Hence in the limit $\Delta \to 0$, $g(x - a) \to \delta(x - a)$.

We return to our expression for $\Phi$ in Equation (1–67c), substitute Equation (1–83c) for the line density of a discrete charge, and use Equation (1–84c). We obtain

$$\Phi(\mathbf{r}) = \frac{q}{4\pi\varepsilon_0} \int \frac{\delta(x' - a)}{|\mathbf{r} - x'\hat{\mathbf{x}}|}\, dx' = \frac{q}{4\pi\varepsilon_0 |\mathbf{r} - a\hat{\mathbf{x}}|} = \frac{q}{4\pi\varepsilon_0 |\mathbf{r} - \mathbf{r}_1|} \tag{1-88}$$

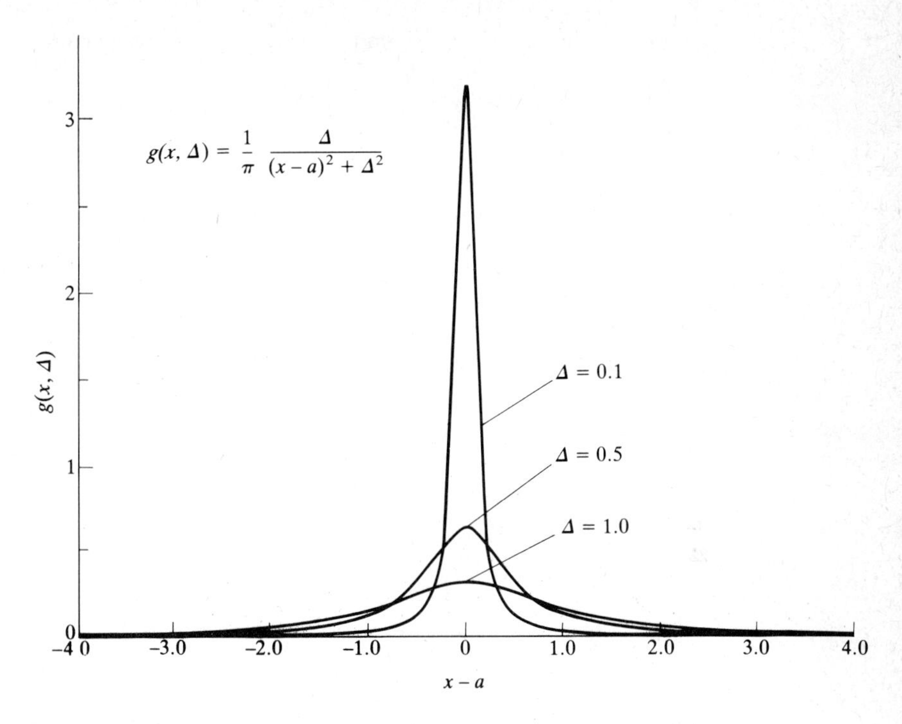


**FIGURE 1-18** Function $g(x, \Delta) = (1/\pi)[\Delta/(x^2 + \Delta^2)]$, which becomes a delta function in the limit as $\Delta$ goes to zero, plotted for various values of $\Delta$

where $\mathbf{r}_1 = a\hat{\mathbf{x}}$. This equation is the usual result for the potential of a point charge.

For a collection of point charges $q_i$ at positions $x_i$, the generalization of the line charge density in Equation (1–83c) is

$$\lambda(x') = \sum_i q_i \delta(x' - x_i) \tag{1-89}$$

For surface or volume charge densities corresponding to discrete charges, we can use delta functions in an analogous fashion to Equation (1–83c). For example, for a surface charge density in the $xy$ plane,

$$\sigma(\mathbf{r}') = q_1 \delta(x' - x_1)\delta(y' - y_1) \tag{1-90a}$$

for a point charge at the location $x_1$, $y_1$.

For a volume charge density,

$$\rho(\mathbf{r}') = q_1 \delta(x' - x_1)\delta(y' - y_1)\delta(z' - z_1) \tag{1-90b}$$

for a point charge at $x_1, y_1, z_1$. From the delta function properties

$$\rho(\mathbf{r}') = 0 \quad \text{if either} \quad x' \neq x_1 \quad \text{or} \quad y' \neq y_1 \quad \text{or} \quad z' \neq z_1 \tag{1-91a}$$

and

$$\int \rho(\mathbf{r}')\, dV' = q_1 \int \delta(x' - x)\delta(y' - y)\delta(z' - z)\, dx'\, dy'\, dz' = q_1 \tag{1-91b}$$

For compactness of notation a standard practice is to write

$$\delta^2(\mathbf{r}' - \mathbf{r}_1) \equiv \delta(x' - x_1)\delta(y' - y_1) \tag{1-92a}$$

$$\delta^3(\mathbf{r}' - \mathbf{r}_1) \equiv \delta(x' - x_1)\delta(y' - y_1)\delta(z' - z_1) \tag{1-92b}$$

With a collection of discrete charges in three dimensions at positions $\mathbf{r}_i$, the volume charge density is

$$\rho(\mathbf{r}') = \sum_i q_i \delta^3(\mathbf{r}' - \mathbf{r}_i) \tag{1-93}$$

Evaluating the potential expression of Equation (1–67a) with the charge density in Equation (1–93), we obtain the result of Equation (1–30) for discrete charges $q_i$:

$$\Phi(\mathbf{r}) = \frac{1}{4\pi\varepsilon_0} \sum_i q_i \int \frac{\delta^3(\mathbf{r}' - \mathbf{r}_i)}{|\mathbf{r} - \mathbf{r}'|}\, dV' = \frac{1}{4\pi\varepsilon_0} \sum_i \frac{q_i}{|\mathbf{r} - \mathbf{r}_i|} \tag{1-94}$$

We will frequently make use of the delta function to facilitate calculations.

A related function is the *theta function* (also called the step function), defined by

$$\theta(z - a) \equiv \begin{cases} 1, & \text{for } z > a \\ 0, & \text{for } z < a \end{cases} \tag{1-95}$$

The theta function is discontinuous at $z = a$. The derivative of the theta function is the delta function:

$$\frac{d}{dz}\theta(z - a) = \delta(z - a) \tag{1-96}$$

This relation can be demonstrated by direct integration or through use of the function in Equation (1–86) in the limit as $\Delta$ goes to zero.

# SUMMARY

## Important Concepts and Equations

$\mathbf{F}_q = q(\mathbf{E} + \mathbf{v} \times \mathbf{B})$ (Lorentz force)

*Force on a charge q moving with velocity* $\mathbf{v}$ *in a magnetic field* $\mathbf{B}$ *and an electric field* $\mathbf{E}$.

$\omega_c = \dfrac{v_\perp}{R} = \dfrac{q}{m} B$ (cyclotron frequency)

*Angular frequency of a particle of mass m and charge q moving in a magnetic field B* ($v_\perp$ *is the velocity component perpendicular to* $\mathbf{B}$, *and R is the radius of the motion*).

$v_D = \dfrac{E}{B}$ (drift velocity)

*The velocity, which is in the* $\mathbf{E} \times \mathbf{B}$ *direction, of the guiding center of a charged particle in crossed* $\mathbf{E}$ *and* $\mathbf{B}$ *fields.*

$\mathbf{F}_q = \dfrac{1}{4\pi\varepsilon_0} \dfrac{qq_1}{R_1^2} \hat{\mathbf{R}}_1$ (Coulomb force)

*The force on charge q due to* $q_1$, *where* $(4\pi\varepsilon_0)^{-1}$ *is a proportionality constant,* $R_1$ *is the distance between the charges, and* $\hat{\mathbf{R}}_1$ *is a unit vector from* $q_1$ *to q.*

$\mathbf{E} = \dfrac{1}{4\pi\varepsilon_0} \displaystyle\sum_{i=1}^{n} \dfrac{q_i \mathbf{R}_i}{R_i^3}$ (Coulomb's law)

*The electric field at a position* $\mathbf{r}$ *due to a distribution of discrete charges at* $\mathbf{r}_1, \mathbf{r}_2, \ldots, \mathbf{r}_n$, *where* $\mathbf{R}_i = \mathbf{r} - \mathbf{r}_i$.

$\mathbf{E} = \displaystyle\sum_i^n \mathbf{E}_i \qquad \mathbf{B} = \sum_i^n \mathbf{B}_i$ (superposition principle)

*The field due to the superposition of sources is the vector sum of the fields of the separate sources.*

Field lines

*Curves drawn tangent everywhere to the field vectors are used to pictorially represent fields. The density of the lines is proportional to the strength of the field. Field lines do not end except on charges.*

Equipotentials

*A spatial surface on which the electric potential is constant. Electric field lines are orthogonal to equipotential surfaces.*

$$\Phi = \frac{1}{4\pi\varepsilon_0} \sum_i \frac{q_i}{R_i} \qquad \text{(electric potential)}$$

*The electrostatic potential at a position* $\mathbf{r}$ *due to a distribution of discrete charges at* $\mathbf{r}_1, \mathbf{r}_2, \ldots,$ *where* $R_i = |\mathbf{r} - \mathbf{r}_i|$.

$$\Phi(\mathbf{r}_b) - \Phi(\mathbf{r}_a) = -\int_{\mathbf{r}_a}^{\mathbf{r}_b} \mathbf{E}(\mathbf{r}') \cdot d\mathbf{r}'$$

*The electric potential difference between two positions is determined by the line integral of the electric field vector.*

$$\mathbf{E} = -\nabla\Phi$$

*The electrostatic field is given by a negative gradient of the scalar potential* $\Phi$.

$$\Phi(\mathbf{r}) = \frac{1}{4\pi\varepsilon_0} \int \frac{\rho(\mathbf{r}')\, dV'}{R} \qquad \text{(electric potential)}$$

*The electrostatic potential at a position* $\mathbf{r}$ *due to a continuous charge distribution* $\rho(\mathbf{r}')$, *where* $\mathbf{R} = \mathbf{r} - \mathbf{r}'$.

$$\mathbf{E}(\mathbf{r}) = \frac{1}{4\pi\varepsilon_0} \int \frac{\rho(\mathbf{r}') R\, dV'}{R^2}$$

*The electric field at* $\mathbf{r}$ *due to a continuous charge distribution* $\rho(\mathbf{r}')$.

$\delta(x' - a)$ (delta function)

*Properties:*

$$\delta(x' - a) = 0 \qquad \text{for} \qquad x' \neq a$$

*For* $x'$ *integration limits that encompass* $a$,

$$\int \delta(x' - a)\, dx' = 1 \qquad \int f(x')\delta(x' - a)\, dx' = f(a)$$

## PROBLEMS

### Introduction: Comparison of Forces

**1-1** In a hydrogen atom in its lowest energy state, the electron and the proton are separated on the average by a distance

$$a_0 = \left(\frac{4\pi\varepsilon_0}{e^2}\right)\left(\frac{h^2}{4\pi^2 m_e}\right) = 0.53 \times 10^{-10} \text{ m}$$

known as the Bohr radius, where $h$ is Planck's constant, $m_e$ is the electron mass, and $e$

is the magnitude of the electron charge. If the electron and proton were bound instead only by the gravitational force, what would their average separation be? Compare your answer with the radius of the observable universe, $R = 15 \times 10^9$ light years.

## Section 1-1: Lorentz Force

**1-2** The Zeeman effect is the change in the frequencies of emission lines of gases due to a magnetic field. As a classical atomic model, assume that electrons are bound to atoms by a simple harmonic force of spring constant $k = m\omega_0^2$ and that the emission lines have a monochromatic frequency $\nu_0 = \omega_0/2\pi$. Find the emission line frequencies when an external magnetic field is applied along the $z$ axis. (In the absence of the external magnetic field the motion can be considered as a superposition of a linear oscillation along the $z$ axis and two circular motions of opposite sense about the $z$ axis, as illustrated in Figure 1-19.)

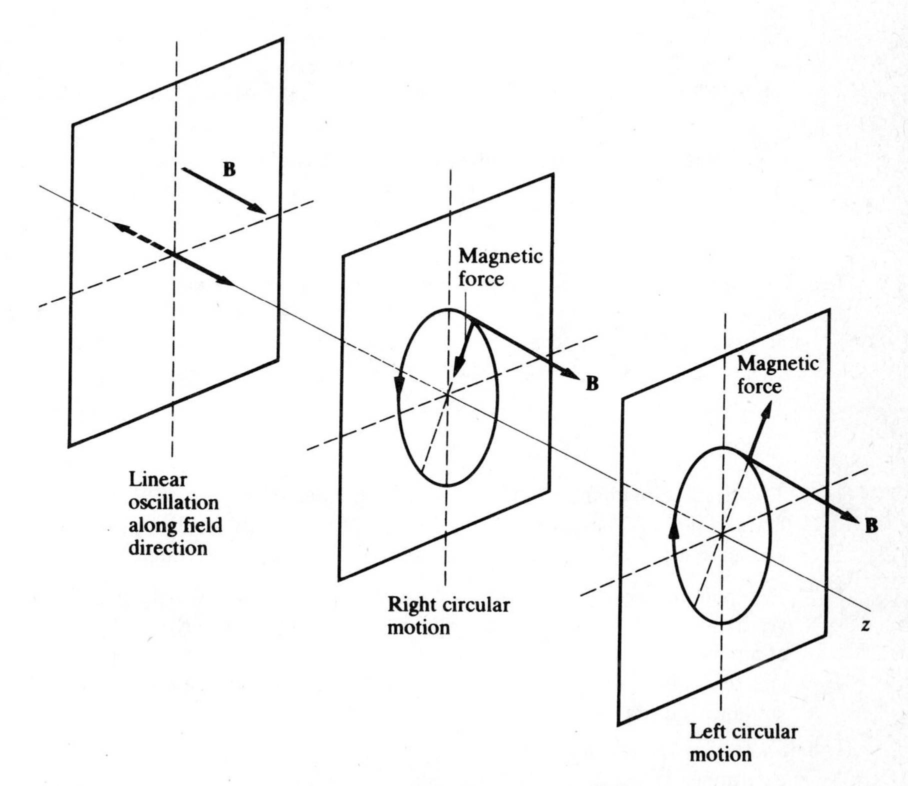


**FIGURE 1-19** Forces acting on an electron, due to a magnetic field along the $z$ axis, for linear motion along the $z$ axis and for right and left circular motion about the axis

## Section 1-2: Motion of Charged Particles

**1-3** An electron passes through an electrostatic deflector of length $l$. The initial velocity $v$ of the electron is parallel to the axis of the deflector. In the deflector the electric field has

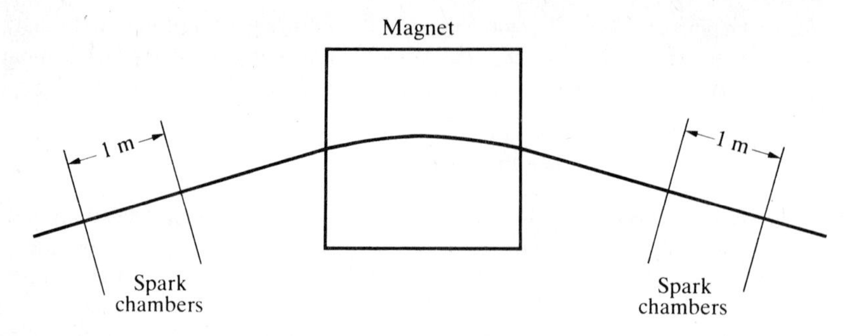


**FIGURE 1-20** Momentum analyzer for charged particles

nearly constant magnitude $E$ and is perpendicular to the axis. Find the angle through which the electron is deflected. Find the deflection angle in a magnetic deflector in which the $B$ field is constant and perpendicular to the initial velocity of the electron.

**1-4** The charge-to-mass ratio of a beam of particles can be measured by first accelerating the particles through a potential difference $V$ and then deflecting these particles in a uniform magnetic field $B$. Find $q/m$ in terms of $V$, $B$, and $R$, where $R$ is the radius of curvature of the beam in the magnetic field $B$. If the kinetic energy $K$ of each particle incident on the collector is also measured, show that $m$ and $q$ can be determined separately.

**1-5** The momentum of a particle is measured by measuring its deflection by a magnet, as shown in Figure 1-20. The spark chambers measure the position of the particle to an accuracy of 0.2 mm and are thin compared with their 1-m separation; and the total angle of deflection is 0.1 rad. To what accuracy $\Delta p/p$ can the momentum be measured?

## Section 1-3: Coulomb's Law

**1-6** Two fixed point charges $+q$ and $-q$ are separated by a distance $2a$. Using Coulomb's law and the superposition principle, find the **E** field in the median plane. Express your answer in terms of the perpendicular distance from the line through the charges.

**1-7** Two point charges are located on the $x$ axis, $q$ at $x = a$ and $-4q$ at $x = -a$. Find a vector expression in Cartesian coordinates for the electric field **E** of the system of charges at any point in the $xy$ plane. Find the $(x, y)$ coordinates of any points at which a test charge would be in equilibrium. Determine whether the equilibrium at any such points is stable or unstable.

**1-8** Show that electric field lines in the $xy$ plane for $\mathbf{E}(x, y) = \hat{\mathbf{x}}E_x(x, y) + \hat{\mathbf{y}}E_y(x, y)$ are determined by the differential equation

$$\frac{dy}{dx} = \frac{E_y(x, y)}{E_x(x, y)}$$

and that the lines of equipotential are solutions of

$$\frac{dy}{dx} = -\frac{E_x(x, y)}{E_y(x, y)}$$

Solve for the field lines and equipotential lines for the electric field

$$\mathbf{E}(x, y) = x\hat{\mathbf{x}} - y\hat{\mathbf{y}}$$

Sketch the solution.

**1-9** Show that the electric field lines in the $r\phi$ plane for $E = \hat{r}E_r + \hat{\phi}E_\phi$ are given by the solutions to the equation

$$\frac{1}{r}\frac{dr}{d\phi} = \frac{E_r}{E_\phi}$$

Solve for the field lines and equipotential lines for the electric field

$$\mathbf{E}(r, \phi) = (2 \cos \phi\hat{r} + \sin \phi\hat{\phi})/r^3. \text{ Sketch the solution.}$$

## Section 1-4: Electric Potential

**1-10** Charges of $+5\ \mu$C are located at points $A$ and $C$ as shown in Figure 1–21. What is the

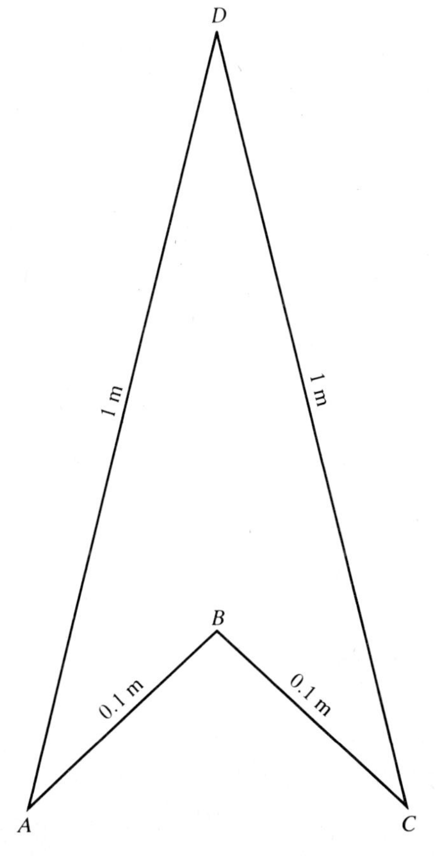


**FIGURE 1-21** Charge released from rest at position $B$. It moves to position $D$ under influence of charges at $A$ and $C$.

velocity of a 15-g mass carrying a charge of $+5\ \mu$C when it passes point $D$ after being released from rest at point $B$?

**1-11** The difference of electrostatic potentials between coaxial cylindrical electrodes of inner radius $a$ and outer radius $b$ is $V$. A uniform magnetic field $B$ is applied to the cylinders along their axis. If $B$ is strong enough (greater than a critical value $B_c$), the electrons emitted with zero velocity from the cathode at the center cannot reach the anode. Find the strength of the critical magnetic field $B_c$. Assume $a \ll b$.

**1-12** Let two coordinate systems be related by a rotation through an angle about the $z$ axis:

$$x' = x\cos\theta + y\sin\theta \qquad y' = -x\sin\theta + y\cos\theta \qquad z' = z$$

Show that the del operator $\boldsymbol{\nabla}$ transforms under this rotation as a vector, that is, that the components $(\partial/\partial x, \partial/\partial y, \partial/\partial z)$ of $\boldsymbol{\nabla}$ transform in the same way as the components $(x, y, z)$ of $\mathbf{r}$.

**1-13** Calculate the electric field for each of the following potentials. Here $k$, $\lambda$, $r_0$, and $p$ are constants.

(i) $\Phi = kxyz$

(ii) $\Phi = \dfrac{-\lambda}{2\pi\varepsilon_0} \ln \dfrac{r}{r_0}$

(iii) $\Phi = \dfrac{p\cos\theta}{4\pi\varepsilon_0 r^2}$

**1-14** Verify the following vector identities, where $s$ and $t$ are scalar fields and $\mathbf{V}$ is a vector field.

(i) $\boldsymbol{\nabla}(st) = s\,\boldsymbol{\nabla}t + t\,\boldsymbol{\nabla}s$
(ii) $\boldsymbol{\nabla}\cdot(s\mathbf{V}) = s\,\boldsymbol{\nabla}\cdot\mathbf{V} + \mathbf{V}\cdot\boldsymbol{\nabla}s$
(iii) $\boldsymbol{\nabla}\times(s\mathbf{V}) = s(\boldsymbol{\nabla}\times\mathbf{V}) - \mathbf{V}\times\boldsymbol{\nabla}s$
(iv) $\boldsymbol{\nabla}\times\boldsymbol{\nabla}s = 0$
(v) $\boldsymbol{\nabla}\cdot\boldsymbol{\nabla}\times\mathbf{V} = 0$

**1-15** From the product rule of differentiation,

$$\boldsymbol{\nabla}\cdot(\mathbf{V}\times W) = \boldsymbol{\nabla}_V\cdot(\mathbf{V}\times\mathbf{W}) + \boldsymbol{\nabla}_W\cdot(\mathbf{V}\times\mathbf{W})$$

$$\boldsymbol{\nabla}\times(\mathbf{V}\times\mathbf{W}) = \boldsymbol{\nabla}_V\times(\mathbf{V}\times\mathbf{W}) + \boldsymbol{\nabla}_W\times(\mathbf{V}\times\mathbf{W})$$

where $\boldsymbol{\nabla}_V$ differentiates only $\mathbf{V}$ and $\boldsymbol{\nabla}_W$ differentiates only $\mathbf{W}$. Using these relations, prove the vector identities

$$\boldsymbol{\nabla}\cdot(\mathbf{V}\times\mathbf{W}) = \mathbf{W}\cdot\boldsymbol{\nabla}\times\mathbf{V} - \mathbf{V}\cdot\boldsymbol{\nabla}\times\mathbf{W}$$

$$\boldsymbol{\nabla}\times(\mathbf{V}\times\mathbf{W}) = (\mathbf{W}\cdot\boldsymbol{\nabla})\mathbf{V} - \mathbf{W}(\boldsymbol{\nabla}\cdot\mathbf{V}) + \mathbf{V}(\boldsymbol{\nabla}\cdot\mathbf{W}) - (\mathbf{V}\cdot\boldsymbol{\nabla})\mathbf{W}$$

[*Hint:* $\mathbf{A}\times(\mathbf{B}\times\mathbf{C}) = \mathbf{B}(\mathbf{A}\cdot\mathbf{C}) - \mathbf{C}(\mathbf{A}\cdot\mathbf{B})$ when $\mathbf{A}$, $\mathbf{B}$, and $\mathbf{C}$ commute.]

**1-16** Which of the following electric fields are not of electrostatic origin?

(i) $\mathbf{E} = c(x - z)^2(\hat{\mathbf{x}} - \hat{\mathbf{z}})$
(ii) $\mathbf{E} = kyz\sin(kxy)\hat{\mathbf{x}} + kxz\sin(kxy)\hat{\mathbf{y}} - \cos(kxy)\hat{\mathbf{z}}$
(iii) $\mathbf{E} = 2xyz\hat{\mathbf{x}} + xz^2\hat{\mathbf{y}} + x^2y\hat{\mathbf{z}}$

## Section 1-6: Electric Lens

**1-17** Sketch the field lines and a typical electron trajectory for the electrostatic lens discussed in the text in the case $V_1 < V_0$.

## Section 1-7: Continuous Charge Distributions

**1-18** Consider a straight-line charge of constant, linear charge density on the $z$ axis. Compute the electric field in cylindrical coordinates ($\imath$, $\phi$, $z$) by direct application of Coulomb's law for a continuous charge distribution in the following cases.

(i) The charge extends from $z = -\infty$ to $z = 0$ only.
(ii) The line charge extends from $z = -\infty$ to $z = +\infty$.

**1-19** A point charge $Q$ is located at a distance $d$ from an infinite line charge with charge density $\lambda$ per unit length; $Q$ and $\lambda$ are related by $Q/\lambda = d$.

(i) Find a vector expression for the electric field of the system in terms of ($x$, $y$, $z$) coordinates. Choose coordinate axes such that the charge $Q$ is located at ($x = d$, $y = 0$, $z = 0$) and the line charge $\lambda$ is along the $z$ axis.
(ii) Find the ($x$, $y$, $z$) coordinates of any points at which a test charge would be in equilibrium.

**1-20** An infinite line charge has density $-\lambda$ for $z < 0$ and $+\lambda$ for $z > 0$. Show that the magnitude of the electric field anywhere is $E = \lambda/(2\pi\varepsilon_0 \imath)$, where $\imath$ is the perpendicular distance from the line charge. Sketch the field lines.

**1-21** Find the electrostatic potential of a line charge density $\lambda$ on the $z$ axis extending from $z_1$ to $z_2$. Calculate the electric field from the potential.

**1-22** Consider two infinitely long, parallel line charges with charges per unit length of $\lambda$ and $-\lambda$, respectively, as shown in Figure 1-22.

(i) Determine the potential $\phi(\imath)$.
(ii) Determine analytically the family of curves that are the equipotential surfaces. (*Hint*: Use a coordinate system centered on the $+\lambda$ charge.)
(iii) Sketch equipotential surfaces for $\Phi > 0$, $\Phi = 0$, $\Phi < 0$.

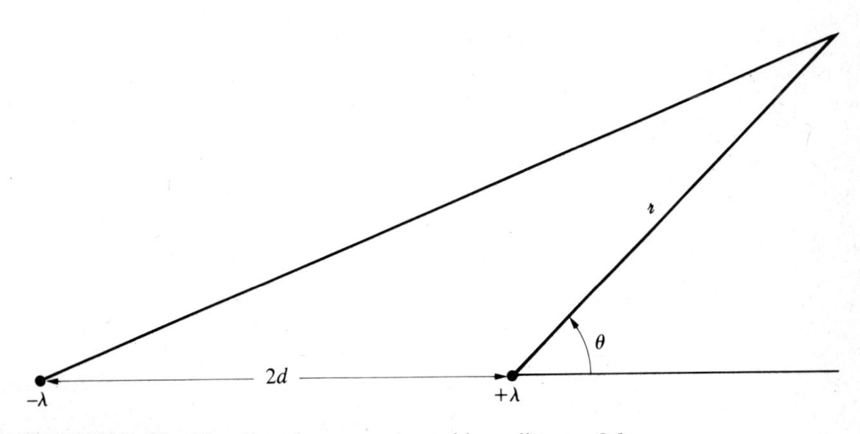


**FIGURE 1-22** Two line charges separated by a distance $2d$

**1-23** A thin disk has uniform surface charge $+\sigma$ coulombs per meter squared ($C/m^2$) and radius $R$. A point charge $q$ is on the axis of symmetry. Derive an expression for the force on the charge as a function of distance $z$ above or below the disk; examine the result in the limits of (i) small $R/z$ and (ii) large $R/z$. Sketch the force versus $z$.

**1-24** A ring of radius $a$ has a total charge $+Q$ uniformly distributed on it.
(i) Calculate the electric field and potential on the axis of the ring.
(ii) Consider a charge $-Q$ constrained to slide along the axis of the ring. Show that the charge will execute simple harmonic motion for small displacements perpendicular to the plane of the ring.

**1-25** A thin circular loop of radius $b$ and linear charge density $\lambda$ lies in the $xy$ plane with its center at the origin. A second loop of radius $a$ and linear charge density $-\lambda$ is parallel to the $xy$ plane with its center at the position $(x = 0, y = 0, z = D)$. Find an expression for the electric potential $\Phi(z)$ on the $z$ axis. Determine the potential difference between the centers of the loops.

**1-26** Two uniform, infinite sheets of electric charge density $+\sigma$ and $-\sigma$ intersect at right angles. Find the magnitude and the direction of the electric field everywhere, and sketch the lines of $\mathbf{E}$.

**1-27** Calculate, by direct integration, the electric field inside and outside an infinitely long circular cylinder that has a uniform charge density $\sigma$ on the surface.

**1-28** A spherical shell of radius $a$ has a charge $Q$ uniformly distributed over its surface. By integration of Coulomb's law, find the potential inside and out. Sketch the potential versus $r/a$. (*Hint*: Use differential rings at constant distance from the field point, and use this distance as your integration variable.)

**1-29** Two large, flat charged sheets of uniform surface charge density $+\sigma$ and $-\sigma$ are separated by a distance $d$. Calculate the electric field both between and outside the plates. Find the potential difference between the plates. Repeat the calculation for surface charge density $+\sigma$ on both plates.

## Section 1-9: Delta Function

**1-30** Prove the property of the delta function $\delta(az) = (1/|a|)\delta(z)$.

**1-31** Using the Dirac delta function representation $g(x)$ discussed in the text, show that $\delta(x) = (d/dx)\theta(x)$, where the $\theta$ step function has the property $\theta(x) = 0$ for $x < 0$ and $\theta(x) = 1$ for $x > 0$.

**1-32** Show that the function $f(x) = (1/\sqrt{\pi}\sigma)e^{-x^2/\sigma^2}$ approaches the delta function in the limit $\sigma \to 0$.

# Gauss's Law, Conductors, and Electrostatics

CHAPTER CONTENTS

Historically, progress in physics has been made by considering an isolated region of space and time on the assumption that we do not need a detailed knowledge of the rest of the universe. In mathematical terms this assumption means that the laws of physics can be expressed as differential equations that can be solved in any space-time region, with the effect of distant regions entering only as boundary conditions. Newton's second law is an example. The acceleration of a point mass depends only on the force at the current location of the mass, although its velocity and position depend on its entire past history.

In electrostatics Coulomb's law yields the electric field at a point if we know the charge distribution everywhere. However, what we would like to have is a local (differential) equation relating the charge density at a point to the electric field at that point, with boundary conditions representing the distant regions. About 1840 an integral relation, variously ascribed to Karl Friedrich Gauss, George Green, and William Thomson (Lord Kelvin), was derived from Coulomb's law. The relation, now known as Gauss's law, is weaker than Coulomb's law but applies more generally, whether charges are at rest or in motion. Recast in differential form, Gauss's law becomes the first of the Maxwell equations. By use of Gauss's law the calculation of electric fields in situations with symmetry can be carried out quite easily. This chapter is primarily devoted to the derivation of Gauss's law and its application.

## 2-1 GAUSS'S ELECTRIC FLUX LAW

For any vector field $\mathbf{V}(\mathbf{r})$ we can define its *flux* $\mathscr{F}_V$ through a surface $S$ as

$$\mathscr{F}_V = \int \mathbf{V}(\mathbf{r}) \cdot d\mathbf{S}(\mathbf{r}) \qquad \textbf{(2-1)}$$

Here $d\mathbf{S} = \hat{\mathbf{n}}\, dS$, with $\hat{\mathbf{n}}$ the unit vector normal to the surface element $dS$. For a closed surface $\hat{\mathbf{n}}$ is chosen by convention to be the outward normal. For an open surface $\hat{\mathbf{n}}$ is taken to be related to the direction of the tangent to the boundary circuit by the right-hand rule; see Figure 2–1. For the electric field $\mathbf{E}$ the electric flux through a surface $S$ is

$$\mathscr{F}_E = \int \mathbf{E}(\mathbf{r}) \cdot d\mathbf{S} \qquad \textbf{(2-2)}$$

In this section we will derive an important relation between the electric flux through a closed surface and the charge enclosed by the surface.

Consider first the case of a point charge $q$ within a closed surface $S$. We choose the origin of coordinates at the position of the charge. Let $d\mathbf{S}$ be an element of the closed surface with outward normal, as illustrated in

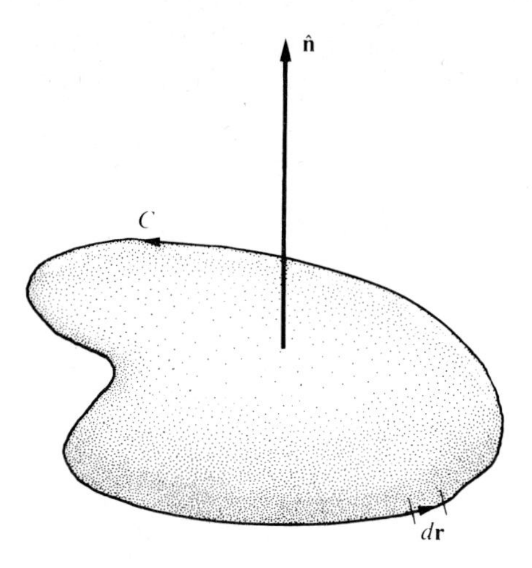


**FIGURE 2–1** Sense of the unit normal vector for an open surface as determined by the right-hand rule

Figure 2–2. Then from Coulomb's law

$$\mathbf{E} \cdot d\mathbf{S} = \frac{q}{4\pi\varepsilon_0} \frac{\hat{\mathbf{r}} \cdot d\mathbf{S}}{r^2} \tag{2-3}$$

The dot product $\hat{\mathbf{r}} \cdot d\mathbf{S}$ is the projected area element on the surface of a sphere of radius $r$. In spherical coordinates we have

$$\hat{\mathbf{r}} \cdot d\mathbf{S} = (r\, d\theta)(r \sin\theta\, d\phi) = r^2\, d\Omega \tag{2-4}$$

where $d\Omega = \sin\theta\, d\theta\, d\phi$ is the *differential solid angle*. Thus Equation (2–3) can be rewritten as

$$\mathbf{E} \cdot d\mathbf{S} = \frac{q}{4\pi\varepsilon_0}\, d\Omega \tag{2-5}$$

The integral of $d\Omega$ over a closed surface is $4\pi$; thus we find

$$\oint \mathbf{E} \cdot d\mathbf{S} = \frac{q}{\varepsilon_0} \tag{2-6}$$

where the circle on the integral sign denotes a closed surface. [For more complex closed surfaces the rays from the position of the charge may intersect the surface more than once. More precisely, they will intersect the surface an odd number of times, and the argument of the next paragraph can be used to show that the result of Equation (2–6) still holds.]

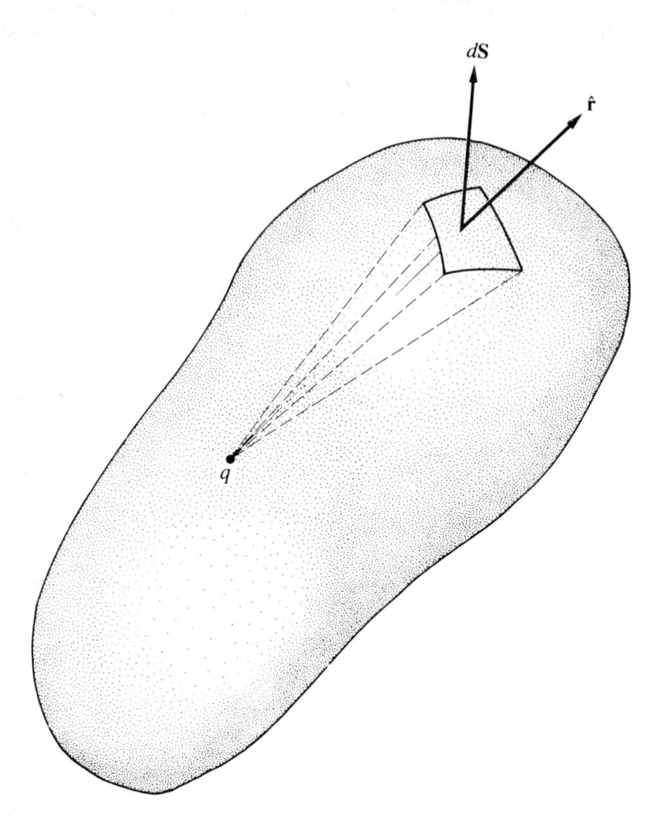


**FIGURE 2-2** Point charge $q$ located at the origin of coordinates within a closed surface $S$

For the case of a charge $q$ outside the closed surface, a cone from $q$ cuts the surface twice, as shown in Figure 2–3. In this case $\hat{\mathbf{r}} \cdot d\mathbf{S}/r^2 = +d\Omega\ (-d\Omega)$ for the farther (nearer) surface element. The contributions from the two surfaces thereby cancel, and hence

$$\oint \mathbf{E} \cdot d\mathbf{S} = 0 \qquad \textbf{(2-7)}$$

for a charge exterior to the closed surface. For a more complex surface the solid-angle rays can intersect regions of the surface an even number of times, but by a suitable modification of the preceding proof, Equation (2–7) continues to hold. Equation (2–6) for a single point charge generalizes, by the superposition principle, to an arbitrary collection of charges, with $q$ of Equation (2–6) becoming the net charge $Q$ inside the surface. The conclu-

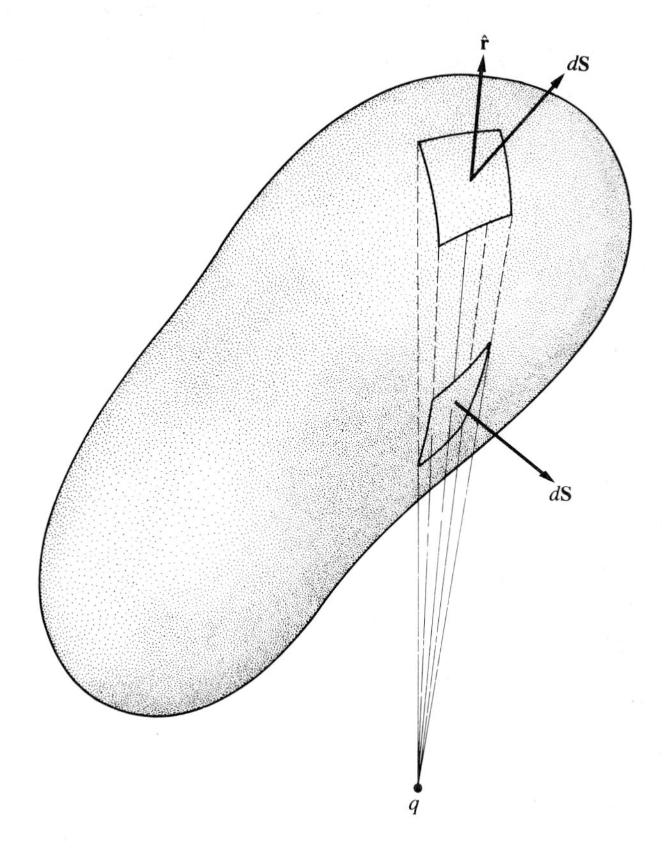


**FIGURE 2–3** Flux cone from a charge located outside a closed surface intersecting the surface twice

sion of Equations (2–6) and (2–7) is that

**Gauss's Law (Integral Form)**

$$\oint \mathbf{E} \cdot d\mathbf{S} = \frac{Q}{\varepsilon_0} \tag{2-8}$$

where $Q$ is the total charge enclosed by the closed surface. The amount of flux through a closed surface thus depends only on the charge enclosed by the surface.

In developing an intuitive picture of electric and magnetic fields, we can often usefully employ the concepts of field lines, as discussed in Section 1–3. For a given vector field **E** we can construct field lines that are everywhere tangent to **E**. Consider a group of neighboring lines that outline a thin tube, as illustrated in Figure 2–4. Assuming that the normal surfaces $S_1$ and $S_2$ cut off a region of the tube containing no charge, the application of Gauss's

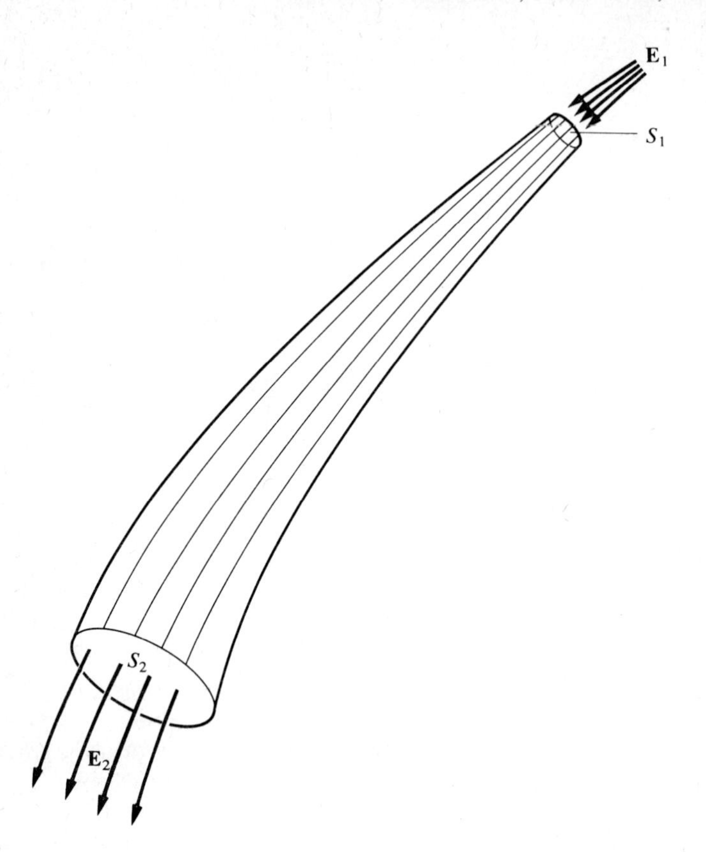


**FIGURE 2-4** Flux tube cut by two normal surfaces $S_1$ and $S_2$ in a region containing no charge

integral law to the closed volume gives

$$E_1 S_1 = E_2 S_2 \tag{2-9}$$

that is, the flux out at end 2 is the same as the flux in at end 1. If we introduce field lines to represent **E**, we can take the density of lines to be proportional to the magnitude of **E** in order to maintain the constancy of the number of lines in a flux tube. The field lines are chosen to point along the direction of **E**, and hence they begin on positive charges and terminate on negative ones. If the total charge of the system is nonzero, some lines must be considered to begin or end at infinity (e.g., the case of a single charge).

## 2-2 DIVERGENCE AND THE DIVERGENCE THEOREM

To cast Gauss's law in differential form, we need a basic theorem of vector calculus known as the divergence theorem. The divergence is a differential operation that plays an important role in the laws of electromagnetism. The divergence $\nabla \cdot \mathbf{E}$ of a vector field $\mathbf{E}$ is the dot product of the del operator in Equation (1–38) with $\mathbf{E}$. In Cartesian coordinates the divergence is given by

$$\nabla \cdot \mathbf{E} = \left(\hat{\mathbf{x}}\frac{\partial}{\partial x} + \hat{\mathbf{y}}\frac{\partial}{\partial y} + \hat{\mathbf{z}}\frac{\partial}{\partial z}\right) \cdot (\hat{\mathbf{x}}E_x + \hat{\mathbf{y}}E_y + \hat{\mathbf{z}}E_z)$$
$$\equiv \frac{\partial E_x}{\partial x} + \frac{\partial E_y}{\partial y} + \frac{\partial E_z}{\partial z} \tag{2-10}$$

In non-Cartesian systems a unit vector will, in general, be a function of the coordinates [e.g., in cylindrical coordinates the direction of $\hat{\mathbf{r}}$ depends on $\phi$; in such cases the differential operation of $\nabla$ on $\mathbf{E}$ must include the derivatives of the unit vectors (see Section 2–5C)].

The *divergence theorem* of vector calculus allows us to relate the integral of any vector function $\mathbf{E}$ through a *closed surface* $S$ to an integral of $\nabla \cdot \mathbf{E}$ throughout the volume enclosed by $S$:

**Divergence Theorem**

$$\oint \mathbf{E} \cdot d\mathbf{S} = \int \nabla \cdot \mathbf{E}\, dV \tag{2-11}$$

where $d\mathbf{S} = \hat{\mathbf{n}}\, dS$ is the vector surface element and $\hat{\mathbf{n}}(\mathbf{r})$ is the *outward* normal unit vector over the closed surface $S$. The circle on the integral sign denotes a closed surface. We will now prove this important theorem.

Consider first the integral $\oint E_x\, dS_x$. A thin prism through the volume, with axis parallel to the $x$ axis and cross section $dy\, dz$, cuts the surface $S$ at two points, $x = x_+$ and $x = x_-$, as shown in Figure 2–5. At $x_+$ the component of the outward-pointing surface element along $x$ is $dS_x|_{x=x_+} = dy\, dz$. Similarly, at $x_-$ we have $dS_x|_{x=x_-} = -dy\, dz$. Thus the integral can be written as

$$\oint E_x\, dS_x = \iint [E_x(x_+, y, z) - E_x(x_-, y, z)]\, dy\, dz$$
$$= \int dy \int dz \int_{x_-}^{x_+} dx \frac{\partial E_x}{\partial x} = \int \frac{\partial E_x}{\partial x}\, dV \tag{2-12}$$

We leave to the reader the elaboration needed if there are more pairs of entrance and exit points. By repeating the above argument with $\oint E_y\, dS_y$ and $\oint E_z\, dS_z$, we obtain

$$\oint (E_x\, dS_x + E_y\, dS_y + E_z\, dS_z) = \int \left(\frac{\partial E_x}{\partial x} + \frac{\partial E_y}{\partial y} + \frac{\partial E_z}{\partial z}\right) dV \tag{2-13}$$

which is the divergence theorem, Equation (2–11).

As a corollary of the divergence theorem, we replace $\mathbf{E}$ in Equation (2–11) by $\mathbf{c} \times \mathbf{E}$, where $\mathbf{c}$ is an arbitrary but constant vector. Using the

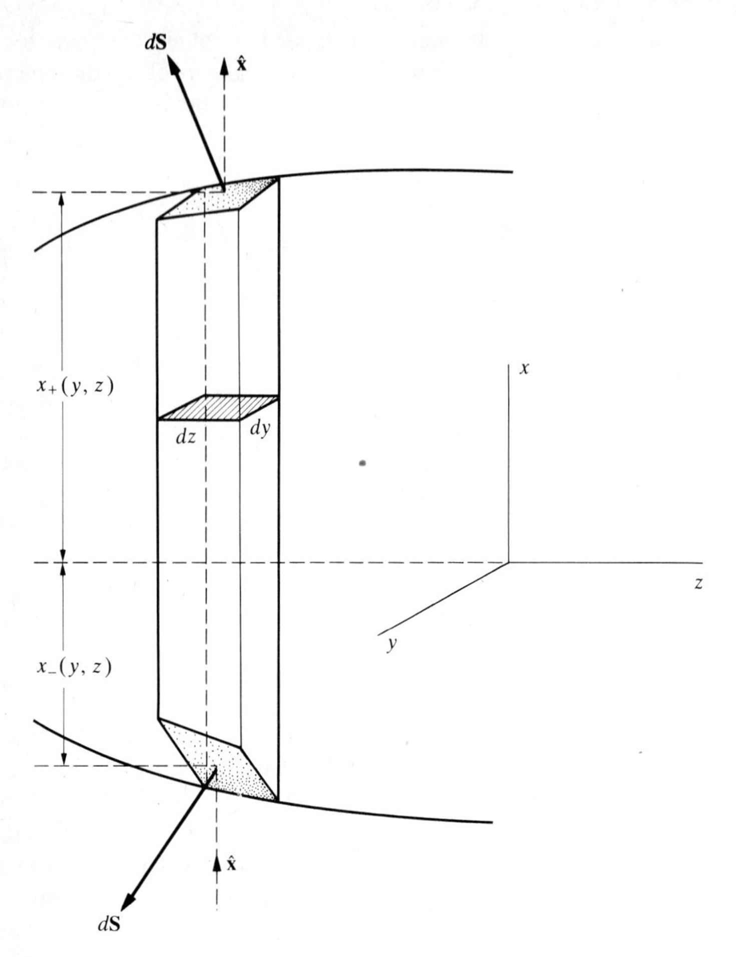


**FIGURE 2-5** Projection of the differential area element $dy\,dz$ onto the closed surface $S$

identity for the divergence of a cross product, we find

$$\oint \mathbf{c} \times \mathbf{E} \cdot d\mathbf{S} = -\mathbf{c} \cdot \int \nabla \times \mathbf{E}\, dV \tag{2-14}$$

Interchanging the dot and cross products and factoring the constant vector **c** from the surface integral gives

**Divergence Theorem, Corollary I**

$$\oint d\mathbf{S} \times \mathbf{E} = \int \nabla \times \mathbf{E}\, dV \tag{2-15}$$

Another corollary is found by replacing **E** by **c**$\Phi$ in Equation (2–11):

**Divergence Theorem, Corollary II**

$$\int \nabla\Phi \, dV = \oint \Phi \, d\mathbf{S} \tag{2-16}$$

The divergence theorem is valid in any coordinate system because of its manifest vector nature. When applied to an infinitesimal volume element, it can be used to evaluate the form of the divergence in a non-Cartesian coordinate system. For an infinitesimal volume element the divergence theorem becomes

$$\nabla \cdot \mathbf{E} = \lim_{\Delta V \to 0} \frac{1}{\Delta V} \oint_{\Delta S} \mathbf{E} \cdot d\mathbf{S} \tag{2-17}$$

For example, in the spherical case illustrated in Figure (1–12), the surface elements are

$$d\mathbf{S} = (r\, d\theta)(r \sin\theta \, d\phi)\hat{\mathbf{r}} + (dr)(r \sin\theta \, d\phi)\hat{\boldsymbol{\theta}} + (dr)(r\, d\theta)\hat{\boldsymbol{\phi}} \tag{2-18a}$$

and the volume element is

$$dV = (r\, d\theta)(r \sin\theta \, d\phi)\, dr \tag{2-18b}$$

The electric field in spherical coordinates is

$$\mathbf{E} = E_r \hat{\mathbf{r}} + E_\theta \hat{\boldsymbol{\theta}} + E_\phi \hat{\boldsymbol{\phi}}$$

From Equations (2–17) and (2–18) we find the divergence expression

$$\begin{aligned} \nabla \cdot \mathbf{E} = \frac{1}{(r\, d\theta)(r \sin\theta \, d\phi)(dr)} & [d\theta \sin\theta \, d\phi (r^2 E_r|_{r+dr} - r^2 E_r|_r) \\ & + dr\, r\, d\phi (\sin\theta E_\theta|_{\theta + d\theta} - \sin\theta E_\theta|_\theta) \\ & + dr\, r\, d\theta (E_\phi|_{\phi + d\phi} - E_\phi|_\phi)] \end{aligned}$$

Using the relation $f(x + dx) - f(x) = (\partial f/\partial x)\, dx$, we obtain the divergence in spherical coordinates:

$$\nabla \cdot \mathbf{E} = \frac{1}{r^2}\frac{\partial}{\partial r}(r^2 E_r) + \frac{1}{r \sin\theta}\frac{\partial}{\partial \theta}(\sin\theta E_\theta) + \frac{1}{r \sin\theta}\frac{\partial E_\phi}{\partial \phi} \tag{2-19}$$

We leave as an exercise an analogous derivation of the divergence in cylindrical coordinates.

## 2-3 DIFFERENTIAL FORM OF GAUSS'S LAW

Combining the divergence theorem of Equation (2–11) for the electric field and Gauss's electric flux law of Equation (2–8), we obtain the equality

$$\int \nabla \cdot \mathbf{E}\, dV = \frac{Q}{\varepsilon_0} \tag{2-20}$$

The net charge $Q$ in the volume $V$ can be expressed in terms of the charge density $\rho(\mathbf{r})$ as the integral

$$Q = \int \rho(\mathbf{r})\, dV \tag{2-21}$$

These equations yield

$$\int \left( \nabla \cdot \mathbf{E} - \frac{\rho}{\varepsilon_0} \right) dV = 0 \tag{2-22}$$

Since the choice of volume is arbitrary, the integrand must vanish, giving

**Gauss's Law (Differential Form)**

$$\nabla \cdot \mathbf{E} = \frac{\rho}{\varepsilon_0} \tag{2-23}$$

This equation is the local form of Gauss's law. It turns out that this relation is valid even in the case of moving charges; Equation (2–23) is one of Maxwell's equations.

Gauss's law is weaker but has greater generality than Coulomb's law. One can show that a vector field, such as $\mathbf{E}$, can be completely specified by its divergence and curl at all points in space. Gauss's law gives only $\nabla \cdot \mathbf{E}$. Coulomb's law specifies in addition that $\nabla \times \mathbf{E} = 0$. We will see later that although Gauss's law holds even when charges are in motion, $\nabla \times \mathbf{E} = 0$ is not generally valid. Hence Gauss's law has general validity, but Coulomb's law is valid only for electrostatics.

## 2-4 ALTERNATIVE DERIVATION OF GAUSS'S LAW

The differential form of Gauss's law can alternatively be derived by directly forming the dot product of $\nabla$ with the $\mathbf{E}$ of Coulomb's law. Some of the methods of this alternative derivation will be useful in later sections. From the Coulomb expression in Equation (1–72a) for the electric field due to a continuous charge density $\rho(\mathbf{r}')$, the divergence of $\mathbf{E}$ is

$$\nabla_{\mathbf{r}} \cdot \mathbf{E} = \frac{1}{4\pi\varepsilon_0} \int \rho(\mathbf{r}') \nabla_{\mathbf{r}} \cdot \left( \frac{\mathbf{R}}{R^3} \right) dV' \tag{2-24}$$

where $\mathbf{R} = \mathbf{r} - \mathbf{r}'$ and the integral is over all space. In Equation (2–24) we have moved the divergence operation inside the integral, since $\rho(\mathbf{r}')$ and the limits do not depend on $\mathbf{r}$. Here we have labeled $\nabla$ by the subscript $\mathbf{r}$ to indicate which coordinate it differentiates. To derive an expression for the factor $\nabla_{\mathbf{r}} \cdot (\mathbf{R}/R^3)$ that appears in the integrand, we first note that for fixed $\mathbf{r}'$, $dx = d(x - x')$, $dy = d(y - y')$, and $dz = d(z - z')$. Consequently, we have $\nabla_{\mathbf{r}} = \nabla_{\mathbf{R}}$. Making use of the vector identity

$$\nabla(s\mathbf{V}) = \mathbf{V} \cdot \nabla s + s\,\nabla \cdot \mathbf{V} \tag{2-25}$$

where $s$ is a scalar and $\mathbf{V}$ is a vector, we can write

$$\nabla_{\mathbf{R}} \cdot \frac{\mathbf{R}}{R^3} = \frac{1}{R^3}\,\nabla_{\mathbf{R}} \cdot \mathbf{R} + \mathbf{R} \cdot \nabla_{\mathbf{R}}\left(\frac{1}{R^3}\right) \tag{2-26}$$

In Equation (2–26) the divergence of the vector $\mathbf{R} = \hat{\mathbf{x}}X + \hat{\mathbf{y}}Y + \hat{\mathbf{z}}Z$ is

$$\nabla_{\mathbf{R}} \cdot \mathbf{R} = \frac{\partial}{\partial X}(X) + \frac{\partial}{\partial Y}(Y) + \frac{\partial}{\partial Z}(Z) = 3 \tag{2-27}$$

and the gradient of $1/R^3$, from Equation (1–71) with $n = 3$, is

$$\nabla_{\mathbf{R}}\left(\frac{1}{R^3}\right) = -\frac{3}{R^5}\,\mathbf{R} \tag{2-28}$$

Hence we obtain

$$\nabla_{\mathbf{R}} \cdot \frac{\mathbf{R}}{R^3} = \frac{3}{R^3} - \frac{3}{R^3} = 0 \tag{2-29}$$

An alternative way to obtain this result is to use Equation (2–19) for the divergence in spherical coordinates, taking $E_r = 1/R^2$. However, Equation (2–29) is not valid at $\mathbf{R} = 0$ where the derivatives are undefined. To treat the $\mathbf{R} = 0$ singular point, we apply the divergence theorem to the vector $\mathbf{R}/R^3$, with the integrations over the coordinates $X$, $Y$, and $Z$ of $\mathbf{R}$:

$$\int \nabla_{\mathbf{R}} \cdot \left(\frac{\mathbf{R}}{R^3}\right) dV = \oint \frac{\mathbf{R} \cdot d\mathbf{S}}{R^3} \tag{2-30}$$

By using a relation analogous to Equation (2–4) for the vector $\mathbf{R}$, this equation becomes

$$\int \nabla_{\mathbf{R}} \cdot \left(\frac{\mathbf{R}}{R^3}\right) dV = \oint d\Omega = 4\pi \tag{2-31}$$

Thus the value of the integral is *independent* of the volume it is integrated over.

Since the volume enclosing the point $\mathbf{R} = 0$ can be made arbitrarily small, we conclude from Equation (2–31) that

$$\nabla_{\mathbf{R}} \cdot \left( \frac{\hat{\mathbf{R}}}{R^2} \right) = 4\pi\delta^3(\mathbf{R}) \tag{2-32}$$

where $\delta^3(\mathbf{R})$ is the three-dimensional delta function. Thus the divergence of $\hat{\mathbf{R}}/R^2$ is singular at $\mathbf{R} = 0$. Substituting this result into Equation (2–24),

$$\nabla \cdot \mathbf{E} = \frac{1}{\varepsilon_0} \int \rho(\mathbf{r}')\delta^3(\mathbf{r} - \mathbf{r}')\, dV' \tag{2-33}$$

we obtain Equation (2–23):

$$\nabla \cdot \mathbf{E} = \frac{\rho}{\varepsilon_0}$$

We will now apply Gauss's law in computing the charge density that produces an electric field

$$\mathbf{E}(\mathbf{r}) = \frac{q}{4\pi\varepsilon_0} f(R) \frac{\hat{\mathbf{R}}}{R^2} \tag{2-34}$$

From Equation (2–23) we have

$$\rho(\mathbf{r}) = \varepsilon_0 \nabla_{\mathbf{r}} \cdot E(\mathbf{r}) = \frac{q}{4\pi} f(R)\, \nabla_{\mathbf{r}} \cdot \left( \frac{\hat{\mathbf{R}}}{R^2} \right) + \frac{q}{4\pi} \frac{\hat{\mathbf{R}}}{R^2} \cdot \nabla_{\mathbf{r}} f(R) \tag{2-35}$$

Since $\mathbf{r}'$ is fixed, $\nabla_{\mathbf{r}} = \nabla_{\mathbf{R}}$; hence using Equation (2–32), we obtain

$$\rho(\mathbf{r}) = qf(0)\delta^3(\mathbf{R}) + \frac{q}{4\pi R^2} \frac{df(R)}{dR} \tag{2-36}$$

A formula similar to Equation (2–32) exists in cylindrical coordinates:

$$\nabla \cdot \left( \frac{\hat{\boldsymbol{\imath}}}{\imath} \right) = 2\pi\delta^2(\boldsymbol{\imath}) \tag{2-37}$$

where $\boldsymbol{\imath} = x\hat{\mathbf{x}} + y\hat{\mathbf{y}}$ is the cylindrical radius vector. To show that $\nabla \cdot (\hat{\boldsymbol{\imath}}/\imath)$ vanishes for $\imath \neq 0$ is straightforward. To derive the singular contribution, we integrate over a cylinder of radius $\imath$ and length $\ell$:

$$\int \nabla \cdot \left( \frac{\hat{\boldsymbol{\imath}}}{\imath} \right) dV = \oint \frac{\hat{\boldsymbol{\imath}}}{\imath} \cdot d\mathbf{S} = 2\pi\ell \tag{2-38a}$$

Since $\hat{\boldsymbol{\imath}}$ does not depend on $z$, we can do the $dz$ part of the volume integral, to obtain

$$\int dx\, dy\, \nabla \cdot \left(\frac{\hat{\boldsymbol{\imath}}}{\imath}\right) = 2\pi \qquad \textbf{(2-38b)}$$

Since this integral is independent of the cylinder radius $\imath$, we conclude that Equation (2–37) holds.

An electric field of the form

$$\mathbf{E}(\mathbf{r}) = \frac{\lambda}{2\pi\varepsilon_0}\frac{\hat{\boldsymbol{\imath}}}{\imath} \qquad \textbf{(2-39)}$$

has a charge density given by

$$\rho = \varepsilon_0\, \nabla \cdot \mathbf{E} = \frac{\lambda}{2\pi}\, \nabla \cdot \left(\frac{\hat{\boldsymbol{\imath}}}{\imath}\right) = \lambda\delta^2(\imath) \qquad \textbf{(2-40)}$$

Hence Equation (2–39) is the electric field from a line charge density $\lambda$.

# 2-5 GENERAL APPLICATION OF GAUSS'S LAW TO SYMMETRIC CHARGE DISTRIBUTIONS

The integral form of Gauss's law provides a simple method of calculating the electric field in three general situations of common interest: (A) infinite plane sheets of charge, (B) spherically symmetric charge distributions, (C) infinitely long, cylindrically symmetric charge distributions. In each of these cases the consideration of symmetry suggests a surface to which the electric field is normal and constant in magnitude at all points. Then the flux integral in Gauss's law of Equation (2–8) can be immediately evaluated. We illustrate in these three cases in turn.

## 2-5A Infinite Charge Sheet

In Section 1–8 the electric field of an infinite uniformly charged sheet was calculated by direct integration of Coulomb's law. Here we show how the result can be more easily obtained from Gauss's law. By the symmetry of the charge distribution, the electric field is necessarily normal to the sheet. For a positive charge density $\sigma$ per unit area, the electric field points outward from the sheet on each side. We imagine an integration surface in the shape of a cylinder, as illustrated in Figure 2–6. Inasmuch as all the flux leaves from

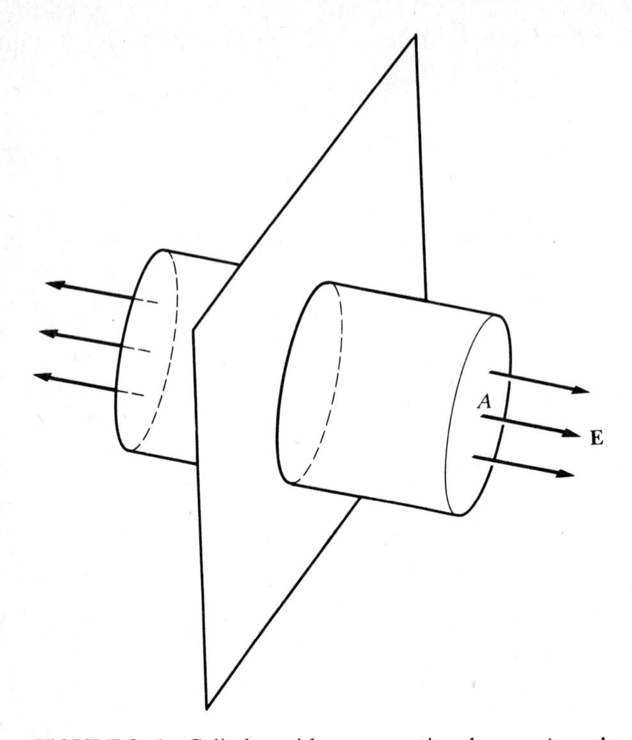


**FIGURE 2-6** Cylinder with cross-sectional area $A$ used as the integration surface to determine the electric field of an infinite charge sheet

the top and bottom of the cylinder, the evaluation of

$$\int \mathbf{E} \cdot d\mathbf{S} = \frac{Q}{\varepsilon_0}$$

over the area $A$ of each end gives

$$2EA = \frac{\sigma A}{\varepsilon_0} \qquad \textbf{(2-41)}$$

Thus

$$\mathbf{E} = \frac{\sigma}{2\varepsilon_0}\,\hat{\mathbf{n}} \qquad \textbf{(2-42)}$$

where $\hat{\mathbf{n}}$ is a unit vector pointing out from the surface. This result agrees with Equations (1-78) and (1-79).

For two or more charged sheets the net electric field is the vector superposition of the fields produced by each sheet.

## 2-5B Spherical Charge Distribution

Next, we consider the case of a spherically symmetric, continuous charge distribution $\rho(r)$. We choose the origin of our coordinate system to be at the center of the distribution. Owing to the symmetry the electric field must be in the radial direction and be a function only of the magnitude of $\mathbf{r}$:

$$\mathbf{E}(\mathbf{r}) = E(r)\hat{\mathbf{r}} \tag{2-43}$$

We choose a spherical surface ($\hat{\mathbf{n}} = \hat{\mathbf{r}}$) of radius $r$ for the surface integral in Equation (2–8). On the surface $E(r)$ is a constant, and we obtain

$$\oint \mathbf{E}\cdot\hat{\mathbf{n}}\,dS = E(r)\oint dS = E(r)4\pi r^2 \tag{2-44}$$

The total charge enclosed by the surface is

$$Q(r) = \int \rho(r')\,dV' = 4\pi\int_0^r \rho(r')r'^2\,dr' \tag{2-45}$$

From Gauss's law the result for the electric field is

$$\mathbf{E}(r) = \frac{Q(r)\hat{\mathbf{r}}}{4\pi\varepsilon_0 r^2} = \frac{\hat{\mathbf{r}}}{\varepsilon_0 r^2}\int_0^r \rho(r')r'^2\,dr' \tag{2-46}$$

Thus the electric field $\mathbf{E}(\mathbf{r})$ of a spherical charge distribution is equivalent to that of a point charge $Q(r)$ located at the center of the distribution: The charges at $r' > r$ do not influence the electric field at $r$. If all the charge is contained within a radius $r = a$, then for $r > a$, $Q(r) = Q(a)$, and the electric field is

$$\mathbf{E}(r) = \frac{Q(a)\hat{\mathbf{r}}}{4\pi\varepsilon_0 r^2}, \qquad \text{for } r > a \tag{2-47}$$

In Newtonian gravitation the same holds: For a spherical mass distribution the gravitational force at the point $r$ acts as if the mass interior to the point $r$ were concentrated at the center and the mass exterior to $r$ had no effect.

As a particular illustration of Equation (2–46), we consider the charge distribution $\rho(r) = (Q/4\pi a^2)\delta(r - a)$, representing a charged shell at radius $r = a$ of total charge $Q$. We obtain

$$\mathbf{E}(\mathbf{r}) = \begin{cases} \dfrac{Q\hat{\mathbf{r}}}{4\pi\varepsilon_0 r^2}, & \text{for } r > a \\ 0, & \text{for } r < a \end{cases} \tag{2-48}$$

which is illustrated in Figure (2–7). The differential of the electric potential defined in Equation (1–31),

$$d\Phi(\mathbf{r}) = -\mathbf{E}(\mathbf{r})\cdot d\mathbf{r} \tag{2-49}$$

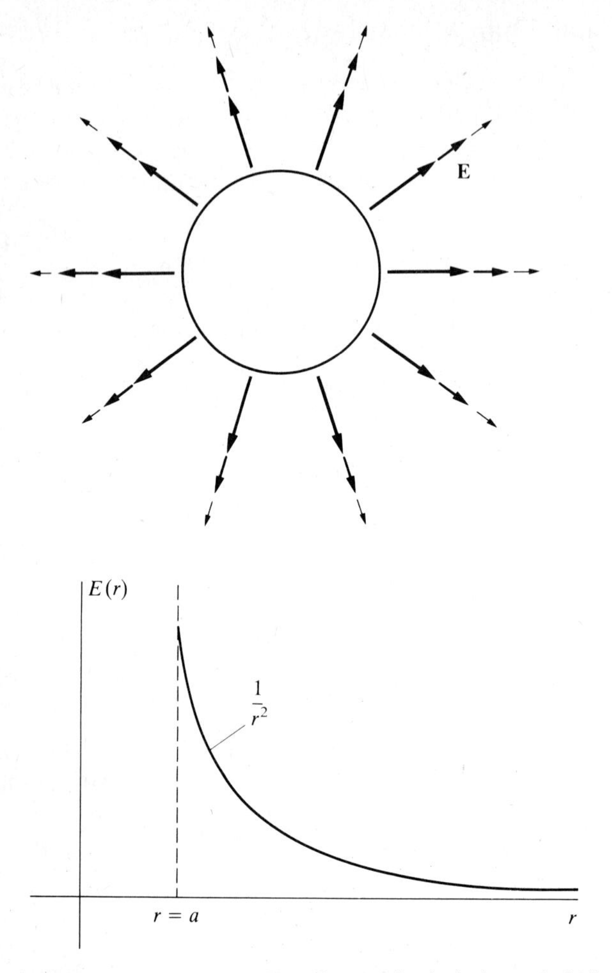


**FIGURE 2-7** Electric field for a spherical charged shell of radius $r = a$. The field falls like $1/r^2$ outside the shell.

is, from Equation (2–48),

$$d\Phi(r) = \begin{cases} -\dfrac{Q}{4\pi\varepsilon_0 r^2}\,dr, & \text{for } r > a \\ 0, & \text{for } r < a \end{cases} \tag{2-50}$$

for the shell. Integrating from $\infty$ to $r$, we get

$$\Phi(r) - \Phi(\infty) = \frac{Q}{4\pi\varepsilon_0}\frac{1}{r} \tag{2-51}$$

The value of $\Phi(\infty)$ can be chosen to be zero. Inside the shell $\mathbf{E} = 0$, so from Equation (2–49) the potential is the same everywhere in the interior. Hence

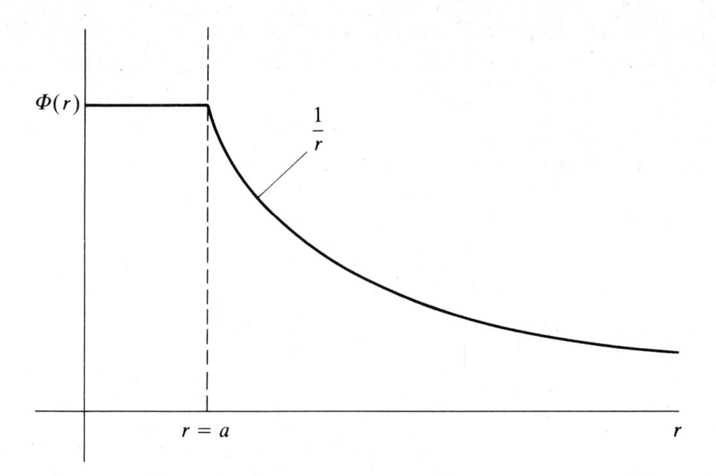


**FIGURE 2-8** Electric potential for a spherical charged shell of radius $r = a$. The potential falls like $1/r$ outside the shell.

the electric potential of the shell is

$$\Phi(r) = \begin{cases} \dfrac{Q}{4\pi\varepsilon_0 r}, & \text{for } r \geqslant a \\ \dfrac{Q}{4\pi\varepsilon_0 a}, & \text{for } r \leqslant a \end{cases} \tag{2-52}$$

as illustrated in Figure 2–8.

## 2-5C Cylindrical Charge Distribution

The integral form of Gauss's law in Equation (2–8) can be applied to cylindrical charge distributions in a similar way as the preceding spherical case. But instead, we pursue a different method here based on the differential form of Gauss's law in Equation (2–23). For this purpose we need the expression for $\nabla \cdot \mathbf{E}$ in cylindrical coordinates $(\imath, \phi, z)$.

We start with the cylindrical component forms of $\nabla$ and $\mathbf{E}$:

$$\begin{aligned} \nabla &= \hat{\boldsymbol{\imath}}\frac{\partial}{\partial \imath} + \hat{\boldsymbol{\phi}}\frac{1}{\imath}\frac{\partial}{\partial \phi} + \hat{\mathbf{z}}\frac{\partial}{\partial z} \\ \mathbf{E} &= \hat{\boldsymbol{\imath}}E_{\imath} + \hat{\boldsymbol{\phi}}E_{\phi} + \hat{\mathbf{z}}E_z \end{aligned} \tag{2-53}$$

The unit vectors in the cylindrical system are related to those in the Cartesian system by

$$\hat{\mathbf{z}} = \hat{\mathbf{z}} \qquad \hat{\boldsymbol{\imath}} = \hat{\mathbf{x}}\cos\phi + \hat{\mathbf{y}}\sin\phi \qquad \hat{\boldsymbol{\phi}} = -\hat{\mathbf{x}}\sin\phi + \hat{\mathbf{y}}\cos\phi \tag{2-54}$$

as can be deduced from the geometry of Figures 1–13 and 2–9. We note that the $z$ and $\imath$ derivatives of the unit vectors vanish but $\phi$ derivatives are non-zero:

$$\frac{\partial \hat{\boldsymbol{\imath}}}{\partial \phi} = -\hat{\mathbf{x}} \sin \phi + \hat{\mathbf{y}} \cos \phi = \hat{\boldsymbol{\phi}}$$
$$\frac{\partial \hat{\boldsymbol{\phi}}}{\partial \phi} = -\hat{\mathbf{x}} \cos \phi - \hat{\mathbf{y}} \sin \phi = -\hat{\boldsymbol{\imath}} \tag{2-55}$$

Hence when we form the dot product of $\nabla$ and $\mathbf{E}$, we obtain

$$\nabla \cdot \mathbf{E} = \frac{\partial E_{\imath}}{\partial \imath} + \frac{1}{\imath}\frac{\partial E_\phi}{\partial \phi} + \hat{\boldsymbol{\phi}} \cdot \frac{\partial \hat{\boldsymbol{\imath}}}{\partial \phi} E_{\imath} + \frac{\partial E_z}{\partial z} + \frac{1}{\imath} \hat{\boldsymbol{\phi}} \cdot \frac{\partial \hat{\boldsymbol{\phi}}}{\partial \phi} E_\phi$$
$$= \frac{\partial E_{\imath}}{\partial \imath} + \frac{1}{\imath} E_{\imath} + \frac{1}{\imath}\frac{\partial E_\phi}{\partial_\phi} + \frac{\partial E_z}{\partial z} \tag{2-56}$$

This expression for the divergence in cylindrical coordinates can be recast in the form

$$\nabla \cdot \mathbf{E} = \frac{1}{\imath}\frac{\partial}{\partial \imath}(\imath E_{\imath}) + \frac{1}{\imath}\frac{\partial E_\phi}{\partial \phi} + \frac{\partial E_z}{\partial z} \tag{2-57}$$

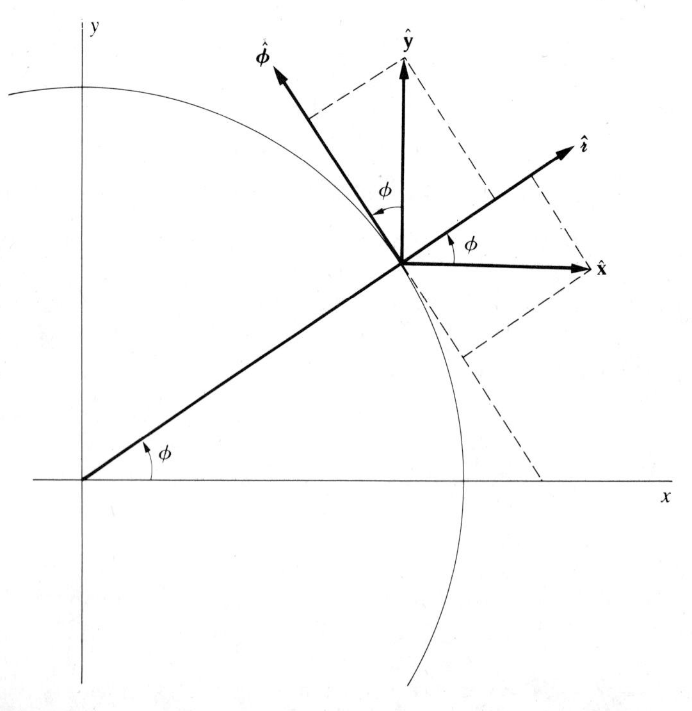


**FIGURE 2-9** Geometry of cylindrical and Cartesian coordinate unit vectors

An alternative method of obtaining this result was outlined in Section 2–2.

We now use Gauss's law in differential form to find the electric field from an infinitely long, continuous charge distribution of cylindrical symmetry. We take the $z$ axis on the line of symmetry. Owing to the symmetry $E_z = E_\phi = 0$ and $E_r$ can be a function only of $r$; and the partial derivative $\partial/\partial r$ becomes a total derivative $d/dr$. Equation (2–23) becomes

$$\frac{1}{r}\frac{d}{dr}[rE_r(r)] = \frac{1}{\varepsilon_0}\rho(r) \tag{2-58}$$

This differential equation can be solved by direct integration:

$$E_r(r) = \frac{[rE_r(r)]_{r=0}}{r} + \frac{1}{\varepsilon_0 r}\int_0^r \rho(r')r'\,dr' \tag{2-59}$$

In the absence of a line charge at $r = 0$, the boundary condition $[rE_r(r)]_{r=0} = 0$ is appropriate since $E_r(r)$ is then a nonsingular function.

The total charge per unit length enclosed in a cylinder of radius $r$ is

$$Q(r) = \int_0^{2\pi} d\phi' \int_0^r r'\,dr'\,\rho(r') \tag{2-60}$$

After doing the trivial integration over $d\phi'$ in Equation (2–60), we recognize that Equation (2–59) can be expressed in terms of $Q(r)$ as

$$E_r(r) = \frac{Q(r)}{2\pi\varepsilon_0 r} \tag{2-61}$$

This expression is just what would be obtained from application of the integral form of Gauss's law, as the reader can verify.

For the case of a distribution amounting to a total charge $\lambda$ per unit length, the electric field at a radius $r$ outside the charge region is

$$E_r(r) = \frac{\lambda}{2\pi\varepsilon_0 r} \tag{2-62}$$

In the case of a *line charge*, where the charge is concentrated at $r = 0$, the differential approach above cannot be used. However, from Equation (2–39) we see that Equation (2–62) still holds. The corresponding potential for the line charge is

$$\Phi(r) = -\int^r E_r(r)\,dr + \text{constant} = -\frac{\lambda}{2\pi\varepsilon_0}\ln r + \text{constant} \tag{2-63}$$

**TABLE 2-1** Summary of electric fields and potentials

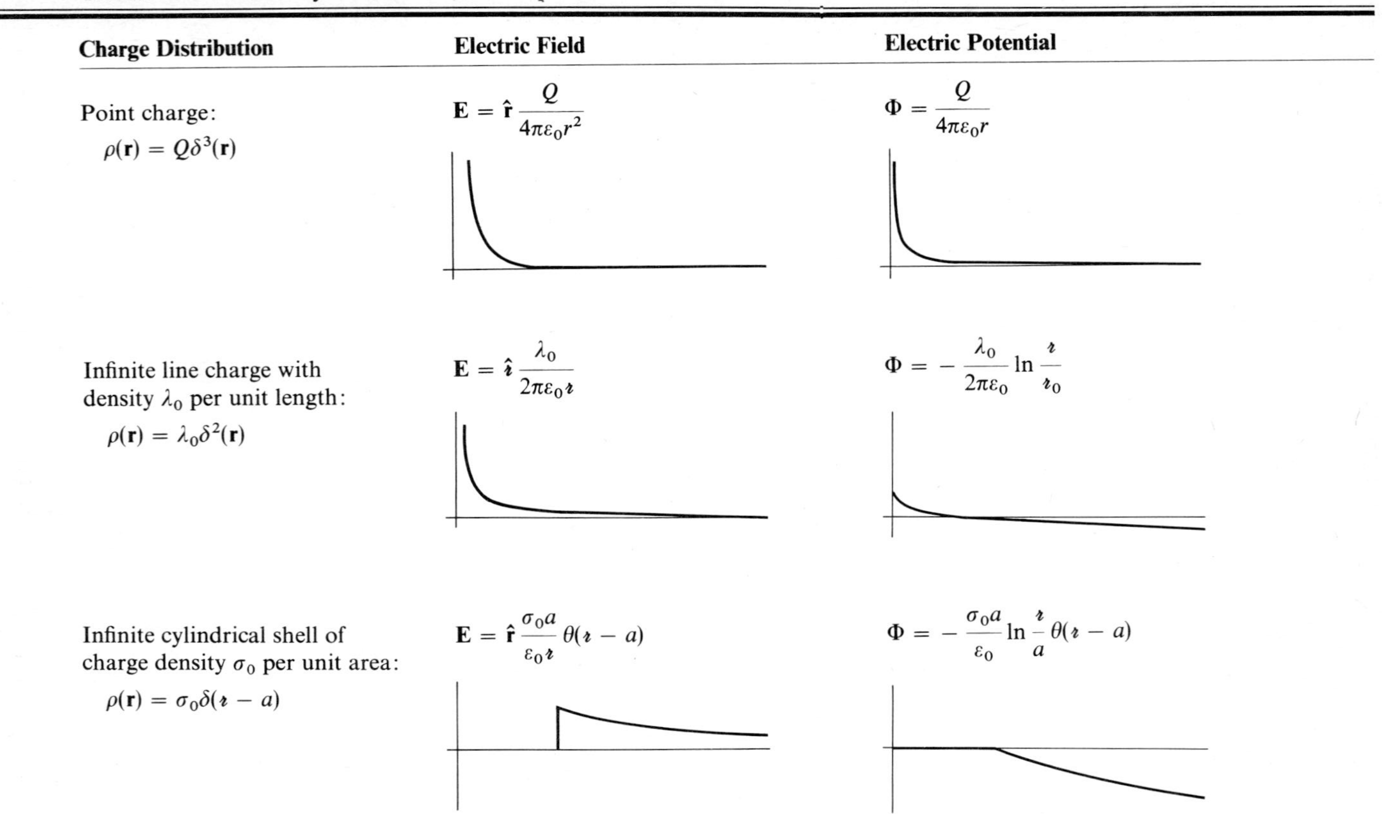

| Charge Distribution | Electric Field | Electric Potential |
|---|---|---|
| Point charge: $\rho(\mathbf{r}) = Q\delta^3(\mathbf{r})$ | $\mathbf{E} = \hat{\mathbf{r}}\dfrac{Q}{4\pi\varepsilon_0 r^2}$ | $\Phi = \dfrac{Q}{4\pi\varepsilon_0 r}$ |
| Infinite line charge with density $\lambda_0$ per unit length: $\rho(\mathbf{r}) = \lambda_0\delta^2(\mathbf{r})$ | $\mathbf{E} = \hat{\mathscr{r}}\dfrac{\lambda_0}{2\pi\varepsilon_0 \mathscr{r}}$ | $\Phi = -\dfrac{\lambda_0}{2\pi\varepsilon_0}\ln\dfrac{\mathscr{r}}{\mathscr{r}_0}$ |
| Infinite cylindrical shell of charge density $\sigma_0$ per unit area: $\rho(\mathbf{r}) = \sigma_0\delta(\mathscr{r} - a)$ | $\mathbf{E} = \hat{\mathbf{r}}\dfrac{\sigma_0 a}{\varepsilon_0 \mathscr{r}}\theta(\mathscr{r} - a)$ | $\Phi = -\dfrac{\sigma_0 a}{\varepsilon_0}\ln\dfrac{\mathscr{r}}{a}\theta(\mathscr{r} - a)$ |

Infinite cylinder of uniform charge density $\rho_0$ per unit volume:

$$\rho(\mathbf{r}) = \rho_0\theta(a - \mathscr{r})$$

$$\mathbf{E} = \hat{\mathscr{r}}\left[\frac{\rho_0 a^2}{2\varepsilon_0 \mathscr{r}}\theta(\mathscr{r} - a) + \frac{\rho_0 \mathscr{r}}{2\varepsilon_0}\theta(a - \mathscr{r})\right]$$

$$\Phi = -\frac{\rho_0 a^2}{2\varepsilon_0}\ln\frac{\mathscr{r}}{a}\theta(\mathscr{r} - a) - \frac{\rho_0(\mathscr{r}^2 - a^2)}{4\varepsilon_0}\theta(a - \mathscr{r})$$

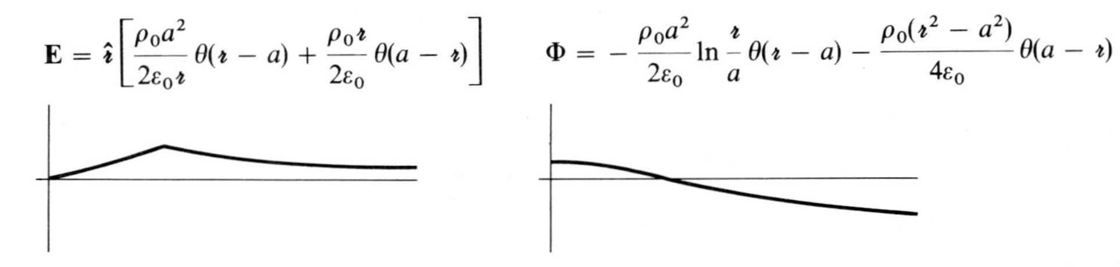

Charged spherical shell:

$$\rho(\mathbf{r}) = \sigma_0\delta(r - a)$$

$$\mathbf{E} = \frac{\sigma_0 a^2}{\varepsilon_0 r^2}\theta(r - a)\hat{\mathbf{r}}$$

$$\Phi = \frac{\sigma_0 a^2}{\varepsilon_0 r}\theta(r - a) + \frac{\sigma_0 a}{\varepsilon_0}\theta(a - r)$$

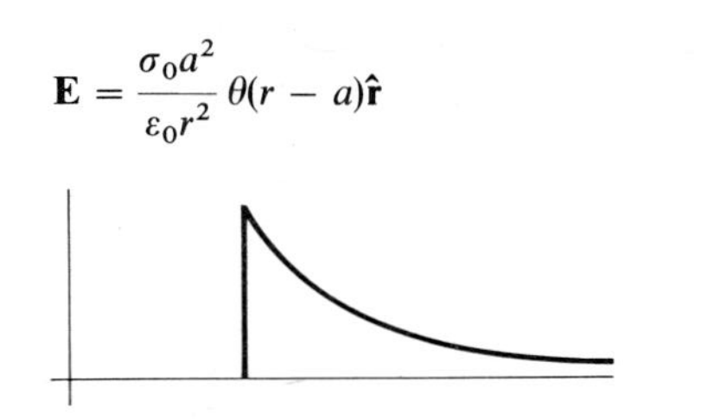

Uniformly charged sphere:

$$\rho(\mathbf{r}) = \rho_0\theta(a - r)$$

$$\mathbf{E} = \frac{\rho_0 a^3}{3\varepsilon_0 r^2}\left[\theta(r - a) + \frac{\rho_0 r}{3\varepsilon_0}\theta(a - r)\right]\hat{\mathbf{r}}$$

$$\Phi = \frac{\rho_0 a^3}{3\varepsilon_0 r}\theta(r - a) + \frac{\rho_0}{6\varepsilon_0}(3a^2 - r^2)\theta(a - r)$$

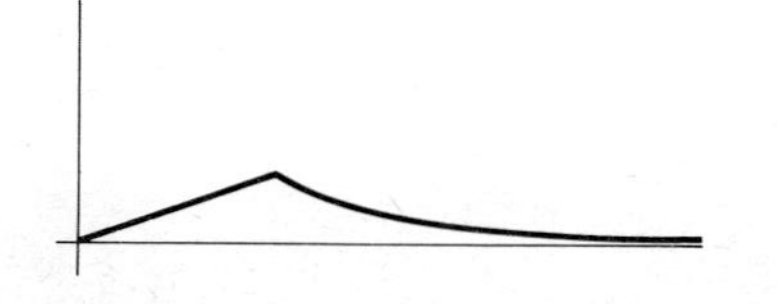

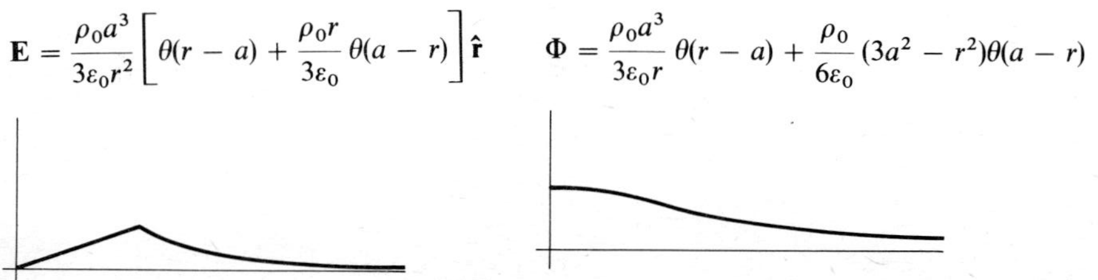

**TABLE 2-1** *Continued*

| Charge Distribution | Electric Field | Electric Potential |
|---|---|---|
| Infinite, planar charged sheet: $\rho(\mathbf{r}) = \sigma_0 \delta(z)$ | $\mathbf{E} = \hat{\mathbf{z}}\left[\frac{\sigma_0}{2\varepsilon_0}\theta(z) - \frac{\sigma_0}{2\varepsilon_0}\theta(-z)\right]$ 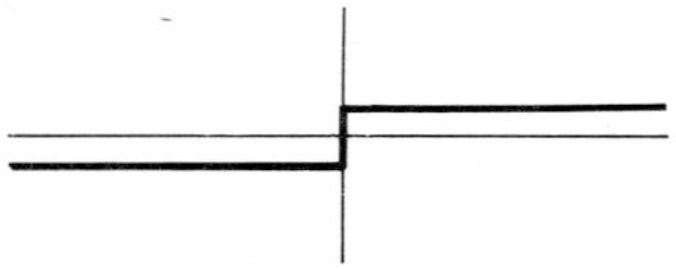 | $\Phi = -\frac{\sigma_0 \lvert z \rvert}{2\varepsilon_0}$ 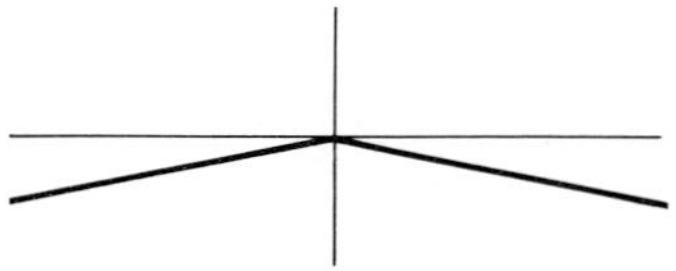 |
| Charged semi-infinite conducting region ($z < 0$): $\rho(\mathbf{r}) = \sigma_0 \delta(z)$ | $\mathbf{E} = \hat{\mathbf{z}}\frac{\sigma_0}{\varepsilon_0}\theta(z)$ 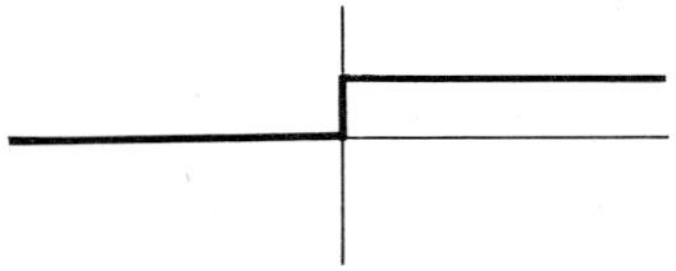 | $\Phi = -\frac{\sigma_0 z}{\varepsilon_0}\theta(z)$ 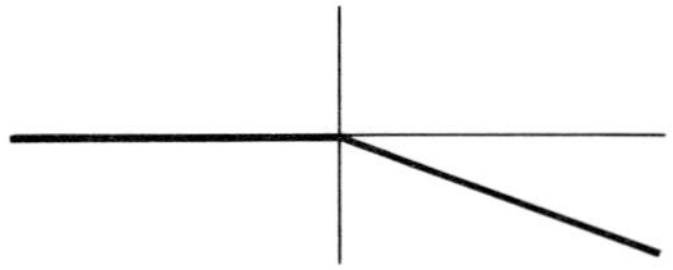 |
| Two infinite, planar charged sheets of opposite charge: $\rho(\mathbf{r}) = \sigma_0[\delta(z + a) - \delta(z - a)]$ | $\mathbf{E} = \hat{\mathbf{z}}\frac{\sigma_0}{\varepsilon_0}[\theta(z + a) - \theta(z - a)]$ 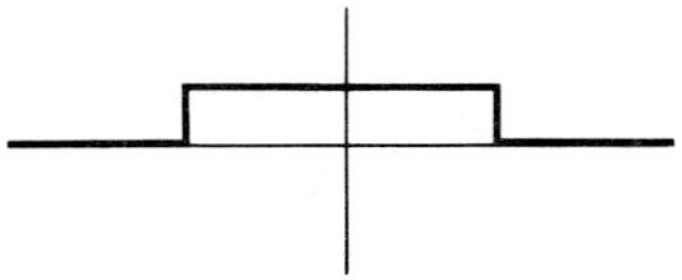 | $\Phi = -\frac{\sigma_0}{\varepsilon_0}[(z + a)\theta(z + a) - (z - a)\theta(z - a)]$ 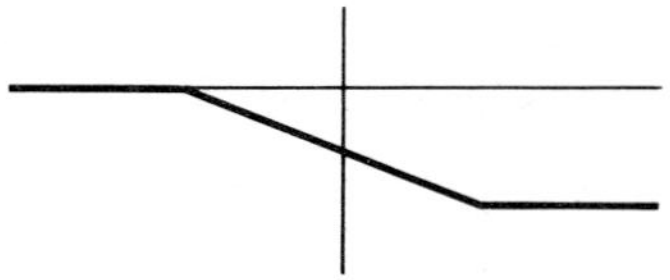 |

To avoid infinities, we cannot choose the reference potential at $r = 0$ or $r = \infty$ in this case. Choosing the zero of the potential at $r = r_0$, we have

$$\Phi(r) = -\frac{\lambda}{2\pi\varepsilon_0} \ln \frac{r}{r_0} \quad \textbf{(2-64)}$$

We can equally well use the differential form of Gauss's law in applications involving planar or spherical symmetry. In planar applications we use the Cartesian form of the divergence in Equation (2–10). For spherical applications the form of the divergence is given in Equation (2–19); the method used above to derive the form of $\nabla \cdot \mathbf{E}$ in cylindrical coordinates can also be used in deriving Equation (2–19).

A summary of electric fields and potentials obtained from the application of Gauss's law to some simple charge distributions is given in Table 2–1.

## 2-6 APPLICATION: ELECTROSTATIC PRECIPITATION

The electrostatic precipitator removes fine suspended material from a gas by charging the particulate material and separating it from the gas with an electric field. The precipitator has assumed great importance in these times of environmental awareness and renewed emphasis on coal-fired electricity generation. For instance, the quantity of coal fly ash that is precipitated annually in the United States alone is nearly $10^8$ tons. Precipitators are used in diverse manufacturing processes and also on a small scale as household air cleaners connected to the central heating and air-conditioning system.

A typical precipitator design is shown in Figure 2–10. The precipitator consists of a grounded outside cylinder with a high-voltage wire running along its axis. The gas enters at one end, the particles are deposited on the grounded cylinder, and the cleaned gas leaves the other end. The device works as follows: The voltage is increased in magnitude until a *corona discharge* forms close to the central wire. In a corona, ions of both signs of charge are created. Ions of one sign are attracted toward the wire. The ions of the other sign are repelled outward. As these latter ions migrate toward the grounded cylinder, they attach themselves to the particulates and carry them to the cylinder. The applied voltage must be high enough to sustain the corona discharge but not so large that electrical breakdown occurs and sparks discharge directly to the outer cylinder. A corona may be produced by an applied voltage of either sign, but a corona with outward-migrating negative ions yields a higher discharge current and also has a limiting breakdown voltage that is higher. The negative corona is initiated by free electrons near the central wire. Imposition of a high electric field near the wire repels the free electrons, which then collide with gas molecules. The collisions produce positive ions and more free electrons. The positive ions are attracted

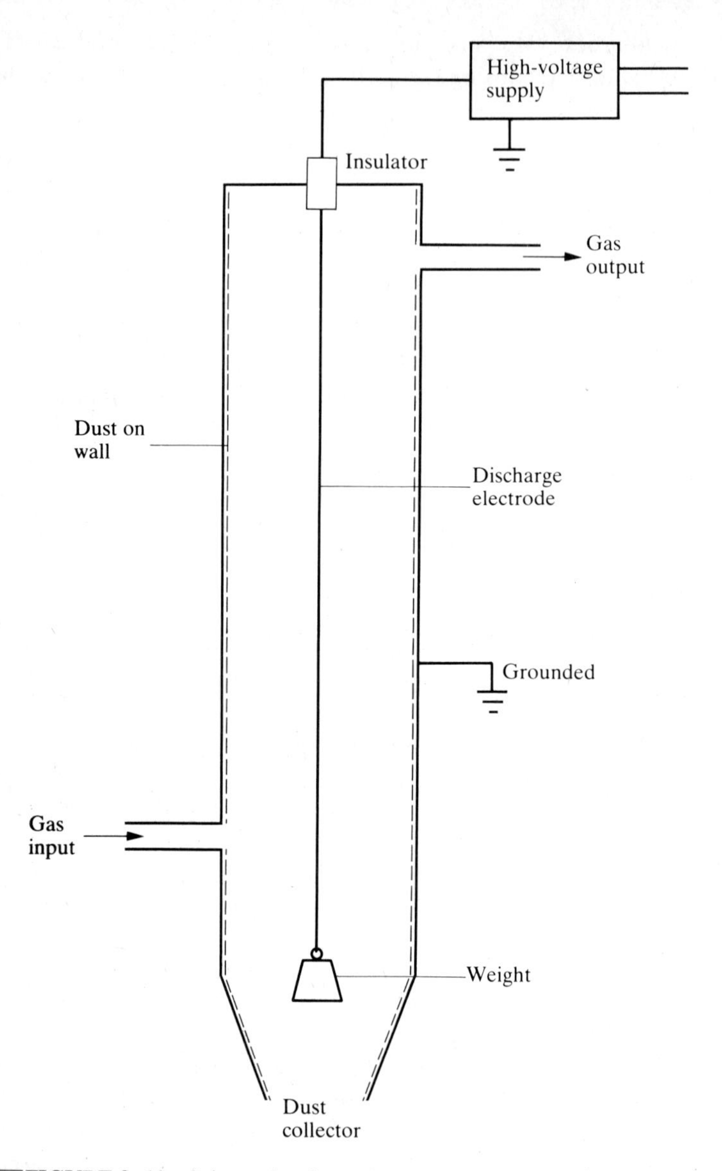


**FIGURE 2-10** Schematic of a typical electrostatic precipitator

to the wire, and the collisions with the wire release additional electrons. As electrons move outward into a region of weaker electric field, they attach to gas molecules to form negative ions. These negative ions fill the region outside the corona, and their movement toward the cylinder produces the corona current.

The electric field in a cylindrical electrostatic precipitator can be easily derived in the region outside the corona where ions of only one sign are present. We start with the differential form of Gauss's law for a cylindrically symmetric field with only a radial component. From Equation (2–58) we have

$$\frac{1}{r}\frac{d}{dr}(rE) = \frac{\rho(r)}{\varepsilon_0} \tag{2-65}$$

where $\rho(r)$ is the space charge, which is largely due to ions. The motion of an ion acted on by the electric field is impeded by repeated collisions with gas molecules in its path. The average drift velocity $v$ of the ion is empirically proportional to the field strength:

$$v = bE \tag{2-66}$$

where $b$ is the mobility constant. The charge of a cylindrical shell of radius $r$, thickness $dr$, and length $\ell$ is

$$dq = \rho\ell(2\pi r\, dr) \tag{2-67}$$

In time $dt$ this amount of charge flows radially through the cylindrical surface at radius $r$, giving rise to a current $i$ per unit length given by

$$i = \frac{1}{\ell}\frac{dq}{dt} = 2\pi r\rho v = 2\pi r\rho bE \tag{2-68}$$

For a steady-state flow charge, $i$ must be constant, independent of $r$. We can eliminate $\rho$ from Equations (2–65) and (2–68), giving

$$\frac{1}{r}\frac{d}{dr}(rE) = \frac{i}{2\pi\varepsilon_0 b}\left(\frac{1}{rE}\right) \tag{2-69}$$

This equation can be rearranged as

$$(rE)\, d(rE) = \frac{i}{2\pi\varepsilon_0 b}\, r\, dr$$

and then integrated from just beyond the corona ($r = r_1$, $E = E_1$) out to a distance $r$. The result is

$$E^2 = E_1^2\left(\frac{r_1}{r}\right)^2 + \frac{i}{2\pi\varepsilon_0 b}\left(1 - \frac{r_1^2}{r^2}\right) \tag{2-70}$$

For sufficiently large $r$ and $i$ the electric field approaches a constant value:

$$E = \left(\frac{i}{2\pi\varepsilon_0 b}\right)^{1/2} \tag{2-71}$$

which is often a good approximation in practical applications.

## 2-7 EARNSHAW'S THEOREM

From Gauss's law we can derive the following useful result of electrostatics known as Earnshaw's theorem.

> **THEOREM:** In a region of zero charge density a test charge cannot be in stable equilibrium under the influence of electrostatic forces only.

We prove the theorem by contradiction, assuming the converse is true at a point *P*, with the following conditions satisfied.

1. There is a small region about *P* containing no charge except a test charge.
2. At *P*, $\mathbf{E} = 0$ since it is a point of equilibrium (i.e., no electric force there).
3. At a small distance away from *P*, **E** points toward *P*, assuring stable equilibrium for the positive test charge at *P*, as illustrated in Figure 2–11. For a negative test charge **E** points away from *P*.

Applying Gauss's integral law to a small region about *P*, the conditions 2 and 3 imply that there must be charge contained within the integration surface. But this contradicts condition 1, thus proving the theorem. (Note that we exclude the trivial case where $\mathbf{E} = 0$ everywhere.) The **E** field

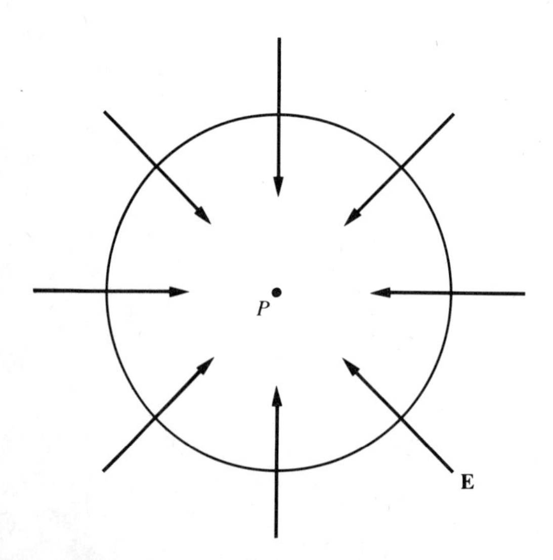


**FIGURE 2–11** Condition for a positive test charge to be in stable equilibrium: The electric field must point radially inward.

lines in Figure 2–11 are those of a point charge at $P$; but since there is no charge at $P$, field lines of this configuration cannot exist. The electric field must point away from $P$ in some directions and toward $P$ in other directions.

Earnshaw's theorem can be restated in terms of the electric potential. Since $\Phi = -\int \mathbf{E} \cdot d\mathbf{r}$ increases or decreases depending on the direction from $P$, the restatement of the theorem is as follows:

> **THEOREM:** There is no maximum or minimum of the potential within a region containing no charge density

(This theorem also follows from $\nabla \cdot \mathbf{E} = 0$ and $\mathbf{E} = -\nabla\Phi$, which gives $\nabla^2\Phi = 0$; see Section 3–3.) In the next section we will apply Earnshaw's theorem in discussing the potential within a conductor.

## 2-8 CONDUCTORS AND GAUSS'S LAW

Until this point, we have mainly considered charges without reference to their response to forces on them. The behavior of matter under an imposed electric field, although complex on the atomic level, can be broadly characterized by whether or not charge carriers move freely. A conductor is a material in which current can flow when an electric field is applied.

### 2-8A Electric Field and Potential of a Conductor

In a static-equilibrium situation where no current is flowing, the electric field must vanish throughout any conductor. Since $\mathbf{E} = 0$ everywhere within the conductor, $\nabla \cdot \mathbf{E} = 0$ internally, and hence by the differential form of Gauss's law $\nabla \cdot \mathbf{E} = \rho/\varepsilon_0$; the charge density inside the conductor vanishes. Thus any charge on the conductor must reside on the surface.

Another consequence of $\mathbf{E} = 0$ in the interior of a conductor in equilibrium is that the potential difference

$$\Phi(\mathbf{r}_b) - \Phi(\mathbf{r}_a) = -\int_{\mathbf{r}_a}^{\mathbf{r}_b} \mathbf{E} \cdot d\mathbf{r} \tag{2-72}$$

between any two points inside the conductor vanishes. The potential is a constant everywhere throughout the conductor—it is an *equipotential region.* In particular, the potential is constant on the conductor's surface, which is thus an *equipotential surface.*

The constancy of $\Phi(\mathbf{r})$ on the surface implies that components of the gradient tangent to the surface must vanish. Hence the tangential component of the electric field on the conductor's surface is zero:

$$\mathbf{E}_t = \hat{\mathbf{n}} \times \mathbf{E} = -\mathbf{n} \times \nabla\Phi = 0 \tag{2-73}$$

This result agrees with the argument that a tangential electric field would cause charges to move and thus violate the equilibrium assumption.

The previous argument does not apply to the normal component of the electric field,

$$E_n = \hat{\mathbf{n}} \cdot \mathbf{E} = -\hat{\mathbf{n}} \cdot \nabla\Phi \tag{2-74}$$

which can be nonvanishing immediately outside the conductor (i.e., discontinuous at the surface). In fact, it is nonvanishing if there is a nonvanishing charge density on the conductor's surface.

The normal electric field can be related to the surface charge density by means of Gauss's law. As a flux integration surface, we take the small prism illustrated in Figure 2–12. The bottom of the prism lies just infinitesimally beneath the conductor's surface and the top just infinitesimally above the surface. The normal area $A$ of the prism is chosen sufficiently small that $\mathbf{E}$ and $\sigma$ are constant over the area A. Application of Gauss's law,

$$\oint \mathbf{E} \cdot d\mathbf{S} = \frac{Q}{\varepsilon_0}$$

gives

$$EA = \frac{\sigma A}{\varepsilon_0} \tag{2-75}$$

since there is no $\mathbf{E}$ field component tangent to the conductor's surface (i.e., no flux leaves the prism from its cylindrical sides). The electric field just

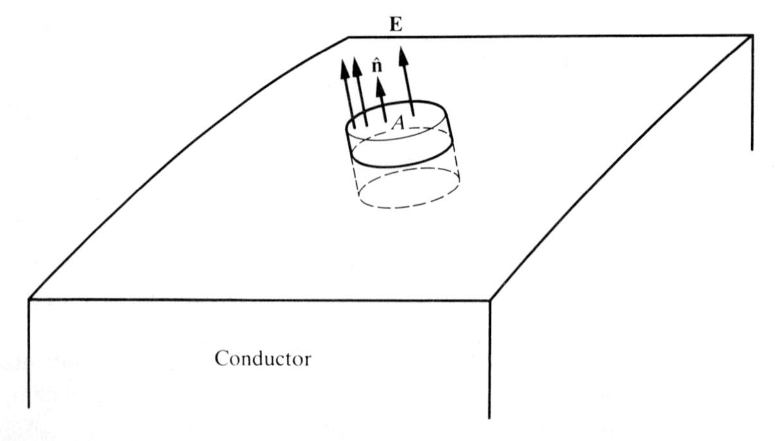


**FIGURE 2–12** Small prism of circular cross-sectional area $A$, used as the flux integration surface to determine the electric field outside the conductor's surface

outside the surface of the conductor is therefore

$$\mathbf{E} = \frac{\sigma}{\varepsilon_0}\hat{\mathbf{n}} \qquad \text{(just outside the conductor's surface)} \tag{2-76}$$

where $\hat{\mathbf{n}}$ is the *outward* normal to the surface.

In summary, we have shown that

$$\mathbf{E}(\mathbf{r}) = 0 \qquad \Phi(\mathbf{r}) = \text{constant} \qquad \text{(inside a conductor)} \tag{2-77}$$

and that

$$\begin{aligned} E_n(\mathbf{r}) &= \frac{\sigma(\mathbf{r})}{\varepsilon_0} \qquad E_t(\mathbf{r}) = 0 \\ \Phi(\mathbf{r}) &= \text{constant} \qquad \text{(just outside the conductor's surface)} \end{aligned} \tag{2-78}$$

## 2-8B Conductor with a Cavity

One of the most surprising and important properties of conductors is that there can be no **E** field inside an empty cavity in a conductor, as illustrated in Figure 2–13, no matter how the conductor is charged externally or what the shape of the conductor or cavity is. To show this property, we need only apply Earnshaw's theorem, proved in Section 2–7, to the cavity. Since the cavity is empty, Earnshaw's theorem specifies that the potential can have no maximum or minimum inside the cavity. The walls of the cavity form an equipotential surface since $\mathbf{E} = 0$ in the conductor. Hence if the potential inside the cavity deviated from its value on the walls, it would somewhere

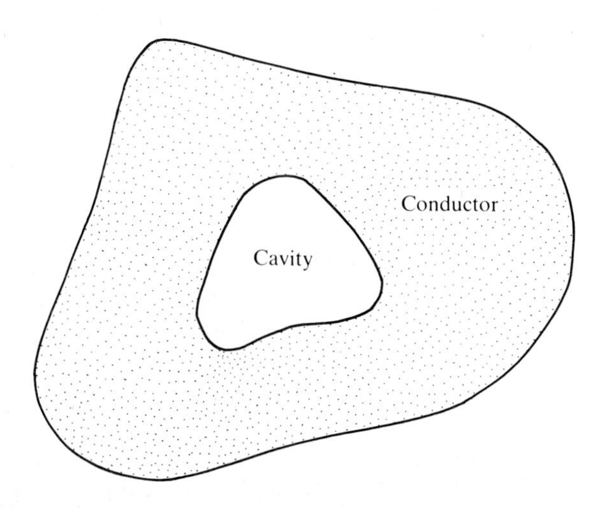


**FIGURE 2-13** Conductor with an arbitrarily shaped cavity

reach a minimum or maximum. Since this condition violates Earnshaw's theorem, the potential must be a constant throughout the cavity. The cavity therefore has no electric field, and the walls of the cavity have no surface charge density. These statements in no way depend on the charge on the conductor, the fields external to the conductor, or the shape of the conductor or cavity.

We have seen that there is no electric field within an empty cavity in a conductor. Surface charges distribute themselves on the outside of the hollow conductor, so the electric field vanishes everywhere within it, including the cavity, no matter what charges might be placed outside the conductor. Suppose now that we place two charged conductors A and B within the cavity of a conductor. There will be a field inside the cavity and a surface charge density induced on the cavity walls. The electric field within the cavity can be regarded as superposition of two parts: (1) $\mathbf{E}_{\text{ext}}$ due to all of the external charges, including those on the outer surface of the conductor, and (2) $\mathbf{E}_{\text{int}}$ due to the charges on A and B as well as those charges induced on the inner cavity surface. Since $\mathbf{E}_{\text{ext}}$ is always zero, we observe that the field in the cavity is exactly the same as if there were no external charge at all. The conductors A and B have been shielded from the effects of external charges. Whenever an electrical apparatus might be adversely affected by external electric fields, it can be enclosed within a conductor, such as wire mesh or aluminum foil, for shielding. A wire mesh is indistinguishable electrically from a solid conductor at distances large compared to the mesh size.

A dramatic demonstration of electrostatic-shielding phenomenon was provided by Michael Faraday in 1838 when he constructed a 4-$\text{m}^2$ metal cage mounted on insulating supports. With Faraday inside the cage was charged to a high voltage, evidenced by large spark discharges to the surroundings. Even with the cage so highly charged, Faraday could experimentally detect no electric field in the interior. As we have seen, all the excess charge would reside on the outside of the cage, and any electric field within the cage would not be affected. For exactly the same reason one of the safest places to be in an electrical storm is inside an automobile—provided, of course, that it is constructed of metal.

The earth is a conductor, albeit not a very good one. Because of its conducting property, the earth can be regarded as an equipotential body. Any conductor placed in contact with the earth will attain the same potential as the earth by the flow of charge; such a conductor is said to be *grounded.* As noted previously, the choice of reference potential is arbitrary. For situations involving grounded conductors the potential of the earth is conventionally chosen to be the reference point from which other potentials are measured; that is, by this convention $\Phi = 0$ on all grounded conductors.

Next, we consider a conductor with a cavity in which charge can be inserted through a very small hole drilled in the conductor, as illustrated in Figure 2–14. We can neglect the effect of the hole on the electric field and the charge distribution if the hole is small compared with the dimensions of the

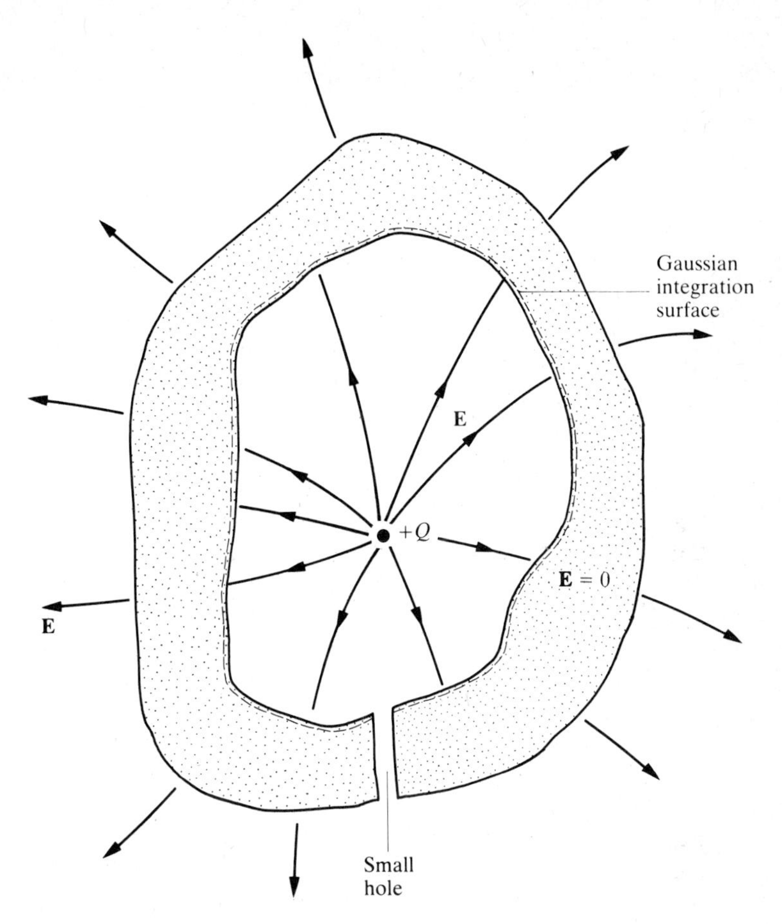


**FIGURE 2-14** Neutral conductor with a cavity and a very small access hole through which charges can be inserted

cavity. Suppose that the conductor is uncharged and that a charge $+Q$ is placed in the cavity. By application of Gauss's law we find that a charge $-Q$ is induced on the cavity walls and a charge $+Q$ appears on the outer surface of the conductor. Electric field lines that originate on the $+Q$ charge in the cavity end on the cavity walls.

If a conducting wire is placed between the interior charge and the walls, current will flow until the cavity and its inside surface have no net charge, while the charge remains on the outside surface of the conductor. This process is the principle of the Van de Graaff generator, devised in 1930, where a moving belt carries charge into a hollow conducting sphere and the charge flows onto the sphere; see Figure 2–15. The conducting sphere can thereby be raised to a very high potential. The maximum obtainable potential in air is that at which the air molecules become ionized by the

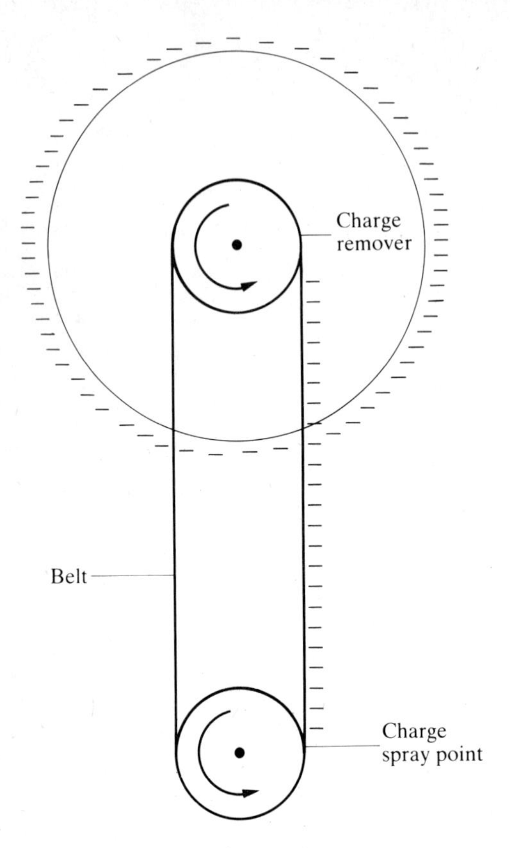


**FIGURE 2-15** Schematic of a Van de Graaff generator. The dome can be charged either positively or negatively.

potential and currents flow through the air. This dielectric breakdown occurs in air for an electric field strength of about $3 \times 10^6$ V/m.

A modern version of the Van de Graaff machine is shown in Figure 2–16. The high voltages are used in the acceleration of ions from a source placed at the sphere. The ions are accelerated down a tube to a target at earth potential. Collisions of the ions with targets provide important information about nuclear properties.

## 2-8C Cavendish Cavity and Faraday Cage Experiments

The history of the discovery of the inverse-square law for electrostatics is rather interesting. The American scientist Benjamin Franklin noticed that charged cork balls placed inside a conducting cup were unaffected by charge placed on the conductor. He felt this result was of great significance and communicated it to his friend Joseph Priestly, of England, in 1766. Priestly

**FIGURE 2–16** Modern version of a Van de Graaff machine

made an analogy with gravitation, where a mass within a massive spherical shell will experience no gravitational force. He concluded that the electrostatic force was also an inverse-square law. However, Priestly's theoretical argument had little impact when he reported his result in 1767.

In 1769 John Robinson of Edinburgh found an $r^{-2.06}$ experimental result for the force between two charged objects. Henry Cavendish then carried out experiments at Cambridge using two concentric charged conductors, to be described later. He determined that the force law must lie between $r^{-1.98}$ and $r^{-2.02}$. Cavendish did not publish his results, and they also had no impact on developments in the field. (A historical search by Maxwell brought the Cavendish notes to light a century later.) The experiment of Charles Augustin Coulomb in 1788–1789, measuring the force between charges, established the $r^{-2}$ behavior less accurately than those of Cavendish but well enough that it became known as Coulomb's law.

The most precise experimental determinations of the exponent in the electrostatic force law are based on Priestly's original idea, which was used in the experiment by Cavendish. A schematic of the experiment is shown in Figure 2–17. Two concentric conducting spheres are connected by a sensitive electrometer, which measures any charge flow between the spheres. The outer sphere is alternately charged to high potential and discharged, and evidence is sought for charge flow between the spheres. If Coulomb's inverse-square force is exact, the inner sphere will always remain exactly at the same potential as the outer sphere (since the potential is constant within a cavity in a conductor), and consequently no charge will flow. In contrast, for a force law $F = kr^{-2+\varepsilon}$, with $\varepsilon \neq 0$, a charge flow will occur

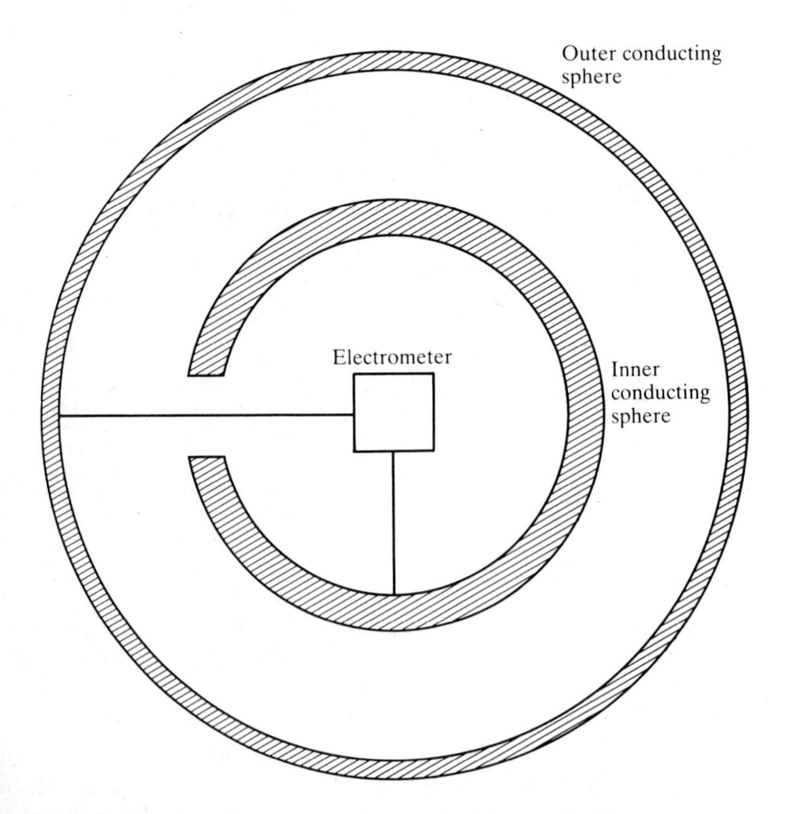


**FIGURE 2-17** Schematic of the experiment used by Cavendish to precisely measure the electrostatic force power law. The electrometer measures the charge transferred between the spheres.

when the potential of the outer sphere is changed. Of course, for $\varepsilon = 0$ no current will flow for any shape of conductor, but the theoretical evaluation of the expected flow for $\varepsilon \neq 0$ is easiest for concentric spheres. A modern version of this experiment [E. R. Williams, J. E. Faller, and H. A. Hill, *Phys. Rev. Lett.*, 26:721 (1971)] establishes that the deviation of $\varepsilon$ from 0 is less than $3 \times 10^{-16}$. This null experiment yields a vastly higher accuracy determination of $\varepsilon$ than a direct force measurement.

The exactness of Coulomb's law is also implied by the success of electromagnetic calculations on the submicroscopic level. The energy levels for hydrogen are accurately predicted by quantum applications of Coulomb's law at distances less than $10^{-10}$ m. Coulomb repulsion energies are successfully calculated in the atomic nucleus at distances of $10^{-15}$ m. At particle accelerators, where distances down to $10^{-18}$ m are probed, no deviations from the predictions of the quantum version of electromagnetic theory have been observed.

## 2-9 ELECTROSTATIC POTENTIAL ENERGY OF A SYSTEM OF DISCRETE CHARGES

For a system of particles that interact via the Coulomb force, the energy of the system is conserved, as shown in Section 1–4. Conservation of energy permits us to relate states of the system at different times without examining the intervening motion.

From Equation (1–34) the potential energy $U$ of a charge $q_1$ is

$$U = q_1\Phi_1 \tag{2-79}$$

where $\Phi_1$ is the electric potential at $\mathbf{r}_1$ due to all other charges. For a system of two charges the potential energy is

$$U = q_1\left(\frac{1}{4\pi\varepsilon_0}\frac{q_2}{R_{12}}\right) = \frac{q_1q_2}{4\pi\varepsilon_0 R_{12}} \tag{2-80}$$

where $R_{12}$ is the distance between $q_1$ and $q_2$. This energy is the amount of work needed to bring the charges from infinity to a separation $R_{12}$. From the superposition principle for $\Phi$, the total potential energy for a system of particles is the sum of potential energies of all possible pairs of particles.

For a system of three particles $U$ is given by

$$U = \frac{1}{4\pi\varepsilon_0}\left(\frac{q_1q_2}{R_{12}} + \frac{q_1q_3}{R_{13}} + \frac{q_2q_3}{R_{23}}\right) \tag{2-81}$$

where $R_{ij}$ denotes the distance between charges $q_i$ and $q_j$. This result can be recast in the form

$$U = \frac{1}{2}(q_1\Phi_1 + q_2\Phi_2 + q_3\Phi_3) \tag{2-82}$$

where

$$\begin{aligned}
\Phi_1 &= \frac{1}{4\pi\varepsilon_0}\left(\frac{q_2}{R_{12}} + \frac{q_3}{R_{13}}\right) \\
\Phi_2 &= \frac{1}{4\pi\varepsilon_0}\left(\frac{q_1}{R_{12}} + \frac{q_3}{R_{23}}\right) \\
\Phi_3 &= \frac{1}{4\pi\varepsilon_0}\left(\frac{q_1}{R_{13}} + \frac{q_2}{R_{23}}\right)
\end{aligned} \tag{2-83}$$

Here $\Phi_i$ is the potential at the charge $q_i$ due to the other two charges. The factor $\frac{1}{2}$ in Equation (2–82) reflects the fact that every pair of charges is counted twice (e.g., as $q_1q_2/R_{12}$ and $q_2q_1/R_{12}$).

The generalization for the potential energy of $N$ charges is

$$U = \frac{1}{2}\sum_{i=1}^{N} q_i\Phi_i \tag{2-84}$$

with

$$\Phi_i = \frac{1}{4\pi\varepsilon_0}\sum_{j\neq i}\frac{q_j}{R_{ij}}$$

With elimination of $\Phi_i$ Equation (2–84) becomes

$$U = \frac{1}{2}\left(\frac{1}{4\pi\varepsilon_0}\right)\sum_{j=1}^{N}\sum_{\substack{i=1 \\ (i\neq j)}}^{N}\frac{q_iq_j}{R_{ij}} \tag{2-85}$$

The $j = i$ term is omitted since it represents the self-energy, or the energy of a charge in its own field. The electrostatic self-energy of a point charge is infinite, since $R_{ii} = 0$. The self-energy is the work required to assemble an individual charge, and since it does not depend on the charge location, it is irrelevant to the interaction potential energy of a system of charges.

The interaction energy of nonoverlapping spherical charge distributions is the same as if the same charges were concentrated at the sphere centers. To establish this result, we recall that the potential $\Phi(r)$ outside a spherical charge distribution is the same as that due to the equivalent point charge at the center. Thus the force on one spherical distribution due to a second spherical distribution is the same as if one of the distributions were replaced by its equivalent point charge at its center. Moreover, since mutual electrostatic forces are equal in magnitude, we can also replace the other spherical distribution by its equivalent point charge in calculating the force. Therefore the interaction energy is given by the point charge result.

## 2-10 APPLICATION: ENERGY OF AN IONIC CRYSTAL

As an application of these ideas to atomic physics, we consider the electrostatic binding energy of ionic crystals. An ionic crystal like sodium chloride consists of a three-dimensional lattice of alternating positive $Na^+$ and negative $Cl^-$, as illustrated in Figure 2–18. The dissociation energy needed to separate the crystal completely into ions can be calculated and compared with the experimentally measured value.

We use the approximation that the charge distribution of each ion is spherical and does not overlap that of neighboring ions. Then the interaction potential energy of an ion and one of its neighbors is just given by

$$U = \frac{\pm e^2}{4\pi\varepsilon_0 R} \tag{2-86}$$

where $R$ is the distance between the ions and $e$ is the electron charge. The positive (negative) sign applies for ion charges of like (unlike) sign. To calculate the total potential energy, we must sum, in Equation (2–86), over all ions in the crystal.

As a simple example of such a calculation, we do a one-dimensional lattice. We first consider a reference $Na^+$ in the crystal (see Figure 2–19) and calculate the potential energy of interaction with all other ions on a line of infinite extent. With a lattice of infinite extent we are neglecting any surface effects. There are two nearest $Cl^-$ at distance $R = a$, two $Na^+$ at distance $R = 2a$, and so on. The energy of this sum is

$$U_{\text{ref}} = -\frac{2e^2}{4\pi\varepsilon_0 a}\left(\frac{1}{1} - \frac{1}{2} + \frac{1}{3} - \frac{1}{4} + \cdots\right) \tag{2-87}$$

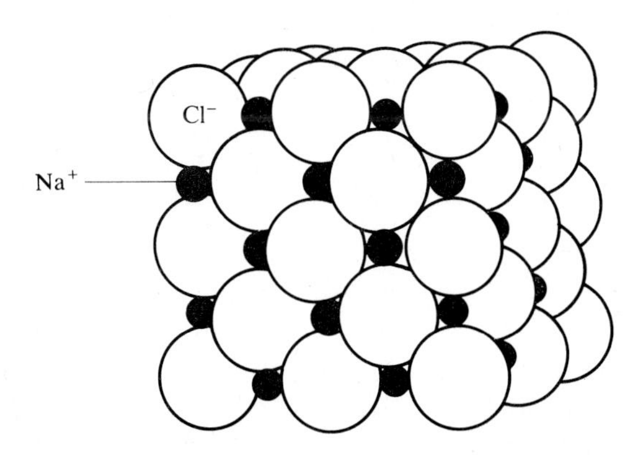


**FIGURE 2–18** Crystal lattice of sodium chloride. The chlorine ions are much larger than the sodium ions and are close-packed.

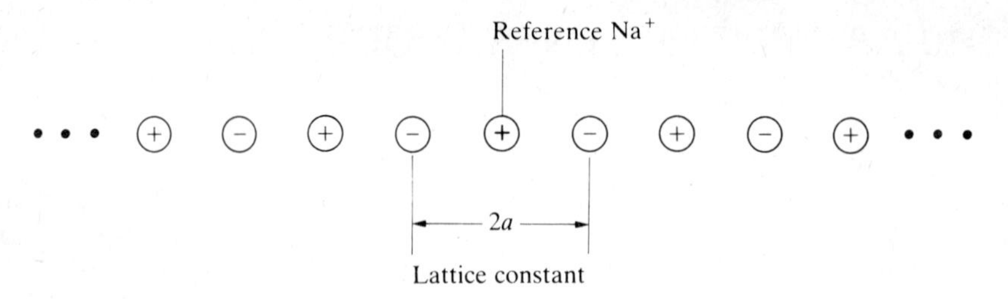


**FIGURE 2-19** Simple one-dimensional lattice of infinite extent

The overall factor of 2 occurs because there are identical ions to the right and left of the reference ion. The series in Equation (2–87) converges slowly. By comparison with the power series expansion,

$$\ln(1 + x) = x - \frac{x^2}{2} + \frac{x^3}{3} - \frac{x^4}{4} + \cdots$$

the complete series in Equation (2–87) can be summed to

$$U_{\text{ref}} = -\frac{e^2 2 \ln 2}{4\pi\varepsilon_0 a} = -1.386\,\frac{e^2}{4\pi\varepsilon_0 a} \tag{2-88}$$

The result for $U_{\text{ref}}$ is the same if our reference ion was a $Cl^-$. A similar calculation of $U_{\text{ref}}$ for a three-dimensional lattice is tedious. The numerical evaluation gives

$$U_{\text{ref}} = -1.7476\,\frac{e^2}{4\pi\varepsilon_0 a} \tag{2-89}$$

Since any position of an $Na^+$ or $Cl^-$ in an infinite lattice is equivalent, the total potential energy is obtained by summing Equation (2–89) for all possible reference ions and dividing by 2 (since such a sum counts the potential energy of each pair twice). If $2N$ is the number of ions ($N$ of $Na^+$ and $N$ of $Cl^-$) in the crystal, the total potential energy is

$$U = \left(\frac{1}{2}\right)(2N)U_{\text{ref}}$$

or

$$\frac{U}{N} = U_{\text{re-}} \tag{2-90}$$

The lattice constant for NaCl is $2a = 5.63 \times 10^{-10}$ m. In electronvolts the theoretical result from Equations (2–89) and (2–90) is

$$\frac{U}{N} = -8.94 \text{ eV} \tag{2-91}$$

Here we have converted the units to electronvolts (eV) from joules by using the relation 1 eV $= 1.6 \times 10^{-19}$ J. The measured energy of dissociation is

$$\frac{U}{N} = -7.93 \text{ eV} \tag{2-92}$$

The small discrepancy with the calculated value of Equation (2–91) is due to the overlapping of ionic-charge distributions and to nonspherical distortions of each ion owing to the nonisotropic distribution of the other ions.

## 2-11 ELECTROSTATIC POTENTIAL ENERGY OF A CONTINUOUS CHARGE DISTRIBUTION

Until now, we have considered the electrostatic energy of an assembly of point charges. To evaluate the corresponding energy of a continuous charge distribution is often useful. To do this evaluation, we replace the point charge $q_i$ by a differential charge $dq = \rho(\mathbf{r})\,dV$ in an infinitesimal volume $dV$ and integrate instead of summing over the charges in Equations (2–84):

$$U = \frac{1}{2}\int \rho(\mathbf{r})\Phi(\mathbf{r})\,dV \tag{2-93}$$

$$\Phi(\mathbf{r}) = \frac{1}{4\pi\varepsilon_0}\int \frac{\rho(\mathbf{r}')}{R}\,dV' \tag{2-94}$$

where $\mathbf{R} = \mathbf{r} - \mathbf{r}'$ as usual. This result for $U$ is valid only for truly continuous charge distributions. If the charge density $\rho(\mathbf{r})$ contains delta functions, then the energy calculated from Equation (2–93) is infinite because of the infinite self-energies of the point charges. These self-energies are the infinities that were omitted in Equation (2–85).

For a continuous charge distribution the electrostatic potential energy can also be written as an integral over the square of the electric field intensity. We use Gauss's law to write the integrand of Equation (2–93) as

$$\rho\Phi = \varepsilon_0\Phi\nabla\cdot\mathbf{E} \tag{2-95}$$

With a vector identity this equation can be expressed in the form

$$\rho\Phi = \varepsilon_0[-\mathbf{E}\cdot\nabla\Phi + \nabla\cdot(\Phi\mathbf{E})] \tag{2-96}$$

Inserting $\nabla\Phi = -\mathbf{E}$ converts the energy expression of Equation (2–93) to

$$U = \frac{1}{2}\varepsilon_0\int E^2\,dV + \frac{1}{2}\varepsilon_0\int \nabla\cdot(\Phi\mathbf{E})\,dV \tag{2-97}$$

Applying the divergence theorem to the second term in Equation (2–97), we obtain

$$U = \frac{1}{2}\varepsilon_0 \int E^2\, dV + \frac{1}{2}\varepsilon_0 \oint \Phi\mathbf{E}\cdot d\mathbf{S} \tag{2-98}$$

If the charge density $\rho(\mathbf{r})$ is nonzero only in a finite volume and if the integration region is taken over all space, the surface integral in Equation (2–98) vanishes. It vanishes because far from a finite charge distribution the electric field approaches that of a point charge; hence

$$\Phi\mathbf{E}\cdot d\mathbf{S} \approx \left(\frac{q}{4\pi\varepsilon_0 r}\right)\left(\frac{q}{4\pi\varepsilon_0 r^2}\right)(r^2\, d\Omega) = \left(\frac{q}{4\pi\varepsilon_0}\right)^2 \frac{d\Omega}{r} \tag{2-99}$$

tends to zero, like $1/r$ for large $r$. The electrostatic energy expression of Equation (2–98) thus reduces to

$$U = \frac{1}{2}\varepsilon_0 \int_{\text{all space}} E^2\, dV \tag{2-100}$$

As an exercise relevant to the application of the next section, we evaluate the electrostatic energy of a uniformly charged sphere of radius $a$ and charge $Q$. The electric field found in Section 2–5 for an arbitrary spherically symmetric charge distribution is

$$\mathbf{E} = \begin{cases} \dfrac{Q(r)\hat{\mathbf{r}}}{4\pi\varepsilon_0 r^2}, & \text{for } r < a \\[2ex] \dfrac{Q\hat{\mathbf{r}}}{4\pi\varepsilon_0 r^2}, & \text{for } r > a \end{cases} \tag{2-101}$$

where $Q(r) = Q(r/a)^3$ is the charge enclosed within a radius $r < a$. We now substitute Equation (2–101) into Equation (2–100) and carry out the integration:

$$\begin{aligned} U &= \frac{\varepsilon_0}{2}\int_0^\infty E_r^2(4\pi r^2\, dr) \\ &= \frac{\varepsilon_0}{2}\left(\frac{Q}{4\pi\varepsilon_0}\right)^2 \left[\int_0^a \left(\frac{r}{a}\right)^6 \frac{4\pi r^2\, dr}{r^2} + \int_a^\infty \frac{4\pi r^2\, dr}{r^4}\right] \\ &= \frac{Q^2}{8\pi\varepsilon_0}\left(\frac{1}{a^6}\int_0^a r^4\, dr + \int_a^\infty \frac{dr}{r^2}\right) \end{aligned}$$

The result is

$$U = \frac{3}{5}\left(\frac{Q^2}{4\pi\varepsilon_0 a}\right) \tag{2-102}$$

for the electrostatic energy of a uniformly charged sphere. We note that in the point charge limit ($a \to 0$) the energy of the charge becomes infinite.

## 2-12 APPLICATION: ENERGY OF NUCLEAR FISSION

From the electrostatic energy of a charged sphere in Equation (2–102), we can estimate the energy yield from nuclear fission. First, we make a brief introduction to nuclear physics.

The nucleus of a heavy atom contains a number $z$ of protons and a somewhat larger number $N$ of neutrons. The nuclear radius is roughly

$$R = (1.2 \times 10^{-15})A^{1/3} \text{ meters} \tag{2-103}$$

where the atomic number $A = Z + N$ is the number of protons and neutrons. This result implies that the nuclear volume is proportional to the number of nucleons, so the nuclear density is the same for any nucleus. Since protons repel each other, there must be an attractive *nuclear force* holding the nucleus together. This force is the same for protons or neutrons and is of short range, about $10^{-15}$ m. A consequence of the short-range force and the universal nuclear density is that potential energy $U_N$ due to the nuclear force alone is approximately proportional to the atomic number $A$. In contrast, the electrostatic energy $U_E$ of a nucleus obtained from Equation (2–102),

$$U_E = \frac{3}{5}\left(\frac{Z^2e^2}{4\pi\varepsilon_0 R}\right) \tag{2-104}$$

increases as the square of $Z$. Since we are interested in the interaction part of the electrostatic energy, we must subtract a term proportional to $Ze^2$ representing the self-energies of the $Z$ protons. We obtain

$$U_E = \frac{3}{5}\frac{Z(Z-1)e^2}{4\pi\varepsilon_0 R} \tag{2-105}$$

which vanishes for a nucleus having a single proton. Note that $\frac{1}{2}Z(Z-1)$ is just the number of pairs of protons in the nucleus.

If a slow neutron is captured by a uranium nucleus, the uranium becomes unstable and fissions into two roughly equal-sized nuclei, as schematically depicted in Figure 2–20. Since the net atomic number is conserved by all reactions, there is little change in $U_N$. However, the electrostatic potential energy of the reaction products is less than the initial electrostatic energy, since separation of like charges occurs in the fission process. The electrostatic energy is converted to kinetic energy.

To estimate the energy release, we calculate the electrostatic potential energy of a uranium nucleus with $Z = 92$ and $A = 235$, using Equation (2–105):

$$U_E({}^{92}U_{235}) = \frac{(\frac{3}{5})(92)(91)(1.6 \times 10^{-19})^2}{(9 \times 10^9)^{-1}(1.2 \times 10^{-15})(235)^{1/3}} \text{ J} \approx 10^9 \text{ eV} \tag{2-106}$$

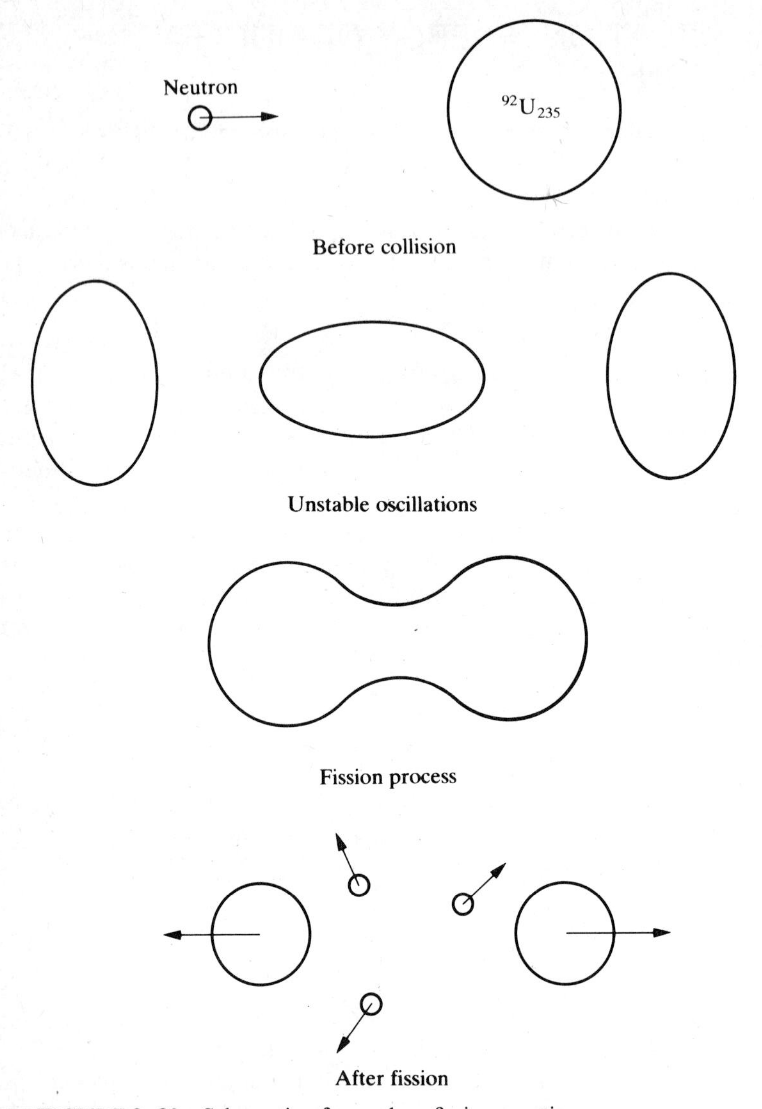


**FIGURE 2-20** Schematic of a nuclear fission reaction

In the last step we have converted the units to electronvolts (eV) from joules, 1 eV $= 1.6 \times 10^{-19}$ J. The two fission fragments each have roughly half the $Z$ and $A$, and their electrostatic energy is

$$U_E(\text{fragments}) = 0.6 \times 10^9 \text{ eV} \tag{2-107}$$

The energy release is

$$\Delta U_E = U_E(^{92}\text{U}_{235}) - U_E(\text{fragments}) \approx 0.4 \times 10^9 \text{ eV} \tag{2-108}$$

This result is an overestimate because those nucleons that reside at the surface do not contribute as fully to the nuclear energy as the interior nucleons. As with other fluids, this feature results in an energy per unit area called surface tension energy. In the fission process the volume of nuclear matter remains the same, but the surface area increases. The change in surface energy is about half the magnitude of the change in electrostatic energy, giving a net energy release, observed experimentally, of

$$\Delta U_{\text{net}} \approx 0.2 \times 10^9 \text{ eV} \tag{2-109}$$

Thus each fission releases roughly $10^8$ times more energy than a chemical reaction, such as burning.

## SUMMARY

### Important Concepts and Equations

$$\mathscr{F}_V = \int \mathbf{V}(\mathbf{r}) \cdot d\mathbf{S}(\mathbf{r})$$

*The flux of a vector field* $\mathbf{V}(\mathbf{r})$ *through a surface is the surface integral of* $\mathbf{V} \cdot d\mathbf{S}$.

$$\oint \mathbf{E} \cdot d\mathbf{S} = \frac{Q}{\varepsilon_0} \qquad \text{(Gauss's law, integral form)}$$

*The electric flux* $\oint \mathbf{E} \cdot d\mathbf{S}$ *through a closed surface depends only on charge* $Q$ *enclosed.*

$$\oint \mathbf{E} \cdot d\mathbf{S} = \int \boldsymbol{\nabla} \cdot dV \qquad \text{(divergence theorem)}$$

*The integral of a vector field over a closed surface is equal to the integral of the divergence* $\boldsymbol{\nabla} \cdot \mathbf{E}$ *of the field over the volume enclosed by the surface.*

$$\boldsymbol{\nabla} \cdot \mathbf{E} = \frac{\rho}{\varepsilon_0} \qquad \text{(Gauss's law, differential form)}$$

*The divergence* $\boldsymbol{\nabla} \cdot \mathbf{E}$ *of the electric field is equal to the charge density* $\rho$ *divided by the permittivity of free space* $\varepsilon_0$.

Earnshaw's theorem in terms of a test charge

*In a region of zero charge density a test charge cannot be in stable equilibrium under the influence of electrostatic forces only.*

Earnshaw's theorem in terms of potential

*There is no maximum or minimum of the potential within a region containing no charge density.*

Electric field of a conductor (in static equilibrium)

*In static equilibrium (i.e., no current flowing) the electric field is zero inside a conductor. From Gauss's law the charge density inside the conductor also vanishes, and hence any charge on the conductor must be on the surface.*

Potential of a conductor (in static equilibrium)

*The surface of a conductor is an equipotential surface.*

Conductor with a cavity

*There is no electric field inside an empty cavity in a conductor.*

$$U = \frac{1}{2}\left(\frac{1}{4\pi\varepsilon_0}\right)\sum_{j=1}^{N}\sum_{\substack{i=1 \\ (i \neq j)}}^{N}\frac{q_i q_j}{R_{ij}} \qquad \text{(electrostatic energy, discrete charges)}$$

*The interaction energy U for a system of N particles is half the sum of the potential energies of all possible pairs of particles. Term $R_{ij}$ is the distance between the pairs; the self-energy must be omitted since the electrostatic self-energy of a point charge is infinite ($R_{ii} = 0$).*

$$U = \frac{1}{2}\int_{\text{all space}} \rho\phi \, dV$$

$$= \frac{1}{2}\varepsilon_0 \int_{\text{all space}} E^2 \, dV \qquad \text{(electrostatic energy, continuous charge distribution)}$$

*For the interaction energy U for a continuous charge distribution, the square of the electric field $E^2$ is integrated over all space. The self-energy is included and must be explicitly subtracted if the charge distribution contains delta functions (discrete charges).*

# PROBLEMS

## Section 2-1: Gauss's Flux Law

**2-1** A line charge $\lambda$ per unit length on the $z$ axis extends from $z = 0$ to $z = +\infty$. Use Coulomb's law to calculate the electric field (see Problem 1–21), and then explicitly verify Gauss's law by direct integration over a cylinder of radius $r_0$ and length $\ell$ with ends at positions $z_1$ and $z_2$, where $z_1$ and $z_2$ may be positive or negative.

## Section 2-2: Divergence Theorem

**2-2** Starting from the divergence theorem, demonstrate the corollary

$$\oint_S f \, d\mathbf{S} = \int_V \nabla f \, dV$$

where $f$ is a scalar field and the surface $S$ bounds the volume $V$.

**2-3** Using the divergence theorem, derive the expression for the divergence in cylindrical coordinates.

## Sections 2-3 and 2-4: Gauss's Differential Law

**2-4** For the electrostatic potential

$$\Phi(r) = \frac{q}{4\pi\varepsilon_0}\frac{e^{-r/r_0}}{r}$$

find the charge density and the total charge.

**2-5** Find the charge densities associated with the following electric fields.
(i) $\mathbf{E} = k(2x\hat{\mathbf{x}} + y\hat{\mathbf{y}} - 3z\hat{\mathbf{z}})$.
(ii) $\mathbf{E} = (\lambda/2\pi\varepsilon_0 \mathscr{r})\hat{\boldsymbol{\mathscr{r}}}$ for $\mathscr{r} > 0$.
(iii) $\mathbf{E} = (Q/4\pi\varepsilon_0 r^2)\hat{\mathbf{r}}$ for $r \geqslant a$ and $\mathbf{E} = 0$ for $r < a$.

**2-6** In a charge-free region, show that if $\mathbf{E}$ is everywhere directed parallel to the $z$ axis, then $\mathbf{E}$ is a constant (i.e., independent of spatial location). (*Hint*: Use $\nabla\cdot\mathbf{E} = \rho/\varepsilon_0$ and $\nabla\times\mathbf{E} = 0$.)

**2-7** A cylindrically symmetric electric field is given by

$$\mathbf{E}(\mathscr{r}) = \frac{\lambda}{2\pi\varepsilon_0 \mathscr{r}} f(\mathscr{r})\hat{\boldsymbol{\mathscr{r}}}$$

where $\mathscr{r}$ is the radius in cylindrical coordinates and $f(\mathscr{r})$ is a specified function with $f(\mathscr{r} = 0) = 1$, $f(\mathscr{r} = \infty) = 0$. Find the charge density for all $\mathscr{r}$ and the total charge.

**2-8** A charge distribution produces an electric field

$$\mathbf{E} = \hat{\mathbf{r}}E_0\theta(a - r)$$

where $E_0$ is constant. Find the charge density $\rho(\mathbf{r})$ in terms of $E_0$ and $a$. Find the total charge. Determine the electric potential $\Phi(\mathbf{r})$ for all $r$.

**2-9** A charge distribution produces an electric field

$$\mathbf{E} = \begin{cases} \dfrac{C}{r^2}\hat{\mathbf{r}}, & \text{for } r > a \\[2ex] \dfrac{Cr}{a^3}\hat{\mathbf{r}}, & \text{for } r \leqslant a \end{cases}$$

where $C$ is a constant.
(i) What charge distribution $\rho(r)$ gives an electric field of this form?
(ii) Determine the electric potential $\Phi(r)$ for all $r$. Take the zero of potential at $r = \infty$.

## Section 2-5: Applications of Gauss's Law

**2-10** The fair-weather electrostatic field above the earth's surface has the empirical form

$$\mathbf{E} = -(ae^{-\alpha z} + be^{-\beta z})\hat{\mathbf{z}}$$

where $a$, $b$, $\alpha$, and $\beta$ are positive constants and $z$ is the height above the surface. Use Gauss's law in differential form to determine the charge density $\rho(z)$ as a function of height. Use Gauss's law in integral form to find the total charge within a vertical column from $z = 0$ to $z = \infty$ with cross-sectional area $A$.

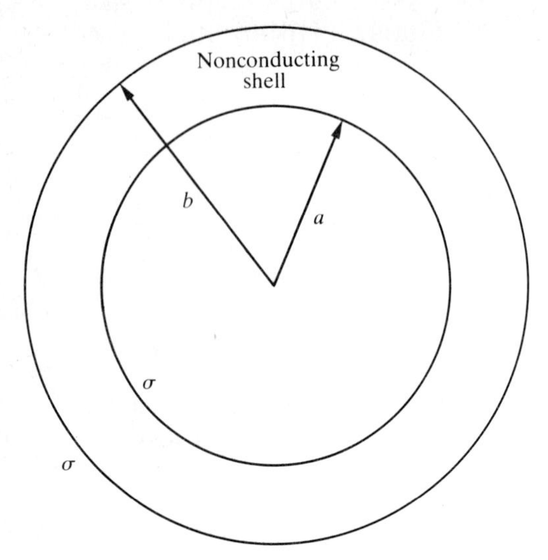


**FIGURE 2-21** Concentric, spherical nonconducting shells of radius $a$ and $b$, both with surface charge density $\sigma$

**2-11** The inner ($r = a$) and outer ($r = b$) surfaces of a spherical nonconducting shell have the same uniform surface charge densities $\sigma$. See Figure 2–21. The charge density elsewhere is zero. Using Gauss's law, find $\mathbf{E}(\mathbf{r})$ for the three regions $r < a$, $a < r < b$, and $r > b$, and find $\Phi(\mathbf{r})$ from $r = 0$ to $\infty$, with $\Phi(\infty) = 0$ everywhere continuous.

**2-12** The plum pudding model of the atom proposed by J. J. Thomson in the early days of atomic theory consisted of a sphere of radius $a$ of positive charge of total value $Ze$, where $Z$ is an integer and $e$ is the fundamental unit of charge. The electrons, of charge $-e$, were considered to be point charges embedded in the positive charge.
(i) Find the force acting on an electron as a function of its distance $r$ from the center of the sphere for the element hydrogen.
(ii) What type of motion does the electron execute?
(iii) Find an expression for the frequency of this motion.

**2-13** Using the principle of superposition, show that the electric field is constant within a charge-free spherical cavity in a uniformly charged sphere.

**2-14** The charge density inside a sphere of radius $a$ is $\rho = cr^{\lambda}$, where $\lambda$ is nonnegative. Determine the potential $\Phi$ for all $r$ in terms of the charge $Q$ of the sphere.

**2-15** In general, when one produces a beam of ions or electrons, the space charge within the beam causes a potential difference between the axis and the surface of the beam. A 10-mA beam of 50-keV protons ($v = 3 \times 10^6$ m/s) travels along the axis of an evacuated beam pipe. The beam has a circular cross section of 1-cm diameter. Calculate the potential difference between the axis and the surface of the beam, assuming that the particles in the beam are uniformly distributed over the beam diameter.

**2-16** Derive the expressions for the electric fields and potentials in Table 2–1 for all cases not explicitly evaluated in the text.

**2-17** Starting from an expression for $\nabla$ in spherical coordinates, derive the expression for the divergence in spherical coordinates (see Section 2–2). The relation between the spherical and Cartesian unit vectors is

$$\hat{\mathbf{r}} = \hat{\mathbf{x}} \sin\theta \cos\phi + \hat{\mathbf{y}} \sin\theta \sin\phi + \hat{\mathbf{z}} \cos\theta$$
$$\hat{\boldsymbol{\theta}} = \hat{\mathbf{x}} \cos\theta \cos\phi + \hat{\mathbf{y}} \cos\theta \sin\phi - \hat{\mathbf{z}} \sin\theta$$
$$\hat{\boldsymbol{\phi}} = -\hat{\mathbf{x}} \sin\phi + \hat{\mathbf{y}} \cos\phi$$

**2-18** Derive expressions for $\nabla \times \mathbf{V}$ in cylindrical and spherical coordinate systems. (*Hint*: See the derivation in Section 2–5C of divergence expressions.)

## Section 2-6: Electrostatic Precipitation

**2-19** A precipitator has outside radius $R$ and length $\ell$. The electric field $E$ outside the corona region is given, to a good approximation, by Equation (2–71). Find the maximum permissible flow rate that allows all the ash to be removed.

## Section 2-7: Earnshaw's Theorem

**2-20** A positive point charge $Q$ is located at a distance $d$ above an infinite planar sheet with positive uniform charge density $\sigma$ per unit area.

(i) Find a vector expression for the electric field of the system of charges in terms of $(x, y, z)$ coordinates. (Choose a coordinate axis such that the charge $Q$ is located at $x = 0$, $y = 0$, and $z = d$ and the conducting sheet is in the $xy$ plane.)

(ii) Find the $(x, y, z)$ coordinates of any points at which a test charge would be in equilibrium. According to Earnshaw's theorem, is the equilibrium at any such points stable or unstable?

**2-21** An electric field is given by $\mathbf{E} = x\hat{\mathbf{x}} - y\hat{\mathbf{y}}$.

(i) Show that this equation could represent an electrostatic field in a charge-free region.

(ii) Find the point of equilibrium. Is it stable, unstable, or neutral?

(iii) Discuss the relevance of Earnshaw's theorem for this electric field.

## Section 2-8: Conductors and Gauss's Law

**2-22** A conductor has a surface charge $\sigma$. Show that the force on a small element of charge $dq$ is radially outward and is given by $dF = \frac{1}{2}E\,dq$, where $E$ is the electric field at the surface of the conductor. (*Hint*: Consider a small disk on the conductor, and find the electric field at points above the disk, on the disk, and below the disk.) Find the force between two parallel-plate conductors of area $A$ and separation $d$ with charges $+Q$ and $-Q$.

**2-23** Two concentric conducting spheres of radii $R_a$ and $R_b$, with $R_a < R_b$, have net charges $Q_a$ and $Q_b$. The inner sphere is grounded; that is, its potential is zero and the potential at infinity is also zero. Determine $Q_a$ in terms of $Q_b$.

**2-24** In an uncharged conductor with a cavity, a point charge $Q$ is inserted into the cavity and fixed there; an additional net charge $Q$ (made up of many small charges) is then placed on the walls of the cavity. See Figure 2–22. After equilibrium is attained, how much charge is on each surface?

**2-25** The force between two point charges is $F = q_1 q_2 / 4\pi\varepsilon_0 r^{2+\varepsilon}$. Calculate the potential difference between two concentric spherical shells by direct integration. For $\varepsilon \ll 1$, show that this potential difference is proportional to $\varepsilon$.

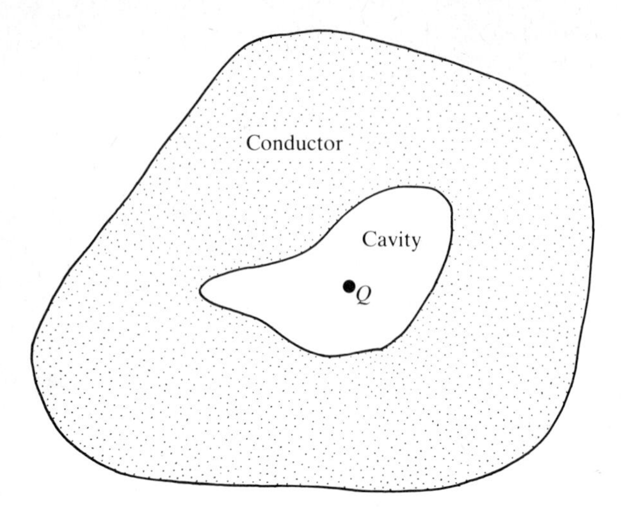


**FIGURE 2-22** Point charge $Q$ inside the cavity of a conductor

## Section 2-9: Electrostatic Energy of Discrete Charges

**2-26** Charges $Q$ are at the corners of a square. A charge $q$ is brought to the center of the square.

(i) Determine $q$ such that there is no force on any of the charges.
(ii) Find the work required to assemble this charge configuration.
(iii) Is this system in stable, unstable, or neutral equilibrium?

## Section 2-11: Electrostatic Energy of Continuous Charge Distributions

**2-27** A charge $+Q$ is distributed with uniform density in a spherical volume having a radius $a$ and centered at the origin of a Cartesian coordinate system. A second charge $-Q$ is also distributed with uniform density in a spherical volume of radius $a$ at a distance $R$ ($R > 2a$) from the origin on the $x$ axis. Determine the work done by the electric field necessary to carry a positive test charge $q$ from the origin ($x = y = z = 0$) to the point $x = R$, $y = 0$, $z = 0$.

**2-28** Consider two conductors of arbitrary shape having charges $+Q$ and $-Q$ and surface potentials $\Phi_1$ and $\Phi_2$. See Figure 2-23. Using the expression $U = \frac{1}{2}\int \rho\Phi\, dV$, show that the potential energy of the charged conductors is $Q^2/2C$, where $C \equiv Q/(\Phi_1 - \Phi_2)$ is the capacitance.

**2-29** Two parallel plates of area $A$ are separated by a distance $d$ and charged to a potential difference $V$.

(i) Find the electrostatic energy stored in the field.
(ii) Suppose the plates are forced apart an infinitesimal distance by a mechanical force. Find the change in electrostatic energy, and from this change, deduce the mechanical force. Assume that the plates are isolated (i.e., not connected to a battery).
(iii) Use the electric field from one plate to directly calculate the force on the other plate.

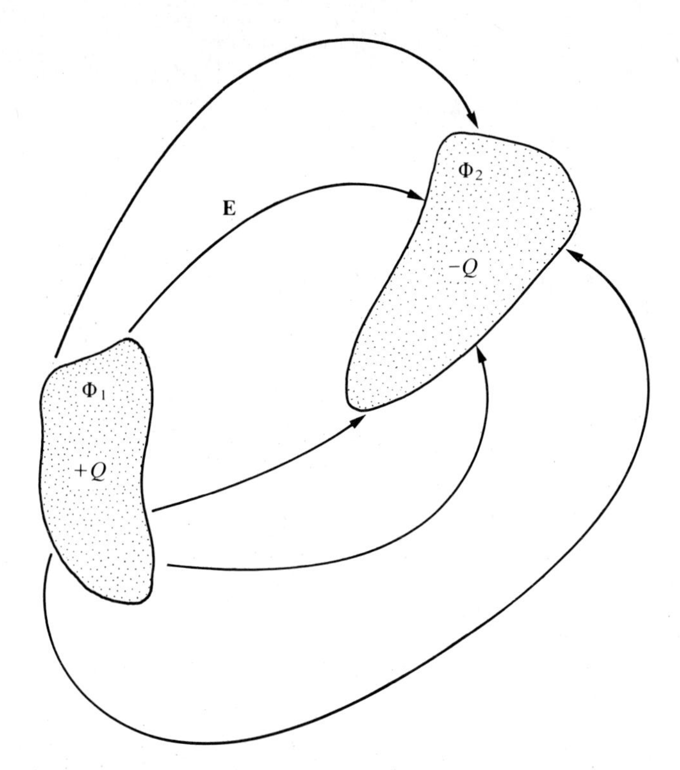


**FIGURE 2-23** Two arbitrary conductors with specified charges and potentials

**2-30** From the expression for the electric field of a uniformly charged sphere, Equation (2.46), find the potential everywhere. Using this result, evaluate the electrostatic potential energy.

**2-31** Calculate the interaction energy of two uniformly charged spheres of charge $Q$ and radius $a$ whose centers are separated by a distance $R > 2a$.

# Boundary Value Methods in Electrostatics

## CHAPTER CONTENTS

When we have a collection of charges at known positions, the electric field can be determined from Coulomb's law. However, many practical problems in electrostatics involve conductors whose surfaces are equipotentials. On the conductor surfaces the potential is known but not the charge distribution. The general strategy in solving for the electric field in such cases is to find an expression for the electric potential that has the correct value on all boundary surfaces and agrees with Gauss's law for the given charge density at all points in the interior of the region. Any expression for $\Phi$ that satisfies these criteria will give the correct solution for the electric field **E**, as we will prove in the following section. A number of different methods can be used to construct such a $\Phi$. The choice of method of solution depends on the nature and the geometry of the particular electrostatic problem to be solved. This chapter is largely devoted to the consideration and applications of a variety of potential methods.

## 3-1 UNIQUENESS OF ELECTROSTATIC SOLUTIONS

In electrostatics the electric field at each point satisfies the equations

$$\nabla \times \mathbf{E}(\mathbf{r}) = 0 \qquad \text{and} \qquad \nabla \cdot \mathbf{E}(\mathbf{r}) = \frac{\rho(\mathbf{r})}{\varepsilon_0}$$

Physical solutions for $\mathbf{E}(\mathbf{r})$ must be unique. For a region with boundaries we need to know what additional constraints must be placed on **E** at the boundary surfaces to ensure that the solution to these differential equations is unique. This question can be most easily addressed by the introduction of the electric potential $\Phi$,

$$\mathbf{E} = -\nabla\Phi$$

Suppose that there are two solutions $\mathbf{E}_1$ and $\mathbf{E}_2$ that have the same divergence $\nabla\cdot\mathbf{E}_1 = \nabla\cdot\mathbf{E}_2 = \rho/\varepsilon_0$ inside the boundary. These solutions can be expressed in terms of their respective potentials as $\mathbf{E}_1 = -\nabla\Phi_1$ and $\mathbf{E}_2 = -\nabla\Phi_2$. We construct the differences $\mathbf{E} = \mathbf{E}_1 - \mathbf{E}_2$ and $\Phi = \Phi_1 - \Phi_2$, and we note that $\nabla\cdot\mathbf{E} = 0$. The divergence of the product $\Phi\mathbf{E}$ is given by

$$\nabla\cdot(\Phi\mathbf{E}) = \Phi\nabla\cdot\mathbf{E} + (\nabla\Phi)\cdot\mathbf{E} = 0 + (-\mathbf{E})\cdot\mathbf{E} = -E^2 \tag{3-1}$$

Applying the divergence theorem to the volume $V$ enclosed by the boundary, we find

$$\int_V \nabla\cdot(\Phi\mathbf{E})\, dV = \oint_S \Phi\mathbf{E}\cdot\hat{\mathbf{n}}\, dS \tag{3-2}$$

where $S$ is the boundary surface. If the surface integral vanishes,

$$\oint_S \Phi \mathbf{E} \cdot \hat{\mathbf{n}}\, dS = 0 \tag{3-3}$$

then Equations (3–1) and (3–2) imply that

$$\int_V E^2\, dV = 0 \tag{3-4}$$

Since the integrand of Equation (3–4) is nonnegative, $\mathbf{E} = 0$ everywhere in $V$, and hence $\mathbf{E}_1 = \mathbf{E}_2$. Thus if the two solutions both satisfy a boundary condition that makes the integral in Equation (3–3) vanish, the solution will be *unique*.

Various ways of specifying physical boundary conditions that yield unique solutions are as follows:

1. The potential is specified on the boundaries. Then $\Phi_1 = \Phi_2$ on the bounding surfaces, and $\Phi = \Phi_1 - \Phi_2 = 0$ in Equation (3–3).
2. The normal component of the electric field is specified on the boundary. Then $\hat{\mathbf{n}} \cdot \mathbf{E}_1 = \hat{\mathbf{n}} \cdot \mathbf{E}_2$ on $S$, which gives $\hat{\mathbf{n}} \cdot \mathbf{E} = \hat{\mathbf{n}} \cdot (\mathbf{E}_1 - \mathbf{E}_2) = 0$ in Equation (3–3).
3. The total charges on conducting boundary surfaces are specified. In this case, since $\Phi = V_i$ is constant on the conducting surface $i$, we can evaluate Equation (3–3) by Gauss's law, as follows:

$$\oint_S \Phi \mathbf{E} \cdot d\mathbf{S} = \sum_i V_i \oint_{S_i} (\mathbf{E}_1 - \mathbf{E}_2) \cdot d\mathbf{S} = \frac{1}{\varepsilon_0} \sum_i V_i (Q_1^{(i)} - Q_2^{(i)}) \tag{3-5}$$

Here $Q_1^{(i)}$ is the total charge on conductor $(i)$ for solution 1. Since $Q_1^{(i)} = Q_2^{(i)}$, the surface integral vanishes as required for a unique solution.

We remark that some of the boundary surfaces may be within the region of interest. For example, conductors may lie in the interior of the region.

In taking boundary values, we must be careful not to overspecify the charge and voltage conditions. For example, if the conductors are maintained at fixed voltages by a set of batteries, we can only take the charges on the conductors as they come. We can, however, specify the charges on some of the conductors and voltages on the others. In fact, the boundary conditions could even be specified differently point by point on the same surface.

The foregoing argument assures us that there is only one solution to the electrostatic differential equations that satisfies the specified boundary conditions. Thus we are free to use any method, including an educated guess, to arrive at a solution. Before going on to an elaboration of these methods, we introduce the concept of capacitance, which is a fundamental geometric property of a system of conductors.

## 3-2 CAPACITANCE

A basic type of electrostatics problem is the following: Find the electrostatic potential $\Phi$ in a charge-free region that is interspersed with conductors, as illustrated in Figure 3–1. Each of the conductors has either a specified potential $\Phi = V_i$ or a specified charge $Q_i$. As we saw in the previous section, the potential is unique if either the $V_i$'s or the $Q_i$'s are known. We will choose a reference potential $\Phi = 0$ at infinity.

We first consider the problem in which the potentials $V_i$ are specified on each of the conducting surfaces $S_i$. A useful technique is to break the problem up into elementary solutions $\Phi_i(\mathbf{r})$, defined to be the solution of the problem in which the conductor $i$ is at unit potential and the other conductors are all grounded (i.e., at zero potential). By superposition, the general potential will be

$$\Phi(\mathbf{r}) = \sum_i V_i \Phi_i(\mathbf{r}) \tag{3-6}$$

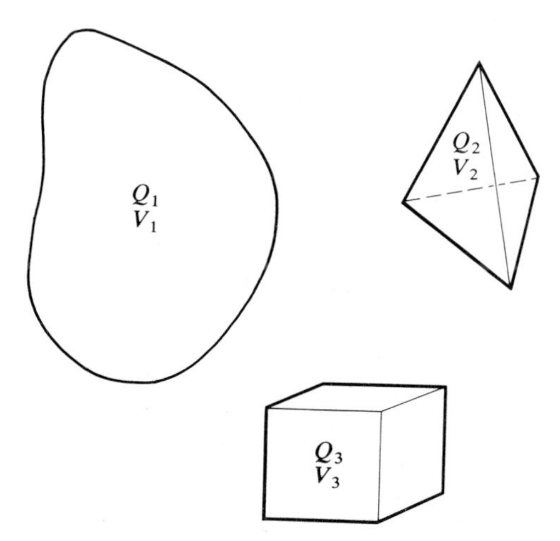


**FIGURE 3-1** Arbitrary charged conductors in a charge-free region

We now define the capacitance—or, more precisely, the *coefficient of capacitance*—$C_{ij}$ as the charge on the $i$th conductor for the elementary solution $\Phi_j(\mathbf{r})$. Physically, this definition means that if we ground all conductors, then raise the potential of the $j$th conductor to unity, the $i$th conductor will have charge $Q_i = C_{ij}$. By superposition, if the $j$th conductor has potential $V_j$, as in Equation (3–6), the charge on the $i$th conductor is

$$Q_i = \sum_j C_{ij} V_j \tag{3-7}$$

The off-diagonal $C_{ij}$ $(i \neq j)$ are called coefficients of electrostatic inductance or mutual capacitances. The unit of capacitance is the farad (F), which is 1 coulomb per volt.

The coefficients of capacitance $C_{ij}$ are *symmetric* in the indices $i$ and $j$,

$$C_{ij} = C_{ji} \tag{3-8}$$

as we will now prove. For the elementary solution $\Phi_j(\mathbf{r})$ we denote the associated charge density at position $\mathbf{r}$ on the boundaries (surfaces of conductors) by $\sigma_j(\mathbf{r})$. The total charge on the $i$th conductor with solution $\Phi_j(\mathbf{r})$ is by definition $C_{ij}$:

$$C_{ij} \equiv \oint_{S_i} \sigma_j(\mathbf{r})\, dS \tag{3-9}$$

The surface integral is over the surface $S_i$ of the $i$th conductor. Since by definition the elementary solution is 1 on the surface $S_i$ and 0 on all others, we can write Equation (3–9) as

$$C_{ij} = \oint \Phi_i(\mathbf{r})\sigma_j(\mathbf{r})\, dS \tag{3-10}$$

The surface integral is over the surface $S_i$ of the $i$th conductor. Since by Coulomb's law, the $i$th elementary solution is given by

$$\Phi_i(\mathbf{r}) = \frac{1}{4\pi\varepsilon_0} \oint \frac{\sigma_i(\mathbf{r}')\, dS'}{|\mathbf{r} - \mathbf{r}'|} \tag{3-11}$$

where the integration is over all the conducting surfaces. Substituting this expression for $\Phi_i(\mathbf{r})$ into Equation (3–10), we obtain

$$C_{ij} = \frac{1}{4\pi\varepsilon_0} \oint\oint \frac{\sigma_i(\mathbf{r}')\, dS'\, \sigma_j(\mathbf{r})\, dS}{|\mathbf{r} - \mathbf{r}'|} \tag{3-12}$$

This expression is manifestly symmetric in indices $i$ and $j$.

Without obtaining a solution, we can determine the signs of the $C_{ij}$ by the following argument. We consider the elementary potential $\Phi_i$, which is unity on conductor $i$ and zero on the other conductors. The field lines of $\mathbf{E}$ are directed from regions of higher to lower potential and hence point away from conductor $i$. Since positive charges are the source of field lines, all the charge density on conductor $i$ must be positive. Because $C_{ii}$ is the integrated charge on conductor $i$, it is therefore positive. By similar reasoning, all the charge density on the conductors at zero potential is negative, and therefore the $C_{ij}$ for $i \neq j$ are negative. In summary, we have the inequalities

$$\begin{aligned} &C_{ii} \geqslant 0 \\ &C_{ij} \leqslant 0, \qquad j \neq i \end{aligned} \tag{3-13}$$

The linear system of equations in (3–7) can be inverted to give the potentials in terms of the charges:

$$V_i = \sum_j P_{ij} Q_j \tag{3-14}$$

The $P_{ij}$ are called the *coefficients of potential*. The *symmetry* $P_{ij} = P_{ji}$ follows from $C_{ij} = C_{ji}$.

From the symmetry of $C_{ij}$ and of $P_{ij}$ we can make the following conclusions:

1. The charge induced on conductor 1 by changing the potential on 2 by $V$ is the same as the charge that would be induced on 2 if the potential on conductor 1 were raised by $V$.
2. The potential to which conductor 1 is raised by putting a charge $Q$ on conductor 2 is the same as the change in potential on conductor 2 that would result from putting the same charge $Q$ on conductor 1.

For a single isolated conductor Equation (3–7) simplifies to

$$Q = CV \tag{3-15}$$

where $C \equiv C_{11}$. As an example, an isolated conducting sphere of radius $R$ has a surface potential $V = Q/(4\pi\varepsilon_0 R)$. Hence its capacitance $C = Q/V$ is

$$C = 4\pi\varepsilon_0 R \tag{3-16}$$

Any isolated conductor of overall size $R$ will have a capacitance comparable to this value.

Of great practical importance is the situation in which two conductors, say 1 and 2, are close to one another but isolated from any others. We consider two conductors roughly of size $R$ separated by a distance $d$, as shown in Figure 3–2. In this case, assuming the effect of other conductors is negligible, Equation (3–7) gives

$$Q_1 = C_{11}V_1 + C_{12}V_2 \qquad Q_2 = C_{21}V_1 + C_{22}V_2 \tag{3-17}$$

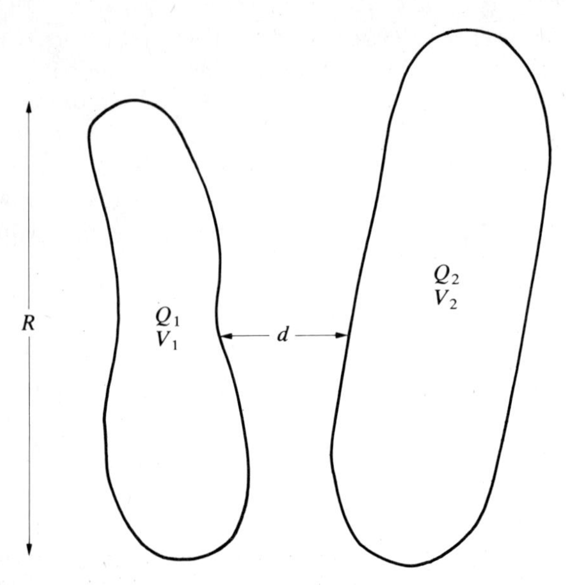


**FIGURE 3-2** Two conductors that are close to one another but isolated from any other conductors. Two such conductors form a capacitor.

We first consider the case in which conductor 1 is held at a fixed potential by a battery and conductor 2 is grounded. Because the potential difference between conductors 1 and 2 is fixed, the smaller the separation $d$ is, the larger is the electric field, and hence the larger is the charge on the conductors. We conclude that $C_{11}$ and $C_{21}$ become large as $d$ gets small. With voltages interchanged we can similarly conclude that $C_{22}$ and $C_{12}$ become large as $d$ gets small. If both conductors are held at the same fixed potential, $Q_1$ and $Q_2$ do not change appreciably as $d$ goes to zero. So from Equations (3–17) we conclude that $C_{11} + C_{12}$ and $C_{21} + C_{22}$ remain finite in the limit of small $d$. Since $C_{21} = C_{12}$, we thereby deduce that for closely spaced conductors

$$C_{11} \approx C_{22} \approx -C_{12} = -C_{21} \tag{3-18}$$

We denote the common value of these coefficients of capacitance by $C$, called the capacitance. From Equations (3–17) and (3–18) we find the approximate relation

$$Q_1 = -Q_2 = C(V_1 - V_2) \tag{3-19a}$$

or equivalently,

$$C = \frac{Q_1}{V_1 - V_2} \tag{3-19b}$$

for a system of two isolated conductors that are close together in comparison with their sizes.

For a parallel-plate capacitor with plate area $A$ and a separation $d$ that is small compared with the size of the plates, the electric field between the plates is nearly uniform and is given by

$$\mathbf{E} = \frac{(Q_1/A)}{\varepsilon_0}\hat{\mathbf{z}} \tag{3-20}$$

where $\hat{\mathbf{z}}$ is directed from plate 1 to plate 2. The electric potential difference is

$$V_2 - V_1 = -\int_0^d E_z\,dz = -\frac{(Q_1/A)d}{\varepsilon_0} \tag{3-21}$$

Hence by Equations (3–19) and (3–21) the capacitance for the parallel-plate configuration is

$$C = \frac{\varepsilon_0 A}{d} \tag{3-22}$$

This result neglects edge effects, which become significant if the separation $d$ is not small compared with the plate size.

For a parallel-plate capacitor with $A = 1\text{ m}^2$ and $d = 10^{-4}$ m, the capacitance is $C \approx 0.1\ \mu$F. By rolling up the pair of plates, a capacitor of these dimensions can be reduced to a volume of order $100\text{ cm}^3$. By placing a dielectric between the plates, the capacitance can be increased by a factor $\kappa$, where $\kappa$ is the dielectric constant.

Capacitors are often used to store energy. A historical example is the Leyden jar. A modern example is the use of capacitors to provide pulses of energy to heat plasmas and to generate the magnetic field confining them. In electric circuits capacitors are used to shift the phase of an alternating current and in conjunction with inductors to make resonant circuits.

## 3-3 LAPLACE EQUATION: SOLUTIONS IN ONE DIMENSION

In electrostatics the differential form of Gauss's law, $\nabla\cdot\mathbf{E} = \rho(\mathbf{r})/\varepsilon_0$, is supplemented by the condition that $\mathbf{E}$ is derivable from a scalar potential, $\mathbf{E} = -\nabla\Phi$. These two equations can be replaced by a single differential equation for $\Phi$, namely,

**Poisson's Equation**

$$\nabla^2\Phi(\mathbf{r}) = \frac{-\rho(\mathbf{r})}{\varepsilon_0} \tag{3-23}$$

where $\nabla^2 = \nabla\cdot\nabla$ is called the Laplacian operator. In regions of space where there are no charges,

**Laplace's Equation**

$$\nabla^2\Phi(\mathbf{r}) = 0 \tag{3-24}$$

For boundary conditions $\Phi(\mathbf{r}) \to 0$ as $r \to \infty$, we already know that the solution to Poisson's equation is

$$\Phi(\mathbf{r}) = \frac{1}{4\pi\varepsilon_0}\int \frac{\rho(\mathbf{r}')\,dV'}{|\mathbf{r} - \mathbf{r}'|}$$

which gives an electric field that satisfies Gauss's law. For boundary conditions that are instead specified in terms of potentials on surfaces enclosing the region of interest, we need to construct a general solution to Poisson's or Laplace's equation satisfying the required boundary values.

The explicit form of the solution to Laplace's equation depends on the choice of coordinates. Symmetries of the boundaries or of the charges usually dictate which coordinate system should be selected and the types of solutions to try. The form of the Laplacian operator $\nabla^2$ in the Cartesian, cylindrical, and spherical coordinate systems can readily be found from our previous results by taking the divergence of the gradient,

$$\nabla^2\Phi = \nabla\cdot(\nabla\Phi)$$

One obtains

$$\nabla^2\Phi = \frac{\partial^2\Phi}{\partial x^2} + \frac{\partial^2\Phi}{\partial y^2} + \frac{\partial^2\Phi}{\partial z^2} \qquad \text{(Cartesian)} \tag{3-25a}$$

$$\nabla^2\Phi = \frac{1}{\mathscr{r}}\frac{\partial}{\partial \mathscr{r}}\left(\mathscr{r}\frac{\partial\Phi}{\partial \mathscr{r}}\right) + \frac{1}{\mathscr{r}^2}\frac{\partial^2\Phi}{\partial\phi^2} + \frac{\partial^2\Phi}{\partial z^2} \qquad \text{(cylindrical)} \tag{3-25b}$$

$$\nabla^2\Phi = \frac{1}{r^2}\frac{\partial}{\partial r}\left(r^2\frac{\partial\Phi}{\partial r}\right) + \frac{1}{r^2\sin\theta}\frac{\partial}{\partial\theta}\left(\sin\theta\frac{\partial\Phi}{\partial\theta}\right) + \frac{1}{r^2\sin^2\theta}\frac{\partial^2\Phi}{\partial\phi^2} \text{(spherical)} \tag{3-25c}$$

A partial differential equation, such as Laplace's equation, has a great variety of solutions. The boundary conditions of a given problem usually dictate which type of solution should be used. In this section we will discuss the simplest solutions of Equation (3–24), which depend on only one coordinate.

We begin with the particularly simple case of an isolated parallel-plate capacitor with plates of infinite extent. We take the $z$ axis normal to the plates. Because of the symmetry in the $xy$ plane, we expect a solution that depends only on $z$. In this circumstance Laplace's equation is

$$\frac{d^2\Phi(z)}{dz^2} = 0 \tag{3-26}$$

which has the solution

$$\Phi(z) = a + bz \tag{3-27}$$

For boundary values $\Phi(0) = V$ and $\Phi(d) = 0$, the solution becomes

$$\Phi(z) = V\left(1 - \frac{z}{d}\right) \tag{3-28}$$

This potential yields the constant electric field

$$E_z(z) = \frac{V}{d} \tag{3-29}$$

For cylindrical boundaries the Laplace equation for a $\Phi$ that depends only on the radial variable $\imath$ is

$$\frac{d}{d\imath}\left[\imath\frac{d\Phi(\imath)}{d\imath}\right] = 0 \tag{3-30}$$

This equation has the general solution

$$\Phi(\imath) = a + b \ln \imath \tag{3-31}$$

Such a potential arises from a line charge; see Section 2–5C.

As an example with spherical symmetry, we consider a region bounded by two concentric, spherical conducting surfaces of radii $a$ and $b$, as shown in Figure 3–3. The Laplace equation for a $\Phi$ that depends only on $r$ is

$$\frac{d}{dr}\left(r^2\frac{d\Phi}{dr}\right) = 0 \tag{3-32}$$

A first integral of this equation yields

$$r^2\frac{d\Phi}{dr} = C_1 \tag{3-33}$$

where $C_1$ is an integration constant. Dividing both sides of Equation (3–33) by $r^2$ and integrating again, we obtain

$$\Phi(r) = -\frac{C_1}{r} + C_2 \tag{3-34}$$

The boundary conditions on $\Phi$ are

$$-\frac{C_1}{a} + C_2 = V_1 \qquad \text{and} \qquad -\frac{C_1}{b} + C_2 = V_2 \tag{3-35}$$

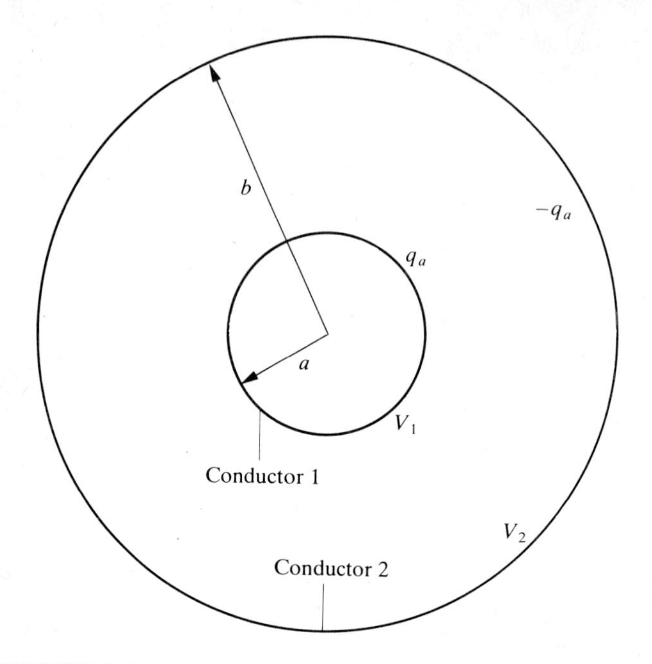


**FIGURE 3-3** Region consisting of a spherical conducting surface of radius $a$ with a charge $+Q$ concentric with a spherical cavity of radius $b$ in another conductor

which determine the integration constants to be

$$C_1 = (V_2 - V_1)\left(\frac{ab}{b-a}\right) \quad \text{and} \quad C_2 = \frac{V_2 b - V_1 a}{b-a} \tag{3-36}$$

Thus the potential in the interior space is

$$\Phi(r) = \frac{-ab(V_2 - V_1)}{r(b-a)} + \frac{V_2 b - V_1 a}{b-a} \tag{3-37}$$

The corresponding electric field is

$$\mathbf{E} = -\frac{\partial \Phi(\mathbf{r})}{\partial r}\hat{\mathbf{r}} = -\left(\frac{V_2 - V_1}{b-a}\right)\left(\frac{ab}{r^2}\right)\hat{\mathbf{r}} \tag{3-38}$$

From Equation (2–75) the charge density on the inner conducting surface is

$$\sigma = \varepsilon_0 E(r = a) = -\varepsilon_0\left(\frac{V_2 - V_1}{b-a}\right)\left(\frac{b}{a}\right) \tag{3-39}$$

The total charge on the surface of the inner conducting surface is

$$q_a = 4\pi a^2 \sigma = -4\pi\varepsilon_0 \left(\frac{V_2 - V_1}{b - a}\right) ab \tag{3-40}$$

Eliminating the potential difference $V_2 - V_1$ between Equations (3–38) and (3–40), we find

$$\mathbf{E}(r) = \frac{q_a}{4\pi\varepsilon_0} \frac{\hat{\mathbf{r}}}{r^2} \tag{3-41}$$

This equation is the result that would be obtained more simply in this example by direct application of the integral form of Gauss's law.

The charge on the outer conducting surface is

$$q_b = -4\pi b^2 \varepsilon_0 E(r = b) = -q_a \tag{3-42}$$

Gauss's law applied to a surface in the interior of the outer conductor, where $\mathbf{E} = 0$, requires the total enclosed charge to be zero, as Equation (3–42) gives.

The concentric spheres form a capacitor. The capacitance of this system is

$$C = \frac{q_a}{V_1 - V_2} = 4\pi\varepsilon_0 \frac{ab}{b - a} \tag{3-43}$$

# 3-4 SEPARATION OF VARIABLES IN CARTESIAN COORDINATES

One of the simplest and most powerful methods of solving Laplace's equation is a separation of variables. In this approach we look for solutions in which the coordinate dependence factors into a product of functions, each of which depends on only one variable. We then try to satisfy the boundary values by a superposition of such terms. We illustrate the method here for Cartesian coordinates in a two-dimensional case.

For a potential that depends only on $x$ and $y$ coordinates, a separated solution to Laplace's equation has the form

$$\Phi(x, y) = X(x)Y(y) \tag{3-44}$$

If the boundary conditions of a given problem can be satisfied with such a form, it constitutes the solution to the problem. If not, we still may be able to satisfy the boundary conditions by a sum over such solutions.

With Equation (3–44) as a trial solution, Laplace's equation becomes

$$Y(y)\frac{d^2X(x)}{dx^2} + X(x)\frac{d^2Y(y)}{dy^2} = 0$$

We can rearrange this equation as

$$\frac{1}{X(x)}\frac{d^2X(x)}{dx^2} = -\frac{1}{Y(y)}\frac{d^2Y(y)}{dy^2} = \text{constant} \tag{3-45}$$

Since the left-hand side is a function only of $x$ and the right-hand side is a function only of $y$, each side must be a constant for the equality to hold. The constant may be positive, negative, or zero; the choice depends on the nature of the solutions that can satisfy the boundary conditions. If the constant is zero, the solutions to Equation (3–45) are

$$X = a + bx \qquad Y = c + dy \tag{3-46}$$

This solution has only very limited application.

We next consider the case of a negative constant ($= -k^2$) in Equation (3–45), for which

$$\frac{1}{X}\frac{d^2X}{dx^2} = -k^2 \qquad \frac{1}{Y}\frac{d^2Y}{dy^2} = k^2 \tag{3-47}$$

The well-known solutions to these differential equations are

$$\begin{aligned} X(x) &= A_k \cos kx + B_k \sin kx \\ Y(y) &= C_k e^{-ky} + D_k e^{ky} \end{aligned} \tag{3-48}$$

as can be verified by differentiation; here $A_k$, $B_k$, $C_k$, and $D_k$ are constant coefficients. The corresponding potential is

$$\Phi(x, y) = (A_k \cos kx + B_k \sin kx)(C_k e^{-ky} + D_k e^{ky}) \tag{3-49}$$

A more general solution of this type can be formed by superposing solutions of different $k$. In the following subsection we examine representations of potential solutions by series expansion in trigonometric functions.

## 3-4A Fourier Expansion

The sine and cosine functions are said to be *orthogonal* in the interval $-a \leqslant x \leqslant a$ because the integrals

$$\begin{aligned} &\int_{-a}^{a} \sin\frac{m\pi x}{a} \sin\frac{n\pi x}{a}\, dx = a\delta_{mn}, \qquad m \text{ or } n \neq 0 \\ &\int_{-a}^{a} \cos\frac{m\pi x}{a} \cos\frac{n\pi x}{a}\, dx = a\delta_{mn}, \qquad m \text{ or } n \neq 0 \\ &\int_{-a}^{a} \sin\frac{m\pi x}{a} \cos\frac{n\pi x}{a}\, dx = 0 \end{aligned} \tag{3-50}$$

vanish for $m \neq n$. (Here $\delta_{mn}$ is the Kronecker delta: $\delta_{mn} = 1$ for $m = n$ and $\delta_{mn} = 0$ for $m \neq n$.) Furthermore, any reasonable function defined on the interval $-a \leqslant x \leqslant a$ can be expressed as an infinite series of sine and cosine terms:

$$f(x) = \sum_{n=0}^{\infty} \left( A_n \cos \frac{n\pi x}{a} + B_n \sin \frac{n\pi x}{a} \right) \tag{3-51}$$

where the $A_n$ and $B_n$ are constant coefficients. (See a mathematics text for a precise definition of *reasonable*.) Hence the cosine and sine functions are said to form a complete set. Another example of a complete set of functions is the Legendre polynomials, which we consider in a later section.

By multiplying Equation (3–51) by $\sin(m\pi x/a)$ or $\cos(m\pi x/a)$ and integrating from $x = -a$ to $x = +a$, we can solve for the coefficients by using the orthogonality properties of Equation (3–50):

$$\begin{aligned} A_0 &= \frac{1}{2a} \int_{-a}^{+a} f(x)\, dx \\ A_n &= \frac{1}{a} \int_{-a}^{+a} f(x) \cos \frac{n\pi x}{a}\, dx, \qquad n \neq 0 \\ B_n &= \frac{1}{a} \int_{-a}^{+a} f(x) \sin \frac{n\pi x}{a}\, dx \end{aligned} \tag{3-52}$$

We note that the $n$th term of Equation (3–51) has the same form as Equation (3–49) at fixed $y$, with $k$ chosen as $k = n\pi/a$. This result suggests that we can write an arbitrary potential in Equation (3–49) as follows for the interval $-a \leqslant x \leqslant +a$:

$$\Phi(x, y) = \sum_{n=0}^{\infty} \left( A_n \cos \frac{n\pi x}{a} + B_n \sin \frac{n\pi x}{a} \right) (C_n e^{-n\pi y/a} + D_n e^{n\pi y/a}) \tag{3-53}$$

## 3-4B Application of Fourier Method

As an illustration, we construct the potential for the interior of the region bounded by the conducting planes in Figure 3–4. The two parallel planes are infinite in the $y$ and $z$ directions and are separated by a distance $a$. These two plates are grounded ($V = 0$). The perpendicular plane is insulated from the parallel planes and maintained at a potential $V_0$. We choose coordinate axes with the parallel plates in the $x = 0$ and $x = a$ planes and the perpendicular plate in the $y = 0$ plane. By symmetry we can expect a solution with no dependence on the variable $z$, and hence we try to satisfy the boundary conditions by using a series solution of the form in Equation (3–53). The boundary conditions $V = 0$ at $x = 0$ and $x = a$ for any $y$ can only be satis-

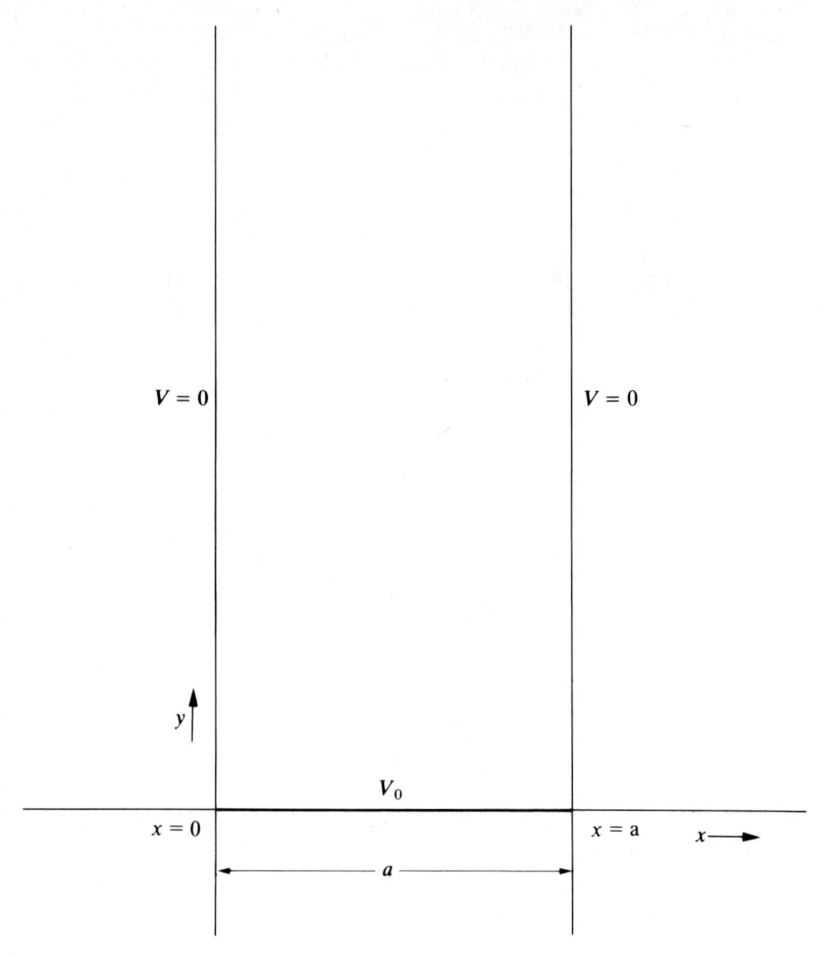


**FIGURE 3-4** Region bounded by two grounded, parallel conducting planes and a conducting plane at a potential $V_0$

fied by taking $A_n = 0$ for all $n$. Thus Equation (3–53) becomes

$$\Phi(x, y) = \sum_{n=1}^{\infty} \sin\frac{n\pi x}{a}\,(C_n e^{-n\pi y/a} + D_n e^{n\pi y/a}) \tag{3-54}$$

where we have rewritten $B_nC_n$ and $B_nD_n$ as $C_n$ and $D_n$, respectively. We have also dropped the $n = 0$ term, which vanishes. Since there is no charge between the parallel plates, we expect the potential to be finite as $y \to \infty$. This boundary condition yields

$$D_n = 0 \qquad \text{all } n \tag{3-55}$$

Finally, we match the solution to $\Phi(x, 0) = V_0$:

$$\sum_{n=1}^{\infty} C_n \sin \frac{n\pi x}{a} = V_0 \tag{3-56}$$

We need to find the set of constants $C_n$ that satisfies this equation. The orthogonality relation in Equation (3–50) can be rewritten as

$$\int_0^a \sin \frac{m\pi x}{a} \sin \frac{n\pi x}{a} = \frac{a}{2} \delta_{mn} \tag{3-57}$$

We multiply Equation (3–56) by $\sin(m\pi x/a)$ and integrate over $x$ from

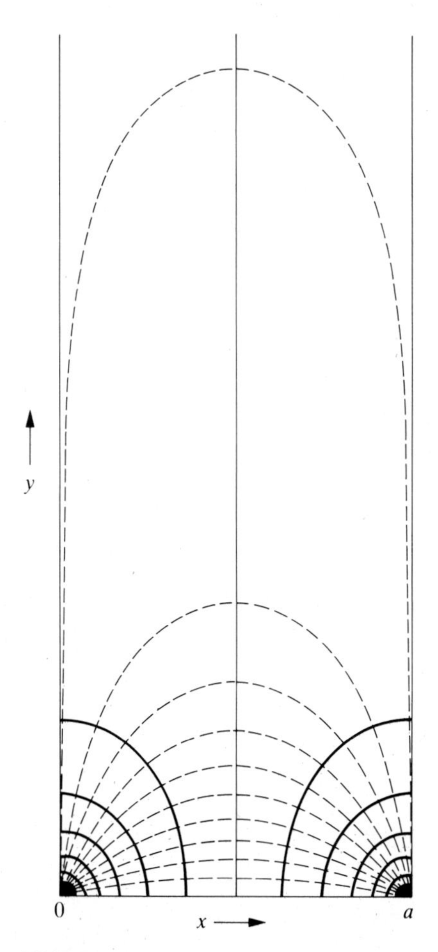


**FIGURE 3-5** Equipotentials (dashed lines) and field lines for the region within the conducting planes of Figure 3–4

$x = 0$ to $x = a$:

$$\sum_{n=1}^{\infty} C_n \int_0^a \sin\frac{m\pi x}{a} \sin\frac{n\pi x}{a}\, dx = V_0 \int_0^a \sin\frac{m\pi x}{a}\, dx \tag{3-58}$$

By Equation (3–57) only one term of the summation is nonzero, namely $n = m$, giving

$$C_m = \frac{2V_0}{a} \int_0^a \sin\frac{m\pi x}{a}\, dx = \frac{2V_0}{m\pi}(1 - \cos m\pi) \tag{3-59}$$

Since $\cos m\pi = (-)^m$, we arrive at (replacing $m$ by $n$)

$$C_n = \frac{4V_0}{n\pi}, \quad n = 1, 3, 5, \ldots; \qquad C_n = 0, \quad n = 2, 4, \ldots \tag{3-60}$$

The final expression for the potential is then

$$\Phi(x, y) = \frac{4V_0}{\pi} \sum_{n\,\mathrm{odd}} \frac{1}{n} \sin\frac{n\pi x}{a}\, e^{-n\pi y/a} \tag{3-61}$$

Some potential and field lines are plotted in Figure 3–5.

## 3-5 SOLUTIONS IN SPHERICAL COORDINATES WITH AXIAL SYMMETRY

The method of separation of variables can also be used in a variety of other two- or three-dimensional coordinate systems. Here we apply this method in spherical coordinates to find solutions to Laplace's equations that depend on radius $r$ and polar angle $\theta$ but not on the azimuthal angle $\phi$. Many physical problems with spherical boundaries have a symmetry axis. Since any rotation about this axis does not change the problem, we know that the solution will not depend on $\phi$.

For a potential that has no $\phi$ dependence, Laplace's equation, (3–25c), becomes

$$\frac{\partial}{\partial r}\left(r^2 \frac{\partial \Phi}{\partial r}\right) + \frac{1}{\sin\theta}\frac{\partial}{\partial \theta}\left(\sin\theta \frac{\partial \Phi}{\partial \theta}\right) = 0 \tag{3-62}$$

We try a solution of factorized form:

$$\Phi(r, \theta) = R(r)\Theta(\theta) \tag{3-63}$$

Substituting this form into Equation (3–62) and dividing by $\Phi = R\Theta$, we obtain

$$\frac{1}{R(r)}\frac{d}{dr}\left(r^2\frac{dR}{dr}\right) = -\frac{1}{\Theta(\theta)}\frac{1}{\sin\theta}\frac{d}{d\theta}\left(\sin\theta\frac{d\Theta}{d\theta}\right) \tag{3-64}$$

Since the left side is a function only of $r$ and the right side only of $\theta$, both sides must be independent of the variables. Equating the left- and right-hand sides to the same constant $C$, we have

$$\frac{d}{dr}\left[r^2\frac{dR(r)}{dr}\right] = CR(r) \tag{3-65}$$

and

$$\frac{1}{\sin\theta}\frac{d}{d\theta}\left[\sin\theta\frac{d\Theta(\theta)}{d\theta}\right] = -C\Theta(\theta) \tag{3-66}$$

We observe that

$$R(r) = r^m \tag{3-67}$$

satisfies Equation (3–65) if

$$m(m+1) = C \tag{3-68}$$

This quadratic equation in $m$ has two solutions. If $m = \alpha$ is one solution, then $m = -\alpha - 1$ is the other solution, as can be verified by substitution in Equation (3–68). Thus the general solution to the radial equation is of the form

$$R(r) = Ar^\alpha + Br^{-\alpha-1} \tag{3-69}$$

where $A$ and $B$ are constants.

To investigate the nature of the angular solutions, we first change the independent variable from $\theta$ to $\cos\theta$, as a convenience, and introduce the notation

$$\mu \equiv \cos\theta \tag{3-70}$$

Since

$$\frac{d}{d\theta} = \frac{d\mu}{d\theta}\frac{d}{d\mu} = -\sin\theta\frac{d}{d\mu} = -\sqrt{1-\mu^2}\frac{d}{d\mu} \tag{3-71}$$

Equation (3–66) becomes, with $C = \alpha(\alpha+1)$ from Equation (3–68),

**Legendre's Equation**

$$\frac{d}{d\mu}\left[(1-\mu^2)\frac{d\Theta}{d\mu}\right] + \alpha(\alpha+1)\Theta = 0 \tag{3-72}$$

This differential equation is known as Legendre's equation. Since it is a second-order differential equation, there are two linearly independent solutions, which are called $P_\alpha(\mu)$ and $Q_\alpha(\mu)$.

The general solution to Equation (3–72) is a linear combination,

$$\Theta(\mu) = C_1 P_\alpha(\mu) + C_2 Q_\alpha(\mu) \tag{3-73}$$

where $C_1$ and $C_2$ are constants. The $Q_\alpha(\mu)$ terms are singular at $\mu = \pm 1$. The $P_\alpha(\mu)$ terms are singular at $\mu = -1$ for noninteger $\alpha$. If $\alpha = n$, a nonnegative integer, then $P_n(\mu)$ is nonsingular at all $\mu$ and is normalized to $P_n(1) = 1$. In fact, $P_n(\mu)$ is a polynomial of order $n$ and is called a *Legendre polynomial.* These polynomials are of importance because any reasonable nonsingular function of $\mu$ can be expressed as an infinite series of Legendre polynomial terms, analogous to the Fourier series that arose in the Cartesian solutions. From the method of separation of variables we are thus led to a general solution $\Phi(r, \theta)$, finite on the polar axis, of the form

$$\Phi(r, \theta) = \sum_n R_n(r) P_n(\mu) = \sum_n (A_n r^n + B_n r^{-n-1}) P_n(\mu) \tag{3-74}$$

We digress now to discuss the Legendre polynomials.

## 3-5A Legendre Polynomials

The Legendre equation, (3–72), can be solved directly by power series development. Instead, we will employ a simple physically motivated method to obtain the Legendre polynomials. Suppose, as in Figure 3–6, we have a point charge of value $q = 4\pi\varepsilon_0$ located at a distance $d = 1$ from $z = 0$ along the $z$ axis. The potential is

$$\Phi = \frac{1}{R} = \frac{1}{|\mathbf{r} - \hat{\mathbf{z}}|} = \frac{1}{\sqrt{1 + r^2 - 2r\mu}} \tag{3-75}$$

where $\mu = \cos\theta$.

This equation presumably has an expansion of the form of Equation (3–74). However, for $r < 1$, we can directly expand this potential as a power series in $r$, using the binomial series

$$(1 + \delta)^{-1/2} = 1 - \frac{\delta}{2} + \frac{1 \cdot 3}{2 \cdot 4}\delta^2 - \frac{1 \cdot 3 \cdot 5}{2 \cdot 4 \cdot 6}\delta^3 + \cdots \tag{3-76}$$

with $\delta = r^2 - 2r\mu$. Grouping the result into powers of $r$, we obtain

$$\Phi = 1 + r\mu + r^2(\tfrac{1}{2})(3\mu^2 - 1) + r^3(\tfrac{1}{2})(5\mu^3 - 3\mu) + \cdots \tag{3-77}$$

Since $\Phi = 1/R$ satisfies the Laplace equation (when $R \neq 0$), we can compare this expansion for $r < 1$ with the expansion in Equation (3–74). There are no

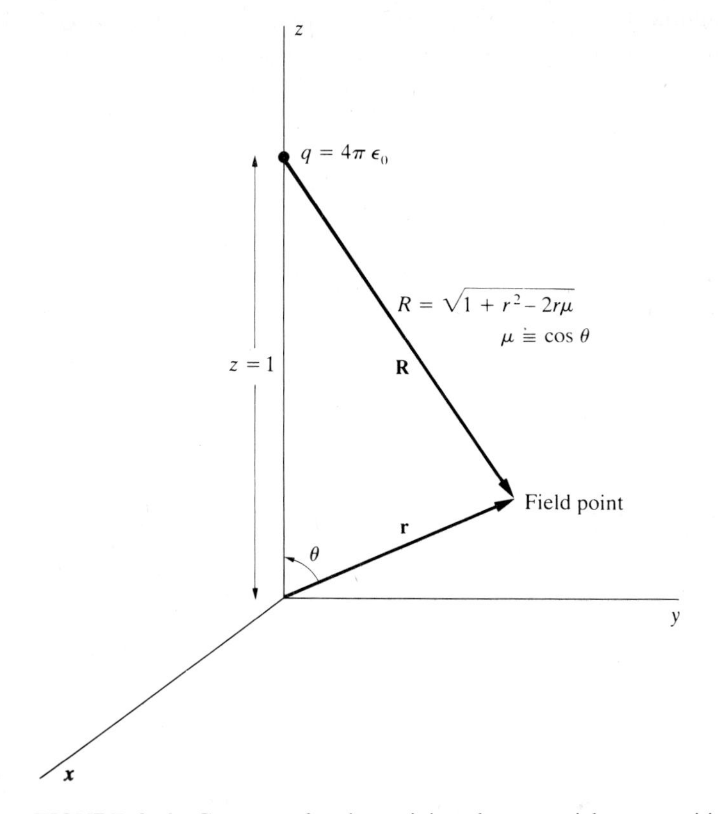


**FIGURE 3-6** Geometry for determining the potential at a position **r** of a point charge $q$ located at (0, 0, 1)

charges and hence no singularities at $r = 0$, so $B_n = 0$ in Equation (3–74). At $\mu = 1$, Equation (3–75) becomes $1/(1 - r) = \sum_{n=0}^{\infty} r^n$, and comparison with Equation (3–74) gives $A_n = 1$, because $P_n(1) = 1$. Then for arbitrary $\mu$

$$\Phi(r, \mu) = \frac{1}{R} = \sum_{n=0}^{\infty} r^n P_n(\mu) \tag{3-78}$$

where the summation is over integral values of $n$. The first few polynomials $P_n(\mu)$ are determined by inspection of Equation (3–77) to be

$$\begin{aligned} P_0(\mu) &= 1 & P_1(\mu) &= \mu \\ P_2(\mu) &= \tfrac{1}{2}(3\mu^2 - 1) & P_3(\mu) &= \tfrac{1}{2}(5\mu^3 - 3\mu) \end{aligned} \tag{3-79}$$

In Equation (3–78),

$$\frac{1}{\sqrt{1 + r^2 - 2r\mu}} \equiv \sum_{n=0}^{\infty} r^n P_n(\mu) \tag{3-80}$$

the function $(1 + r^2 - 2r\mu)^{-1/2}$ is called the *generating function* of Legendre polynomials. From this formula we see that $P_n(\mu)$ is an even or odd function of $\mu$, according to whether $n$ is even or odd. Hence, in particular, $P_n(-1) = (-1)^n$.

Explicit expressions for higher-order polynomials can be evaluated as above, but with increasing difficulty. A more efficient approach is to use a recursion relation, which can be derived from the generating function. To obtain it, we differentiate Equation (3–80) with respect to $r$ and divide by the original equation. By subsequent cross multiplication and recombination of all coefficients of the power $r^n$, we obtain

**Legendre Recursion Relation**

$$(n + 1)P_{n+1}(\mu) = (2n + 1)\mu P_n(\mu) - nP_{n-1}(\mu) \tag{3-81}$$

Starting with $P_0(\mu) = 1$ and $P_1(\mu) = \mu$, any Legendre polynomial can be found by iteration, using the recursion relation. Several low-order polynomials are plotted in Figure 3–7.

A property of particular importance is the orthogonality of the Legendre polynomials. To demonstrate the orthogonality and deduce the normalization, we proceed as follows. First, square the generating function of Equation (3–80):

$$\sum_{m,n=0}^{\infty} r^n s^m P_n(\mu)P_m(\mu) = \frac{1}{\sqrt{1 + r^2 - 2r\mu}}\,\frac{1}{\sqrt{1 + s^2 - 2s\mu}} \tag{3-82}$$

Then integrate over the physical range of $\mu$, $-1 \leqslant \mu \leqslant +1$:

$$\begin{aligned}\sum_{m,n=0}^{\infty} r^n s^m \int_{-1}^{+1} P_n(\mu)P_m(\mu)\,d\mu &= \int_{-1}^{+1} \frac{d\mu}{\sqrt{(1 + r^2 - 2r\mu)(1 + s^2 - 2s\mu)}} \\ &= \frac{1}{\sqrt{rs}} \ln \frac{1 + \sqrt{rs}}{1 - \sqrt{rs}} \\ &= \frac{2}{\sqrt{rs}}\left[\sqrt{rs} + \frac{1}{3}(\sqrt{rs})^3 + \cdots\right] = 2\sum_{n=0}^{\infty} \frac{(rs)^n}{2n + 1}.\end{aligned} \tag{3-83}$$

Comparing powers of $r$ and $s$, we obtain the orthogonality and normalization property

$$\int_{-1}^{+1} P_n(\mu)P_m(\mu)\,d\mu = \frac{2}{2n + 1}\delta_{mn} \tag{3-84}$$

where $\delta_{mn}$ is the Kronecker delta.

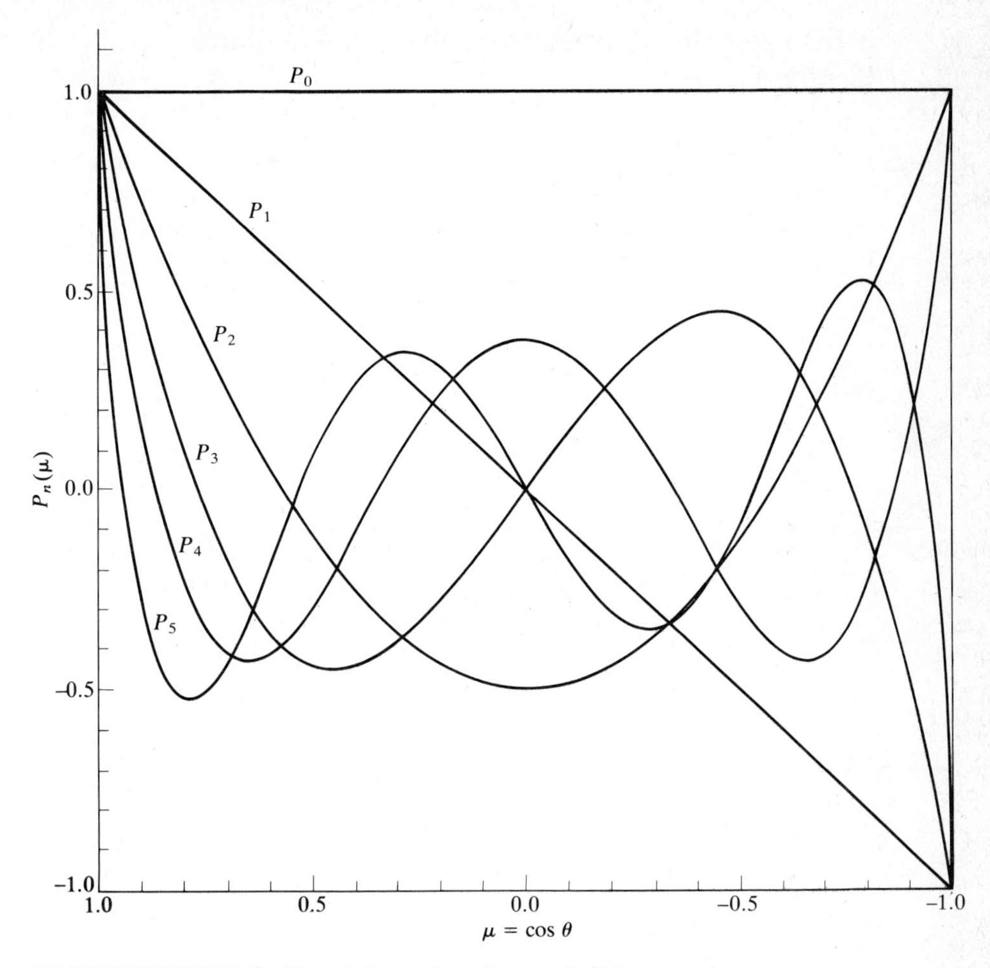


**FIGURE 3-7** First six Legendre polynomials $P_n(\mu)$ versus $\mu$

## 3-5B Legendre Expansion

Any reasonable function in the range $-1 \leqslant \mu \leqslant +1$ can be expressed as a superposition of Legendre polynomials

$$f(\mu) = \sum_{n=0}^{\infty} a_n P_n(\mu) \tag{3-85}$$

The coefficients $a_n$ can be calculated from $f(\mu)$ by use of the orthogonality property of Legendre polynomials. If we multiply Equation (3–85) by $P_m(\mu)$ and integrate over $\mu$ from $-1$ to $+1$, the $a_m$ term gives the only non-vanishing contribution, and we thereby deduce that

$$a_n = \frac{2n+1}{2} \int_{-1}^{+1} d\mu\, f(\mu) P_n(\mu) \tag{3-86}$$

## 3-5C Example: Conducting Sphere in a Uniform Electric Field

As an illustration of these spherical solutions, we will solve for the electric potential for a conducting sphere of radius $a$ that is located in a uniform electric field. (This condition means, of course, uniform in the *absence* of the conducting sphere, which in turn means uniform far away from the sphere.) In this case the axis of symmetry is the electric field direction, which we take to be the $z$ axis, as in Figure 3–8. The form of the solution in Equation (3–74) is

$$\Phi(r, \mu) = \left(A_0 + \frac{B_0}{r}\right) P_0(\mu) + \left(A_1 r + \frac{B_1}{r^2}\right) P_1(\mu) + \left(A_2 r^2 + \frac{B_2}{r^3}\right) P_2(\mu) + \cdots \tag{3-87}$$

The conducting sphere is to be an equipotential, that is, $\Phi(a, \mu) = V$. By the orthogonality property, Equation (3–86), each coefficient of the Legendre polynomials in Equation (3–87), except the $n = 0$ coefficient, must vanish at $r = a$; that is,

$$\begin{aligned} A_n &= -\frac{B_n}{a^{2n+1}}, \qquad n > 0 \\ A_0 + \frac{B_0}{a} &= V, \qquad n = 0 \end{aligned} \tag{3-88}$$

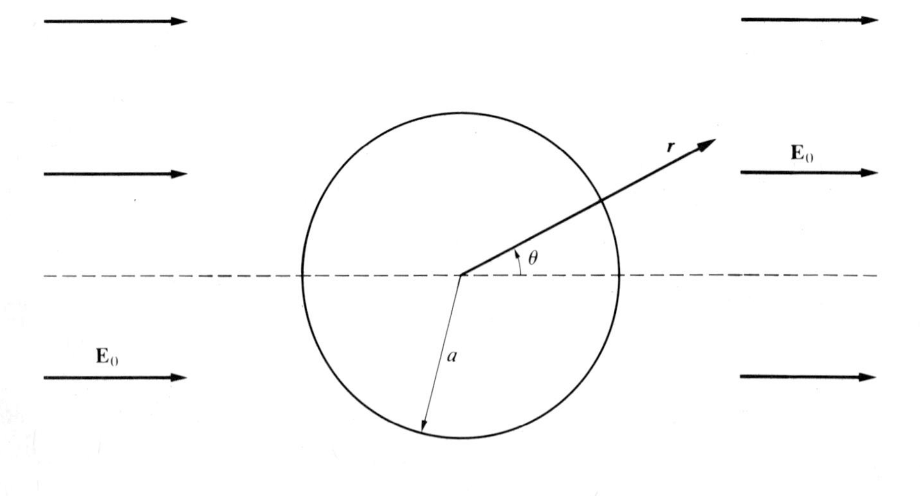


**FIGURE 3–8** Conducting sphere of radius $a$ in an initially uniform electric field

The condition for the field at large distance $r$ being the uniform field $\mathbf{E} = E_0\hat{\mathbf{z}}$ is that

$$\lim_{r \to 0} \Phi(r, \cos\theta) = -E_0 z = -E_0 r P_1(\cos\theta) \tag{3-89}$$

Comparing this asymptotic limit with the solution in Equation (3–87), we see that

$$A_{n>1} = 0 \qquad A_1 = -E_0 \tag{3-90}$$

Referring to Equation (3–88), we conclude that

$$\begin{aligned} B_n &= 0, \qquad n \geqslant 2 \\ B_1 &= -a^3 A_1 = a^3 E_0 \end{aligned} \tag{3-91}$$

The potential is therefore

$$\Phi(r, \cos\theta) = \left(A_0 + \frac{B_0}{r}\right) + E_0\left(-r + \frac{a^3}{r^2}\right)\cos\theta \tag{3-92}$$

The charge density on the sphere [see Equation (2–74)] is

$$\begin{aligned} \sigma &= \varepsilon_0 E_r(r = a) = -\varepsilon_0 \left.\frac{\partial\Phi}{\partial r}\right|_{r=a} \\ &= -\varepsilon_0\left[-\frac{B_0}{a^2} - E_0\left(1 + \frac{2a^3}{a^3}\right)\mu\right] = \frac{\varepsilon_0 B_0}{a^2} + 3\varepsilon_0 E_0 \mu \end{aligned} \tag{3-93}$$

From a surface integral of $\sigma(\mu)$ we find that the total charge $Q$ on the sphere is

$$Q = \int_{-1}^{+1} \sigma(\mu)(2\pi a^2\, d\mu) = 4\pi\varepsilon_0 B_0 \tag{3-94}$$

This derivation determines all the constants in Equation (3–92) in terms of observable quantities. The final solution is then

$$\Phi(r, \cos\theta) = V - \frac{Q}{4\pi\varepsilon_0 a} + \frac{Q}{4\pi\varepsilon_0 r} - E_0 r\cos\theta + \frac{E_0 a^3}{r^2}\cos\theta \tag{3-95}$$

The $1/r$ term in Equation (3–95) is the potential due to an isolated charged sphere, and the term linear in $r$ is the potential of the uniform field if the sphere were absent. The $1/r^2$ term describes the effect of charge distribution induced on the sphere by the external field and is an electric dipole field, as will be discussed in the next chapter. The net induced charge is zero. The equipotentials and electric field lines are shown in Figure 3–9(a) for an uncharged sphere and in Figure 3–9(b) for a sphere with net positive charge.

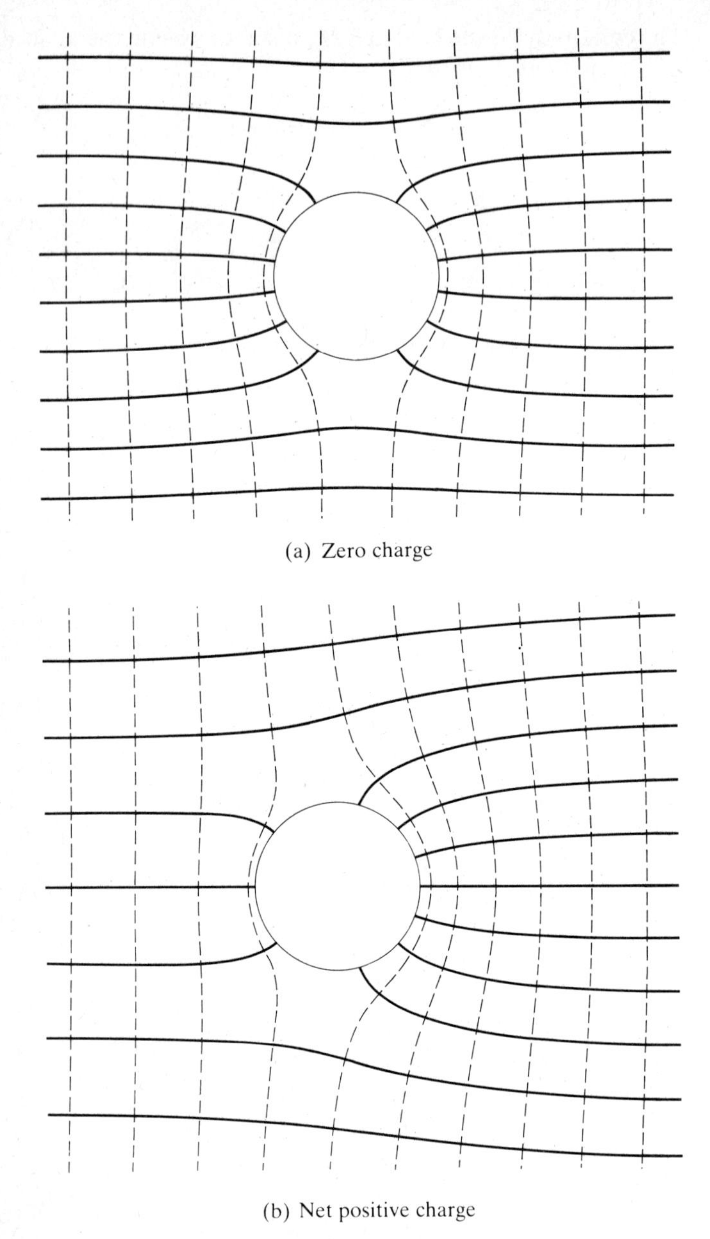


**FIGURE 3-9** Equipotentials (dashed lines) and field lines for (a) a conducting sphere carrying zero charge and (b) a conducting sphere with net positive charge, placed in an initially uniform electric field

# 3-6 ANALYTIC-FUNCTION METHODS

In two-dimensional electrostatic problems the properties of analytic functions of a complex variable can sometimes be used in constructing the potential solution. We introduce the complex variable

$$w = x + iy \tag{3-96}$$

where $x$ and $y$ are Cartesian coordinates. A complex function $f(w)$ is said to be analytic if a unique derivative

$$\frac{df(w)}{dw} = \lim_{\Delta w \to 0} \frac{f(w + \Delta w) - f(w)}{\Delta w} \tag{3-97}$$

exists for any complex increment $\Delta w$. This condition is very easy to satisfy; any differentiable function $f(x)$, taken as a function of $w = x + iy$, $f(w)$, satisfies Equation (3–97). However, remarkable things follow.

If we take $\Delta w = \Delta x$ in Equation (3–97), we find

$$\frac{df(w)}{dw} = \frac{\partial f}{\partial x} \tag{3-98}$$

But with $\Delta w = i\,\Delta y$, we obtain

$$\frac{df(w)}{dw} = -i\frac{\partial f}{\partial y} \tag{3-99}$$

Equating these expressions for $df/dw$ yields

$$\frac{\partial f}{\partial x} = -i\frac{\partial f}{\partial y} \tag{3-100}$$

If we write $f(w)$ in terms of real functions $\Phi$ and $\Psi$ as

$$f(w) = \Phi(x, y) + i\Psi(x, y) \tag{3-101}$$

then the real and imaginary parts of Equation (3–100) are

**Cauchy-Riemann Equations**

$$\frac{\partial \Phi}{\partial x} = \frac{\partial \Psi}{\partial y} \qquad \frac{\partial \Psi}{\partial x} = -\frac{\partial \Phi}{\partial y} \tag{3-102}$$

These equations, which the parts $\Phi$ and $\Psi$ of an analytic function $f$ satisfy, are called the Cauchy-Riemann equations. Differentiating the first relation with respect to $x$ and the second with respect to $y$ and then subtracting the result, we find

$$\nabla^2\Phi(x, y) = 0 \tag{3-103}$$

Reversing the order of the preceding operations gives

$$\nabla^2\Psi(x, y) = 0 \tag{3-104}$$

Hence both real and imaginary parts of an analytic function satisfy the Laplace equation, and so either can represent an electrostatic potential in a charge-free region. Furthermore, the curves defined by $\Phi$ = constant and $\Psi$ = constant are mutually orthogonal, as we now verify. The normals to these curves are along $\nabla\Phi$ and $\nabla\Psi$, respectively, and the dot product of these vectors is

$$\nabla\Phi \cdot \nabla\Psi = \left(\hat{\mathbf{x}}\frac{\partial\Phi}{\partial x} + \hat{\mathbf{y}}\frac{\partial\Phi}{\partial y}\right)\cdot\left(\hat{\mathbf{x}}\frac{\partial\Psi}{\partial x} + \hat{\mathbf{y}}\frac{\partial\Psi}{\partial y}\right) = \frac{\partial\Phi}{\partial x}\frac{\partial\Psi}{\partial x} + \frac{\partial\Phi}{\partial y}\frac{\partial\Psi}{\partial y} \tag{3-105}$$

This product vanishes by the Cauchy-Riemann equations. Hence if the curves $\Phi(x, y)$ = constant represent equipotentials, then the curves $\Psi(x, y)$ = constant will correspond to the electric field lines, and vice versa.

To demonstrate the analytic-function method, we consider

$$f(w) = Cw^2 \tag{3-106}$$

with $C$ a real constant, for which

$$\Phi(x, y) = C(x^2 - y^2) \qquad \Psi(x, y) = 2Cxy \tag{3-107}$$

The function $\Phi(x, y)$ could represent the potential in the space between four conductors whose surfaces are hyperbolas (hyperbolic cylinders in three dimensions) maintained at constant potential. For example,

$$\Phi(x, y) = \frac{V_0}{a^2}(x^2 - y^2) \tag{3-108}$$

represents the potential between four conducting surfaces at potentials $\pm V_0$ located at $x^2 - y^2 = \pm a^2$, as illustrated in Figure 3–10. This arrangement is called an electric quadrupole lens. The electric field for the lens is

$$\mathbf{E} = -\nabla\Phi = \frac{2V_0}{a^2}(-x\hat{\mathbf{x}} + y\hat{\mathbf{y}}) \tag{3-109}$$

For positively charged particles moving in the $z$ direction, the lens will focus in the $x$ direction and defocus in the $y$ direction. The field lines are shown in Figure 3–10.

Another example is

$$f(w) = C \ln w \tag{3-110}$$

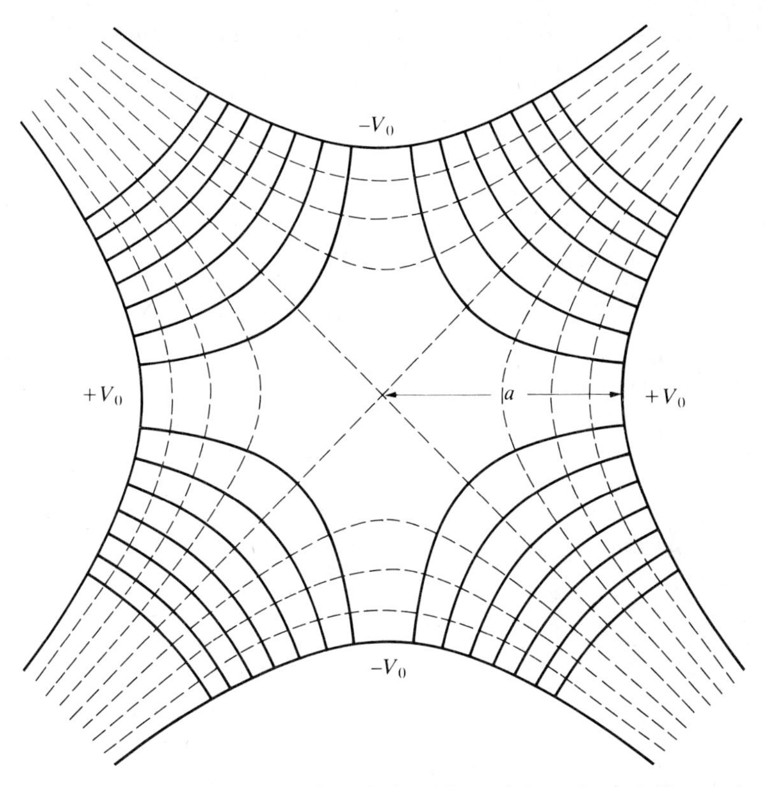


**FIGURE 3-10** Electric quadrupole lens formed from hyperbolic conductors. The equipotentials (dashed lines) and the field lines are shown.

which is analytic for $w \neq 0$. In cylindrical coordinates

$$w = \imath e^{i\phi} \tag{3-111}$$

$$\Phi = C \ln \imath \tag{3-112}$$

$$\Psi = C\phi \tag{3-113}$$

The potential $\Phi$ is that of a line charge at $\imath = 0$.

Since any analytic function gives a solution to Laplace's equation, one can write a large number of solutions for electric potentials, some of which correspond to physically interesting problems.

## 3-7 METHOD OF IMAGES

A typical electrostatics problem consists of a set of charges and conductors with specified potentials. The solution consists of a potential function $\Phi(\mathbf{r})$ that is constant on the conducting surfaces and satisfies Poisson's equation at all points outside the conductor. Once $\Phi(\mathbf{r})$ is known, we can compute the electric field and hence the induced charge density on the conductors.

Let us now consider a related problem. Starting with the same solution $\Phi(\mathbf{r})$, we now imagine that the conductors vanish. Where the conducting surfaces used to be, $\Phi(\mathbf{r})$ still assumes the same constant values, but now the lines of $\mathbf{E}$ pass right through. These $\mathbf{E}$ lines end on a charge distribution contained within the former conducting surfaces, which in some cases is a set of point charges called *image charges*. The net image charge is exactly equal to the net induced charge for each conductor.

The method of images reverses this sequence. We guess at an image charge distribution that in conjunction with the outside charges will yield the desired equipotential surfaces and hence solve the conductor problem in the region outside the conductors. In practice, the only equipotential shapes that have simple enough charge distributions to be treated by the method of images are composed of planes, spheres, and cylinders.

We illustrate the method of images for three simple cases. Other electrostatic problems can sometimes be solved by use of these basic image results.

## 3-7A Point Charge and Infinite Conducting Plane

Figure 3–11 illustrates a point charge $q$ located a distance $h$ above an infinite, flat conducting surface that is grounded. We choose the origin of the coordinate system on the conductor at the point vertically beneath the point charge. We know that for a system of two equal and opposite charges separated by a distance $2h$, $\Phi(\mathbf{r})$ vanishes at all points equidistant from the charges, that is, on the median plane. Thus this two-charge system gives the solution for a point charge at a distance $h$ from a grounded conducting surface; in this application charge $-q$ in the conducting region is fictitious and is an image charge. The potential outside the conductor calculated from the real and image charges is

$$\Phi(\mathbf{r}) = \frac{q}{4\pi\varepsilon_0}\left(\frac{1}{R_+} - \frac{1}{R_-}\right) \tag{3–114}$$

where

$$R_+ = [(z - h)^2 + \imath^2]^{1/2} \qquad R_- = [(z + h)^2 + \imath^2]^{1/2} \tag{3–115}$$

The electric field can be found by differentiation of Equation (3–114) or by direct construction from Coulomb's law. From the geometry of Figure 3–11 the $\mathbf{E}$ field on the plane is

$$\mathbf{E} = -\hat{\mathbf{z}}\frac{q}{4\pi\varepsilon_0}\left(\frac{2\cos\theta}{R^2}\right) \tag{3–116}$$

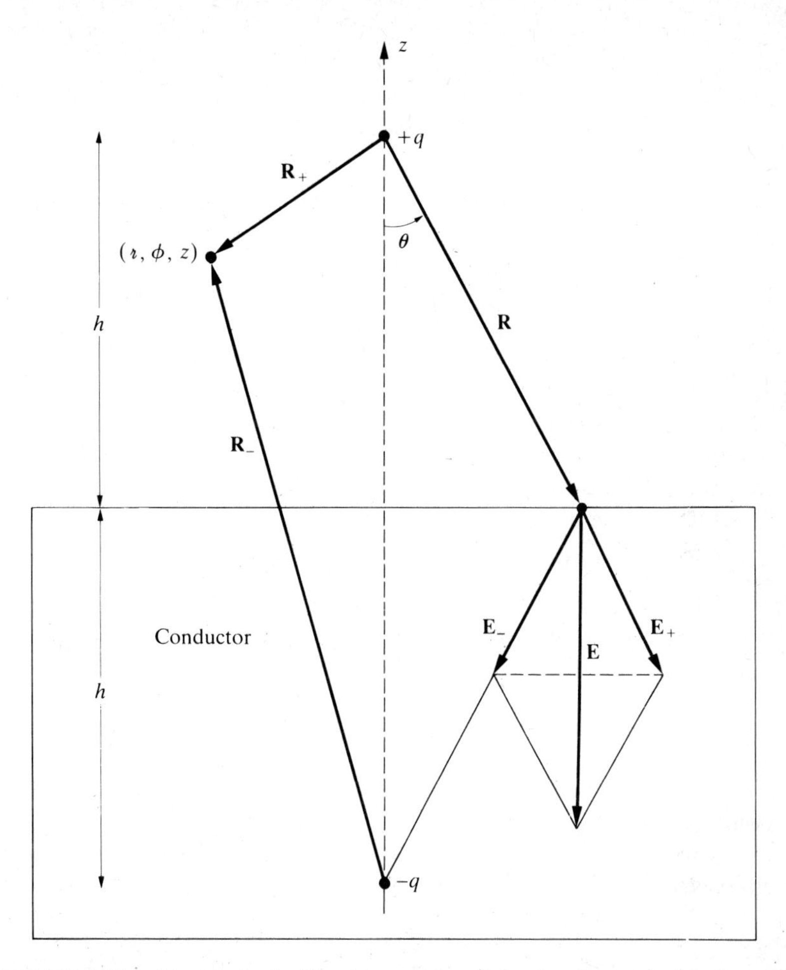


**FIGURE 3-11** Method of images used in solving for the potential of a point charge $q$ above an infinite conducting plane. The image charge $-q$ and the field contributions of the real and image charges are shown.

where

$$R = (h^2 + \imath^2)^{1/2} \qquad \cos\theta = \frac{h}{R} \tag{3-117}$$

Hence

$$E_z = -\frac{qh}{2\pi\varepsilon_0 R^3} \tag{3-118}$$

From the Gauss's law result of Equation (2–74), the surface charge density on the conductor is

$$\sigma = \varepsilon_0 E_z = -\frac{qh}{2\pi(h^2 + r^2)^{3/2}} \tag{3-119}$$

Integrating over the entire conducting plane, we find the net induced charge on the conductor:

$$q_{\text{induced}} = -\frac{qh}{2\pi}\int_0^{2\pi} d\phi \int_0^{\infty} \frac{r\,dr}{(h^2 + r^2)^{3/2}} = -qh\int_h^{\infty} \frac{R\,dR}{R^3} = -q \tag{3-120}$$

Thus the induced charge is the same as the image charge: The electric field due to the induced charge on the surface of the conductor is correctly represented by an equivalent image charge at a distance $h$ below the surface, as illustrated in Figure 3–12.

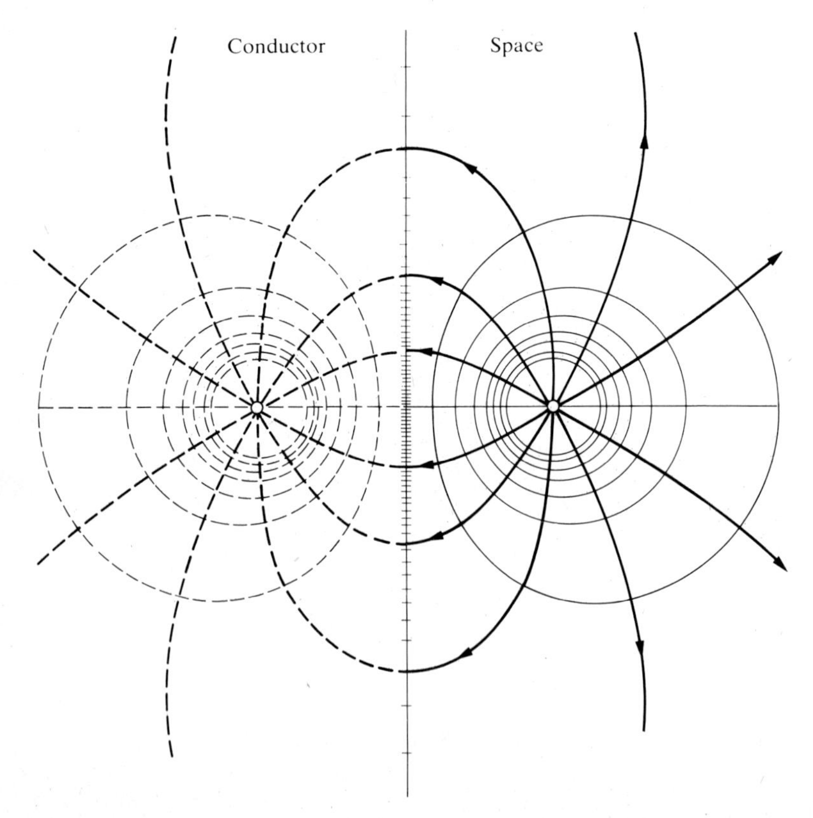


**FIGURE 3-12** Equipotentials and field lines (with arrows) for a point charge outside an infinite conducting plane. The tick marks represent the density of the induced charge on the conductor. The dashed lines show the image equipotentials and field lines.

The force on the point charge $q$ is

$$\mathbf{F} = q\mathbf{E}$$

where $\mathbf{E}$ is the electric field due to the induced charges on the conducting plane. We can calculate $\mathbf{F}$ from the image charge, obtaining

$$\mathbf{F} = -\frac{q^2}{4\pi\varepsilon_0(2h)^2}\,\hat{\mathbf{z}} \tag{3-121}$$

## 3-7B Point Charge Outside a Conducting Sphere

In this problem, illustrated in Figure 3–13, we want to find the electric potential in the region outside a conducting sphere in the presence of a point charge. We take the origin of the coordinate system, $\mathbf{r} = 0$, at the center of the sphere. The point charge $q$ is located at $\mathbf{r} = r_q\hat{\mathbf{z}}$. The boundary condition on the potential is $\Phi = 0$ on the surface of the grounded sphere, $r = a$. We try to satisfy the boundary condition by using an image charge $q_{\mathrm{I}}$ at $r = r_{\mathrm{I}}$, on a line between the source and the origin. The potential due to the source $q$ and the image $q_{\mathrm{I}}$ is

$$4\pi\varepsilon_0\Phi(\mathbf{r}) = \frac{q}{|\mathbf{r} - r_q\hat{\mathbf{z}}|} + \frac{q_{\mathrm{I}}}{|\mathbf{r} - r_{\mathrm{I}}\hat{\mathbf{z}}|} \tag{3-122}$$

This equation can be rewritten as

$$4\pi\varepsilon_0\Phi(\mathbf{r}) = \frac{q/r}{\sqrt{1 + (r_q/r)^2 - 2(r_q/r)(\cos\theta)}} + \frac{q_{\mathrm{I}}/r_{\mathrm{I}}}{\sqrt{1 + (r/r_{\mathrm{I}})^2 - 2(r/r_{\mathrm{I}})(\cos\theta)}} \tag{3-123}$$

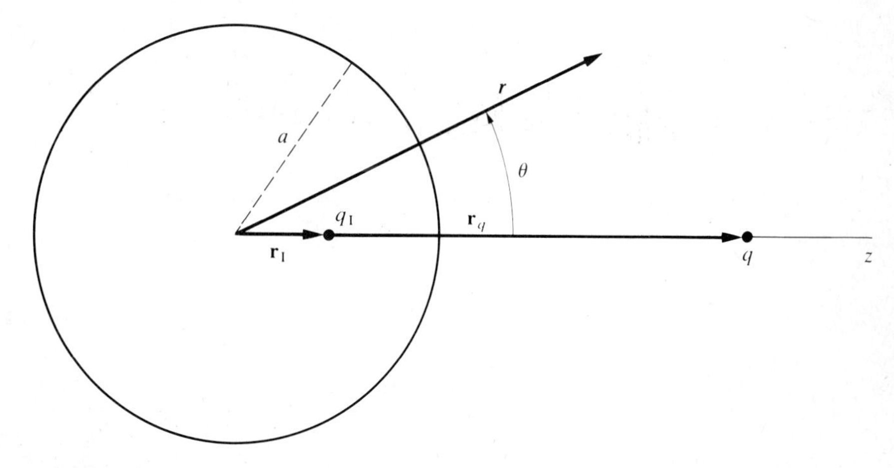


**FIGURE 3–13** Geometry for the image charge $q_{\mathrm{I}}$ associated with a point charge $q$ outside a conducting sphere of radius $a$

where $\cos\theta \equiv \hat{\mathbf{r}} \cdot \hat{\mathbf{z}}$. We can get $\Phi = 0$ at $r = a$ for all $\theta$ by the choices

$$\frac{q_{\mathrm{I}}}{r_{\mathrm{I}}} = -\frac{q}{a} \tag{3-124a}$$

and

$$\frac{r_q}{a} = \frac{a}{r_{\mathrm{I}}} \tag{3-124b}$$

Thus an image charge

$$q_{\mathrm{I}} = -\left(\frac{a}{r_q}\right)q \tag{3-125a}$$

located on a line between $q$ and the origin at a distance

$$r_{\mathrm{I}} = \frac{a^2}{r_q} \tag{3-125b}$$

can be used to solve the problem of a point charge outside a grounded sphere. An example of the field lines and the equipotentials for this solution is shown in Figure 3–14. Conversely, the problem of a point charge at $r_{\mathrm{I}}$ inside a spherical cavity in a conductor is solved by use of an image charge at the point $r_q$ defined by Equation (3–125b) outside the cavity.

We can easily generalize the calculation above to the case of an ungrounded spherical conductor in the vicinity of a point charge. We solve this problem as a superposition of the potentials of (1) the grounded sphere plus external point charge $q$ and (2) the ungrounded sphere with net charge $q'$ and no external point charge. The potential for (2) in the region $|\mathbf{r}| \geqslant a$ is

$$4\pi\varepsilon_0\Phi(\mathbf{r}) = \frac{q'}{|\mathbf{r}|} \tag{3-126}$$

The complete potential is given by the sum of Equations (3–122) and (3–126). With the external point charge present, the net charge $Q$ on the sphere is given by

$$Q = q' + q_{\mathrm{I}} \tag{3-127}$$

where $q_{\mathrm{I}}$ is the image charge of Equation (3–125a). Hence the charge $q'$ appearing in Equation (3–126) is

$$q' = Q - q_{\mathrm{I}} \tag{3-128}$$

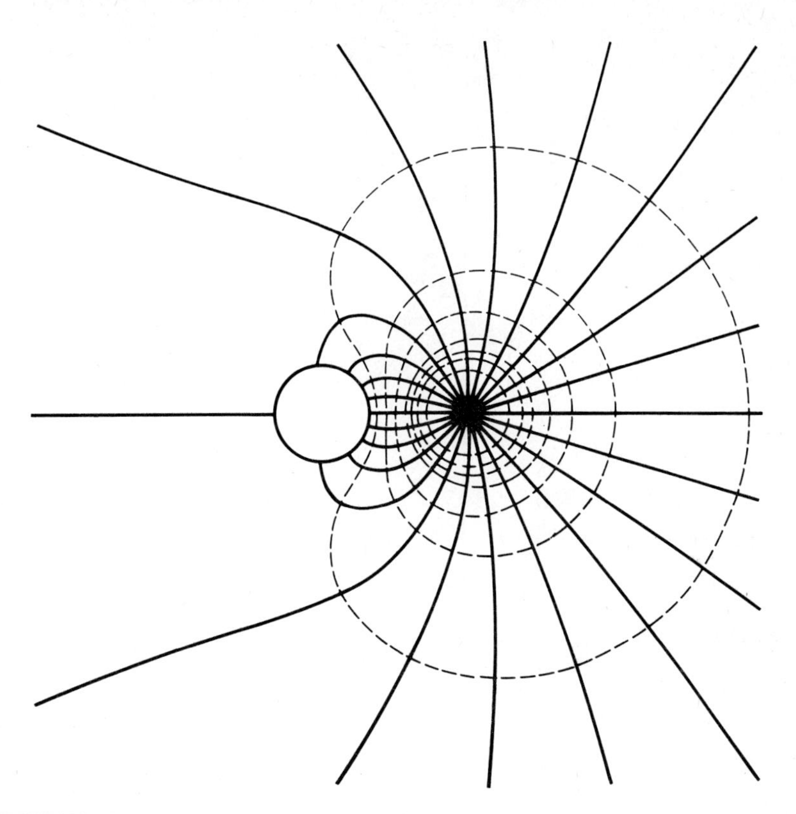

**FIGURE 3-14** Equipotentials (dashed lines) and field lines for a point charge outside a conducting sphere

On the surface of the ungrounded conductor the potential is

$$\Phi(r = a) = \frac{Q - q_I}{4\pi\varepsilon_0 a} \tag{3-129}$$

## 3-7C Line Charge Image Methods

A similar image technique can be used in solving problems with cylindrical symmetry.

**Line Charge and Conducting Plane.** We first consider the case of a line charge outside an infinite grounded conducting plane, as illustrated in Figure 3–15. The true line charge $\lambda$ per unit length is a distance $D/2$ from the plane. We place an image line charge $\lambda_I = -\lambda$ symmetrically below the surface. Denoting by $\mathscr{r}_\pm$ the radial distances from the $\pm\lambda$ line charges to an observation point at $\mathscr{r}$, we obtain the potential due to the line charge and its

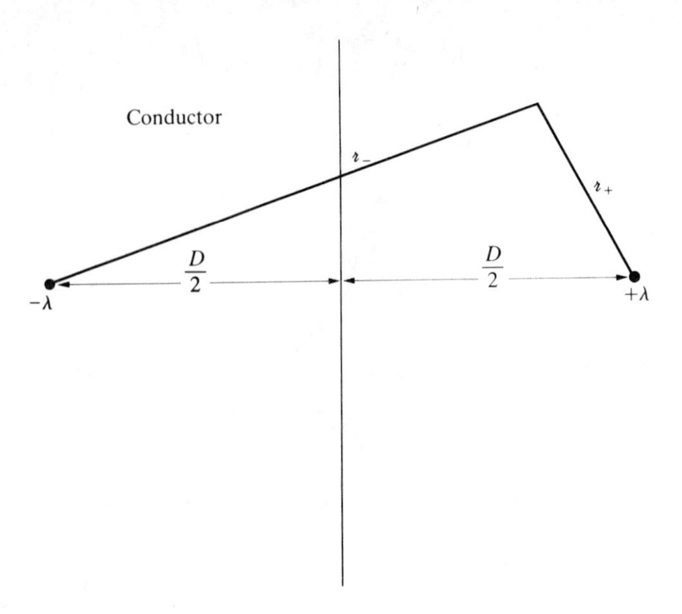


**FIGURE 3-15** Method of images used in solving the problem of a line charge outside an infinite, conducting grounded plane

image, from Equation (2–62):

$$\Phi(\imath) = -\frac{\lambda}{2\pi\varepsilon_0}\left(\ln\frac{\imath_+}{\imath_0} - \ln\frac{\imath_-}{\imath_0}\right) = -\frac{\lambda}{2\pi\varepsilon_0}\ln\frac{\imath_+}{\imath_-} \tag{3-130}$$

On the conducting plane $\imath_+ = \imath_-$ and $\Phi = 0$. This value is the appropriate boundary value for the grounded conductor, so Equation (3–130) is the unique potential outside the conductor, $\imath_+ < \imath_-$.

In the coordinate system of Figure 3–16, with origin at a distance $p$ to the left of the image line charge, the distances $\imath_+$ and $\imath_-$ can be expressed

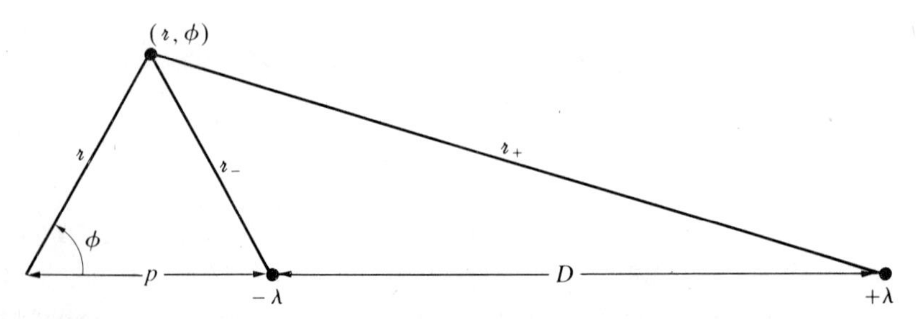


**FIGURE 3-16** Geometry for the potential at a point $(\imath, \phi)$ for line charges of $-\lambda$ and $+\lambda$ located at distances $p$ and $p + D$ from the origin, respectively

in polar coordinates ($\imath$, $\phi$) by the law of cosines as

$$\imath_+^2 = \imath^2 + (D + p)^2 - 2(D + p)\imath \cos \phi$$
$$\imath_-^2 = \imath^2 + p^2 - 2\imath p \cos \phi \tag{3–131}$$

The equipotentials of Equation (3–130) are specified by

$$\frac{\imath_+}{\imath_-} = c \tag{3–132}$$

where $c$ is a positive constant. Combining Equations (3–131) and (3–132), we can express the equipotential condition as

$$\imath^2(1 - c^2) - 2(D + p - c^2 p)\imath \cos \phi = c^2 p^2 - (D + p)^2 \tag{3–133}$$

By judicious choice of our origin of coordinates at location

$$p = \frac{D}{c^2 - 1} \tag{3–134}$$

this equation simplifies to

$$\imath = \left|\frac{cD}{c^2 - 1}\right| \tag{3–135}$$

Thus all the equipotentials are circles, unlike the spherical case where only one equipotential was spherical.

**Line Charge and Conducting Cylinder.** We can immediately apply the preceding solution—for the potential due to two line charges—to the problem of a line charge and a conducting cylinder, illustrated in Figure 3–17. If we place the axis of symmetry at the origin used in Figure 3–16, and let $-\lambda$ be the image line charge, then the surface of the conductor is an equipotential of Equation (3–130). The distance $R$ from the true line charge to the cylinder axis is

$$R = D + p = \frac{c^2 D}{c^2 - 1} \tag{3–136}$$

From Equation (3–135) the cylinder radius $a$ is given by

$$a = \frac{cD}{c^2 - 1} \tag{3–137}$$

From Equations (3–134), (3–136), and (3–137) we obtain

$$p = \frac{a^2}{R} \tag{3–138}$$

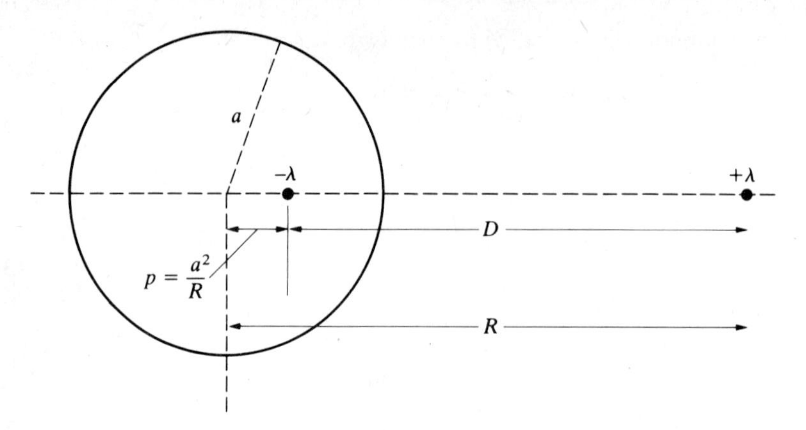


**FIGURE 3-17** Image charge of $-\lambda$ located at $p = a^2/R$ and used in determining the potential of a line charge $+\lambda$ located a distance $R$ from the center of a cylinder of radius $a$

and

$$c = \frac{R}{a} \tag{3-139}$$

The image charge location is thus the same as in the spherical case; see Equation (3–125b).

From Equations (3–130), (3–132), and (3–139) the potential on the surface of the conducting cylinder is

$$\Phi(r = a) = -\frac{\lambda}{2\pi\varepsilon_0}\ln\frac{R}{a} \tag{3-140}$$

The net charge on the cylinder surface in this example is $-\lambda$ per unit length. In the more general circumstance of a line charge $\lambda$ per unit length outside a conducting cylinder having net charge $\Lambda$ per unit length, we must add a term,

$$\Phi(r) = -\frac{(\Lambda - \lambda_{\mathrm{I}})}{2\pi\varepsilon_0}\ln\frac{r}{r_0} \tag{3-141}$$

to the potential of Equation (3–130). Here $\lambda_{\mathrm{I}} = -\lambda$ is the original image line charge, and $\Lambda - \lambda_{\mathrm{I}}$ is a new image charge at $r = 0$, required to maintain the conducting surface as an equipotential with net charge $\Lambda$ per unit length.

**Two Oppositely Charged, Parallel Conducting Cylinders.** Another application of the potential in Equation (3–130) from opposite line charges is that of two parallel conducting cylinders of opposite charge, as illustrated in

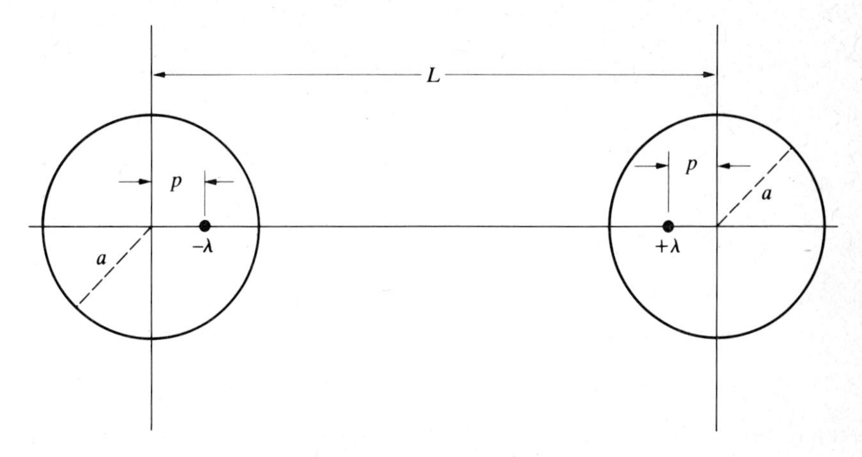


**FIGURE 3–18** Two nonoverlapping, oppositely charged, parallel conducting cylinders. The potential outside the conducting cylinders can be solved with two image line charges

Figure 3–18. The equipotential and electric field lines for Equation (3–130) are plotted in Figure 3–19. If we choose the locations of the conductors to coincide with equipotential circles, then Equation (3–131) represents the unique solution to the two-parallel-conductor problem with net charges $\pm\lambda$ per unit length on the conductors. Both the line charges represent images in this case. The distance $L$ between the axes of the cylinders is

$$L = D + 2p = R + p = R + \frac{a^2}{R} \tag{3-142}$$

The potentials on the two conducting surfaces are, from Equation (3–140),

$$\Phi = \frac{\pm\lambda}{2\pi\varepsilon_0} \ln \frac{R}{a} \tag{3-143}$$

The capacitance per unit length is

$$C = \frac{\lambda}{\Delta\Phi} = \frac{\pi\varepsilon_0}{\ln R/a} \tag{3-144}$$

With Equation (3–142) this capacitance can be reexpressed in terms of $L$ as

$$C = \frac{\pi\varepsilon_0}{\cosh^{-1}(L/2a)} \tag{3-145}$$

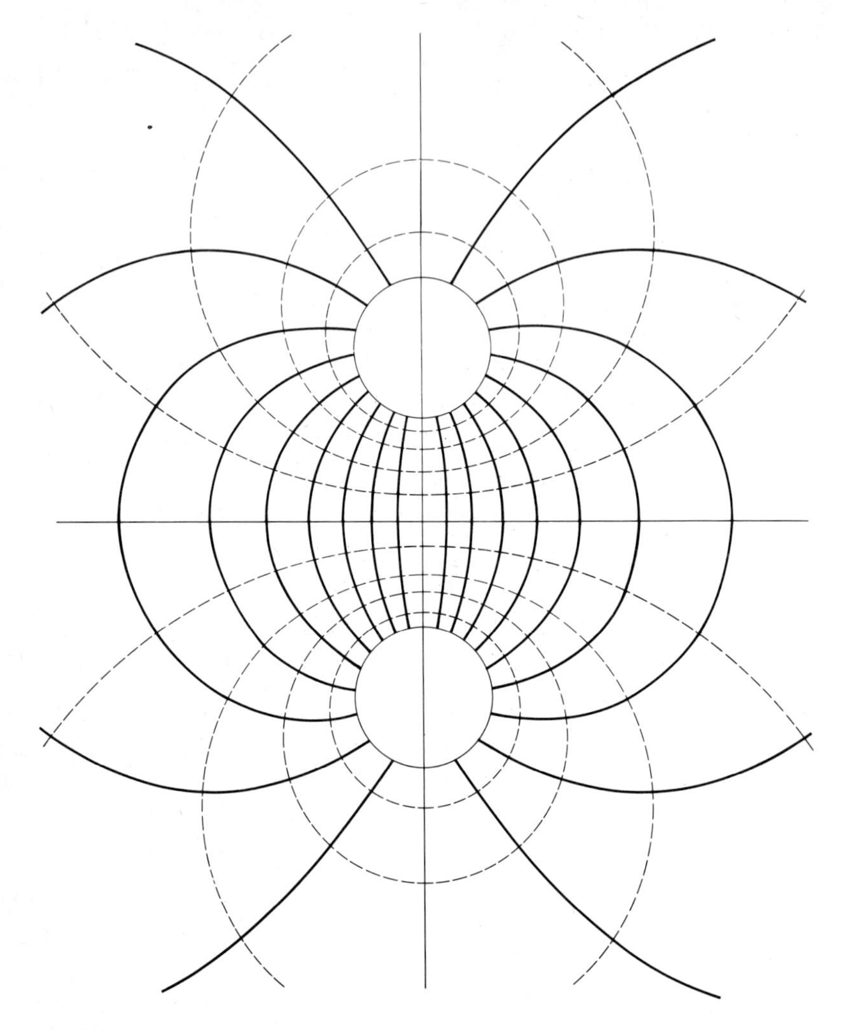

**FIGURE 3-19** Equipotentials (dashed lines) and field lines for two oppositely charged, parallel conducting cylinders

## 3-8 NUMERICAL SOLUTION OF THE LAPLACE EQUATION

Analytic solutions to the Laplace equation are possible only for physical problems with simple geometries. However, for any boundary value problem in electrostatics the potential can be determined to any desired accuracy by numerical solution of Laplace's equation. With calculators or computers a numerical approach to electrostatic problems is quite practical. In fact, numerical solutions are sometimes preferable to complicated series solutions, which may not be rapidly convergent.

In the numerical approach the continuous spatial dimensions are replaced by a lattice system with finite spacings. The derivative of the potential is approximated by the difference between potentials at adjacent lattice points divided by the lattice separation. Similar approximations are used for higher derivatives. The Laplace equation is thereby represented by coupled algebraic equations that can be solved by iteration or by matrix inversion. The finer the lattice, the better the approximation becomes.

As a simple illustration of numerical methods, we solve Laplace's equation in two Cartesian dimensions,

$$\frac{\partial^2 \Phi}{\partial x^2} + \frac{\partial^2 \Phi}{\partial y^2} = 0 \tag{3-146}$$

for an arbitrary boundary surface on which the potential has specified values. The spatial region of interest is spanned by a rectangular lattice whose points are separated by a length $\Delta$. The $x$ and $y$ coordinates of the lattice points are specified by

$$x_i = i\Delta \qquad y_j = j\Delta \tag{3-147}$$

where $i$ and $j$ are integers. The potential at the site $(x_i, y_j)$ is denoted by $\Phi_{i,j}$. The first derivative with respect to $x$ is

$$\frac{\partial \Phi}{\partial x}(x_i^+, y_j) \approx \frac{\Phi_{i+1,j} - \Phi_{i,j}}{\Delta} \tag{3-148a}$$

where $x_i^+$ denotes the midpoint of the interval $x_i$ to $x_{i+1}$. The corresponding result for the interval $i - 1$ to $i$ is

$$\frac{\partial \Phi}{\partial x}(x_i^-, y_j) \approx \frac{\Phi_{i,j} - \Phi_{i-1,j}}{\Delta} \tag{3-148b}$$

The second derivative with respect to $x$ evaluated at $(x_i, y_j)$ is then

$$\frac{\partial^2 \Phi}{\partial x^2}(x_i, y_j) \approx \frac{(\partial\Phi/\partial x)(x_i^+, y_j) - (\partial\Phi/\partial x)(x_i^-, y_j)}{\Delta}$$

$$\approx \frac{\Phi_{i-1,j} - 2\Phi_{i,j} + \Phi_{i+1,j}}{\Delta^2} \tag{3-149}$$

Similarly, the second derivative with respect to $y$ is given by

$$\frac{\partial^2 \Phi}{\partial y^2}(x_i, y_j) = \frac{\Phi_{i,j-1} - 2\Phi_{i,j} + \Phi_{i,j+1}}{\Delta^2} \tag{3-150}$$

Consequently, the lattice approximation to the Laplace equation is

$$\nabla^2\Phi_{ij} = \frac{1}{\Delta^2}\left[(\Phi_{i-1,j} - 2\Phi_{i,j} + \Phi_{i+1,j}) + (\Phi_{i,j-1} - 2\Phi_{i,j} + \Phi_{i,j+1})\right]$$
$$= 0 \tag{3-151}$$

Introducing the average potential for the four adjacent sites surrounding the $(i, j)$ site by

$$\bar{\Phi}_{ij} \equiv \tfrac{1}{4}(\Phi_{i-1,j} + \Phi_{i+1,j} + \Phi_{i,j-1} + \Phi_{i,j+1}) \tag{3-152}$$

we have the lattice version of the Laplace equation:

$$\Phi_{ij} = \bar{\Phi}_{ij} \tag{3-153}$$

The Laplace equation is thereby equivalent to the requirement that the potential at a lattice point be given by the average of the potentials at the closest adjoining sites. Equation (3–153) is a set of coupled homogeneous algebraic equations that can be solved by matrix methods. An alternative is to solve Equation (3–153) by iteration; this technique is known as the relaxation method. We illustrate the iterative method of solution in the following example.

Consider a square with sides of unit length that are charged to potentials of 0, 1, 2, and 3 V. We divide the square into a six-by-six lattice. The lattice points on the boundary of the square are assigned the values of the boundary potential. The potentials at the corner lattice points are ambiguous, but their values do not enter in the calculation of potential averages at interior lattice points. At the interior points any initial values can be adopted for the potential, but faster convergence of the iterative process is achieved if the starting values roughly interpolate the boundary potentials. We take the initial potential values given in Table 3–1.

For the first iteration we start at $i = 1$ and $j = 1$ and calculate $\Phi_{1,1}$, using the boundary values and the assumed interior potential values

$$\bar{\Phi}_{1,1} = \frac{3 + 0 + 1 + 2}{4} = 1.5$$

This result is assigned as the new value of $\Phi_{1,1}$. We now proceed to the point $i = 2$ and $j = 1$ and calculate

$$\bar{\Phi}_{2,1} = \frac{1.5 + 0 + 1 + 1}{4} = 0.88$$

which is assigned as the value of $\Phi_{2,1}$ in this iteration. Similarly, the potential at $i = 3$ and $j = 2$ becomes

$$\bar{\Phi}_{3,1} = \frac{0.88 + 0 + 1 + 1}{4} = 0.72$$

**TABLE 3–1** Iterations for a numerical solution to the Laplace equation for the case of a two-dimensional rectangular boundary with specified potentials

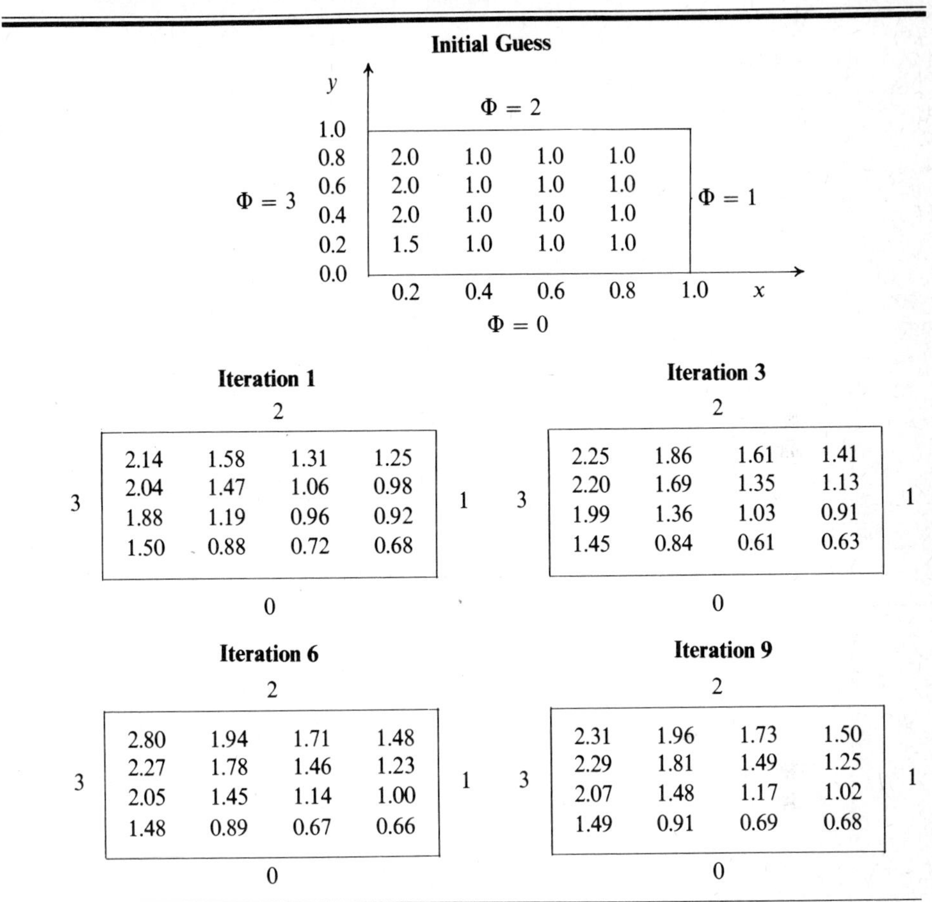

**Initial Guess**

$\Phi = 2$ (top), $\Phi = 3$ (left), $\Phi = 1$ (right), $\Phi = 0$ (bottom)

| y \ x | 0.2 | 0.4 | 0.6 | 0.8 |
|---|---|---|---|---|
| 1.0 | | | | |
| 0.8 | 2.0 | 1.0 | 1.0 | 1.0 |
| 0.6 | 2.0 | 1.0 | 1.0 | 1.0 |
| 0.4 | 2.0 | 1.0 | 1.0 | 1.0 |
| 0.2 | 1.5 | 1.0 | 1.0 | 1.0 |
| 0.0 | | | | |

(x axis: 0.2, 0.4, 0.6, 0.8, 1.0)

**Iteration 1** (boundaries: top 2, left 3, right 1, bottom 0)

| | | | |
|---|---|---|---|
| 2.14 | 1.58 | 1.31 | 1.25 |
| 2.04 | 1.47 | 1.06 | 0.98 |
| 1.88 | 1.19 | 0.96 | 0.92 |
| 1.50 | 0.88 | 0.72 | 0.68 |

**Iteration 3** (boundaries: top 2, left 3, right 1, bottom 0)

| | | | |
|---|---|---|---|
| 2.25 | 1.86 | 1.61 | 1.41 |
| 2.20 | 1.69 | 1.35 | 1.13 |
| 1.99 | 1.36 | 1.03 | 0.91 |
| 1.45 | 0.84 | 0.61 | 0.63 |

**Iteration 6** (boundaries: top 2, left 3, right 1, bottom 0)

| | | | |
|---|---|---|---|
| 2.80 | 1.94 | 1.71 | 1.48 |
| 2.27 | 1.78 | 1.46 | 1.23 |
| 2.05 | 1.45 | 1.14 | 1.00 |
| 1.48 | 0.89 | 0.67 | 0.66 |

**Iteration 9** (boundaries: top 2, left 3, right 1, bottom 0)

| | | | |
|---|---|---|---|
| 2.31 | 1.96 | 1.73 | 1.50 |
| 2.29 | 1.81 | 1.49 | 1.25 |
| 2.07 | 1.48 | 1.17 | 1.02 |
| 1.49 | 0.91 | 0.69 | 0.68 |

By using the new value of $\Phi_{2,1}$ in the determination of $\bar{\Phi}_{3,1}$, we accelerate the convergence of the iterations.

We proceed counterclockwise around the lattice, calculating the averages on the outer points first and then the inner ones. The results of this first iteration are tabulated in Table 3–1. Any path that covers all lattice points could have been used. The results for different paths may differ in early iterations but will eventually converge to a unique solution. The results of further iterations following the original path are also shown in Table 3–1. The solution converges reasonably rapidly, as can be seen by comparing the results of the latter iterations. The numerical solution based on ten iterations is compared with an exact analytical solution to this particular problem in Figure 3–20. The numerical values are within 1% of the exact calculation. More sophisticated numerical methods can be used

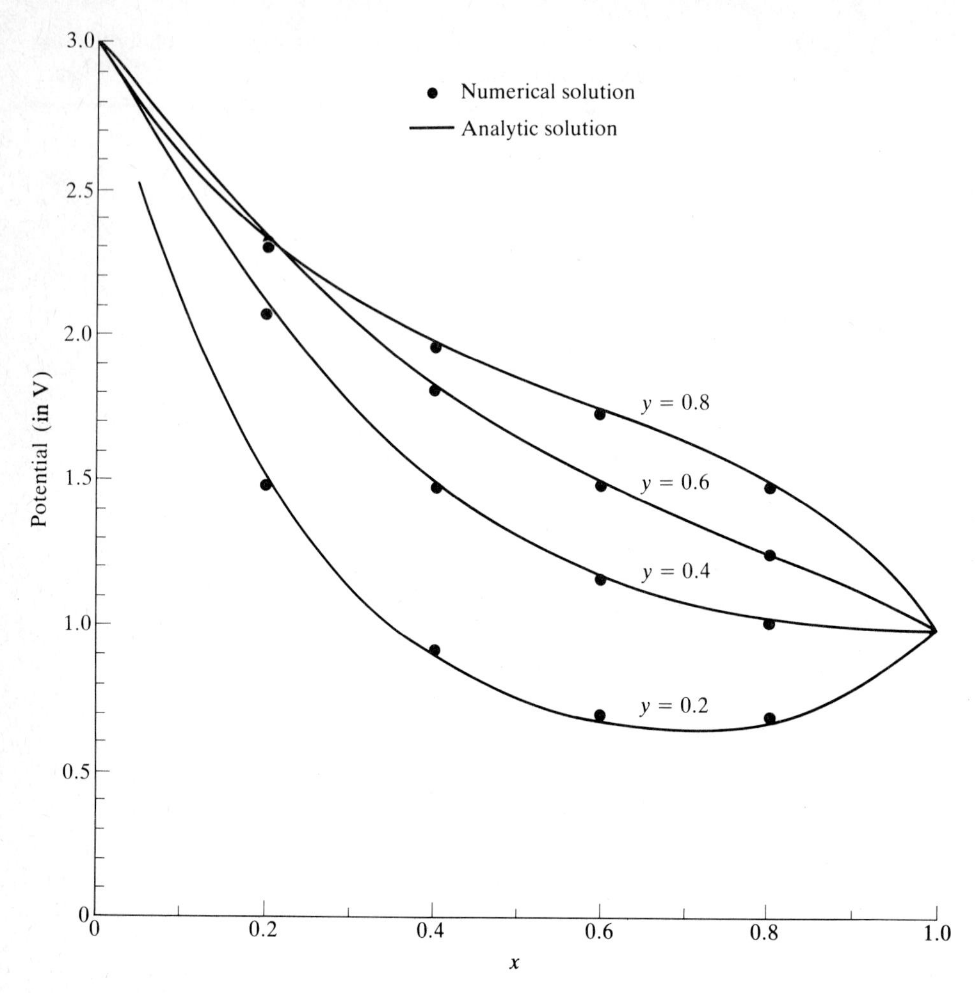


**FIGURE 3-20** Comparison of analytic and numerical solutions for the electric potential within a square whose sides are charged to potentials of 0, 1, 2, and 3 V

to accelerate convergence. Higher accuracy for the potential can be realized with more lattice points. This iterative method relies on the uniqueness theorem—the boundary values of the potential fix the potential everywhere.

## SUMMARY

### Important Concepts and Equations

Uniqueness of electrostatic solutions

*There is a unique solution to the electrostatic differential equations, which satisfies the boundary conditions.*

Coefficients of capacitance $C_{ij}$

*The coefficient of capacitance $C_{ij}$ is the charge on the ith conductor in a charge-free region, where the jth conductor has a potential of unity and all other conductors are grounded. The off-diagonal $C_{ij}$ $(i \neq j)$ are called the coefficients of electrostatic inductance or the mutual capacitances.*

Properties of coefficients of capacitance

*The coefficients of capacitance $C_{ij}$ are symmetric in the indices i and j: $C_{ij} = C_{ji}$. The charge induced on conductor 1 by changing the potential on conductor 2 by V is the same as the charge that would be induced on conductor 2 if the potential on conductor 1 were raised by V.*

$$Q_i = \sum_j C_{ij} V_j$$

*The charge on the ith conductor $Q_i$ in a system of conductors; $C_{ij}$ is the coefficient of capacitance, and $V_j$ is the potential of the jth conductor.*

Coefficients of potential

*The coefficient of potential $P_{ij}$ is the potential on the ith conductor in a charge-free region, where the jth conductor has a charge of unity and all the other conductors are uncharged.*

Properties of coefficients of potential

*The coefficients of potential $P_{ij}$ are symmetric in the indices i and j: $P_{ij} = P_{ji}$. The potential to which conductor 1 is raised by putting a charge Q on conductor 2 is the same as the change in potential on conductor 2 that would result from putting the same charge Q on conductor 1.*

$$V_i = \sum_j P_{ij} Q_j$$

*The potential $V_i$ on the ith conductor in a system of conductors in a charge-free region; $P_{ij}$ is the coefficient of potential, and $Q_j$ is the charge on the jth conductor.*

$$\nabla^2 \Phi = -\frac{\rho}{\varepsilon_0} \qquad \text{(Poisson's equation)}$$

*The Laplacian $\nabla^2 \equiv \mathbf{\nabla} \cdot \mathbf{\nabla}$ of the potential $\Phi$ is specified by the charge density $\rho$.*

$$\nabla^2 \Phi = 0 \qquad \text{(Laplace's equation)}$$

*Poisson's equation in regions of space where there are no charges.*

$$\nabla^2 \Phi = \frac{\partial^2 \Phi}{\partial x^2} + \frac{\partial^2 \Phi}{\partial y^2} + \frac{\partial^2 \Phi}{\partial z^2}$$

*Laplacian operator in Cartesian coordinates.*

$$\nabla^2\Phi = \frac{1}{\imath}\frac{\partial}{\partial \imath}\left(\imath\frac{\partial\Phi}{\partial \imath}\right) + \frac{1}{\imath^2}\frac{\partial^2\Phi}{\partial\phi^2} + \frac{\partial^2\Phi}{\partial z^2}$$

*Laplacian operator in cylindrical coordinates.*

$$\nabla^2\Phi = \frac{1}{r^2}\frac{\partial}{\partial r}\left(r^2\frac{\partial\Phi}{\partial r}\right) + \frac{1}{r^2\sin\theta}\frac{\partial}{\partial\theta}\left(\sin\theta\frac{\partial\Phi}{\partial\theta}\right) + \frac{1}{r^2\sin^2\theta}\frac{\partial^2\Phi}{\partial\phi^2}$$

*Laplacian operator in spherical coordinates. Note that*

$$\frac{1}{r^2}\frac{\partial}{\partial r}\left(r^2\frac{\partial\Phi}{\partial r}\right) = \frac{1}{r}\frac{\partial^2}{\partial r^2}(r\Phi)$$

Separation of variables

*A technique for solving a partial differential equation by looking for solutions in which the coordinate dependence factors into a product of functions, each of which depends only on one variable.*

$$f(x) = \sum_{n=0}^{\infty}\left(A_n\cos\frac{n\pi x}{a} + B_n\sin\frac{n\pi x}{a}\right) \qquad \text{(Fourier expansion)}$$

*Any reasonable function $f(x)$ can be expressed as an infinite series of sine and cosine terms. Here*

$$A_0 = \frac{1}{2a}\int_{-a}^{+a} f(x)\,dx$$

$$A_n = \frac{1}{a}\int_{-a}^{+a} f(x)\cos\frac{n\pi x}{a}\,dx, \qquad n > 0$$

$$B_n = \frac{1}{a}\int_{-a}^{+a} f(x)\sin\frac{n\pi x}{a}\,dx$$

$$\Phi = \sum_n (A_n r^n + B_n r^{-n-1})P_n(\mu)$$

*The solution to Laplace's equation in spherical coordinates for a potential with no $\phi$ dependence. Here $A_n$ and $B_n$ are arbitrary constants determined by boundary conditions, the $P_n(\mu)$ are Legendre polynomials of argument $\mu \equiv \cos\theta$.*

$P_0(\mu) = 1 \qquad P_1(\mu) = \mu$

$P_2(\mu) = \frac{1}{2}(3\mu^2 - 1) \qquad P_3(\mu) = \frac{1}{2}(5\mu^3 - 3\mu)$

*The first four Legendre polynomials.*

$$(n+1)P_{n+1}(\mu) = (2n+1)\mu P_n(\mu) - nP_{n-1}(\mu) \qquad \text{(Legendre recursion relation)}$$

*A recursive relation for generating the Legendre polynomials for $n > 1$ from $P_0(\mu)$ and $P_1(\mu)$.*

$$\frac{\partial \Phi}{\partial x} = \frac{\partial \Psi}{\partial y} \qquad \frac{\partial \Psi}{\partial x} = -\frac{\partial \Phi}{\partial y} \qquad \text{(Cauchy-Riemann equations)}$$

*A given function $f = \Phi(x, y) + i\Psi(x, y)$ is analytic if its real and imaginary parts satisfy the Cauchy-Riemann equations. The real and imaginary parts of every analytic function are solutions of Laplace's equation. The curves corresponding to $\Phi$ = constant and $\Psi$ = constant are mutually orthogonal and hence represent field lines and equipotentials.*

Method of images

*The solution to some boundary value problems in electrostatics can be found by assuming a distribution of image charges, which, together with real charges, yield the specified equipotential surfaces. The electric field and the potential can then be calculated outside the conductors, ignoring the conducting regions.*

## PROBLEMS

### Section 3-3: Laplace Equation in One Dimension

**3-1** An air-spaced coaxial cable has an inner conductor 0.5 cm in diameter and an outer conductor 1.5 cm in diameter. The inner conductor is at a potential of +8000 V with respect to the grounded outer conductor.
(i) Find the charge per meter on the inner conductor.
(ii) Find the electric field intensity at $r$ = 1 cm.

**3-2** An infinitely long cylindrical capacitor has an inner conductor of radius $r_1$ and an outer conductor of radius $r_2$. The outer conductor is grounded, and the inner conductor is charged to a positive potential $V_0$.
(i) In terms of $V_0$, $r_1$, and $r_2$, find the electric field and the electric potential at radius $r$, where $r_1 < r < r_2$.
(ii) If the outer radius $r_2$ is fixed and the inner radius $r_1$ is varied, show that the electric field at $r_1$ is a minimum when $r_2/r_1 = e$, where $\ln e = 1$.
(iii) Find the capacitance per unit length.
(iv) If $r_2 = r_1 + d$, show that for $d \ll r_1$, the capacitance of the cylindrical capacitor reduces to that for a parallel-plate capacitor of the same surface area.

**3-3** A capacitor is made of two concentric, conducting spherical shells. Initially, the inner shell (of radius $a$) is grounded, and the outer shell (of radius $b$) is at potential $V$. The ground of the inner shell is then removed and replaced by an electrical connection between the shells.
(i) What is the final potential of the outer shell?
(ii) What is the change of energy of the system?

**3-4** Two concentric spherical conductors with radii $a$ and $b$ are arranged as a capacitor. The outer sphere is grounded, and the inner is held at $V_0$. Find a relation between $a$ and $b$ such that the electric field at $r = a$ is a minimum.

## Section 3-4: Separation of Variables in Cartesian Coordinates

**3-5** A square prism of infinite length has one side grounded and the opposite side maintained at voltage $V_0$. The voltage on the other sides varies linearly from zero to $V_0$. Find the potential and the electric field within the prism. Hint: Use zero separation constant.

**3-6** Solve for the potential outside a grounded cylinder in a uniform electric field that is perpendicular to the cylinder axis. Use the method of separation of variables to solve the Laplace equation in coordinates $r$ and $\phi$.

**3-7** Use the method of separation of variables in two rectangular dimensions to find a series expression for the potential in a square of unit sides with three sides that are grounded and one side that is at unit potential.

## Section 3-5: Spherical Solutions

**3-8** The potential on the surface of a nonconducting sphere of radius $R$ is given by $\Phi = VP_2(\cos\theta)$, where $\theta$ is the polar angle in spherical coordinates. Find the potential and the electric field external to the sphere.

**3-9** A grounded conducting plane with a hemispherical boss of radius $a$ (see Figure 3–21) is placed in an electric field that is asymptotically uniform and perpendicular to the plane. Find the potential everywhere outside the conductor, and determine the induced charge density everywhere on the conducting surface.

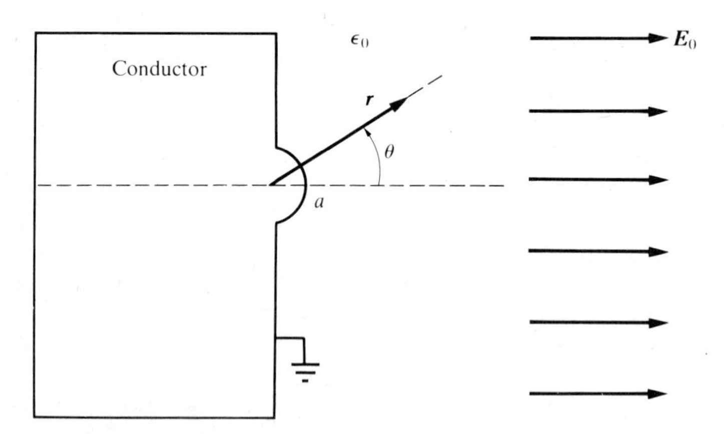


**FIGURE 3-21** Conductor with a hemispherical boss of radius $a$ placed in an asymptotically uniform electric field

## Section 3-6: Analytic-Function Methods

**3-10** For the function $f(w) = C \ln w$, $w = x + iy$ and $C$ real, propose an electrostatics problem for which $\Psi = \text{Im}\, f$ are the equipotentials.

## Section 3-7A: Images—Plane Conductors

**3-11** Two equal charges of opposite sign are located on a vertical line above a grounded conducting plane, with the $+Q$ charge at a height $h_+$ and the $-Q$ charge at a distance $h_- < h_+$ above the plane. Calculate the electric field and the charge density on the plane. At what radius from the vertical line does $E$ on the plane vanish? This model is a realistic description of a thundercloud above the surface of the earth. For typical values of $Q = 40$ C at heights $h_+ = 8$ km and $h_- = 3$ km, calculate the numerical value of $E$ at the earth's surface as a function of distance from the charge axis.

**3-12** What image charges are needed to solve the problem of a charge $q$ arbitrarily placed between two grounded conducting planes making an angle of 90°? Calculate the electric field, and sketch the field lines.

**3-13** Two semi-infinite, grounded conducting planes make an angle of 60°. One plane coincides with the positive $x$ axis. A single point charge $+Q$ is located between the planes, closer to one plane than the other, at coordinates $x = b$, $y = a$. Find the image charges that satisfy the boundary conditions.

**3-14** Charges $+q$ and $-q$ lie at the points $(x, y, z) = (a, 0, a)$, $(-a, 0, a)$ above a grounded conducting plane at $z = 0$. What image charges are needed to determine the potential in the region $z > 0$? What is the electric potential at the point $(x, y, z) = (0, 0, a)$?

**3-15** A charge $q$ is located at the origin between two infinite conducting planes at $x = a$ and $x = -a$. The planes are both at zero potential. Find the image charges and their locations to calculate the potential between the planes. (The number of image charges need not be finite.) Obtain a series expression for the potential on the $x$ axis for $-a \leqslant x \leqslant a$.

## Section 3-7B: Images—Spherical Conductors

**3-16** Find the charge density induced on a grounded conducting sphere of radius $a$ by a point charge $q$ at a distance $b$ in terms of $a$, $b$, and the distance $R$ from $q$ to the location on the sphere.

**3-17** Use the image charge method to solve for the electric potential inside a conducting sphere of radius $a$ containing a point charge $q$ at a distance $rq < a$ from the center.

**3-18** A point charge $q$ is located at a distance $b$ from the center of a conducting sphere of radius $a$. Calculate the force on the charge for (i) a grounded sphere, (ii) an uncharged sphere, and (iii) a sphere charged to a potential $V_0$.

## Section 3-7C: Images—Cylindrical Conductors

**3-19** An infinite straight wire of radius $a$, carrying a charge per unit length $\lambda$, is placed at a distance $b > a$ from an infinite-plane conducting surface, whose potential is maintained at zero. For this system, determine the capacitance per unit length of wire.

**3-20** Two line charges $\pm\lambda$ are located at distance $d$ on opposite sides of an uncharged conducting cylinder of radius $a$. Use the method of images to find the electric potential. Take the limit $d \to \infty$ with $\lambda/d$ finite to obtain the result for an uncharged conducting cylinder in a uniform electric field $E_0$.

## Section 3-8: Numerical Methods

**3-21** Using numerical methods, find the potential inside a square of unit sides when two opposite sides are maintained at potentials of 1 and $-1$ V and the other two sides are grounded. Use a six-by-six lattice. Determine the potential to two significant figures.

**3-22** Use the solution to Problem 3–7 to compute the analytical potential at the lattice points of Problem 3–21. Compare the numerical and analytic results.

**3-23** Solve Problem 3–5 by using numerical iteration with a six-by-six lattice (i.e., 16 interior lattice points). Take the side of the square prism to be 5 m, and take $V_0 = 1$ V. Start with initial values of 0.5 for $\Phi$ at all interior lattice points and do ten iterations. Compare the final numerical solution at the lattice points with the exact solution.

# Electric Fields in Matter

CHAPTER CONTENTS

Matter in its usual state appears to be electrically neutral. In fact, it is composed of equal numbers of positive and negative charges arranged within molecular structures in such a way that the average electric field inside the material is usually zero. As a consequence, the electric field outside the material vanishes. However, when external **E** and **B** fields act on the matter, they change the charge and current densities. This change results in the production of fields inside and outside the matter by the charges and currents in the matter. Refractive indices and permanent magnetism are two examples of the resulting phenomena. In this chapter we discuss the effect of imposed electric fields on dielectric materials. The macroscopic phenomenology will be described first; the macroscopic electric field contributions associated with matter are represented to an excellent approximation by the leading nonvanishing term (namely, the electric dipole) in a power series expansion in inverse distance from the molecules. We then investigate the microscopic origins of the molecular electric dipole moments.

# 4-1 ELECTRIC DIPOLE AND MULTIPOLES

The charge density $\rho(\mathbf{r}')$ in an atom or molecule vanishes outside a region typically $10^{-10}$ m in size. For our purposes, it is somewhat advantageous to take the origin of the coordinate system somewhere within the molecular charge distribution. The distance from the origin to a point in the molecule is denoted by $\mathbf{r}'$, and the distance from the origin to the observation point is denoted by $\mathbf{r}$, as shown in Figure 4–1. At the observation point $\mathbf{r}$ the electric potential is

$$\Phi(\mathbf{r}) = \frac{1}{4\pi\varepsilon_0} \int_V \frac{\rho(\mathbf{r}')\, dV'}{|\mathbf{r} - \mathbf{r}'|} \tag{4-1}$$

We can write $R^2 = |\mathbf{r} - \mathbf{r}'|^2$ as

$$\begin{aligned} |\mathbf{r} - \mathbf{r}'|^2 &= (\mathbf{r} - \mathbf{r}')\cdot(\mathbf{r} - \mathbf{r}') = r^2 - 2\mathbf{r}\cdot\mathbf{r}' + r'^2 \\ &= r^2\left(1 - \frac{2\mathbf{r}\cdot\mathbf{r}'}{r^2} + \frac{r'^2}{r^2}\right) \end{aligned} \tag{4-2}$$

If the observation point is far enough away from the charge distribution so that $r' \ll r$ for all $r'$, we can make a rapidly convergent power series expansion of $1/|\mathbf{r} - \mathbf{r}'|$. Using

$$(1 + \beta)^{-1/2} = 1 - \frac{1}{2}\beta + \frac{3}{8}\beta^2 + \cdots \tag{4-3}$$

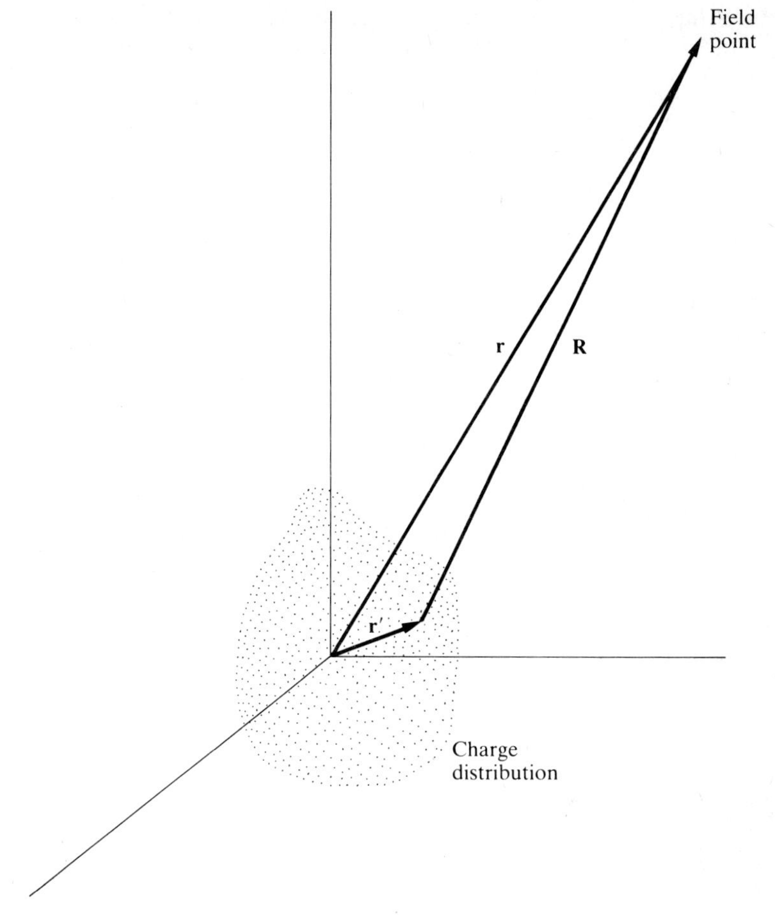


**FIGURE 4-1** Charge distribution localized to some region around the origin; **r** is the vector from the origin to the field point, **r**′ is the vector from the origin to a point in the charge distribution, and **R** is the vector from the point in the charge distribution to the field point

we have

$$\frac{1}{|\mathbf{r}-\mathbf{r}'|} = \frac{1}{r}\left(1 - \frac{2\mathbf{r}\cdot\mathbf{r}'}{r^2} + \frac{r'^2}{r^2}\right)^{-1/2}$$
$$= \frac{1}{r}\left[1 - \frac{1}{2}\left(-\frac{2\mathbf{r}\cdot\mathbf{r}'}{r^2} + \frac{r'^2}{r^2}\right) + \frac{3}{8}\left(-\frac{2\mathbf{r}\cdot\mathbf{r}'}{r^2} + \frac{r'^2}{r^2}\right)^2 + \cdots\right] \tag{4-4}$$

Collecting terms of the same order in $1/r$, we have the expansion

$$\frac{1}{|\mathbf{r}-\mathbf{r}'|} = \frac{1}{r} + \frac{\mathbf{r}\cdot\mathbf{r}'}{r^3} + \frac{1}{2}\frac{[3(\mathbf{r}\cdot\mathbf{r}')^2 - r'^2r^2]}{r^5} + \cdots \tag{4-5}$$

Inserting this result in Equation (4–1), we obtain

$$\Phi(\mathbf{r}) = \frac{1}{4\pi\varepsilon_0}\left\{\frac{1}{r}\int \rho(\mathbf{r}')\,dV' + \frac{1}{r^3}\int \mathbf{r}\cdot\mathbf{r}'\rho(\mathbf{r}')\,dV' + \frac{1}{r^5}\int\left[\frac{3(\mathbf{r}\cdot\mathbf{r}')^2 - r^2 r'^2}{2}\right]\rho(\mathbf{r}')\,dV'\right\} + \cdots \quad \text{(4–6)}$$

This expansion of the potential in powers of $1/r$ is called a multipole expansion. The coefficients of the expansion terms are known as moments. The coefficient of the $1/r$ term is the monopole moment or total charge,

**Electric Monopole Moment**

$$q = \int \rho(\mathbf{r}')\,dV' \quad \text{(4–7)}$$

For a neutral molecule the monopole term of the expansion is absent. The dipole moment is the coefficient of the $1/r^2$ term:

**Electric Dipole Moment**

$$\mathbf{p} = \int \mathbf{r}'\rho(\mathbf{r}')\,dV' \quad \text{(4–8)}$$

The coefficient of the $1/r^3$ term is known as the quadrupole moment:

**Electric Quadrupole Moment**

$$Q_{ij} = \int (3x_i'x_j' - \delta_{ij}r'^2)\rho(\mathbf{r}')\,dV' \quad \text{(4–9)}$$

where $\delta_{ij}$ is the Kronecker delta, $\delta_{ij} = 1$ for $i = j$ and $\delta_{ij} = 0$ for $i \neq j$, and $(x_1, x_2, x_3)$ denotes $(x, y, z)$. The quadrupole moment is a tensor, with nine components (of which only five are independent since $Q_{ij}$ is a symmetric and traceless tensor). In terms of these moments Equation (4–6) becomes

$$\Phi(\mathbf{r}) = \frac{q}{4\pi\varepsilon_0 r} + \frac{\mathbf{p}\cdot\mathbf{r}}{4\pi\varepsilon_0 r^3} + \sum_{i,j}\frac{\frac{1}{2}Q_{ij}x_i x_j}{4\pi\varepsilon_0 r^5} + \cdots \quad \text{(4–10)}$$

Very far away from a charge distribution only the first nonvanishing moment gives an appreciable contribution to the electric potential. As we move closer, the detailed distribution of the charges becomes increasingly important. Equation (4–10) shows that the relative contributions of higher moments increase as $r$ becomes smaller.

A simple example of a charge distribution in which the first nonzero term in $\Phi$ is a dipole is shown in Figure 4–2: two point charges of opposite sign separated by a distance $d$. The potential of this charge arrangement is

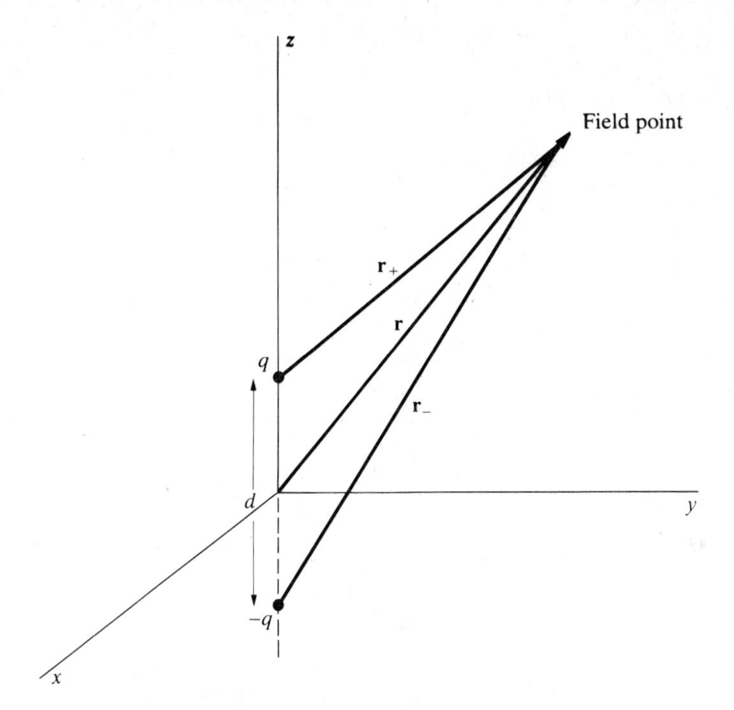


**FIGURE 4-2** Simple dipole consisting of two oppositely charged point charges separated by a distance $d \ll r$

$$\Phi = \frac{q}{4\pi\varepsilon_0}\left(\frac{1}{r_+} - \frac{1}{r_-}\right) \tag{4-11}$$

where

$$r_+^2 = x^2 + y^2 + \left(z - \frac{d}{2}\right)^2 \qquad r_-^2 = x^2 + y^2 + \left(z + \frac{d}{2}\right)^2 \tag{4-12}$$

To calculate $\Phi$ for $r \gg d$, we write

$$\frac{1}{r_+} - \frac{1}{r_-} = \frac{r_- - r_+}{r_+ r_-} = \frac{r_-^2 - r_+^2}{r_+ r_-(r_+ + r_-)} \tag{4-13}$$

The numerator is $r_-^2 - r_+^2 = 2zd$. We are interested only in the lowest-order contribution in $d/r$, so we can approximate $r_+ \approx r_- \approx r$ in the denominator; thus we find

$$\Phi(\mathbf{r}) \approx \frac{(qd)z}{4\pi\varepsilon_0 r^3} = \frac{\mathbf{p}\cdot\mathbf{r}}{4\pi\varepsilon_0 r^3} \tag{4-14}$$

where $\mathbf{p}$, the dipole moment, is given by

$$\mathbf{p} = p\hat{\mathbf{z}} \equiv (qd)\hat{\mathbf{z}} \qquad \textbf{(4-15)}$$

Equation (4–15) for $\mathbf{p}$ can also be found from the general expression (4–8) for $\mathbf{p}$ by taking

$$\rho(\mathbf{r}') = q\left[\delta^3\left(\mathbf{r}' - \frac{d}{2}\hat{\mathbf{z}}\right) - \delta^3\left(\mathbf{r}' + \frac{d}{2}\hat{\mathbf{z}}\right)\right] \qquad \textbf{(4-16)}$$

The electric field at $\mathbf{r} \neq 0$ due to a dipole is

$$\mathbf{E}(\mathbf{r}) = -\nabla_r \Phi(\mathbf{r}) = -\frac{1}{4\pi\varepsilon_0}\left[\mathbf{r}\cdot\mathbf{p}\,\nabla\left(\frac{1}{r^3}\right) + \frac{1}{r^3}\nabla(\mathbf{r}\cdot\mathbf{p})\right]$$

or

**Electric Dipole Field**

$$\mathbf{E}(\mathbf{r}) = \frac{1}{4\pi\varepsilon_0}\left[\frac{3(\mathbf{r}\cdot\mathbf{p})\mathbf{r} - r^2\mathbf{p}}{r^5}\right] \qquad \textbf{(4-17)}$$

The singular point $\mathbf{r} = 0$ can be treated by the divergence theorem (see Problem 4–4). The complete result for any $\mathbf{r}$ is

$$\mathbf{E}(\mathbf{r}) = \frac{1}{4\pi\varepsilon_0}\left[\frac{3\mathbf{r}(\mathbf{r}\cdot\mathbf{p}) - r^2\mathbf{p}}{r^5} - \frac{4\pi}{3}\mathbf{p}\delta^3(\mathbf{r})\right] \qquad \textbf{(4-18)}$$

In spherical coordinates (see Figure 4–3) with $\mathbf{p} = \mathbf{p}\hat{\mathbf{z}}$, the dipole potential is

$$\Phi(\mathbf{r}) = \frac{p\cos\theta}{4\pi\varepsilon_0 r^2} \qquad \textbf{(4-19)}$$

For $\hat{\mathbf{z}} = \hat{\mathbf{r}}\cos\theta - \hat{\boldsymbol{\theta}}\sin\theta$ the electric dipole field in Equation (4–17) is given in spherical coordinates by

$$E_r = \frac{p}{4\pi\varepsilon_0}\frac{2\cos\theta}{r^3} \qquad E_\theta = \frac{p}{4\pi\varepsilon_0}\frac{\sin\theta}{r^3} \qquad E_\phi = 0 \qquad \textbf{(4-20)}$$

Alternatively, we could obtain Equation (4–20) by using the gradient in spherical coordinates to operate on Equation (4–19). The electric dipole field lines of Equation (4–20) and the equipotentials of Equation (4–19) are illustrated in Figure 4–4. The field lines and equipotentials of the two-charge system in Equation (4–11), before we take the limit $r \gg d$, are shown in Figure 4–5. For large distances the fields resemble the true dipole fields of Figure 4–4, but for small distances $r \ll d$ the fields actually reverse. This

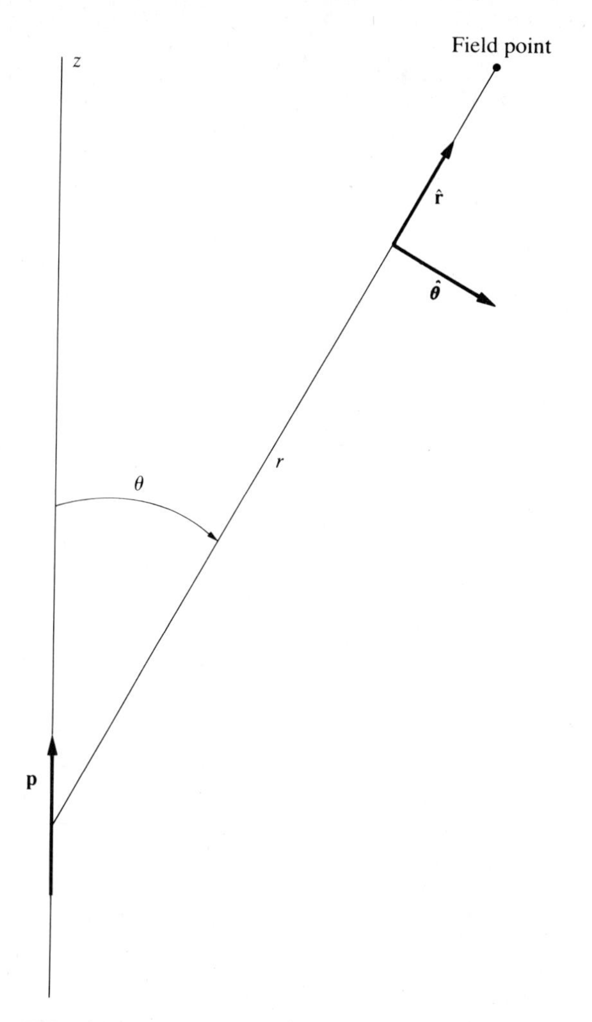


**FIGURE 4–3** Electric dipole aligned along the positive $z$ axis in a spherical coordinate system

behavior for small distances is the reason for the delta function at the origin in the exact dipole expression in Equation (4–18).

The electric dipole moment also appears in the expression for the force and torque on a charge system due to an external field. The *torque* on a charge distribution due to a constant **E** field is

$$\mathbf{N} = \int \mathbf{r}' \times [\rho(\mathbf{r}')\mathbf{E}]\, dV' = [\int \mathbf{r}'\rho(\mathbf{r}')\, dV'] \times \mathbf{E} = \mathbf{p} \times \mathbf{E} \qquad \textbf{(4–21)}$$

The relative orientations of **p**, **E**, and **N** are illustrated in Figure 4–6. The action of the torque is to align **p** with **E**. We introduce the angle $\theta$ that **p** makes with **E**. In terms of a unit vector $\hat{\boldsymbol{\theta}}$ normal to the **pE** plane and in the

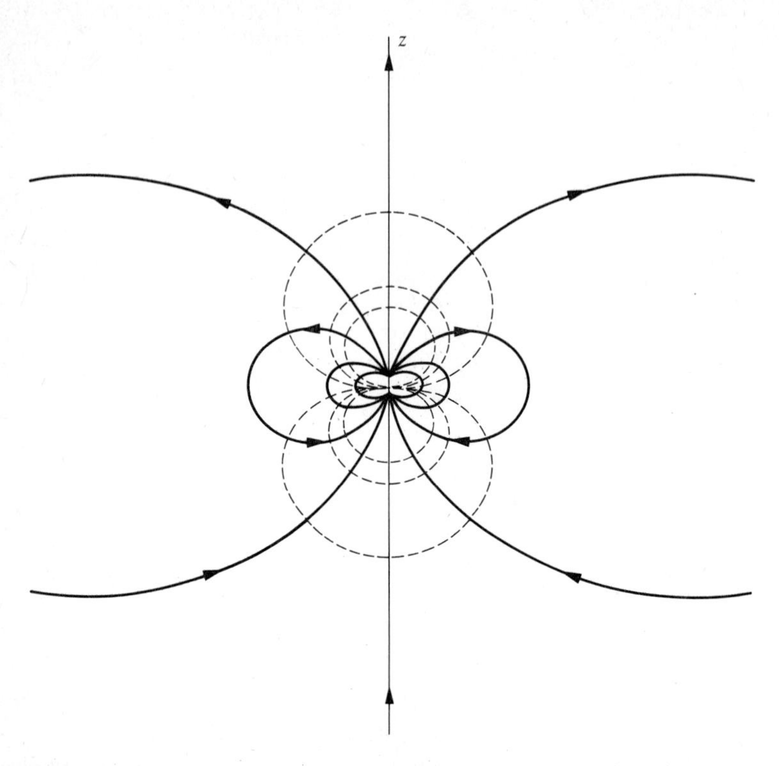


FIGURE 4-4 Equipotentials (dashed lines) and field lines for an electric dipole at $z = 0$ aligned along the positive $z$ axis

sense of increasing $\theta$, we have

$$\mathbf{N} = -pE \sin \theta \hat{\boldsymbol{\theta}} \tag{4-22}$$

The *work* done by the torque in rotating the charge distribution in the plane of $p$ and $E$ through an angle $d\theta$ is

$$dW = \mathbf{N} \cdot (\hat{\boldsymbol{\theta}}\, d\theta) = -pE \sin \theta \, d\theta \tag{4-23}$$

The net work done by the electric field in rotating the system from $\theta_0$ to $\theta$ is

$$W = -pE \int_{\theta_0}^{\theta} \sin \theta \, d\theta = pE \cos \theta \Big|_{\theta_0}^{\theta} = pE(\cos \theta - \cos \theta_0) \tag{4-24}$$

The increase in potential energy $U$ of the system is the negative of the work done by the electric field. We take the zero of potential energy at $\theta_0 = \pi/2$,

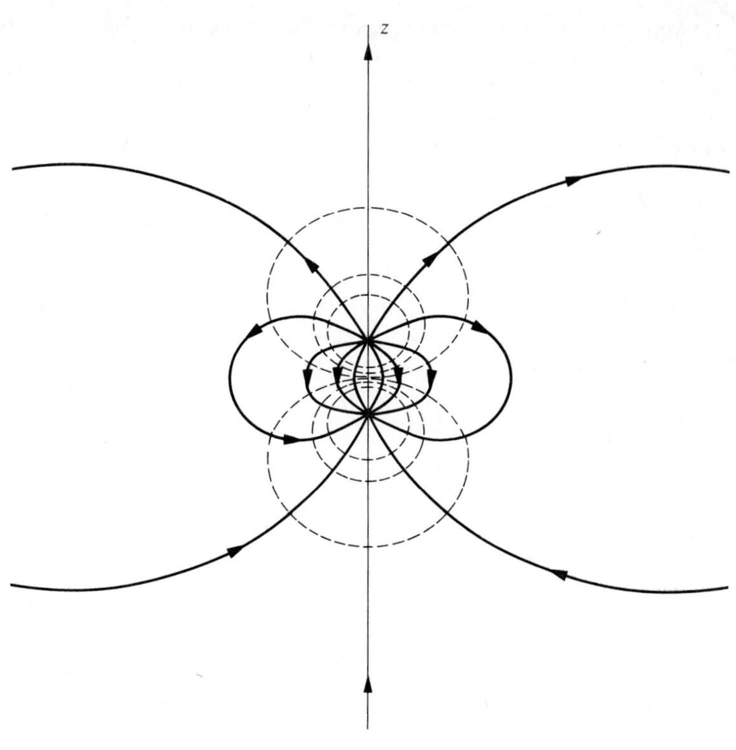


**FIGURE 4–5** Equipotentials (dashed lines) and field lines for two oppositely charged point charges $+q$ at $z = d/2$ and $-q$ at $z = -d/2$

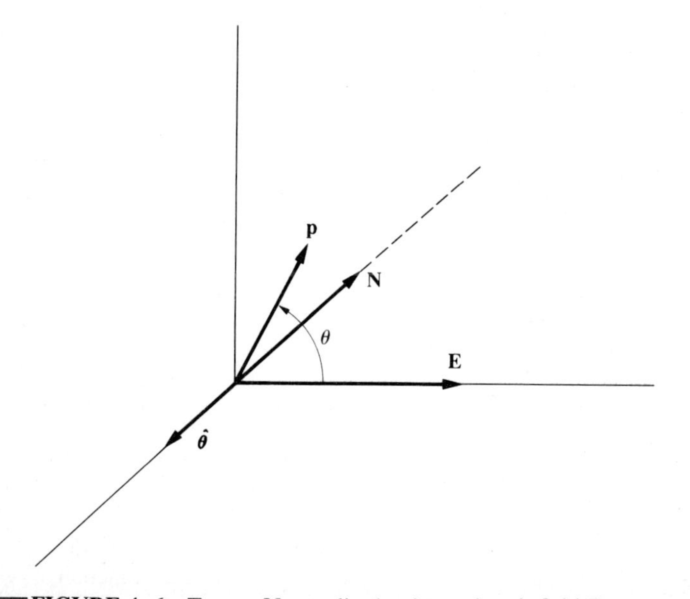


**FIGURE 4–6** Torque **N** on a dipole **p** in an electric field $E$

for convenience. The *potential energy* of the dipole in the electric field is thus

$$U = -pE\cos\theta = -\mathbf{p}\cdot\mathbf{E} \tag{4-25}$$

A plot of this dipole energy is shown in Figure 4–7. The minimum energy occurs when $\theta = 0$ and $\mathbf{p}$ is parallel to $\mathbf{E}$.

When an external electric field acts on a sample of matter consisting of neutral molecules, the mean positions of the electrons are displaced from the ions, thus inducing dipole and higher moments. To investigate the relative importance of the various multipoles, we consider a cube of side $\ell$ consisting of $N \approx (\ell/a_0)^3$ close-packed molecules, where $a_0 \approx 10^{-10}$ m is the molecular size. Under the influence of an external electric field $E$ each molecule acquires a dipole moment of order $p \approx 4\pi\varepsilon_0 a_0^3 E$, as we will demonstrate in Section 4–5A. With Equation (4–20) the electric dipole field at a distance $\ell$ from a single molecule is of order

$$|\mathbf{E}_D| \sim \frac{p}{4\pi\varepsilon_0 \ell^3} \sim \left(\frac{a_0}{\ell}\right)^3 E \tag{4-26}$$

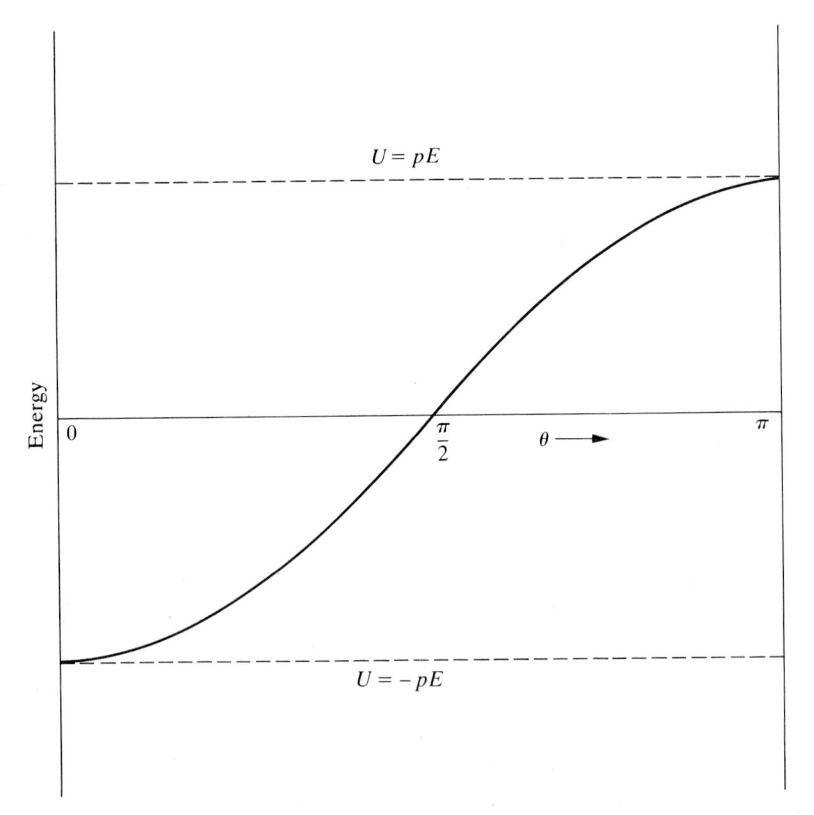


**FIGURE 4–7** Potential energy of a dipole in an electric field as a function of the angle between the dipole and the electric field

For a cube of $N$ molecules the order of magnitude of the induced dipole field at a distance $\ell$ from the center of the cube is

$$|\mathbf{E}^{p}_{\text{induced}}| \sim N|\mathbf{E}_{\text{D}}| \sim E \tag{4-27}$$

The induced dipole field is thus of the same order of magnitude as the external electric field and therefore certainly not negligible.

With Equation (4–10) the quadrupole electric field at a distance $\ell$ is of the order of

$$|\mathbf{E}_{Q}| \sim \frac{Q}{4\pi\varepsilon_0 \ell^4} \tag{4-28}$$

Since by Equation (4–9) the quadrupole moment $Q$ has two coordinate factors, we expect, by analogy to the dipole result, that $Q \sim 4\pi\varepsilon_0 a_0^4 E$. Hence the total induced quadrupole electric field due to a cube of molecules is of order

$$|\mathbf{E}^{Q}_{\text{induced}}| \sim N \frac{4\pi\varepsilon_0 a_0^4}{4\pi\varepsilon_0 \ell^4} E \approx \frac{a_0}{\ell} E \tag{4-29}$$

Thus outside the sample the quadrupole and the higher-moment fields can be neglected. Similar arguments apply for fields in cavities within the sample. Only when one considers the local field on an atomic scale within the material do the higher moments of the molecular charge distribution become significant.

## 4-2 MACROSCOPIC DIELECTRIC THEORY

We consider a parallel-plate capacitor with plate separation $d$ and area $A$, as illustrated in Figure 4–8(a). The potential difference $V$ imposed by a battery is related to the electric field between the plates by $V = Ed$. The charge density on the plates is $\sigma_0 = \varepsilon_0 E$, and the total charge on a plate is $Q_0 = \sigma_0 A$.

When a dielectric is inserted to fill the space between the plates, as in Figure 4–8(b), additional charge flows onto the plates. For example, with pure water between the plates, a charge increase of about eighty times is observed. The potential difference $V$ across the capacitor is maintained at a constant value by the battery. Thus the increase in charge $Q$ on the plates implies an increase in the capacitance $C = Q/V$.

The factor of increase of the charge or of the capacitance due to the dielectric is called the *dielectric constant* $\kappa$. The additional charge density that flows onto the plates is

$$\Delta\sigma = (\kappa - 1)\sigma_0 \tag{4-30}$$

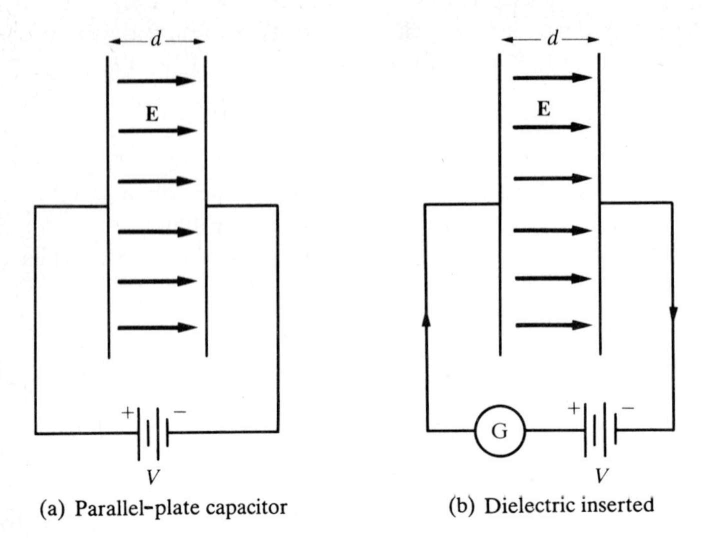


**FIGURE 4-8** Parallel-plate capacitor with separation distance $d$ connected to a battery with potential difference $V$. (a) Original circuit. (b) With dielectric inserted.

Since the electric field $E = V/d$ in the dielectric is unchanged, a negative surface charge density must appear on the dielectric surface that exactly compensates $\Delta\sigma$. This quantity is called the *polarization charge density*,

$$\sigma_P = -(\kappa - 1)\sigma_0 \tag{4-31}$$

Since charges cannot freely move within the dielectric, this apparent charge flow onto the dielectric surfaces must arise from charge separations within the dielectric molecules caused by the applied electric field, as illustrated in Figure 4–9 for a single molecule.

A microscopic picture of how the polarization charge density originates on the dielectric surface is given in Figure 4–10. Before the external field is applied, the matter is electrically neutral. When the electric field is applied, positive and negative charges in the molecules are displaced in opposite directions. As a consequence, there is a net charge flow through any plane in the material parallel to the capacitor plates. This charge flow is called the *polarization current*. The result is that charge accumulates at the surfaces of the material that face the capacitor plates, with positive charge facing the negative plate and vice versa. Notice the difference of this kind of charge flow from that occurring in a conductor. This surface charge density is bound to the molecules, as distinguished from the free charge associated with the motion of conduction electrons in conductors.

An alternative way of expressing the dielectric constant $\kappa$ is, by Equation (4–31),

$$\sigma_P = -\left(\frac{\chi}{\varepsilon_0}\right)\sigma_0 \tag{4-32a}$$

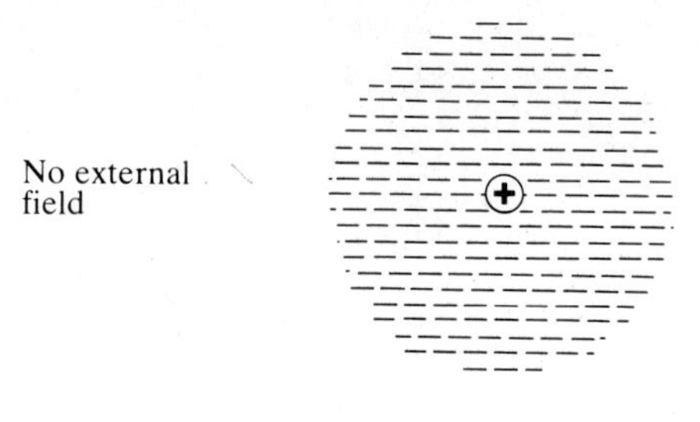


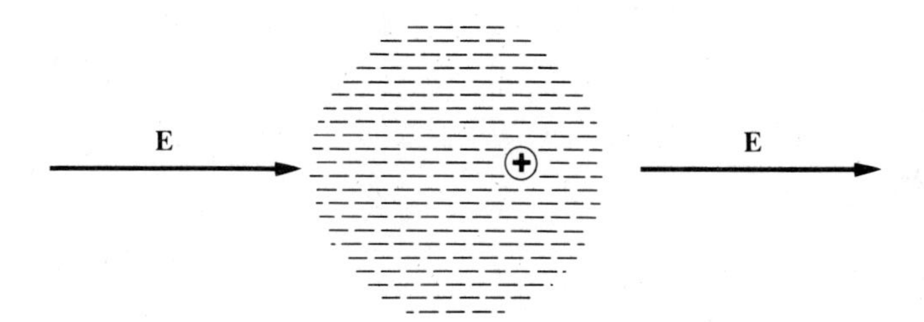


**FIGURE 4-9** Schematic representation of the charge separation within a dielectric molecule caused by the applied electric field

where the *electric susceptibility* $\chi$ is

$$\chi = (\kappa - 1)\varepsilon_0 \tag{4-32b}$$

The *electric permittivity* $\varepsilon$ is defined as

$$\varepsilon = \varepsilon_0 \kappa \tag{4-33}$$

In terms of the susceptibility the permittivity is given by

$$\varepsilon = \varepsilon_0 + \chi \tag{4-34}$$

The above simple case, while instructive, is not very general. The ideas presented can, however, be readily generalized. We have observed that when an **E** field is turned on, the dielectric will polarize, yielding polarization charge densities that modify the net electric field. The molecular dipoles, when averaged over a volume $\Delta V$, give an average dipole moment per unit volume,

$$\mathbf{P} = \frac{1}{\Delta V} \sum_i \mathbf{p}_i \tag{4-35}$$

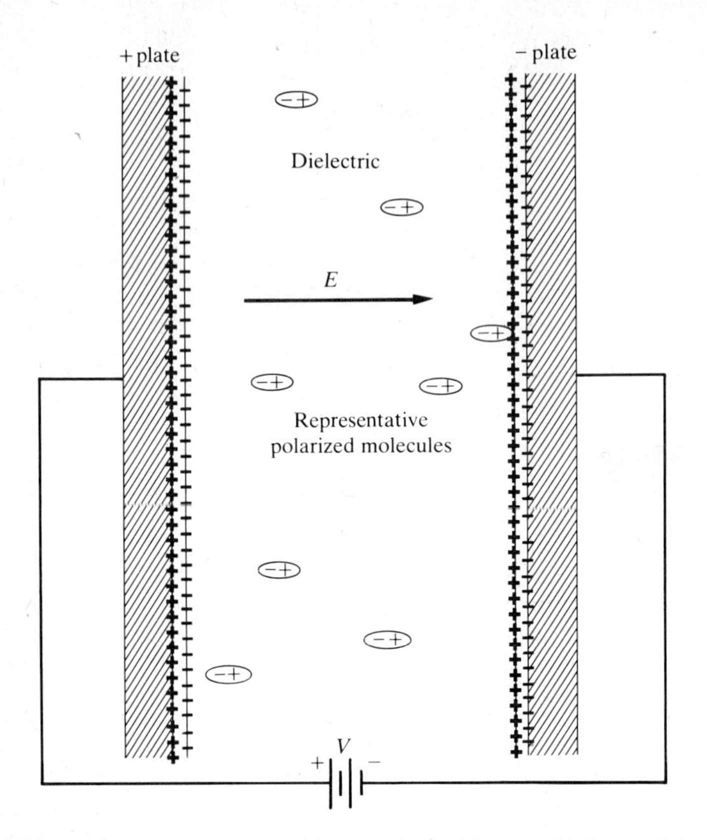


**FIGURE 4-10** Schematic diagram of the microscopic picture of the origin of polarization charge density on a dielectric surface in an electric field

where the summation includes all of the molecular dipoles within $\Delta V$. Even if $\Delta V$ is very small, it can still contain many dipoles; for example, a cube with sides of one wavelength of visible light ($5 \times 10^{-7}$ m) will contain about $10^{11}$ molecular dipoles. We can therefore consider **P** as dipole moment volume density, and the dipole moment of a volume element $dV'$ is thus

$$d\mathbf{p}(\mathbf{r}') = \mathbf{P}(\mathbf{r}')\, dV' \tag{4-36}$$

With Equation (4–10) the electric potential at **r** due to this dipole moment element at **r**′ is

$$d\Phi = \frac{1}{4\pi\varepsilon_0} \frac{d\mathbf{p} \cdot \mathbf{R}}{R^3} = \frac{1}{4\pi\varepsilon_0} \frac{\mathbf{P} \cdot \mathbf{R}\, dV'}{R^3} \tag{4-37}$$

where $\mathbf{R} = \mathbf{r} - \mathbf{r}'$; see Figure 4–11. Integrating over a volume gives the potential due to dipoles within the volume:

$$\Phi(\mathbf{r}) = \frac{1}{4\pi\varepsilon_0}\int \frac{\mathbf{P}\cdot\mathbf{R}}{R^3}\,dV' = \frac{1}{4\pi\varepsilon_0}\int \mathbf{P}\cdot\nabla_{\mathbf{r}'}\left(\frac{1}{R}\right)dV' \tag{4-38}$$

Using the identity $\mathbf{A}\cdot\nabla B = \nabla\cdot(\mathbf{A}B) - B\,\nabla\cdot\mathbf{A}$ in Equation (4–38), we obtain

$$\Phi(\mathbf{r}) = \frac{1}{4\pi\varepsilon_0}\int \nabla_{\mathbf{r}'}\cdot\left(\frac{\mathbf{P}}{R}\right)dV' + \frac{1}{4\pi\varepsilon_0}\int \frac{(-\nabla_{\mathbf{r}'}\cdot\mathbf{P})}{R}\,dV' \tag{4-39}$$

The divergence theorem applied to the first term of Equation (4–39) then yields the result

$$\Phi(\mathbf{r}) = \frac{1}{4\pi\varepsilon_0}\int \frac{\mathbf{P}\cdot\hat{\mathbf{n}}}{R}\,dS' + \frac{1}{4\pi\varepsilon_0}\int \frac{(-\nabla\cdot\mathbf{P})}{R}\,dV' \tag{4-40}$$

Thus the potential due to the dipole density within a volume is the same as the potential due to a charge distribution that has two parts, a polarization surface charge density,

$$\sigma_P = \mathbf{P}\cdot\hat{\mathbf{n}} \tag{4-41}$$

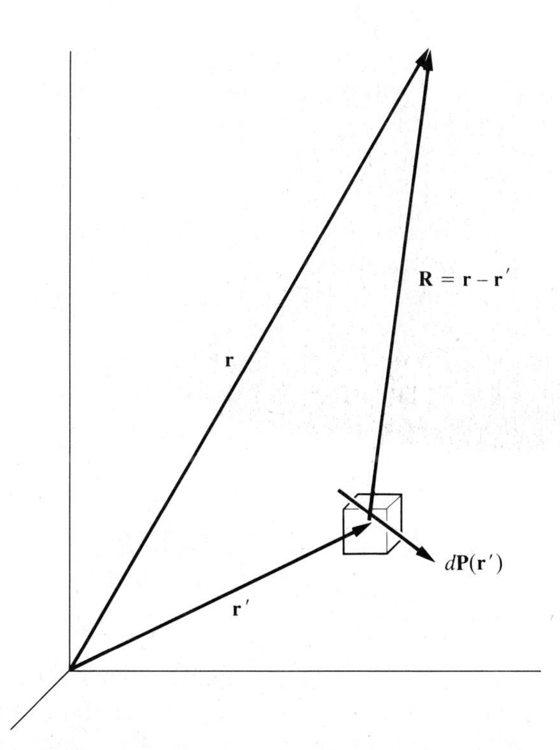


**FIGURE 4–11** Differential $d\mathbf{P}(\mathbf{r}')$ of the polarization vector at $\mathbf{r}'$ as the source of the electric field at $\mathbf{r}$

and a polarization volume charge density,

$$\rho_P = -\nabla \cdot \mathbf{P} \tag{4-42}$$

Observe that if we had chosen to integrate over a volume larger than the region occupied by the dipoles, then the surface term of Equation (4–40) would have vanished and its contribution would have been represented by a surface delta function in $\nabla \cdot \mathbf{P}$, owing to the discontinuity in $\mathbf{P}$ at the surface of the dielectric.

In the parallel-plate capacitor the electric field (and hence $\mathbf{P}$) is uniform in the dielectric volume, and thus $\rho_P = 0$. The surface charge density is $\sigma_P = -P$ on the left-hand plate and $+P$ on the right. Since $E = \sigma_0/\varepsilon_0$ for the left plate ($E = -\sigma_0/\varepsilon_0$ for the right plate), we conclude from Equation (4–32a) that

$$P = \chi E \tag{4-43}$$

In general, for a static electric field the $i$ component of $\mathbf{P}$ is related to the field by

$$P_i = \sum_{j=1}^{3} \chi_{ij} E_j \tag{4-44}$$

where $\chi_{ij}$ is the susceptibility tensor. For an isotropic (no preferred directions) dielectric, the off-diagonal elements of $\chi_{ij}$ are zero, and the diagonal elements are equal. Hence Equation (4–44) can be written as

$$\mathbf{P}(\mathbf{r}) = \chi(E)\mathbf{E}(\mathbf{r}) \tag{4-45}$$

If the applied electric field is considerably smaller than the internal field of the molecule, $\chi(E)$ can be approximated by a constant, which is the reason for the name *dielectric constant*. Dielectrics in which the susceptibility is nearly constant are called *linear*.

For time-dependent fields a relation like Equation (4–45) holds for each frequency component of the wave,

$$\mathbf{P}(\mathbf{r}, \omega) = \chi(\omega)\mathbf{E}(\mathbf{r}, \omega) \tag{4-46}$$

## 4-3 GAUSS'S LAW IN DIELECTRIC MATTER AND BOUNDARY CONDITIONS AT A DIELECTRIC INTERFACE

Gauss's law states that the divergence of $\mathbf{E}$ is proportional to the total charge density $\rho$. In dielectrics the total charge density is the sum of the free charge density $\rho_f$ and the polarization charge density $\rho_P$. By use of Equation (4–42) Gauss's law becomes

$$\nabla \cdot \mathbf{E} = \frac{\rho}{\varepsilon_0} = \frac{1}{\varepsilon_0}(\rho_{\mathrm{f}} + \rho_{\mathrm{P}}) = \frac{1}{\varepsilon_0}(\rho_{\mathrm{f}} - \nabla \cdot \mathbf{P})$$

or

$$\nabla \cdot \left(\frac{\mathbf{E} + \mathbf{P}}{\varepsilon_0}\right) = \frac{\rho_{\mathrm{f}}}{\varepsilon_0} \tag{4-47}$$

We define the *displacement field* **D** as

$$\mathbf{D} \equiv \varepsilon_0 \mathbf{E} + \mathbf{P} = \varepsilon \mathbf{E} \tag{4-48}$$

to obtain

**Gauss's Law in Dielectric Matter**

$$\nabla \cdot \mathbf{D} = \rho_{\mathrm{f}} \tag{4-49}$$

The integral form of Equation (4–49) is then

$$\oint_S \mathbf{D} \cdot d\mathbf{S} = Q_{\mathrm{f}} \tag{4-50}$$

where $Q_{\mathrm{f}}$ is the total free charge enclosed by $S$.

With Equation (4–50) and $\oint \mathbf{E} \cdot d\mathbf{r} = 0$, which was derived in Section 1–4, the boundary conditions at the interface of two dielectrics can be determined. First, we apply Equation (4–50) to an infinitesimal prism centered in the interface, as shown in Figure 4–12(a). In the limit that the height of the prism vanishes, we obtain the relation

**Boundary Condition on Normal D**

$$(\mathbf{D}_2 - \mathbf{D}_1) \cdot \hat{\mathbf{n}} = \sigma_{\mathrm{f}} \tag{4-51}$$

which implies that the normal component of **D** changes by the free surface charge density at the interface. Next, we apply $\oint \mathbf{E} \cdot d\mathbf{r} = 0$ to an infinitesimal rectangular circuit, as shown in Figure 4–12(b). In the limit that the width of the rectangle goes to zero, we obtain the relation

**Boundary Condition on Tangential E**

$$(\mathbf{E}_2 - \mathbf{E}_1) \cdot \hat{\mathbf{t}} = 0 \tag{4-52}$$

where $\hat{\mathbf{t}}$ is the unit tangent vector. This relation implies that the tangential component of **E** is continuous across the interface.

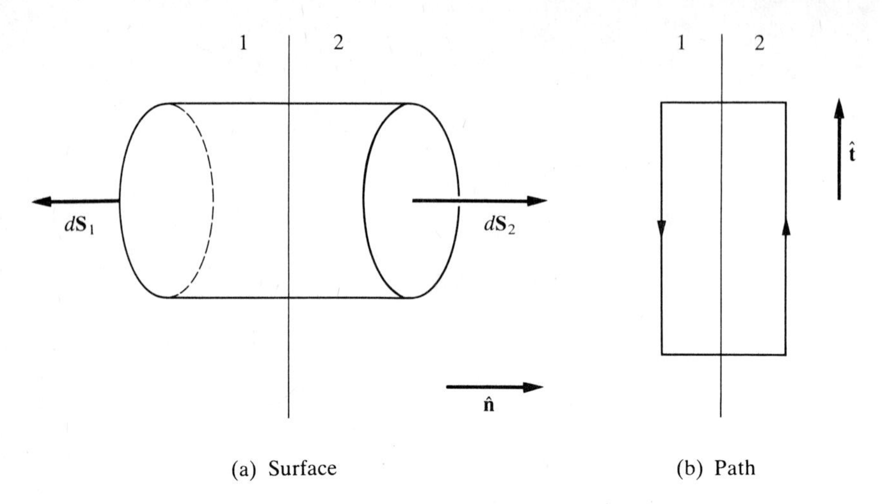


**FIGURE 4-12** Boundary conditions at the interface of two dielectrics. The boundary conditions are determined by applying Gauss's law to the surface in (a) and integrating $\mathbf{E} \cdot d\mathbf{r}$ around the path in (b). The unit vectors $\hat{\mathbf{n}}$ and $\hat{\mathbf{t}}$ are normal and tangent, respectively, to the dielectric interface.

In addition to the above boundary conditions on **D** and **E**, the electrostatic potential $\Phi$ must be continuous

**Boundary Condition on Potential**

$$\Phi_2 = \Phi_1 \tag{4-53}$$

This boundary condition must hold unless the radial component of the electric field is infinite at the dielectric boundary, a situation not ordinarily present.

## 4-3A Parallel-Plate Capacitor with a Dielectric

As an application of Equation (4–50), we calculate the capacitance of the parallel-plate capacitor of Figure 4–13, which has a dielectric medium that does not completely fill the space between the plates. Surrounding the inside surface of the left plate by a Gaussian prism, we deduce that the $D$ field in region 1 is

$$D_1 = \sigma \tag{4-54a}$$

since both **E** and **D** vanish inside the conducting plate. Here $\sigma$ is the charge density on the conducting plate. Since there are no free charges within the capacitor, the use of Gaussian prisms that enclose the interfaces between different regions leads to

$$\sigma = D_1 = D_2 = D_3 \tag{4-54b}$$

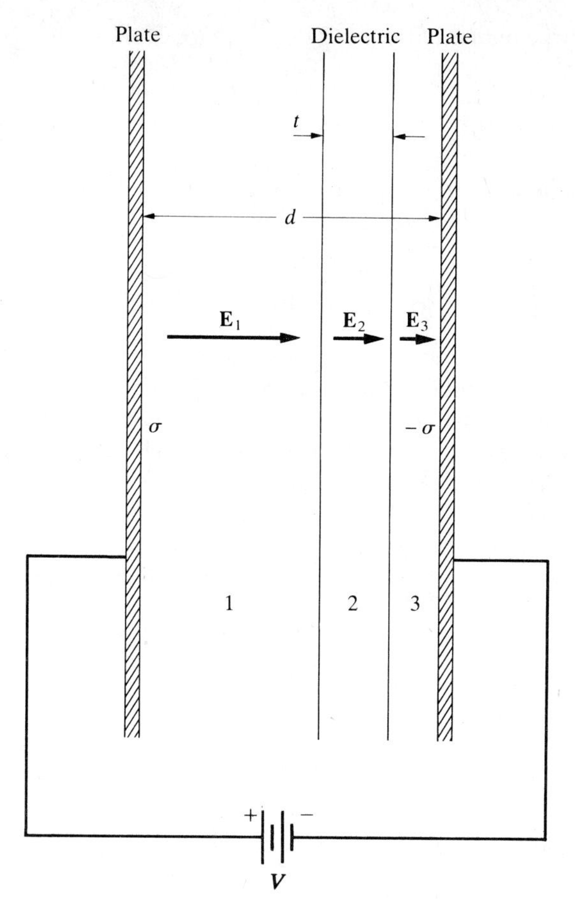


**FIGURE 4-13** Parallel-plate capacitor with a dielectric that does not completely fill the space between the plates

Equivalently,

$$\sigma = \varepsilon_0 E_1 = \varepsilon E_2 = \varepsilon_0 E_3 \tag{4-55}$$

from which we conclude that

$$E_3 = E_1 \qquad E_2 = \frac{\varepsilon_0}{\varepsilon} E_1 \tag{4-56}$$

The potential difference across the capacitor is

$$\begin{aligned} V &= (d - t)E_1 + tE_2 = \left(\frac{d - t}{\varepsilon_0} + \frac{t}{\varepsilon}\right)\sigma \\ &= \left(\frac{d - t}{\varepsilon_0} + \frac{t}{\varepsilon}\right)\frac{Q}{A} \end{aligned} \tag{4-57}$$

where $Q$ is the charge on the left plate. Thus the capacitance $C = Q/V$ is

$$C = \frac{A}{(d - t)/\varepsilon_0 + t/\varepsilon} \tag{4-58}$$

The practical advantage of the displacement field is that only the free charges need be explicitly considered in the calculation.

## 4-3B Dielectric Shielding of a Point Charge

Another application of Equation (4–50) is the calculation of the electric field due to a point charge $q$ imbedded in a linear dielectric of infinite extent. Because of the symmetric nature of the problem, $\mathbf{D}$ must be purely radial, just as the electric field is purely radial for a point charge in a vacuum. Assuming the dielectric has an electric permittivity $\varepsilon$, a simple application of Equation (4–50) gives

$$\mathbf{D} = \frac{q}{4\pi r^2}\,\hat{\mathbf{r}} \tag{4-59}$$

The quantities $\mathbf{E}$ and $\mathbf{P}$ can be obtained from Equation (4–59) and Equation (4–48):

$$\mathbf{E} = \frac{q}{4\pi\varepsilon r^2}\,\hat{\mathbf{r}} \tag{4-60}$$

$$\mathbf{P} = \frac{(\varepsilon - \varepsilon_0)q}{4\pi\varepsilon r^2}\,\hat{\mathbf{r}} \tag{4-61}$$

Comparing Equation (4–60) with $\mathbf{E}$ for a point charge in empty space, we see that the electric field in the dielectric is reduced by the factor $\kappa = \varepsilon/\varepsilon_0$.

To see why the electric field is diminished in the dielectric, consider determining the electric field by an application of Gauss's law. The total charge $Q$ within the dielectric is the sum of the point charge $q$ and the polarization charge $q_P$, which can be obtained from Equations (4–42) and (4–61):

$$q_P = \int \rho_P\, dV = -\int \nabla\cdot\mathbf{P}\, dV = -\frac{(\varepsilon - \varepsilon_0)}{4\pi\varepsilon}q(4\pi)\int \delta^3(\mathbf{r})\, dV$$

$$= -\frac{(\varepsilon - \varepsilon_0)}{\varepsilon}q \tag{4-62}$$

$$Q = q + q_P = q - \frac{(\varepsilon - \varepsilon_0)}{\varepsilon}q = \frac{1}{\kappa}q \tag{4-63}$$

Hence the electric field in the dielectric is reduced by the factor $\kappa$ owing to the polarization charge contribution.

## 4-4 RELATION BETWEEN MACROSCOPIC AND MICROSCOPIC POLARIZATIONS

At the atomic level matter is not electrically neutral since atoms consist of a positively charged nucleus and a negatively charged electron cloud. Within matter there are strong electromagnetic fields that are not readily observed in ordinary laboratory experiments. Laboratory instruments are normally sensitive only to the fields averaged over regions much larger than atomic dimensions. These macroscopic fields are averages of the real fields at the atomic level, which are referred to as *microscopic fields.*

The molecular electric field $\mathbf{E}_M$ is the field "seen" by the molecule. More precisely, it is the part of the microscopic field at the location of the molecule that is due to all charges except those in the molecule. The field $\mathbf{E}_M$ differs from the macroscopic field $\mathbf{E}$ in two ways. First, $\mathbf{E}$ results from all charges including those in the molecule. Second, the molecule may be in a special spatial relationship to nearby ions or polarized molecules (e.g., in a crystal lattice) so that the field at its location is different from the average $\mathbf{E}$ field in the material.

To find the molecular field $\mathbf{E}_M$, we must sum over the electric field contributions due to all charges outside the molecule. For charges farther away than a distance $R$, we approximate the discrete charges by a continuous distribution of charge and dipole densities. The contributions of those charges closer than the distance $R$ will be calculated exactly. The results will be more accurate the larger $R$ is taken to be but, in fact, will be found to vary little once $R$ is larger than a few atomic spacings. This distance can be taken small enough so that the macroscopic field is nearly uniform over the distance $R$.

The macroscopic electric field $\mathbf{E}$ at a point is the average of the field over a local region whose size is large compared with the molecular dimensions. This averaging of the field is equivalent to the averaging of its sources (i.e., smearing out the charge distributions). The distant sources contribute the same to the macroscopic field $\mathbf{E}$ and the molecular field $\mathbf{E}_M$. Only the sources close to the molecule give a difference between $\mathbf{E}_M$ and $\mathbf{E}$; that is, only sources within a radius $R$ contribute to $\mathbf{E}_M - \mathbf{E}$. We denote the field contributions at the center of the sphere due to sources within $R$ by $\mathbf{E}'_M$ for the molecular field and $\mathbf{E}'$ for the macroscopic field; then the field acting on the molecule is

$$\mathbf{E}_M \approx \mathbf{E} + \mathbf{E}'_M - \mathbf{E}' \tag{4-64}$$

For a material with no macroscopic charge density, the sources of $\mathbf{E}'_M$ and $\mathbf{E}'$ are electric dipoles. To calculate $\mathbf{E}'_M$, we sum the dipole fields of all molecules in the sphere, omitting the dipole of the molecule at $r = 0$. To calculate $\mathbf{E}'$, we use the macroscopic polarization vector $\mathbf{P}$ for the region $r < R$.

Since the macroscopic distribution of the charge inside the sphere is described by $\mathbf{P}$, nearly uniform inside the sphere, the volume polarization charge is zero:

$$\rho_{\mathrm{P}} = -\nabla \cdot \mathbf{P} \approx 0 \tag{4-65}$$

And the polarization surface charge density is (see Figure 4–14)

$$\sigma_{\mathrm{P}} = \mathbf{P} \cdot \hat{\mathbf{n}} = P \cos \theta \tag{4-66}$$

The field at the center of the sphere is

$$\mathbf{E}' = -\frac{1}{4\pi\varepsilon_0} \int \frac{\sigma_{\mathrm{P}}(\mathbf{r}')\hat{\mathbf{n}}\, dS'}{R^2} \tag{4-67}$$

We choose the $z$ axis along $\mathbf{P}$; then $E'_z$ will be the only nonvanishing component of $\mathbf{E}'$. Using the relations

$$\begin{aligned} \hat{\mathbf{n}} \cdot \hat{\mathbf{z}} &= \cos \theta \\ dS' &= R^2 \sin \theta \, d\theta \, d\phi = -R^2 \, d(\cos \theta) \, d\phi \end{aligned} \tag{4-68}$$

in Equation (4–67), we obtain

$$\begin{aligned} E'_z &= -\frac{1}{4\pi\varepsilon_0} \int \frac{(P \cos \theta)(\cos \theta)[-R^2 \, d(\cos \theta) \, d\phi]}{R^2} \\ &= \frac{P}{2\varepsilon_0} \int_{+1}^{-1} \cos^2 \theta \, d(\cos \theta) = -\frac{P}{3\varepsilon_0} \end{aligned} \tag{4-69}$$

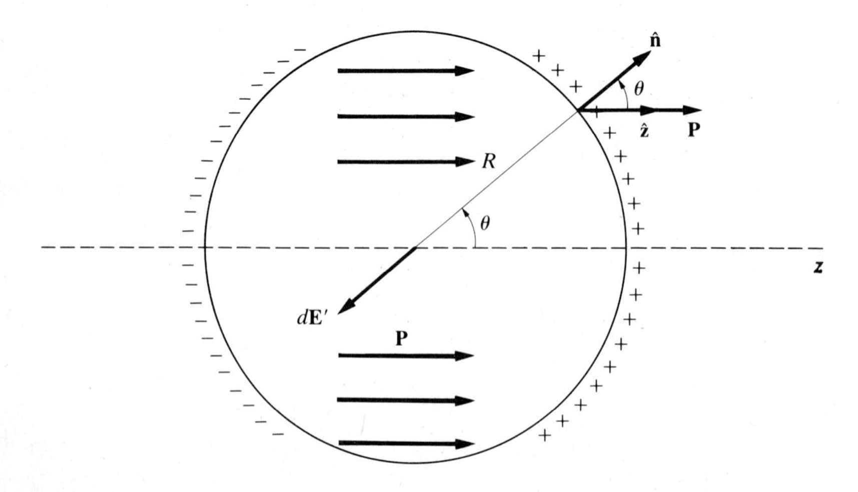


**FIGURE 4-14** Polarization surface charge density on a uniformly polarized dielectric sphere: $\mathbf{P} \cdot \hat{\mathbf{n}}$. The electric field at the center of the sphere due to this polarization surface charge is obtained by integrating the elements $d\mathbf{E}'$ over the surface of the sphere.

Thus the contribution of the charge inside the sphere to **E** at the molecule is

$$\mathbf{E}' = -\frac{\mathbf{P}}{3\varepsilon_0} \tag{4-70}$$

Since this result is independent of the sphere radius $R$, we can think of the source of $\mathbf{E}'$ as the delta function, $-\mathbf{P}/(3\varepsilon_0)\delta^3(\mathbf{r})$, integrated over a volume including the origin. From Equation (4–18) this delta function is equivalent to a dipole located at the origin having moment $\mathbf{p} = \mathbf{P}$. The average polarization effect is thus the volume-averaged field of a dipole having moment **P**.

The field at the molecule $\mathbf{E}'_{\mathrm{M}}$ due to all charge inside the sphere, except in the molecule itself, is not so generally calculable; it depends on the microscopic structure of the material. For a fluid medium the distribution of molecules around the molecule centered at $\mathbf{r} = 0$ is isotropic; that is, a molecule at a distance $r$ is equally likely to be at any point on a sphere of radius $r$. In this case $\mathbf{E}'_{\mathrm{M}}$ vanishes, as we now show.

The electric field at $\mathbf{r} = 0$ due to a dipole $\mathbf{p}$ at $\mathbf{r}'$ is obtained from Equation (4–17) by the replacement of $\mathbf{r}$ by $-\mathbf{r}'$:

$$\mathbf{E}_{\mathrm{D}}(\mathbf{r} = 0) = \frac{1}{4\pi\varepsilon_0}\left[\frac{3\mathbf{r}'(\mathbf{r}' \cdot \mathbf{p}) - \mathbf{r}'^2\mathbf{p}}{r'^5}\right] \tag{4-71}$$

We assume that the direction of $\mathbf{p}$ points along the macroscopic field **E**, which we take as the $z$ direction. The field due to the dipole at $\mathbf{r}'$ is proportional to

$$3\mathbf{r}'z' - r'^2\hat{\mathbf{z}} = \hat{\mathbf{x}}(3x'z') + \hat{\mathbf{y}}(3y'z') + \hat{\mathbf{z}}(2z'^2 - x'^2 - y'^2) \tag{4-72}$$

These terms all vanish when averaged over the direction of $\mathbf{r}'$. Likewise, the field at $\mathbf{r} = 0$ due to any higher multipole carried by molecules isotropically distributed about $\mathbf{r} = 0$ vanishes.

We can similarly demonstrate that $E'_{\mathrm{M}}$ vanishes if the molecules are on a cubic lattice (simple, face-centered, or body-centered). In these cases molecules are located at distances

$$\mathbf{r}_{lmn} = d(\hat{\mathbf{x}}l + \hat{\mathbf{y}}m + \hat{\mathbf{z}}n)$$

where $l$, $m$, $n$ are positive or negative integers and $d$ is the lattice constant, usually of the order of $10^{-10}$ m. From Equations (4–71) and (4–72) the electric dipole field $\mathbf{E}'_{\mathrm{M}}$ is then

$$E'_{\mathrm{M}} = \frac{p}{4\pi\varepsilon_0 d^3}\sum_{lmn}\frac{3ln\hat{\mathbf{x}} + 3mn\hat{\mathbf{y}} + (2n^2 - l^2 - m^2)\hat{\mathbf{z}}}{(l^2 + m^2 + n^2)^5} \tag{4-73}$$

where we sum over dipoles within the sphere excluding $l = m = n = 0$. If the dipole location $(l, m, n)$ lies within the sphere, then $(-l, m, n)$ does also. By Equation (4–73) the $x$ component contributions cancel pairwise; likewise for the $y$ components. All permutations of $(l, m, n)$ have the same distance from the origin, and the quantity $2n^2 - l^2 - m^2$ vanishes when summed over all permutations since each component value appears three times.

For such isotropic materials $\mathbf{E}'_\mathrm{M} = 0$, and so

$$\mathbf{E}_\mathrm{M} = \mathbf{E} - \mathbf{E}'$$

or

$$\mathbf{E}_\mathrm{M} = \mathbf{E} + \frac{\mathbf{P}}{3\varepsilon_0} \tag{4-74}$$

We should emphasize that the result $\mathbf{E}'_\mathrm{M} = 0$ is only an important special case. For a general crystal $\mathbf{E}'_\mathrm{M}$ does not vanish but can be computed in analogy to Equation (4–73). In the two calculations of $\mathbf{E}'_\mathrm{M}$ considered here, we have assumed that each dipole moment is parallel and aligned along the electric field $\mathbf{E}_\mathrm{M}$. If the molecules have intrinsic electric dipole moments, such as in the case of water, $\mathbf{p}$ is not parallel to $\mathbf{E}_\mathrm{M}$. Such molecules are known as polar molecules and will be discussed further in Section 4–5D. The calculation of $\mathbf{E}'_\mathrm{M}$ for such polar molecules is complex and depends on the temperature. For a sufficiently tenuous vapor the effects of short-range order will be negligible, and the random approximation will be accurate. Of course, the polarization effect itself for a vapor will be small.

For the remainder of this section we assume that induced moment $\mathbf{p}$ is parallel to $\mathbf{E}_\mathrm{M}$ and that $\mathbf{E}'_\mathrm{M}$ vanishes. The molecular field acting on each molecule induces a dipole moment

$$\mathbf{p} = \alpha \mathbf{E}_\mathrm{M} \tag{4-75}$$

The proportionality constant $\alpha$ is called the *polarizability*; it parameterizes the microscopic response of the molecule to the field. If there are $n_\mathrm{V}$ molecules per unit volume, the dipole moment per unit volume $\mathbf{P}$ is

$$\mathbf{P} = n_\mathrm{V} \alpha \mathbf{E}_\mathrm{M} \tag{4-76}$$

Inserting Equation (4–74) in Equation (4–76), we find

$$\mathbf{P} = \frac{n_\mathrm{V}\alpha}{1 - n_\mathrm{V}\alpha/(3\varepsilon_0)} \mathbf{E}$$

Equating this expression for $\mathbf{P}$ to the macroscopic formula,

$$\mathbf{P} = \varepsilon_0(\kappa - 1)\mathbf{E}$$

we obtain the Clausius-Mossotti (CM) relation:

**Clausius-Mossotti Relation**

$$\kappa = \frac{1 + 2n_V\alpha/(3\varepsilon_0)}{1 - n_V\alpha/(3\varepsilon_0)} \tag{4-77}$$

This relation specifies the macroscopic parameter $\kappa$ in terms of the microscopic parameter $\alpha$.

The number of molecules per unit volume $n_V$ is given in terms of the mass density $\rho$, Avogadro's number $N_A$, and the molecular weight $A$ by

$$n_V = \frac{\rho N_A}{A} \tag{4-78}$$

For a gas $(n_V\alpha)/(3\varepsilon_0) \ll 1$, $\mathbf{P}/\varepsilon_0$ is small compared with $\mathbf{E}$, and $\mathbf{E}_M \approx \mathbf{E}$. From Equation (4–77) the dielectric constant for a gas is

$$\kappa \approx 1 + \frac{n_V\alpha}{\varepsilon_0} \tag{4-79}$$

The CM relation in Equation (4–77) forms a bridge between the macroscopic description of a dielectric and theories of microscopic behavior that describe the action of an electric field on a molecule. This relation is correct as long as $\mathbf{E}'_M$ can be neglected. In the next section we will discuss some simple models for the polarizability $\alpha$. Although the only completely valid theory for microscopic quantities, such as the polarizability, is the quantum theory, many results obtained by classical reasoning are at least qualitatively correct and provide considerable insight into the observed behavior.

# 4-5 MICROSCOPIC POLARIZABILITY MODELS

The microscopic polarizability of most materials can be described in one or the other of the following two ways. First, the applied electric field can separate the average positive- and negative-charge positions in the atom, molecule, or crystal, giving rise to a net dipole moment. This induced polarization can be due to electron movement or ionic movement. Second, each individual molecule may possess an intrinsic electric dipole moment. In the absence of an external field no net dipole moment is observed because the molecular moments are randomly oriented. An applied field acts to align these moments and yields a net orientational polarization.

## 4-5A Induced Static Polarizability

The quantum mechanical picture of an atom consists of a nearly point nuclear charge $Ze$ surrounded by a cloud of $Z$ electrons. Most of these electrons are tightly bound and generally remain close to the nucleus. A few of the outer electrons, called *valence electrons*, are more loosely bound and participate easily in chemical reactions and interatomic processes. The size of any atom is roughly $a_0 \approx 10^{-10}$ m, nearly independent of $Z$.

When an electric field $\mathbf{E}_M$ acts on an atom, the nucleus and the electrons are displaced in opposite directions, producing a dipole moment. The tightly bound electrons follow the nucleus, and only the valence electrons contribute significantly to the polarizability. If there is a single valence electron, we see from Figure 4–15 that the electron will be displaced to the left of the positive ion, and hence the distance from the ion to the electron, $z$, is negative. The resulting induced dipole moment is then

$$p = -ez \tag{4-80}$$

Since the atom is a stable system, the force of attraction between the ion and the valence electron gives rise to a restoring force on the electron,

$$F = -Kz \tag{4-81}$$

where $K$ is the effective spring constant.

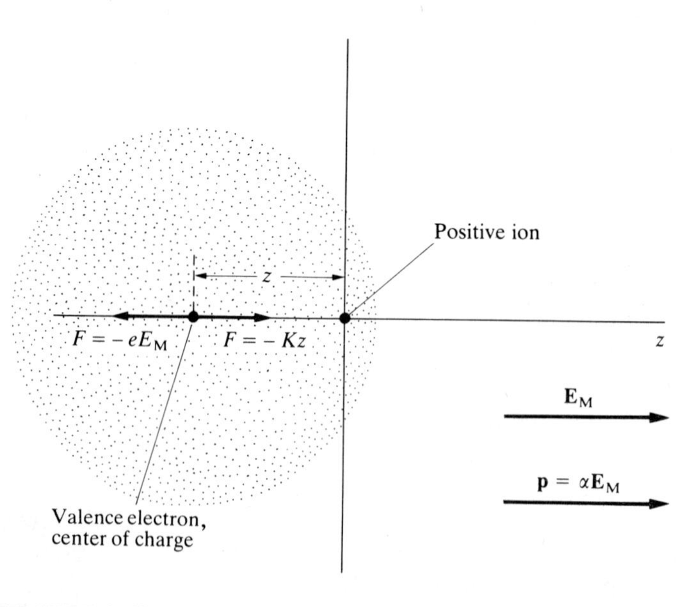


**FIGURE 4-15** Forces on the valence electron due to the electric field $\mathbf{E}_M$. The shaded sphere represents the valence electron cloud, and the large dot at the origin is the ionic core. The displacement $z$ is greatly exaggerated.

To calculate the spring constant $K$, we must know something of the detailed atomic structure; furthermore, the calculation must be done quantum mechanically. However, we can obtain a rough idea of the correct result with the following simple classical model. We assume that the charge density of the single valence electron is spread uniformly in a sphere of radius $a_0$ much larger than the ion radius. The electric field due to the valence electron charge at a distance $r$ from the cloud center can be calculated by using Gauss's law:

$$E(4\pi r^2) = -\frac{e}{\varepsilon_0}\frac{(\frac{4}{3}\pi r^3)}{(\frac{4}{3}\pi a_0^3)}$$

which gives

$$E = -\frac{er}{4\pi\varepsilon_0 a_0^3} \tag{4-82}$$

The restoring force on the nucleus plus the $Z - 1$ closely bound electrons is

$$F_{\text{ion}} = \frac{e^2 z}{4\pi\varepsilon_0 a_0^3} \tag{4-83}$$

where $z = -r$. By Newton's third law the corresponding force on the valence electron is given by

$$F = \frac{-e^2 z}{4\pi\varepsilon_0 a_0^3} = -Kz \tag{4-84}$$

In static equilibrium the applied force on the electron, $-eE_{\text{M}}$, plus the restoring force on the electron, $-Kz$, must add to zero,

$$Kz = -eE_{\text{M}} \tag{4-85}$$

From Equations (4–80) and (4–85) we obtain

$$p = \left(\frac{e^2}{K}\right)E_{\text{M}} = (4\pi\varepsilon_0 a_0^3)E_{\text{M}} \tag{4-86}$$

Then the static polarizability $\alpha_0$, defined by $p = \alpha_0 E_{\text{M}}$, is

**Static Polarizability**

$$\alpha_0 = 4\pi\varepsilon_0 a_0^3 \tag{4-87}$$

The numerical value of this static polarizability is

$$\alpha_0 \approx \frac{(10^{-10})^3}{9 \times 10^9} \approx 10^{-40}\ \text{C}^2\text{m/N} \tag{4-88}$$

From Equations (4–84) and (4–85) the charge displacement $z$ in the molecule relative to the atomic dimension $a_0$ is

$$\frac{|z|}{a_0} = \frac{eE_M}{a_0 K} = \frac{4\pi\varepsilon_0 a_0^2}{e} E_M \tag{4-89}$$

For a field $E_M$ comparable to or less than the largest laboratory fields ($E \leqslant 10^7$ V/m), we find that

$$\frac{z}{a_0} \leqslant 10^{-4} \tag{4-90}$$

## 4-5B Induced Dynamic Polarizability

If the applied field $E(t)$ varies rapidly enough with time, the polarization vector $\mathbf{P}$ will lag or lead the electric field because of damping and inertia effects. To describe the dynamic case, we assume that the charge separation $z$ obeys the forced–harmonic oscillation equation

$$m_e \ddot{z} = -b\dot{z} - Kz + eE_M(t) \tag{4-91}$$

where $m_e$ is the electron mass and the damping term $b\dot{z}$ is due to radiation from the oscillating electron. We write this equation in the standard form

$$\ddot{z} + 2\gamma\dot{z} + \omega_0^2 z = -\frac{e}{m_e} E_M(t) \tag{4-92}$$

where

$$\omega_0^2 = \frac{K}{m_e} = \frac{e^2}{4\pi\varepsilon_0 a_0^3 m_e} = \frac{e^2}{m_e \alpha_0} \qquad \gamma = \frac{b}{2m_e} \tag{4-93}$$

For a sinusoidal molecular field of frequency $\omega$

$$E_M(t) = \mathscr{E} \cos \omega t = \mathscr{E}\, \mathrm{Re}(e^{-i\omega t}) \tag{4-94}$$

The solution to Equation (4–92) can be found by standard methods discussed in classical mechanics textbooks. The steady-state solution gives a dipole moment of the form

$$p(t) = -ez(t) = \mathrm{Re}[\alpha(\omega)\mathscr{E} e^{-i\omega t}] \tag{4-95}$$

where the complex polarizability $\alpha(\omega)$,

$$\alpha(\omega) = \frac{e^2/m_e}{\omega_0^2 - \omega^2 - 2i\gamma\omega} \tag{4-96}$$

is found by substitution of Equation (4–95) into Equation (4–92). Because of the complex nature of $\alpha(\omega)$, the sinusoidal time dependence of $p(t)$ is shifted from that of $E_M(t)$.

Multiplying both the numerator and denominator of $\alpha(\omega)$ by $\omega_0^2 - \omega^2 + 2i\gamma\omega$, we can find the real and imaginary parts of $\alpha(\omega)$:

$$\operatorname{Re} \alpha(\omega) = \frac{(e^2/m_e)(\omega_0^2 - \omega^2)}{(\omega_0^2 - \omega^2)^2 + 4\gamma^2\omega^2} \tag{4-97a}$$

$$\operatorname{Im} \alpha(\omega) = \frac{2\gamma\omega e^2/m_e}{(\omega_0^2 - \omega^2)^2 + 4\gamma^2\omega^2} \tag{4-97b}$$

These quantities are plotted in Figure 4–16. The dynamic polarizability exhibits *resonant* structure; the resonant frequency, from Equation (4–93), is

$$\omega_0 = \left(\frac{e^2}{m_e\alpha_0}\right)^{1/2} = \left[\frac{(1.6 \times 10^{-19})^2}{(9 \times 10^{-31})(10^{-40})}\right]^{1/2} \approx 2 \times 10^{16} \text{ rad/s} \tag{4-98}$$

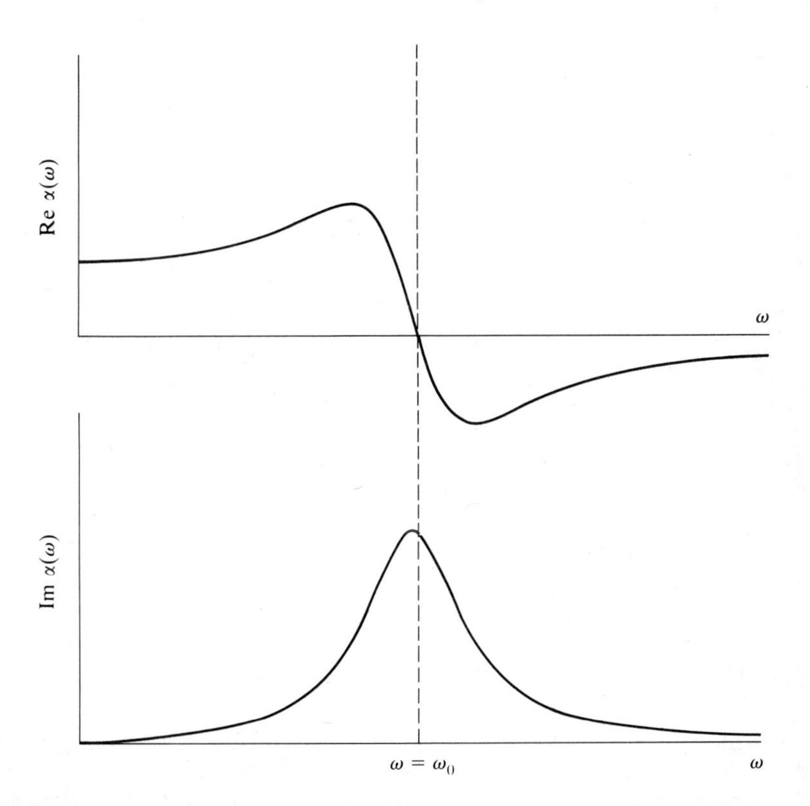


**FIGURE 4–16** Polarizability curve for a substance in a rapidly time-varying electric field (induced dynamic polarizability). When the applied field frequency equals the natural resonance frequency, the imaginary part of the polarizability becomes large.

The corresponding wavelength $\lambda_0 = 2\pi c/\omega_0$ is

$$\lambda_0 \approx 10^{-7}\ \mathrm{m} = 1000\ \text{Å} \tag{4-99}$$

This wavelength is in the ultraviolet range. For waves propagating in a dielectric, strong absorption occurs at the resonance frequency. This phenomenon will be discussed further in Chapter 9. There we will derive the relation $n = \sqrt{\kappa}$ between the index of refraction $n$ and the dielectric constant $\kappa$.

For many substances the polarizability can be described by simple resonance models, with several resonant frequencies that are the atomic spectral line frequencies. With quantum theory the resonant frequencies and damping terms can be calculated from first principles.

## 4-5C Induced Ionic Polarizability

An applied electric field generally induces charge separations in any bound-charge distribution. For example, with ionically bound molecules of a crystal lattice the restoring forces on the ions are similar to the restoring forces for the valence electrons in the atomic case, and the static polarizabilities are

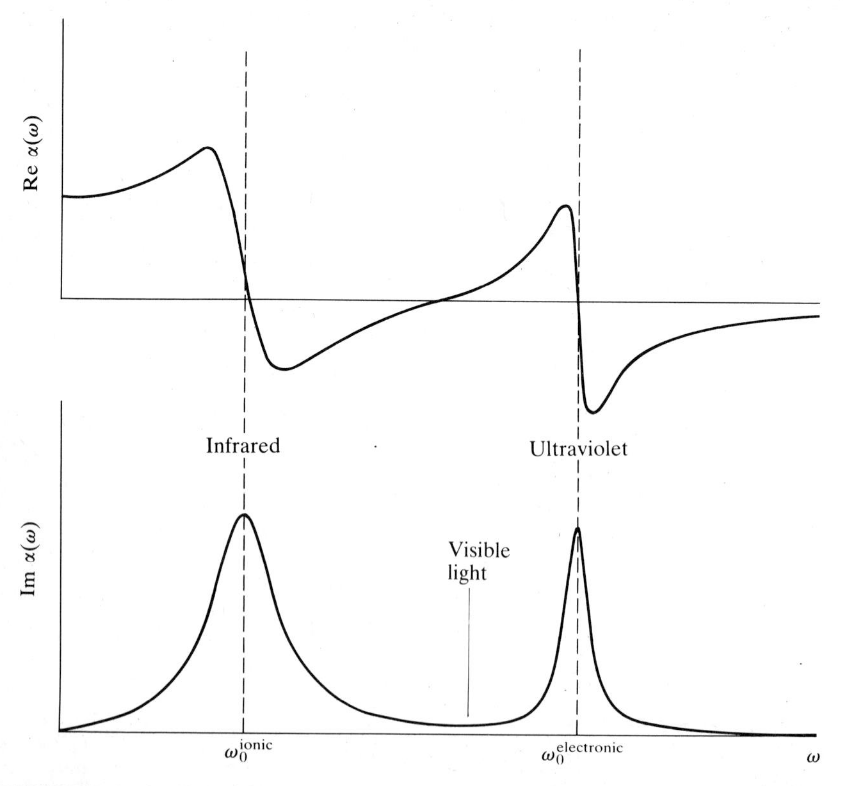


**FIGURE 4-17** Polarizability curve for a transparent substance (e.g., glass) that exhibits only electronic and ionic polarizabilities. The substance is transparent to visible light because the imaginary part of $\alpha(\omega)$ is small in the visible-frequency range.

therefore similar. For the dynamic polarizability there is a difference in that the reduced mass of the ionic pair is much larger than the electron's mass. For an NaCl (common salt) crystal the reduced mass is $M = 25$ proton masses or $M = 5 \times 10^4 m_e$. From Equation (4–98) the corresponding ionic resonant frequency is thus reduced to

$$\omega_0^{\text{ionic}} = \sqrt{\frac{m_e}{M}}\, \omega_0^{\text{electronic}} \approx \frac{\omega_0^{\text{electronic}}}{200} \approx 10^{14} \text{ rad/s} \tag{4-100}$$

The wavelength associated with this angular frequency is

$$\lambda^{\text{ionic}} = \frac{2\pi c}{\omega_0^{\text{ionic}}} = 2 \times 10^{-5} \text{ m} = 200{,}000 \text{ Å} \tag{4-101}$$

which lies in the infrared spectral region.

A typical polarizability curve characteristic of many substances that have only electronic and ionic polarizabilities is shown in Figure 4–17.

## 4-5D Orientational Polarizability

Some substances such as ordinary water exhibit a very large static dielectric constant. For water $\text{Re}(\kappa)$ rises from a value of 1.8 in the optical region to a value $\kappa \approx 80$ in the static case (i.e., $\omega = 0$). Also, the large static polarizability has a strong temperature dependence, as illustrated in Figure 4–18. The data indicate a temperature-independent component that is consistent with the electric and ionic contributions and a temperature-dependent part proportional to $T^{-1}$ that can be understood as a statistical phenomenon. The water molecule consists of an oxygen atom bound to two hydrogen atoms, as schematically shown in Figure 4–19. A water molecule has an intrinsic electric dipole moment of $p = 0.62 \times 10^{-29}$ C·m. The intrinsic moment occurs because of the bonding angle and the fact that the electrons of the hydrogen atoms spend a fraction of the time near the oxygen atom. This intrinsic dipole is comparable to the dipole moment,

$$p = ea_0 = (1.6 \times 10^{-19} \text{ C})(10^{-10} \text{ m}) = 1.6 \times 10^{-29} \text{ C·m}$$

of a charge separation $e$ over the typical molecular dimension $a_0$, which is $10^4$ larger than large induced dipole moments.

If it were not for thermal motion, a small applied electric field would cause all of the permanent dipoles to line up parallel with the field because of the torque $\mathbf{N} = \mathbf{p} \times \mathbf{E}_M$. This alignment would result in an infinite polarizability (i.e., a finite polarization would be produced by an arbitrarily small electric field). When there is no applied field, collisions due to thermal motion cause the dipoles to point randomly, giving no net polarization. The application of a weak field causes a small shift in the average orientation and results in a small net dipole moment.

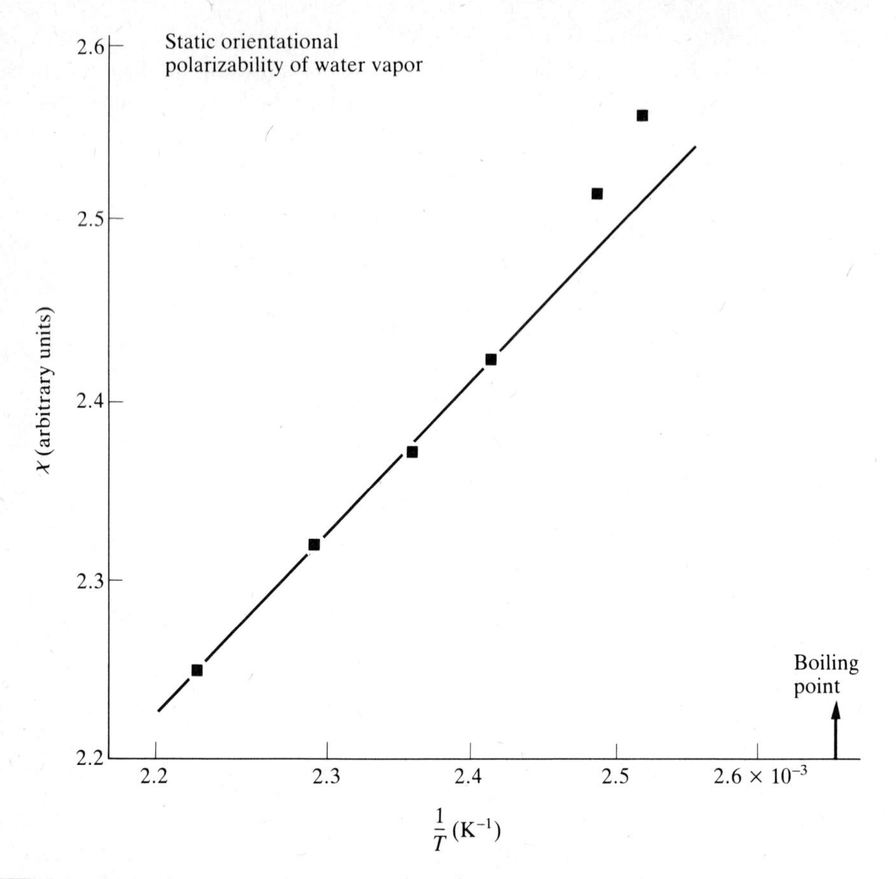


**FIGURE 4-18** Observed susceptibility of water, which exhibits $1/T$ temperature dependence and thus indicates a permanent electric dipole. Extrapolation to infinite temperature gives the induced dipole contribution.

From the Boltzmann statistics of thermal motion, the probability that a molecule has an energy $U$ is proportional to the Boltzmann factor $e^{-U/kT}$, where $k$ is the Boltzmann constant, $k = 1.38 \times 10^{-23}$ J/K, and the temperature $T$ is measured in degrees Kelvin. The orientation energy of an electric dipole in an $\mathbf{E}_\mathrm{M}$ field is, by Equation (4–25),

$$U = -\mathbf{p} \cdot \mathbf{E}_\mathrm{M}$$

Thus

$$\frac{-U}{kT} = \frac{pE_\mathrm{M} \cos\theta}{kT} \tag{4-102}$$

where $\theta$ is the angle between $\mathbf{p}$ and $\mathbf{E}_\mathrm{M}$. The probability that a given molecular dipole moment points into a given solid-angle element $d\Omega$ is proportional to

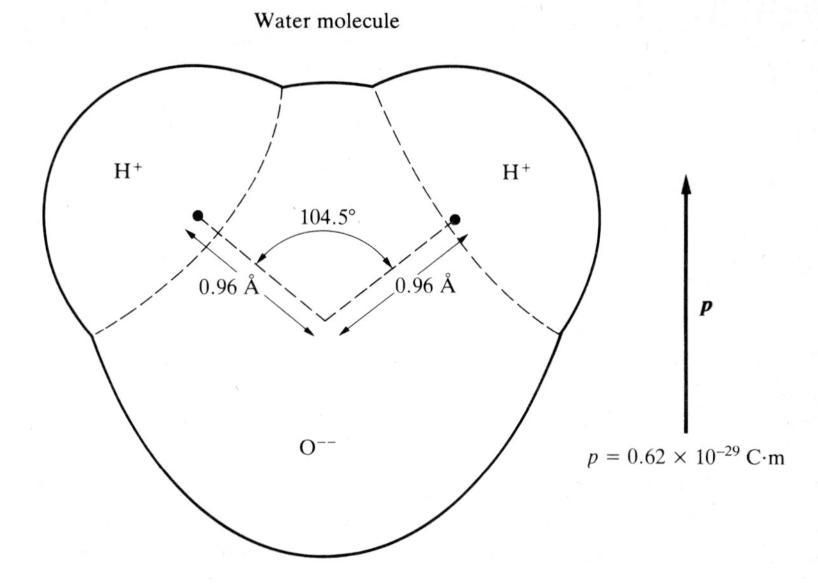


**FIGURE 4–19** Schematic of the permanent dipole moment of a water molecule

$e^{-U/kT}$. Since all dipole orientations are equally probable, the average energy $\langle U \rangle$ is given by the angular average

$$\langle U \rangle = \frac{\int d\Omega \, U e^{-U/kT}}{\int d\Omega \, e^{-U/kT}} \tag{4-103}$$

Since the energy $U$ is proportional to $\cos\theta$, the average $\cos\theta$, from Equation (4–102), is

$$\langle \cos\theta \rangle = \frac{\int_{-1}^{+1} d(\cos\theta) \cos\theta \, e^{(pE_M/kT)\cos\theta}}{\int_{-1}^{+1} d(\cos\theta) e^{(pE_M/kT)\cos\theta}} \tag{4-104}$$

For a typical intrinsic dipole moment ($p = 10^{-29}$ C·m) at room temperature ($T = 300$ K), the quantity $|U/kT|$ in the exponentials is much less than 1, even for a large electric field ($E_M \approx 10^6$ V/m):

$$\left|\frac{U}{kT}\right| = \frac{(10^{-29})(10^6)}{(1.4 \times 10^{-23})(3 \times 10^2)} \approx 2 \times 10^{-3} \tag{4-105}$$

Thus we can make the approximation

$$e^{-U/kT} \approx 1 + \frac{pE_M \cos\theta}{kT} \tag{4-106}$$

in Equation (4–104), which leads to

$$\langle \cos\theta \rangle \approx \frac{pE_M}{3kT} \tag{4-107}$$

The corresponding polarization per unit volume $P$ is

$$P = n_V p\langle \cos\theta \rangle = n_V \frac{p^2}{3kT} E_M \tag{4-108}$$

Thus the orientational polarizability is

**Orientational Polarizability**

$$\alpha_{or} = \frac{p^2}{3kT} \tag{4-109}$$

The proportionality of $\alpha_{or}$ to $1/T$ explains the observed temperature dependence of the static polarizability; see Figure 4–18. The higher the temperature, the more the alignment is reduced by thermal agitation. At room temperature

$$\alpha_{or} = \frac{(10^{-29})^2}{3(1.4 \times 10^{-23})(3 \times 10^2)} \approx 10^{-38}\ \mathrm{C^2{\cdot}m/N} \tag{4-110}$$

This static orientational polarizability is much larger than the induced electronic or ionic polarizabilities of $10^{-40}$, in agreement with observation.

If the applied electric field varies slowly with time compared with the frequency of molecular collisions, the time-averaged dipole angle

$$\langle \cos\theta(t) \rangle = \frac{pE_M(t)}{3kT} \tag{4-111}$$

also oscillates with time. For a sufficiently rapid time variation of $E_M(t)$, the oscillating dipoles will encounter rotational friction and will begin to lag behind the applied field.

## SUMMARY

### Important Concepts and Equations

$$\Phi = \frac{q}{4\pi\varepsilon_0 r} + \frac{\mathbf{p}\cdot\mathbf{r}}{4\pi\varepsilon_0 r^3} + \sum_{i,j} \frac{\frac{1}{2}Q_{ij}x_i x_j}{4\pi\varepsilon_0 r^5} + \cdots$$

*Multipole expansion of a charge distribution.*

$$q = \int \rho(\mathbf{r}')\, dV'$$

*Electric monopole moment.*

$$\mathbf{p} = \int \mathbf{r}'\rho(\mathbf{r}')\,dV'$$

*Electric dipole moment.*

$$Q_{ij} = \int (3x_i'x_j' - \delta_{ij}r'^2)\rho(r')\,dV'$$

*Electric quadrupole moment.*

$$\mathbf{E} = \frac{1}{4\pi\varepsilon_0}\left[\frac{3(\mathbf{r}\cdot\mathbf{p})\mathbf{r} - r^2\mathbf{p}}{r^5}\right]$$

*The field at* $\mathbf{r} \neq 0$ *of an electric dipole at* $\mathbf{r} = 0$.

$$\mathbf{N} = \mathbf{p} \times \mathbf{E}$$

*The torque* **N** *on a charge distribution in a static electric field is given by the cross product of the electric dipole moment* **p** *and the electric field* **E**.

$$U = -p\cdot\mathbf{E}$$

*The potential energy U of a dipole in an electric field is equal to the negative of the dot product of the electric dipole moment* **p** *and the electric field* **E**.

$$\mathbf{P} = \frac{1}{\Delta V}\sum_i \mathbf{p}_i$$

*The average dipole moment per unit volume* **P** *is the sum of the individual dipoles* $\mathbf{p}_i$ *in a small volume* $\Delta V$ *divided by* $\Delta V$.

$$\sigma_\mathrm{P} = \mathbf{P}\cdot\hat{\mathbf{n}}$$

*The polarization surface charge density* $\sigma_\mathrm{P}$ *is equal to the dot product of* **P** *and the unit vector normal to the surface.*

$$\rho_\mathrm{P} = -\nabla\cdot\mathbf{P}$$

*The polarization volume charge density* $\rho_\mathrm{P}$ *is the negative of the divergence of the average dipole moment per unit volume* **P**.

$$\mathbf{P} = \chi\mathbf{E}$$

*The average dipole moment per unit volume* **P** *is equal to the product of the electric susceptibility* $\chi$ *and the electric field* **E** *for a linear* ($\chi$ = *constant*) *isotropic (no preferred direction) dielectric. More generally, for a static electric field*

$$P_i = \sum_{j=1}^{3} \chi_{ij}E_j$$

*where the susceptibility tensor* $\chi_{ij}$ *may depend on the electric field.*

$$\mathbf{D} = \varepsilon_0\mathbf{E} + \mathbf{P} \equiv \varepsilon\mathbf{E}$$

*The displacement field* **D** *is defined to be the sum of the average dipole moment* *per unit volume* **P** *and* $\varepsilon_0$ *times the electric field* **E**. *The displacement field is also equal to electric field* **E** *multiplied by the electric permittivity* $\varepsilon$. *The* **D** *field is introduced so that polarization charges will not appear in Gauss's law.*

$$\varepsilon = \kappa\varepsilon_0 = \varepsilon_0 + \chi \qquad \kappa = \frac{\varepsilon}{\varepsilon_0} \qquad \chi = (\kappa - 1)\varepsilon_0$$

*The basic quantities used to describe dielectric phenomena: the electric permittivity* $\varepsilon$, *the dielectric constant* $\kappa$, *and the electric susceptibility* $\chi$, *and their relationships.*

$\nabla \cdot \mathbf{D} = \rho_f$ (Gauss's law in dielectric matter, differential form)

*The divergence of the displacement field* $\nabla \cdot \mathbf{D}$ *is equal to the free charge density* $\rho_f$.

$\oint_S \mathbf{D} \cdot d\mathbf{S} = Q_f$ (Gauss's law in dielectric matter, integral form)

*The integral of the displacement field* **D** *over a closed surface S is equal to the total free charge* $Q_f$ *enclosed by S.*

$$\left.\begin{aligned}(\mathbf{D}_2 - \mathbf{D}_1)\cdot\hat{\mathbf{n}} &= \sigma_f \\ (\mathbf{E}_2 - \mathbf{E}_1)\cdot\hat{\mathbf{t}} &= 0\end{aligned}\right\} \quad \text{(boundary conditions for dielectric interface)}$$

*The normal component of* **D** *changes by the free surface charge density at an interface. The tangential component of* **E** *is continuous across the interface.*

$\mathbf{P} = n_V \alpha \mathbf{E}_M$

*The dipole moment per unit volume* **P** *is equal to the product of number density (molecules per unit volume),* $n_V$, *the polarizability* $\alpha$, *which parameterizes the microscopic response of the molecule to the field, and the electric field* $\mathbf{E}_M$ *acting on a given molecule.*

$$\kappa = \frac{1 + 2n_V\alpha/(3\varepsilon_0)}{1 - n_V\alpha/(3\varepsilon_0)} \qquad \text{(Clausius-Mossotti relation)}$$

*The CM relation specifies the macroscopic dielectric constant* $\kappa$ *in terms of the microscopic polarizability* $\alpha$. *This relation holds as long as* $\mathbf{E}'_M$, *the net electric field due to nearby molecules, is negligible.*

Induced static polarizability

*When a static electric* $\mathbf{E}_M$ *acts on an atom, the nucleus and the electrons are displaced in opposite directions, inducing a static dipole moment.*

$\alpha_0 = 4\pi\varepsilon_0 a_0^3$

*The static polarizability* $\alpha_0$ *is approximately equal to* $4\pi\varepsilon_0$ *times the cube of the atomic radius* $a_0$, *which is approximately the volume.*

Induced dynamic polarizability

$$\alpha(\omega) = \frac{e^2/m_e}{\omega_0^2 - \omega^2 - 2i\gamma\omega} \qquad \mathbf{P} = n_V \operatorname{Re}[\alpha(\omega)\mathscr{E}e^{-i\omega t}] \qquad \omega_0 = \left(\frac{e^2}{m_e\alpha_0}\right)^{1/2}$$

*When a time-varying electric field* $\mathbf{E}_M(t) = \operatorname{Re}(\mathscr{E}e^{-i\omega t})$ *acts on an atom, a time-varying dipole moment is induced. If the applied field* $\mathbf{E}_M(t)$ *varies rapidly enough with time, the polarization vector* $\mathbf{P}$ *will lag the electric field. Here* $\omega_0$ *is the resonant frequency, and* $\gamma$ *is the damping constant. The resonant wavelength* $\lambda = 2\pi c/\omega_0$ *is of order* $10^{-7}$ m.

Induced ionic polarizability

$$\omega_0^{\text{ionic}} = \left(\frac{e^2}{M\alpha_0}\right)^{1/2}$$

*A static electric field applied to an ionically bonded molecule causes the positive and negative ions to move in opposite directions, inducing a dipole moment. Here* $M$ *is the reduced mass of the ionic pair. The resonant wavelength is of order* $10^{-5}$ m *for* NaCl.

Orientational polarizability

*Polar molecules have permanent dipole moments that an electric field partially aligns, thus leading to orientational polarizability. A small applied field cannot completely align the dipoles because of random thermal motions, which increase with temperature.*

$$\alpha_{or} = \frac{p^2}{3kT}$$

*Orientational polarizability, where* $k = 1.38 \times 10^{-23}$ *J/K is Boltzmann's constant and* $T$ *is the absolute temperature.*

## PROBLEMS

### Section 4-1: Electric Dipoles and Multipoles

**4-1** A system of static charges is distributed along the $x$ axis in the interval $-a \leqslant x_1 \leqslant a$. The charge density is

$$\begin{array}{lll} \rho(x_1) & \text{for} & |x_1| \leqslant a \\ 0 & \text{for} & |x_1| > a \end{array}$$

(i) Write an expression for the electrostatic potential $\Phi(x)$ at a point $x$ on the axis in terms of $\rho(x_1)$.

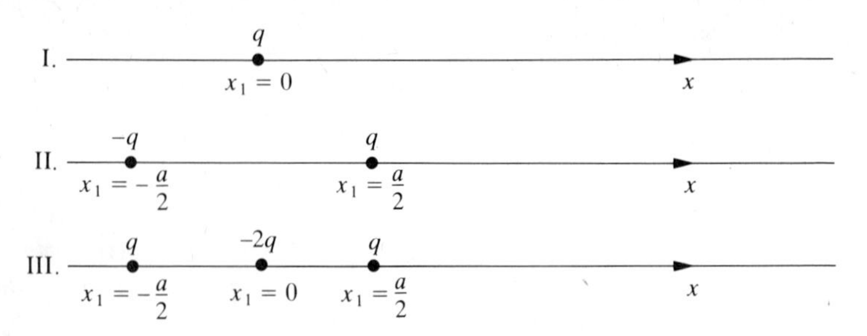


**FIGURE 4-20** Linear configurations of point charges

(ii) Derive a multipole expansion for the potential valid for $x > a$. (Expand your expression for the potential in powers of $1/x$.)

(iii) Consider each charge configuration illustrated in Figure 4-20.
- (a) Find the total charge $Q = \int \rho \, dx$.
- (b) Find the dipole moment $p = \int x\rho \, dx$.
- (c) Find the quadrupole moment $Q_{xx} = 2 \int x^2 \rho dx$.
- (d) Find the leading term (in powers of $1/x$) in the potential $\Phi$ at a point $x \gg a$.

**4-2** For a system of charges, show that the dipole moment $\mathbf{p} = \int \mathbf{r}\rho(\mathbf{r})\, dV$ is independent of the location of the coordinate system origin if the total charge of the system is zero. As an application, consider a ring of negative charge $-q$ of radius $a$ that lies in the $xy$ plane centered at the origin. A charge $+q$ lies inside the ring at $x = +d$. Find the electric monopole and dipole moments of this system, and give an expression for the electric potential to this order.

**4-3** A charge $-4q$ is uniformly distributed on the surface of a sphere of radius $a$ with center at the origin. Two point charges, each $+q$, are located at ($x = a/2$, $y = 0$, $z = 0$) and ($x = 0$, $y = a/2$, $z = 0$), respectively. Find the electric monopole and dipole moments of the system about the origin. Write the expression for the electric potential due to these moments.

**4-4** Show that the electric dipole field must have the delta function contribution at $\mathbf{r} = 0$ given by Equation (4-18). (*Hint:* Recall that $\mathbf{E} = -\nabla\Phi$, and apply the divergence theorem corollary $\int \mathbf{E}\, dV = -\oint \Phi \, d\mathbf{S}$ to a spherical region. In doing the surface integral, remember that the unit vector $\hat{\mathbf{r}}$ changes direction over the surface of the sphere. If the dipole lies along the $z$ axis, only the $z$ component of $\mathbf{E}$ will be nonzero.)

**4-5** Consider an electric dipole of magnitude $p$ situated a distance $L$ from a grounded sphere of radius $a$ and directed radially outward. Show that the image charges necessary to satisfy the boundary conditions are a dipole of magnitude $(a/L)^3 p$ and a point charge $ap/L^2$, both located at a distance $a^2/L$ from the center of the sphere on a radius through the dipole.

**4-6** A dipole of moment $p$ is at the center of a spherical cavity of radius $a$ in a conductor. Find the electric field inside the sphere due to the induced charges on the cavity walls.

**4-7** An infinite sheet of electric dipoles has a dipole moment per unit area $\mathbf{p}$ directed along the normal. Find the potential and the electric field everywhere in space. What is the result if the dipole density is directed uniformly along an axis in the sheet?

## Section 4-2: Macroscopic Dielectric Theory — Energy and Dielectrics

**4-8** For a uniform dielectric, show that the polarization volume charge density $\rho_P$ vanishes outside a charge distribution that causes the polarization.

**4-9** Show that the electrostatic energy density with polarizable material present is $\frac{1}{2}\mathbf{E} \cdot \mathbf{D}$. (*Hint:* Start with $U = \frac{1}{2} \int \rho\Phi \, dV$ and proceed as in the derivation in Section 2–11.)

**4-10** Find the electric force on a dielectric slab of thickness $t$ partially inserted between the plates of an isolated parallel-plate capacitor of separation $t$. (*Hint:* Using the result of the previous problem, compute the field energy change when the slab moves a distance $dx$ into the capacitor. The change in field energy is due to mechanical work done on the slab.)

**4-11** Show that a dielectric object is attracted to a point charge. This demonstration explains why pieces of paper are attracted to a charged rod. (*Hint:* Use an energy argument based on the change of field energy $\frac{1}{2} \int \mathbf{E} \cdot \mathbf{D} \, dV$; see Problem 4–9.)

## Section 4-3: Gauss's Laws in Dielectric Matter

**4-12** Show that the average (macroscopic) electric field in a dielectric is obtained by a measurement of the electric field in a longitudinal slot (the field direction lies in the plane of the slot). Show that the displacement field is determined by measuring the electric field in a transverse slot (the field direction is perpendicular to the slot).

**4-13** A parallel-plate capacitor of plate separation $d_1$ is filled with a solid dielectric material of permittivity $\varepsilon$. The capacitor is charged to voltage $V_1$ and then *disconnected* from the supply and pulled apart, so the plate separation becomes $d_2$ with a dielectric-free gap of distance $d_2 - d_1$. Assuming the lateral dimensions of the plate and the dielectric are large compared with both $d_1$ and $d_2$, compute the voltage ($V_2$) after the capacitor is pulled apart.

**4-14** An isolated parallel-plate capacitor has separation $d$ and is charged to a potential difference $V_0$. A dielectric of permittivity $\varepsilon$ and thickness $d/2$ is inserted. Find the potential difference after the dielectric is in place. What is the smallest possible potential difference that can be obtained by varying $\varepsilon$?

**4-15** Consider a neutral, hollow dielectric sphere, with outer radius $2R$, inner radius $R$, and dielectric constant $\kappa > 1$, which surrounds a point charge $Q$ at its center.
(i) Find the surface charge density on the inner surface of the dielectric.
(ii) Find the potential at a point located at $r = \frac{1}{2}R$.
(iii) How would the previous results differ if the point charge were replaced by a conducting sphere of radius $R$ and charge $Q$?

**4-16** The boundary at $z = 0$ of two dielectrics (dielectric constants $\kappa_1$ for $z < 0$ and $\kappa_2$ for $z > 0$) is free of surface charge. The displacement field vector $\mathbf{D}$ in region 1 is given by

$$\mathbf{D}_1 = 5x^2\hat{\mathbf{x}} + 5y\hat{\mathbf{y}} + (3 + z)\hat{\mathbf{z}}$$

Find $\mathbf{D}_2$ and $\mathbf{E}_2$ at the boundary ($z = 0$) of region 2.

**4-17** A total charge $Q$ is placed on a hollow metal sphere floating in a liquid of dielectric constant $\kappa$. The sphere is half-submerged. How does the charge distribute itself on the sphere? (*Hint:* This problem is not difficult if you make a simple assumption about the electric field.)

**4-18** A charge $q_1$ is located in air a height $a$ above a semi-infinite slab of dielectric constant $\kappa$. At a depth $a$ in the dielectric, on a line normal to the surface and passing through $q_1$, is located a second charge $q_2$. Find the force on the charge $q_1$. (*Hint:* Use the method of images and apply the proper boundary conditions.)

**4-19** A sphere of dielectric constant $\kappa$ is placed in an initially uniform electric field $\mathbf{E}_0$.
(i) Show that the induced surface charge density is

$$\sigma_P = \frac{3\varepsilon_0(\kappa - 1)}{\kappa + 2} E_0 \cos\theta$$

where $\theta$ is measured from the $\mathbf{E}_0$ direction.
(ii) Show that the induced field outside the sphere is that of a dipole, and find the magnitude of the dipole moment.

**4-20** A dielectric sphere of radius $a$ has a permanent radial polarization (i.e., dipole moment per unit volume) of

$$\mathbf{P} = f(r)\hat{\mathbf{r}}$$

where $\mathbf{r}$ is the spherical radius vector from the center of the sphere.
(i) Use Gauss's law for $\mathbf{D}$ to determine the displacement field both inside and outside the sphere.
(ii) Using the result of (i), find $\mathbf{E}$ inside and outside the sphere.
(iii) Determine the polarization surface charge density $\sigma_P$ on the sphere.

**4-21** A sheet of dielectric, of thickness $2a$, has infinite extent in the $xy$ plane and a permanent, uniform dipole moment density $\mathbf{P} = P\hat{\mathbf{z}}$. (*Note:* A permanent dipole moment is a dipole moment that is present in the absence of an applied $\mathbf{E}$ field.)
(i) Find the polarization surface and volume charge densities, $\sigma_P$ and $\rho_P$.
(ii) Find $\mathbf{E}$ and $\mathbf{D}$ everywhere.
(iii) Find the electrostatic potential $\Phi$ everywhere.

**4-22** A long cylindrical conductor of radius $a$, with charge $\lambda$ per unit length, is immersed in a dielectric medium of constant permittivity $\varepsilon$. Use Gauss's law for $\mathbf{D}$ to find the electric field at distance $r > a$ from the axis of the cylinder.

**4-23** Find the electric field inside and outside a dielectric sphere of radius $R$ that has a uniform polarization vector $\mathbf{P}$.

**4-24** The space between two long, thin metal cylinders is filled with a material with dielectric constant $\varepsilon$. The cylinders have radii $a$ and $b$.
(i) What is the charge per unit length on the cylinders when the potential between them is $V$, with the outer cylinder at the higher potential?
(ii) What is the electric field between the cylinders?

## Section 4-5: Microscopic Polarizability Models

**4-25** Compute the polarizability of a conducting sphere of radius $a_0$. Compare your result with the static dielectric polarizability obtained in the text.

**4-26** Derive the expression $\varepsilon(\omega) = 1 - \omega_p^2/\omega^2$; $\omega_p^2 = n_v e^2/\varepsilon_0 m$ for the dielectric constant as a function of $\omega$ for a free-electron gas of number density $n_v$. Then, show that metals are opaque to light for which $\omega$ is less than $\omega_p$. Finally, calculate the wavelength cutoff for Na metal where the volume per free electron is $35 \times 10^{-30}$ m$^3$.

# Magnetic Fields and Vector Potential from Steady Currents

CHAPTER CONTENTS

Magnetism and electricity were considered to be unrelated subjects until the discovery by Hans Christian Oersted in 1820 that an electric current through a wire deflected a magnetic compass needle. The force acting on a magnet due to a current element was deduced very shortly thereafter by Jean Baptiste Biot and Félix Savart from measurements near current-carrying wires. Upon hearing of Oersted's discovery, André Marie Ampère began a series of experiments on the force between wires carrying electric currents and determined the force law between two current elements. At the time the Newtonian approach to physics prevailed, in which all interactions were expressed in terms of forces acting at a distance. The concept of an intermediating electromagnetic field, and its description in terms of field lines, came many years later from the English physicist Michael Faraday. Eventually, the results of Ampère and Biot-Savart were expressed in terms of the magnetic field due to a current element and the force on a current element due to the magnetic field; the magnetic field formula became known as the Biot-Savart law. Actually, the work of Ampère had far greater impact in the subsequent developments in electromagnetic theory, because only Ampère's results were widely known outside France.

The study of magnetic fields due to steady currents is known as magnetostatics and is the central theme of this chapter. After discussing the equation of continuity and the fundamental laws of magnetostatics, we consider current distributions that produce some magnetic fields of practical interest. We introduce the vector potential **A** from which a magnetostatic field **B** can be derived. From the Biot-Savart law we obtain the curl of **B** in terms of the current density **J**. This relation is then recast in an integral form, known as Ampère's circuital law. Steady current flow in a conductor is discussed as an example of a continuity equation for electric charge and current density.

## 5-1 MAGNETIC FORCE ON CURRENTS AND THE CONTINUITY EQUATION

In ordinary matter the charge density and the fields vary dramatically over distances of the order of $10^{-10}$ m. But the observations of fields and sources in the laboratory often reflect the combined effects of a large number of charges and not the details of the actual charge distributions and fields on a microscopic (molecular) scale. Most measuring-instrument probes are macroscopic; they are far larger than molecules and measure averages of microscopic quantities. A macroscopic quantity $\mathbf{A}(\mathbf{r})$ is defined as the space average of the corresponding microscopic quantity $A_{\mathrm{m}}(\mathbf{r})$ over a macroscopic volume $\Delta V$,

$$\mathbf{A}(\mathbf{r}) \equiv \frac{1}{\Delta V} \int_{\Delta V} A_{\mathrm{m}}(\mathbf{r} - \mathbf{r}')\, dV' \qquad \textbf{(5-1)}$$

The volume $\Delta V$ is taken to be small by macroscopic standards but large enough to contain many molecules. (In ordinary matter a sphere having the radius of a single wavelength of visible light will contain billions of molecules.)

The total electromagnetic force $\mathbf{F}$ on a collection of point charges $q_i$ at positions $\mathbf{r}_i(t)$ and velocities $\mathbf{v}_i = \dot{\mathbf{r}}_i$ is, by Equation (1–1),

$$\mathbf{F} = \sum_i q_i[\mathbf{E}_m(\mathbf{r}_i) + \mathbf{v}_i \times \mathbf{B}_m(\mathbf{r}_i)] \tag{5-2}$$

where the fields acting on the charges are the microscopic fields. We can write Equation (5–2) as

$$\mathbf{F} = \int [\rho_m(\mathbf{r}, t)\mathbf{E}_m(\mathbf{r}) + \mathbf{J}_m(\mathbf{r}, t) \times \mathbf{B}_m(\mathbf{r})]\, dV \tag{5-3}$$

where the microscopic charge and current densities, $\rho_m$ and $\mathbf{J}_m$, are defined as

$$\begin{aligned} \rho_m(\mathbf{r}, t) &= \sum_i q_i \delta^3[\mathbf{r} - \mathbf{r}_i(t)] \\ \mathbf{J}_m(\mathbf{r}, t) &= \sum_i q_i \mathbf{v}_i \delta^3[\mathbf{r} - \mathbf{r}_i(t)] \end{aligned} \tag{5-4}$$

The macroscopic charge density $\rho(\mathbf{r})$ is defined, à la Equation (5–1), as the average of the microscopic charge density defined in Equation (5–4),

$$\rho(\mathbf{r}) = \frac{1}{\Delta V}\int_{\Delta V} \sum_i q_i \delta(\mathbf{r} - \mathbf{r}' - \mathbf{r}_i)\, dV' = \frac{\sum_i q_i}{\Delta V} = \frac{\Delta q}{\Delta V} \tag{5-5}$$

where $\Delta q$ is the total charge contained in the volume $\Delta V$. This macroscopic charge density is the same as that introduced in Equation (1–64). Similarly, the macroscopic *current density* $\mathbf{J}(\mathbf{r})$ is given by

$$\mathbf{J}(\mathbf{r}) = \frac{1}{\Delta V}\sum_i q_i \mathbf{v}_i \tag{5-6}$$

By separating the contributions to $\rho$ and $\mathbf{J}$ of electrons and ions, we can express them as

$$\rho(\mathbf{r}) = \rho_e + \rho_{ion} \qquad \mathbf{J}(\mathbf{r}) = \rho_e \mathbf{v}_e + \rho_{ion}\mathbf{v}_{ion} \tag{5-7}$$

where the electron and positive-ion charge densities are

$$\rho_e = -\frac{N_e e}{\Delta V} \qquad \rho_{ion} = \frac{N_{ion} e}{\Delta V} \tag{5-8}$$

and the average velocities are defined as

$$\mathbf{v}_e = \frac{1}{N_e}\sum_{i(\text{electrons})} \mathbf{v}_i \qquad \mathbf{v}_{ion} = \frac{1}{N_{ion}}\sum_{i(\text{ions})} \mathbf{v}_i \tag{5-9}$$

Here $N_e$ and $N_{ion}$ are the number of electrons and ions, respectively, contained in $\Delta V$. (In materials such as ionic aqueous solutions there are also negative ions, whose contributions to $\rho$ and **J** must be added.) For a metallic conductor the average ion velocity is just the overall conductor velocity, since the ions have little relative motion. Since an ion is much more massive than an electron, the ions carry almost all of the momentum. The average electron velocity is a superposition of the conductor velocity and the drift velocity responsible for the conduction current in the metallic conductor. Conduction will be further discussed in Section 5–11.

To give additional physical interpretation to **J**, we derive a differential equation relating $\rho$ and **J**. The partial derivative of $\rho_m$ with respect to time is found from chain differentiation:

$$\frac{\partial \rho_m}{\partial t} = \sum_i q_i \left( \frac{dx_i}{dt}\frac{\partial}{\partial x_i} + \frac{dy_i}{dt}\frac{\partial}{\partial y_i} + \frac{dz_i}{dt}\frac{\partial}{\partial z_i} \right) \delta^3[\mathbf{r} - \mathbf{r}_i(t)]$$

$$= \sum_i q_i \mathbf{v}_i \cdot \nabla_{\mathbf{r}_i} \delta^3(\mathbf{r} - \mathbf{r}_i) \quad \textbf{(5–10)}$$

In deriving this result, we have assumed that the charges $q_i$ of the particles are not time-dependent. Since the argument of the delta function is the difference $\mathbf{r} - \mathbf{r}_i$, we can write Equation (5–10) as

$$\frac{\partial \rho_m}{\partial t} = -\sum_i q_i \mathbf{v}_i \cdot \nabla_{\mathbf{r}} \delta^3(\mathbf{r} - \mathbf{r}_i) \quad \textbf{(5–11)}$$

Comparison of this equation with the divergence of $\mathbf{J}_m$,

$$\nabla_{\mathbf{r}} \cdot \mathbf{J}_m = \sum_i q_i \mathbf{v}_i \cdot \nabla_{\mathbf{r}} \delta^3(\mathbf{r} - \mathbf{r}_i) \quad \textbf{(5–12)}$$

leads to the continuity equation:

$$\nabla \cdot \mathbf{J}_m + \frac{\partial \rho_m}{\partial t} = 0 \quad \textbf{(5–13)}$$

The continuity equation is a consequence of charge conservation, that is, $dq_i/dt = 0$. By space-averaging Equation (5–13), we see that a similar equation holds for the macroscopic quantities:

**Macroscopic Continuity Equation**

$$\nabla \cdot \mathbf{J} + \frac{\partial \rho}{\partial t} = 0 \quad \textbf{(5–14)}$$

Integrating the continuity equation over a volume $V$ and using the divergence theorem gives

$$\int \nabla \cdot \mathbf{J}\, dV = \oint \mathbf{J} \cdot d\mathbf{S} = -\frac{\partial}{\partial t} \int \rho(\mathbf{r}, t)\, dV \quad \textbf{(5–15)}$$

Since the total charge within $V$ is

$$Q = \int \rho(\mathbf{r}, t)\, dV \tag{5–16}$$

we can rewrite Equation (5–15) as

$$\oint \mathbf{J} \cdot d\mathbf{S} = -\frac{dQ}{dt} \tag{5–17}$$

From this relation we conclude that the left-hand side of this equation represents the rate at which charge leaves the volume. Thus the physical interpretation of **J** is the charge flow per unit normal area per unit time.

The integral of **J** over a surface [either closed, as in Equation (5–17), or open] is the rate at which charge is flowing through the surface, in other words, the *current* $I$ through the surface,

$$\int \mathbf{J} \cdot d\mathbf{S} = \frac{dq}{dt} = I \tag{5–18}$$

as illustrated in Figure 5–1. We note that for a closed surface the charge $dq$ passing out through the surface $S$ corresponds to a decrease $dQ = -dq$ of the charge contained in the volume bounded by $S$. Although **J** is called *current density*, note that it is current per unit *area*.

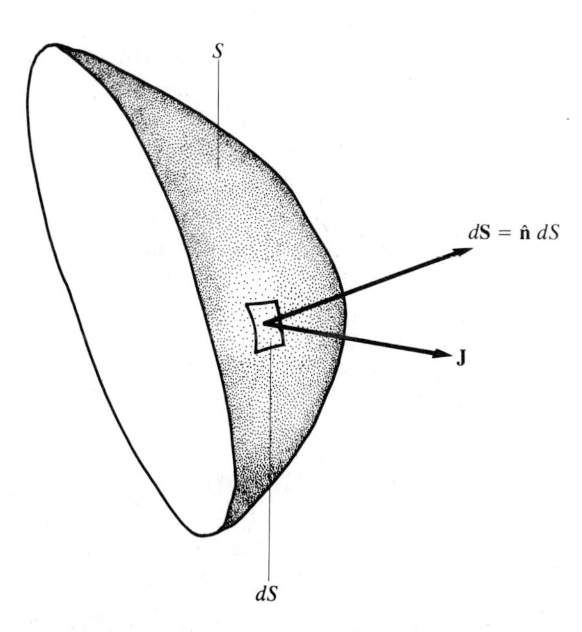


**FIGURE 5–1** Current through a surface $S$. The current is equal to the integral of the current density **J** dotted into the differential surface area $d\mathbf{S}$.

As we mentioned, the fields and densities appearing in the Lorentz force expression of Equation (5–3) are microscopic. If we write the microscopic quantities $\rho_{\text{m}}$, $\mathbf{E}_{\text{m}}$, $\mathbf{J}_{\text{m}}$, and $\mathbf{B}_{\text{m}}$ as the sum of the microscopic (space-averaged) values and fluctuations generically as

$$A_{\text{m}}(\mathbf{r}) = A(\mathbf{r}) + [A_{\text{m}}(\mathbf{r}) - A(\mathbf{r})] \tag{5-19}$$

then the *Lorentz force* acting on $\Delta V$ becomes

$$\begin{aligned}\Delta\mathbf{F} = (\rho\mathbf{E} + \mathbf{J}\times\mathbf{B})\,\Delta V &+ \int_{\Delta V} [(\rho_{\text{m}} - \rho)\mathbf{E} + (\mathbf{J}_{\text{m}} - \mathbf{J})\times\mathbf{B}]\,dV \\ &+ \int_{\Delta V} [\rho(\mathbf{E}_{\text{m}} - \mathbf{E}) + \mathbf{J}\times(\mathbf{B}_{\text{m}} - \mathbf{B})]\,dV \\ &+ \int_{\Delta V} [(\rho_{\text{m}} - \rho)(\mathbf{E}_{\text{m}} - \mathbf{E}) + (\mathbf{J}_{\text{m}} - \mathbf{J})\times(\mathbf{B}_{\text{m}} - \mathbf{B})]\,dV\end{aligned} \tag{5-20}$$

The first term in Equation (5–20) involves only macroscopically measurable quantities. The second and third terms vanish when integrated over the microscopic volume element $\Delta V$. The last term depends on microscopic interactions between charges. In some circumstances this term is as important as the purely macroscopic first term. Consider, for example, a steady conduction current through a wire. The battery provides a constant macroscopic electric field, and the associated force on the electrons given by the first term of Equation (5–20), with $\rho = \rho_{\text{e}}$, accelerates the electrons. The microscopic interactions of the conduction electrons with the ions in the lattice cause the electrons to locally deviate from straight trajectories. The corresponding averaged force from the last term in Equation (5–20) decelerates the electrons. Since, on the average, the conduction electrons are unaccelerated (i.e., the current is constant), the first and last terms of Equation (5–20) must cancel exactly.

The purely macroscopic differential force of Equation (5–20) is

$$d\mathbf{F}(\mathbf{r}) = (\rho\mathbf{E} + \mathbf{J}\times\mathbf{B})\,dV \tag{5-21}$$

For charges confined to a surface the corresponding result is

$$d\mathbf{F} = (\sigma\mathbf{E} + \mathbf{j}\times\mathbf{B})\,dS \tag{5-22}$$

where $\sigma$ is the surface charge density of Equation (1–65) and $\mathbf{j}$ is the surface current density. The surface current density $\mathbf{j}$ is the current passing through a curve lying in the conducting surface but perpendicular to $\mathbf{j}$. The surface current density is thus the current per unit length. Finally, for charges confined to a line (not necessarily straight), the current density has a component only along the line since charge is moving only in this direction. For a differential volume $dV = dS\,dr$, shown in Figure 5–2, the quantity $\mathbf{J}\,dV$ for charges

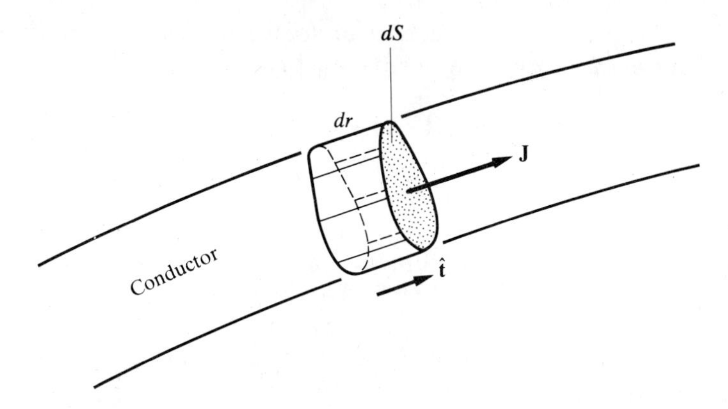


**FIGURE 5-2** Differential volume element for charges confined to a thin wire

on a line becomes

$$\mathbf{J}\,dV = (J\hat{\mathbf{t}})\,dS\,dr = I\,dr\,\hat{\mathbf{t}} = I\,d\mathbf{r} \tag{5-23}$$

where $I$ is the current along the line, that is, charge per unit time passing a point on the line, and $\hat{\mathbf{t}}$ is a unit vector tangent to the wire. Consequently, the force on a differential charge element of a line is

$$d\mathbf{F} = \lambda\mathbf{E}\,dr + I\,d\mathbf{r} \times \mathbf{B} \tag{5-24}$$

where $\lambda$ is the line charge density of Equation (1–66).

## 5-2 BIOT-SAVART LAW FOR MAGNETIC FIELD

The Lorentz force law gives the electromagnetic force on a charged particle in terms of **E** and **B** fields. Coulomb's law provides an expression for **E** in terms of the *electric charges*, if the charge distribution is static. In a similar way, the Biot-Savart law provides an expression for **B** in terms of the *electric currents*, if the current is steady (i.e., the current is time-independent) around a closed-circuit loop. Since there is no convincing evidence for the existence of magnetic charge analogous to electric charge, magnetic forces exist only when there is a current due to electric charge motion.

To have magnetic forces without electric forces is possible. For instance, current flow occurs in conductors even when the charge density in the conductor is zero. Any macroscopic volume in the conductor contains as many electrons as positive ions, resulting in electric charge neutrality. The positive ions forming the crystal lattice of the conductor are fixed, while the conduction electrons flow as a result of an applied external electric field.

The expression of Biot and Savart for the magnetic field produced by a steady current $I$ flowing in a closed circuit is

$$\mathbf{B}(\mathbf{r}) = \frac{\mu_0}{4\pi}\oint (I\, d\mathbf{r}') \times \frac{\hat{\mathbf{R}}}{R^2} \tag{5-25}$$

Here $\mu_0/4\pi$ is a constant of proportionality for empty space, and the line integral is around the circuit loop; see Figure 5–3. For magnetic fields in matter, as will be shown in Section 7–5, $\mu_0$ is replaced by $\mu$, an empirically determined parameter called the *permeability* that depends on the magnetic properties of matter. Compare Equation (5–25) with the expression for **E** from a line charge:

$$\mathbf{E}(\mathbf{r}) = \frac{1}{4\pi\varepsilon_0}\int \lambda(\mathbf{r}')\, dr' \frac{\hat{\mathbf{R}}}{R^2} \tag{5-26}$$

The generalization of the Biot-Savart law from line currents to general steady current distributions is obtained by replacing $I\, d\mathbf{r}$ in Equation (5–25) by $\mathbf{J}\, dV$ [see Equation (5–23)]:

**Biot-Savart Law**

$$\mathbf{B}(\mathbf{r}) = \frac{\mu_0}{4\pi}\int \mathbf{J}(\mathbf{r}') \times \frac{\hat{\mathbf{R}}}{R^2}\, dV' \tag{5-27}$$

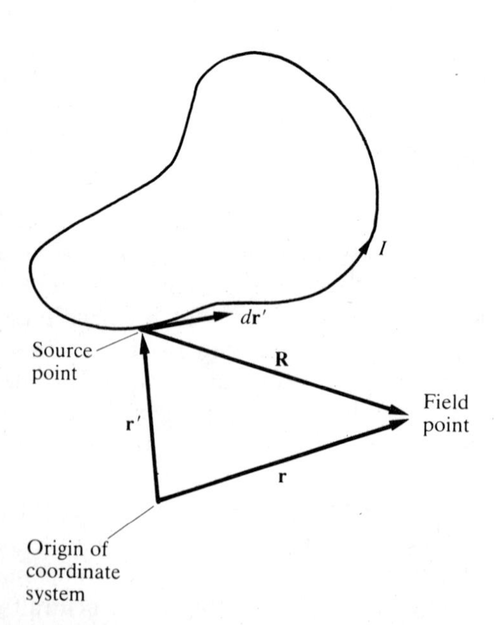


**FIGURE 5-3** Magnetic field at a position **r** due to a closed current-carrying circuit at **r**′ as determined from the Biot-Savart law

## 5-3 APPLICATION: MAGNETIC FIELD OF A STRAIGHT WIRE AND THE FORCE BETWEEN PARALLEL WIRES

As an example, we use Equation (5–25) to calculate the **B** field due to an infinitely long straight wire. Use of cylindrical coordinates is advantageous:

$$\mathbf{r} = (\imath, \phi, z)$$

where $\imath$ is the perpendicular distance to the field point from the wire; see Figure 5–4. We have

$$\mathbf{B}(\mathbf{r}) = \frac{\mu_0}{4\pi} \int (I\, d\mathbf{r}') \times \frac{\hat{\mathbf{R}}}{R^2}$$

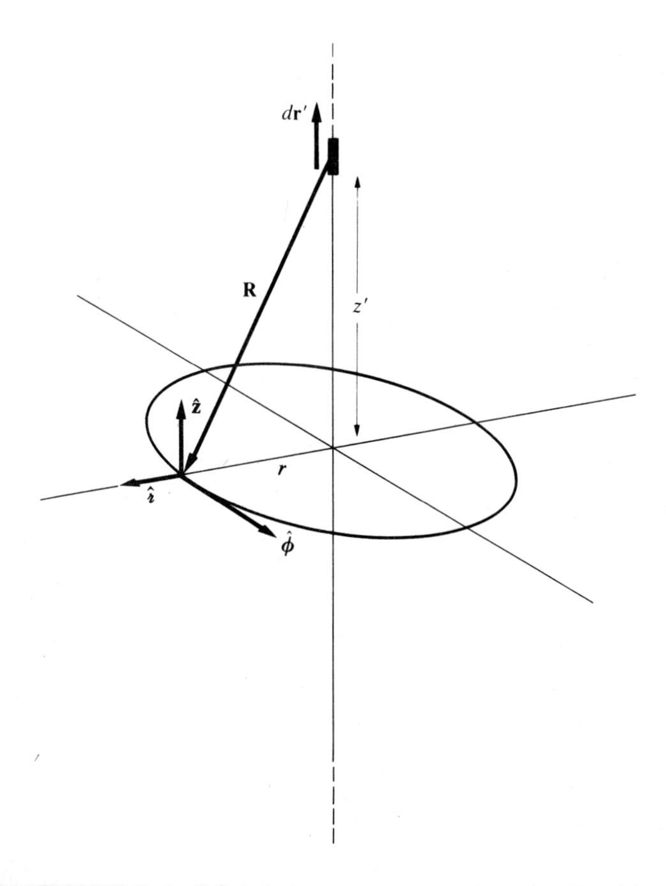


**FIGURE 5–4** Infinitely long, current-carrying wire. The Biot-Savart law is used to determine its magnetic field.

where

$$R^2 = \imath^2 + z'^2 \qquad d\mathbf{r}' = \hat{\mathbf{z}}\, dz' \qquad d\mathbf{r}' \times \hat{\mathbf{R}} = \hat{\boldsymbol{\phi}} \frac{\imath}{R} dz' \tag{5-28}$$

Thus **B** is given by

$$\mathbf{B}(\mathbf{r}) = \frac{\mu_0 I}{4\pi} \hat{\boldsymbol{\phi}} \int_{-\infty}^{+\infty} \frac{\imath\, dz'}{(\imath^2 + z'^2)^{3/2}} = \frac{\mu_0 I}{4\pi} \hat{\boldsymbol{\phi}} \left. \frac{z'}{\imath(\imath^2 + z'^2)^{1/2}} \right|_{z'=-\infty}^{z'=+\infty}$$

or

$$\mathbf{B}(\mathbf{r}) = \frac{\mu_0 I}{2\pi \imath} \hat{\boldsymbol{\phi}} \tag{5-29}$$

The magnetic field around the wire is azimuthal, with the directions of **B** and the current related by the right-hand rule, as shown in Figure 5–5.

We can use Equation (5–29) to calculate the force per unit length between two long current-carrying wires; see Figure 5–6. From Equation (5–24) the electromagnetic force on wire 2 due to wire 1 is

$$\mathbf{F}_{21} = \int (I_2\, d\mathbf{r}_2) \times \mathbf{B}(\mathbf{r}_2) \tag{5-30}$$

where the field at $\mathbf{r}_2$ due to $I_1$ is given by

$$\mathbf{B}(\mathbf{r}_2) = \frac{\mu_0 I_1}{2\pi \imath} \hat{\boldsymbol{\phi}} \tag{5-31}$$

where $\imath$ is a cylindrical radius vector from wire 1 to wire 2. Since $d\mathbf{r}_2 \times \hat{\boldsymbol{\phi}} = -\hat{\boldsymbol{\imath}}\, dz_2$, we obtain

$$\mathbf{F}_{21} = -\frac{\mu_0 I_1 I_2}{2\pi \imath} \hat{\boldsymbol{\imath}} \int dz_2 \tag{5-32}$$

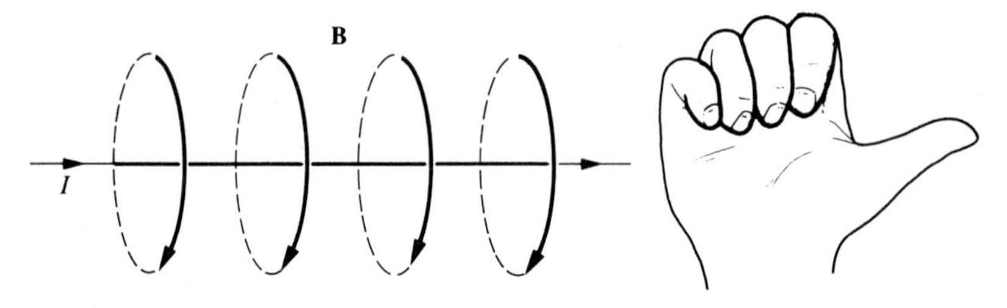


**FIGURE 5-5** Right-hand rule for the direction of the magnetic field through a closed, current-carrying circuit. With the thumb in the direction of the current flow, the fingers of the right hand curl in the direction of the magnetic field.

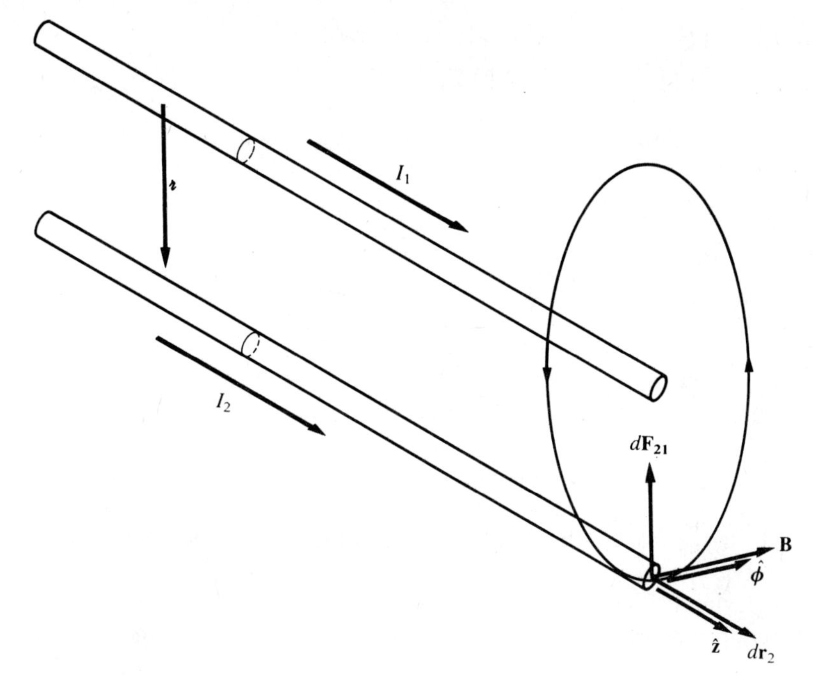


FIGURE 5-6 Geometry for calculating the force between two long current-carrying wires

and the force per unit length on wire 2 due to wire 1 is

**Force Per Unit Length Between Parallel Wires**

$$\mathbf{F}_{21} = -\frac{\mu_0 I_1 I_2}{2\pi \imath}\hat{\imath} \tag{5-33}$$

If the currents are in the same direction, the wires attract each other; wires with currents in opposite directions repel.

The general expression for the force between two closed circuits $C_1$ and $C_2$ carrying steady current $I_1$ and $I_2$ is obtained from Equations (5–25) and (5–30):

**Ampère's Force Law**

$$\mathbf{F}_{21} = \mu_0 \frac{I_1 I_2}{4\pi} \oint_{C_2} \oint_{C_1} \frac{d\mathbf{r}_2 \times (d\mathbf{r}_1 \times \hat{\mathbf{R}})}{R^2} \tag{5-34}$$

where $\mathbf{R} = \mathbf{r}_2 - \mathbf{r}_1$. This relation is Ampère's force law; it was in this form that Ampère presented his magnetic results in 1822. Ampère's result may be viewed as the combination of the Biot-Savart result and the magnetic term of the Lorentz force law.

## 5-4 UNITS OF ELECTRICAL AND MAGNETIC QUANTITIES

The mechanical quantities in the metric system are as follows:

| Quantity | SI System | cgs System |
|---|---|---|
| Length | Meter (m) | Centimeter (cm) |
| Mass | Kilogram (kg) | Gram (g) |
| Time | Second (s) | Second (s) |

In mechanics all other physical quantities can be expressed in terms of these units. For example, from $\mathbf{F} = m\ddot{\mathbf{r}}$ the SI unit of force is 1 kg·m/s$^2$, which is called 1 newton (N).

When electromagnetic forces are involved, the introduction of a new unit is required. In the SI system it has been chosen to be the unit of electric current—the ampere (A). The ampere is defined in terms of the force between two infinite parallel wires 1 meter apart in empty space that are carrying the same current. Specifically, a current of 1 ampere is the current that produces a force per meter of wire equal to $2 \times 10^{-7}$ newtons per meter.

The force per unit length between parallel wires each carrying current $I$ (see Section 5–3) is

$$|\mathbf{F}| = \frac{\mu_0 I^2}{2\pi \mathscr{r}}$$

From the definition of the ampere and this equation, we can solve for the numerical value of the constant $\mu_0$:

$$\frac{\mu_0}{4\pi} = \frac{|\mathbf{F}|\mathscr{r}}{2I^2} = \frac{(2 \times 10^{-7}\ \text{N/m})(1\ \text{m})}{2(1\ \text{A})^2} = 10^{-7}\ \text{N/(A)}^2 \qquad \textbf{(5-35)}$$

in SI units.

The unit of charge in SI units is given in terms of the units of current and time [see Equation (5–18)] as

$$1\ \text{ampere-second} \equiv 1\ \text{coulomb (C)}$$

That is, a charge of 1 coulomb passes a given point in 1 second in a wire carrying a current of 1 ampere.

Once the unit of charge is given in terms of the ampere, the proportionality constant $\varepsilon_0$ in Coulomb's law,

$$F = \frac{1}{4\pi\varepsilon_0}\frac{qq_1}{R^2}$$

must be chosen so that the electric force in newtons between two known charges agrees with the measured force. In SI units the numerical value of $\varepsilon_0$, to four significant figures, is

$$\varepsilon_0 = 8.854 \times 10^{-12} \text{ C}^2/(\text{N}\cdot\text{m}^2) \tag{5-36}$$

Actually, $\varepsilon_0$ is known to an accuracy of one part in $10^9$, not directly from electrostatic measurements but from measurements of the speed of light. From Maxwell's theory the speed of light $c$ in a vacuum is

$$c = \frac{1}{\sqrt{\varepsilon_0 \mu_0}}$$

as will be discussed later. For convenience, we frequently will use the approximation

$$\frac{1}{4\pi\varepsilon_0} \approx 9 \times 10^9 \text{ N}\cdot\text{m}^2/\text{C}^2 \tag{5-37}$$

which is good to 0.2%.

In summary, the equations of electromagnetism introduce one new unit and one fundamental constant. In the SI system the unit is the ampere and the fundamental constant is $\varepsilon_0$.

From the definition of the electric field through the Lorentz force law,

$$\mathbf{F} = q\mathbf{E}$$

the units of **E** are newtons per coulomb. From the formula for potential difference in a static **E** field,

$$d\Phi = -\mathbf{E} \cdot d\mathbf{r}$$

or just from the fact that $\Phi$ is potential energy per unit charge, the units of $\Phi$ are specified as newton-meter per coulomb, which is called a volt (V):

$$1 \text{ newton-meter per coulomb} \equiv 1 \text{ volt}$$

Normally, electric field strength is quoted in volts per meter. The unit of capacitance is the farad (F):

$$1 \text{ coulomb per volt} \equiv 1 \text{ farad}$$

From the magnetic part of the Lorentz force,

$$\mathbf{F} = I\, d\mathbf{r} \times \mathbf{B}$$

the SI unit of the magnetic field is

$$1 \text{ newton per (ampere-meter)} \equiv 1 \text{ tesla (T)}$$

**TABLE 5-1** Representative electrical quantities

|  |  |
|---|---|
| **Typical Currents** | |
| High-voltage transmission line (500,000 V) | $10^2$ A |
| House (220 V) | 50 A |
| 100-W light bulb | 1 A |
| Flashlight | $10^{-3}$ A |
| Human nerve fiber | $10^{-6}$ A |
| **Typical Charges** | |
| Top of thundercloud | 40 C |
| Van de Graaff static-electricity generator | $10^{-6}$ C |
| Electron | $-1.6 \times 10^{-19}$ C |
| **Typical Electric Fields** | |
| Rotating neutron star (pulsar) | $10^{14}$ V/m |
| Hydrogen atom | $5 \times 10^{11}$ V/m |
| Capacitor | $10^6$ V/m |
| Region between thundercloud and earth | $10^4$ V/m |
| **Typical Magnetic Fields** | |
| Neutron star | $10^8$ T |
| White dwarf star | $10^3$ T |
| Superconducting magnet | 5 T |
| Cupboard magnetic latch | 0.1 T |
| Earth | $10^{-4}$ T = 1 G |
| Galactic field | $10^{-10}$ T |

In practice, magnetic fields are often expressed in a smaller unit:

$$10^{-4} \text{ tesla} = 1 \text{ gauss (G)}$$

Magnetic flux $d\mathscr{F} = \mathbf{B} \cdot d\mathbf{S}$, where $d\mathbf{S}$ is a differential area element, has the SI unit

$$1 \text{ tesla-meter}^2 \equiv 1 \text{ weber (Wb)}$$

Table 5–1 lists some representative electrical and magnetic quantities and their SI units.

## 5-5 MAGNETIC FIELDS IN THE LABORATORY AND IN NATURE

Wherever we look in the universe, we find magnetic fields. A vast range of magnetic field strengths is seen, ranging from zero inside a superconductor, to a few times $10^{-10}$ T for the galactic field, to more than $10^8$ T near pulsars. While there are cosmic magnetic fields, there are no appreciable cosmic electric fields. There is a simple explanation for this phenomenon. When electric fields exist, free charge flows to reduce the field; but there are no magnetic charges that can flow.

Astronomical magnetic fields are usually measured by observing spectral line splittings (the Zeeman effect; see Problem 1–2). Of known magnetic fields, the one with the largest spatial dimension is the galactic field; its average strength is about $3 \times 10^{-10}$T and is approximately azimuthal in the galactic plane. This weak but widely extended field is able to confine the very energetic particles of the cosmic rays to our galaxy. Since our galaxy has a radius of about 50,000 light years, we find, from Equation (1–15), that cosmic rays having a single electronic charge can have momenta as high as

$$pc = (erB_\phi)c \approx 2 \times 10^{19}\ \text{eV}$$

where $c$ is the speed of light. Very energetic rays entering the earth's atmosphere initiate extensive showers of particles on the earth's surface. Some showers have been observed exceeding the above momentum. Such cosmic rays may be extragalactic in origin.

The earth and several other planets have magnetic fields with strengths of $10^{-4}$ to $10^{-3}$ T at their surfaces. The sun also has an average field of this order, but much larger fields are observed in sunspots. Other normal stars probably have surface magnetic fields ranging from $10^{-4}$ to $10^{-2}$ T. These magnetic fields are not static. The earth's field reverses its direction every $10^5$ to $10^6$ years, while the sun's dipole field changes sign every 11 years and is related to the sunspot cycle of the same length.

The existence of the magnetic field of the earth and its variation with time presents a profound problem. Permanent magnetization of the iron core of the earth is not a possible explanation because the core temperature is far above the highest temperature (Curie temperature) at which permanent magnetization occurs in any known material. Hence the magnetic field must be due to electric currents, but any electric currents even in the metallic core would decay in the order of $10^4$ to $10^5$ years through resistive dissipation. Direct evidence from rock magnetism shows that earth has had a magnetic field of roughly the present strength, though of varying direction, for at least the past $10^8$ years.

The crucial features in the accepted explanation of the earth's field is that the core is rotating (with the rest of the earth), has an energy source, is molten, and is highly conductive. Other astronomical bodies with magnetic fields—the sun, Jupiter, the stars, and so on—share these features. The source of the field is a type of self-excited dynamo: The convective motion of the fluid conductor through the magnetic field induces electric currents, which in turn produce a magnetic field.

To develop a feeling for this mechanism, consider a self-excited Faraday disk generator as shown in Figure 5–7, consisting of a conducting disk on a metal axle that is driven by an external torque. Sliding contacts provide a conducting path from the axle through a coaxial coil and back to

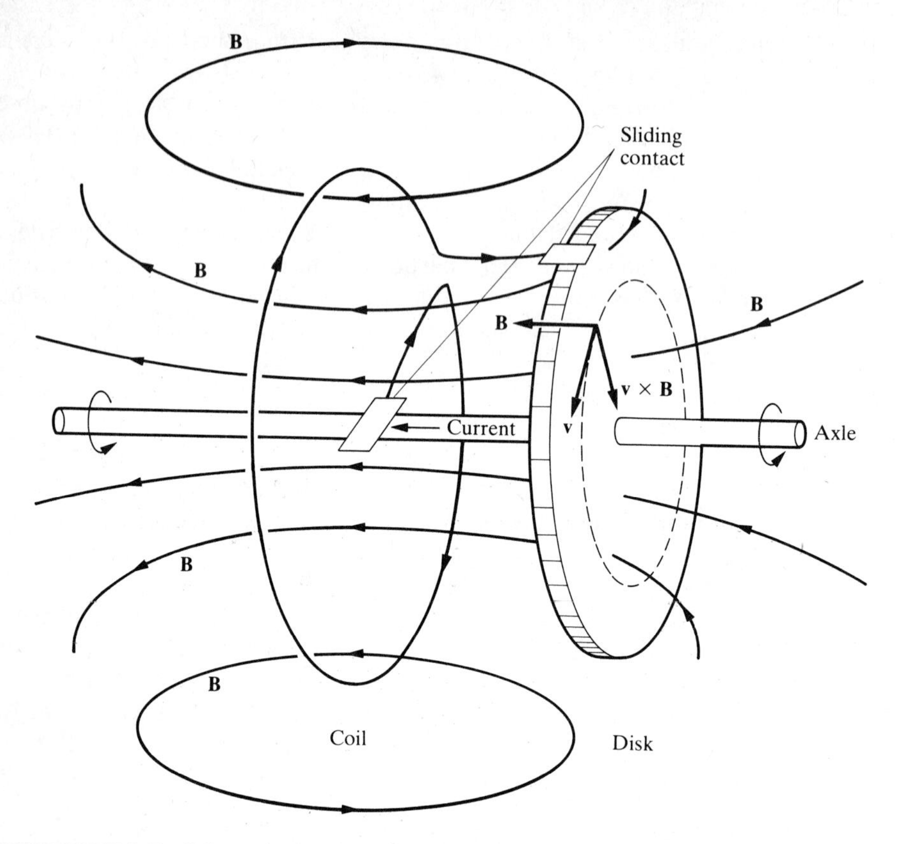


**FIGURE 5-7** Schematic drawing of a self-excited Faraday disk dynamo

the rim of the disk. Suppose there is initially a small axial magnetic field. As the disk rotates, charge carriers in the disk experience a radial Lorentz force. The resulting current in the loop completes the circuit and increases the magnetic field. The device is thus unstable at zero magnetic field; a small initial field will grow.

In the laboratory, magnetic fields of up to about 3 T are not difficult to produce by direct-current electromagnets with iron cores, but ferromagnetic effects do not help in reaching higher fields. By means of superconductors and/or inputting megawatts of power, steady fields up to 25 T have been achieved. Higher fields can only be produced for short periods of time. Pulsed fields of $10^2$ T are achieved typically for times less than 1 ms. At $10^2$ T the magnetic energy density is comparable to the energy density of high explosives and must be accompanied by currents of roughly $10^6$ A. One useful way of reaching still higher fields is known as flux compression. An initial axial magnetic field of a few tesla is established inside a hollow conducting cylinder. The cylinder is then radially compressed by explosives, and the magnetic flux trapped inside the conductor is compressed, producing a very great magnetic field strength for a period of about a microsecond. Fields of about $10^3$ T have been produced in this way.

The highest magnetic fields known in nature are at the surface of neutron stars. These fields are of the order of $10^8$ T.

## 5-6 APPLICATION: MAGNETIC FIELD ON THE AXIS OF A CURRENT LOOP

The magnetic field associated with a current flowing in a circular loop can be used in determining the field of more complex circular current configurations via the superposition principle. We now calculate the **B** field on the $z$ axis of a current loop of radius $a$ in the $xy$ plane and centered at the origin, as illustrated in Figure 5–8. In the Biot-Savart law, (5–25), we substitute

$$d\mathbf{r}' = a\,d\phi'\hat{\boldsymbol{\phi}}', \qquad R^2 = z^2 + a^2 \tag{5-38}$$

with $\phi'$ the azimuthal angle of the current element and $\hat{\boldsymbol{\phi}}'$ the unit vector tangent to the loop. The magnetic field of the loop is

$$\mathbf{B} = \frac{\mu_0 I a}{4\pi(z^2 + a^2)} \oint d\phi'\hat{\boldsymbol{\phi}}' \times \hat{\mathbf{R}} \tag{5-39}$$

After integration over $\phi'$, the only surviving field component will be $B_z$; the magnetic field on the $z$ axis cannot have any $(x, y)$ components because of the circular symmetry around the $z$ axis. To determine $B_z$, we note that

$$\hat{\mathbf{z}} \cdot \hat{\boldsymbol{\phi}}' \times \hat{\mathbf{R}} = \hat{\mathbf{z}} \times \hat{\boldsymbol{\phi}}' \cdot \hat{\mathbf{R}} = (-\hat{\boldsymbol{\imath}}') \cdot \hat{\mathbf{R}} = \cos\alpha = \frac{a}{R}$$

Hence

$$B_z = \frac{\mu_0 I a^2}{4\pi(a^2 + z^2)^{3/2}} \int_0^{2\pi} d\phi'$$

and we obtain the result

$$B_z = \frac{\mu_0 I a^2}{2(a^2 + z^2)^{3/2}} \tag{5-40}$$

A laboratory example of a current loop is illustrated in Figure 5–9. The leads give a negligible contribution to **B** because they represent equal and opposite currents at nearly the same position.

For a tightly wound coil of $N$ turns carrying current $I$, the magnetic

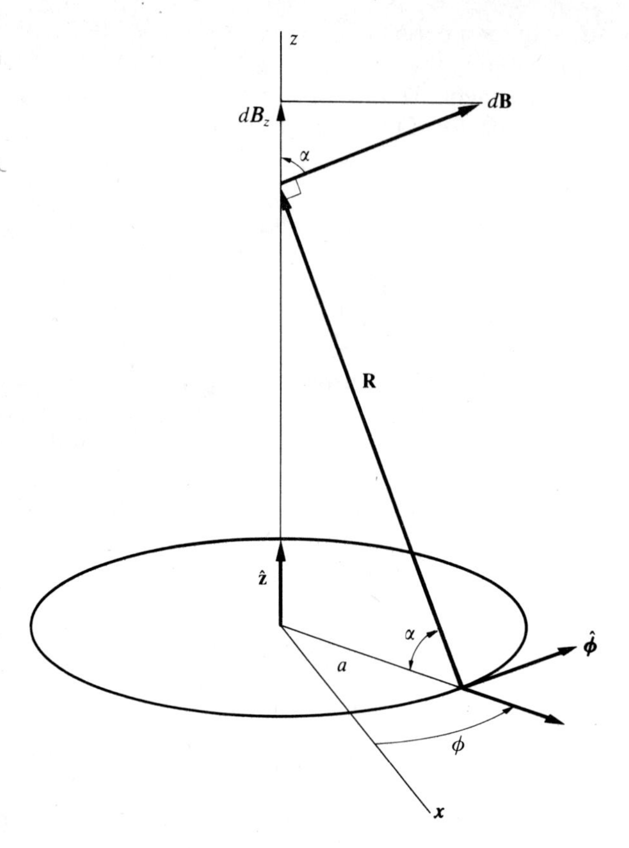


**FIGURE 5-8** Magnetic field on the axis of a current loop in the $xy$ plane centered at the origin

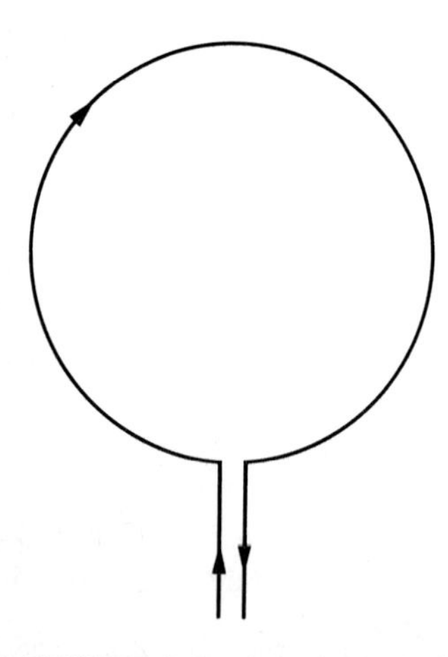

**FIGURE 5-9** Actual circuit that can be approximated as a current loop

field on the axis is just $N$ times the single-loop result, giving

**Magnetic Field on Axis of a Current Loop**

$$B_z = \frac{\mu_0 N I a^2}{2(a^2 + z^2)^{3/2}} \tag{5-41}$$

In this case, even if the leads are not coincident, their magnetic field contribution will be negligible in comparison with that of the coil if $N$ is large.

## 5-7 APPLICATION: HELMHOLTZ COIL

When each of the two identical coils in Figure 5–10 has $N$ loops carrying current $I$ in the same sense, the magnetic field along their common axis is

$$B_z(z) = \frac{\mu_0 N I a^2}{2} [D_+^{-3} + D_-^{-3}] \tag{5-42}$$

where

$$D_+ = \left[a^2 + \left(z + \frac{\ell}{2}\right)^2\right]^{1/2} \qquad D_- = \left[a^2 + \left(z - \frac{\ell}{2}\right)^2\right]^{1/2} \tag{5-43}$$

and $z$ is the distance along the axis measured from the midpoint, $\ell$ is the separation of the coils, and $a$ is their radii. If the two coils are closely spaced,

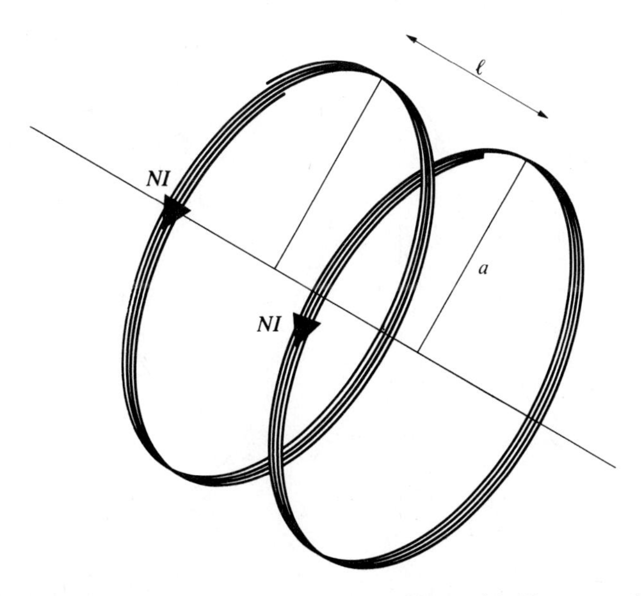


**FIGURE 5-10** Two coaxial current loops with $N$ turns and radii $a$ separated by a distance $\ell$

their magnetic field has nearly the same shape as that of a single coil, as illustrated in Figure 5–11, and the field on the axis has its maximum at the midpoint. If the two coils are widely spaced, their magnetic field has a minimum of the midpoint, as illustrated in Figure 5–11. At intermediate separation of the coils the field at the midpoint will have a vanishing second derivative and therefore be nearly constant in the region near the midpoint.

To investigate $B_z$ in the vicinity of the midpoint, we make a Taylor series expansion about $z = 0$,

$$B_z(z) = B_z(0) + \left.\frac{dB_z}{dz}\right|_0 z + \frac{1}{2!}\left.\frac{d^2B_z}{dz^2}\right|_0 z^2 + \frac{1}{3!}\left.\frac{d^3B_z}{dz^3}\right|_0 z^3 + \cdots \tag{5-44}$$

The derivatives calculated from Equation (5–42) are

$$\begin{aligned}\frac{dB_z}{dz} &= \frac{\mu_0 N I a^2}{2}\left[-3\left(z+\frac{\ell}{2}\right)D_+^{-5} - 3\left(z-\frac{\ell}{2}\right)D_-^{-5}\right]\\ \frac{d^2B_z}{dz^2} &= \frac{\mu_0 N I a^2}{2}\left[-3(D_+^{-5}+D_-^{-5})\right.\\ &\quad \left.+ 15\left(z+\frac{\ell}{2}\right)^2 D_+^{-7} + 15\left(z-\frac{\ell}{2}\right)^2 D_-^{-7}\right]\end{aligned} \tag{5-45}$$

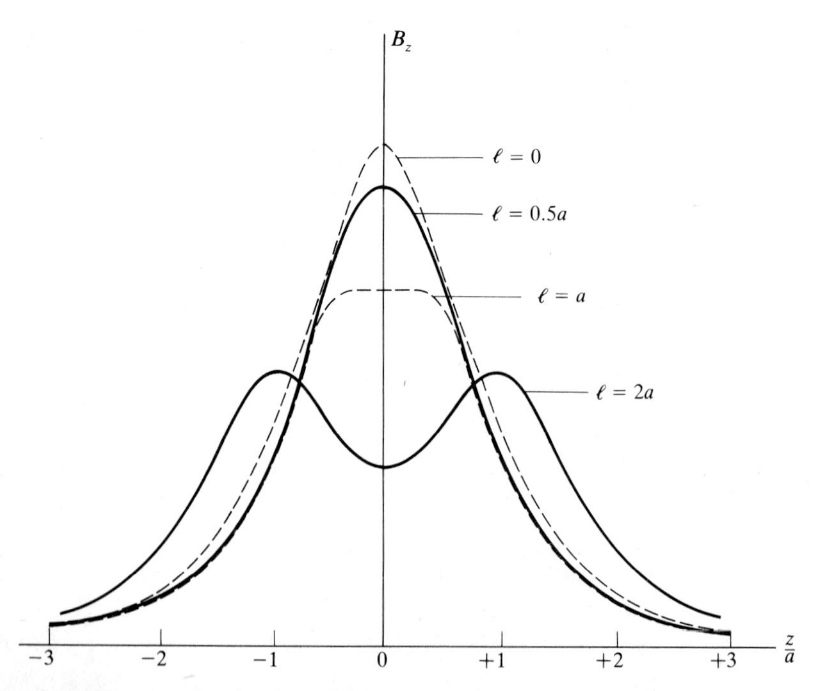


**FIGURE 5-11** Magnetic field along the axis of the coils for various coil spacings. Note that for $\ell = a$ the magnetic field is nearly uniform between the coils. This arrangement is known as a Helmholtz coil

In terms of $D = (a^2 + \ell^2/4)^{1/2}$, the values at $z = 0$ are

$$B_z(0) = \mu_0 N I a^2 D^{-3} \qquad \left.\frac{dB_z}{dz}\right|_0 = 0$$

$$\left.\frac{d^2B_z}{dz^2}\right|_0 = \mu_0 N I a^2 D^{-5}\left(-3 + \frac{15}{4}\frac{\ell^2}{D^2}\right) \tag{5-46}$$

By symmetry,

$$\left.\frac{d^3B_z}{dz^3}\right|_0 = 0$$

For a coil separation equal to the radius of the coil (i.e., $\ell = a$), $D = \sqrt{5}a/2$ and $(d^2B_z/dz^2)|_{z=0} = 0$. Then the constant is the only surviving term in Equation (5–42) through order $z^3$, and

**Magnetic Field on Axis of a Helmholtz Coil**

$$B_z(z) \approx B_z(0) = \frac{\mu_0 N I}{a}\left(\frac{4}{5}\right)^{3/2} \tag{5-47}$$

Consequently, for $z \ll a$ the field is nearly constant. This arrangement is called a *Helmholtz coil*.

Figure 5–12 shows the variation of $B_z(z)$ on the axis as a function of $z$ for the Helmholtz coil, with the shaded regions representing the locations of the two coils. Off the axis the magnetic field has a cylindrical radial com-

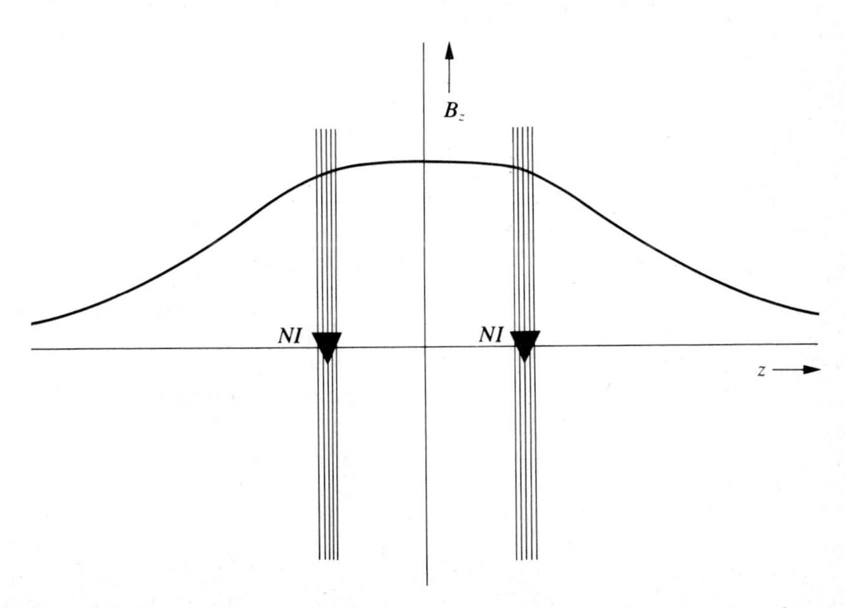


**FIGURE 5–12** Magnetic field on the axis of a Helmholtz coil as a function of $z$

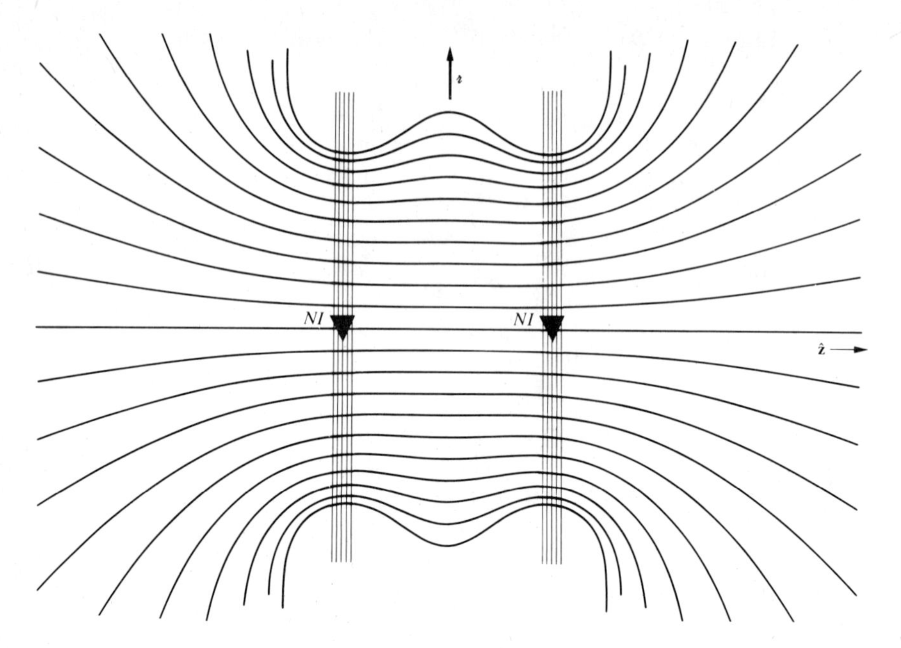


**FIGURE 5-13** Magnetic field lines of a Helmholtz coil in the $(\mathscr{r}, z)$ plane

ponent $B_{\mathscr{r}}$ as well. The magnetic field lines in the $(\mathscr{r}, z)$ plane are shown in Figure 5–13.

A Helmholtz coil is often used when a uniform magnetic field over a large spatial region is required. For example, it can be used to cancel the earth's magnetic field for experiments that require a magnetic field-free region.

## 5-8 APPLICATION: SOLENOID

The *solenoid* is a common element of electric circuits. It consists of wire closely wound in the form of a helix, as illustrated in Figure 5–14. We will calculate the magnetic field of a current-carrying solenoid in the approximation that the helix is represented by a sequence of circular current loops. In a laboratory solenoid the wires are wound in layers of helices, with the pitch of the windings in adjacent layers in opposite directions, so that the component of the current along the axis of the solenoid vanishes.

For a single loop of radius $a$ at coordinate $z'$ on the solenoid axis, the magnetic field on the axis at coordinate $z$ is, by Equation (5–41),

$$B_z(z, z') = \frac{\mu_0 I a^2}{2[a^2 + (z' - z)^2]^{3/2}} \tag{5-48}$$

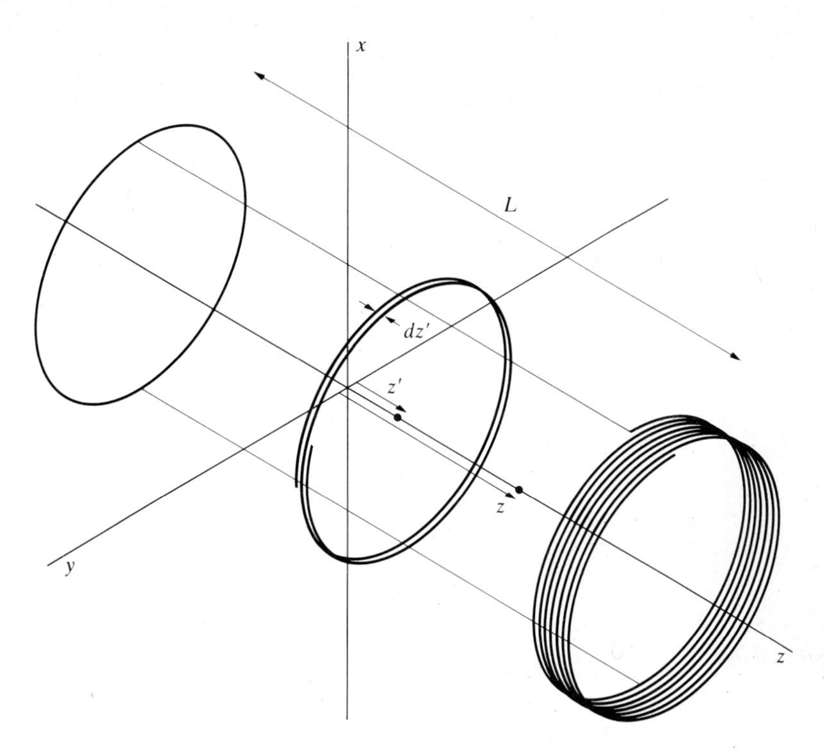


FIGURE 5-14 Solenoid, which consists of wire closely wound in the form of a helix

For a solenoid of $N$ turns and length $L$ the number of loops in the length $dz'$ is $(N/L)\,dz'$, and the magnetic field due to the loops in element $dz'$ is

$$dB_z(z, z') = \frac{\mu_0 N I a^2}{2L} \frac{dz'}{[a^2 + (z' - z)^2]^{3/2}} \tag{5-49}$$

Integrating $dz'$ over the length of the coil, from $-L/2$ to $+L/2$,

$$\begin{aligned} B_z(z) &= \frac{\mu_0 N I a^2}{2L} \int_{-L/2}^{L/2} \frac{dz'}{[a^2 + (z' - z)^2]^{3/2}} \\ &= \frac{\mu_0 N I a^2}{2L} \left. \frac{z' - z}{[a^2 + (z' - z)^2]^{1/2}} \right|_{-L/2}^{L/2} \end{aligned} \tag{5-50}$$

we obtain, for the solenoid field,

$$B_z(z) = \frac{\mu_0 N I}{2L} \left\{ \frac{L/2 - z}{[a^2 + (L/2 - z)^2]^{1/2}} + \frac{L/2 + z}{[a^2 + (L/2 + z)^2]^{1/2}} \right\} \tag{5-51}$$

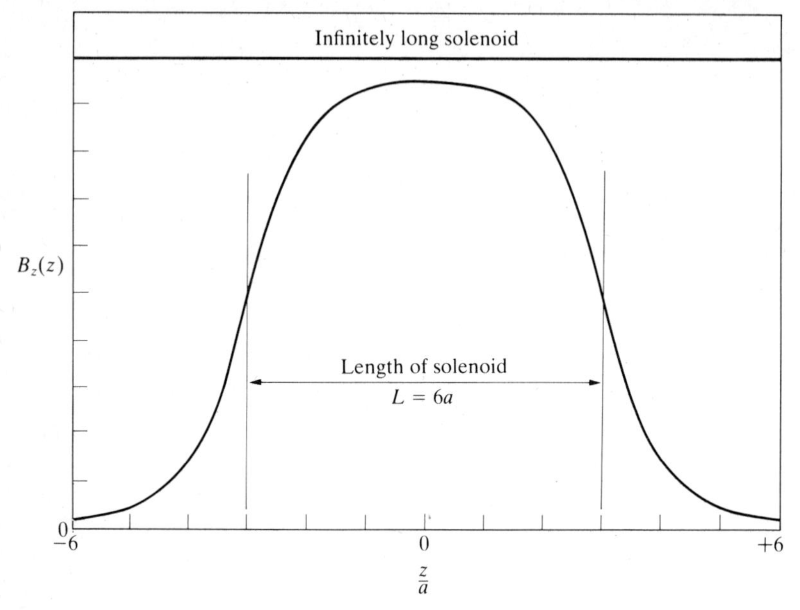


**FIGURE 5-15** Magnetic field on the axis of a solenoid

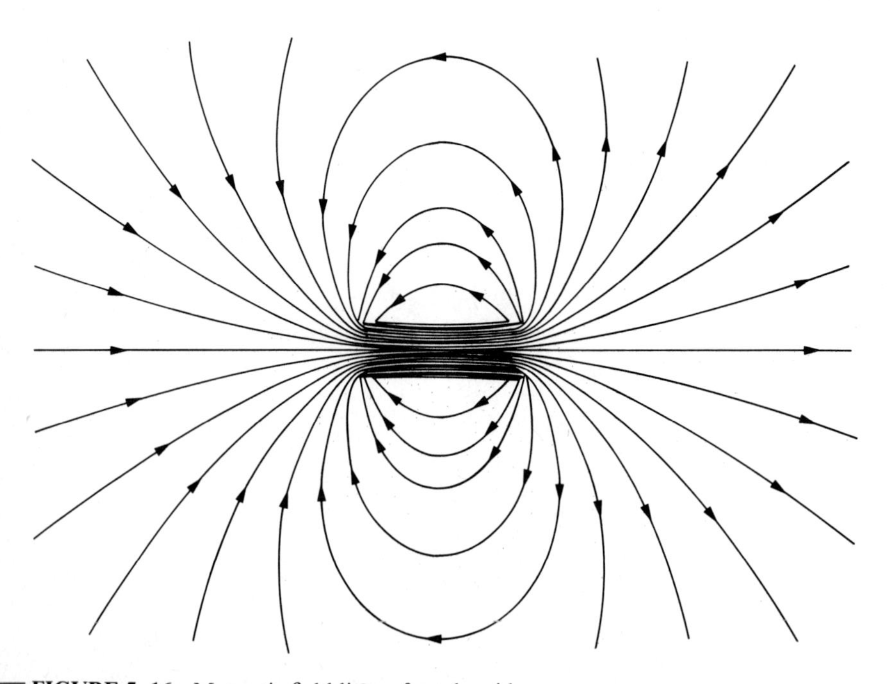

**FIGURE 5-16** Magnetic field lines of a solenoid

Figure 5–15 shows $B_z$ versus $z/a$ for a solenoid for which $L = 6a$. Note that $B_z(z)$ is approximately constant in the interior of the solenoid. In particular, for $a \ll L$ and $|z| \ll L/2$ (i.e., far from the ends of a long solenoid), the approximately constant value of the magnetic field on the solenoid axis is

**Magnetic Field on the Axis of a Long Solenoid**

$$B_z = \frac{\mu_0 NI}{L} \tag{5-52}$$

The magnetic field of an infinitely long solenoid with $n = N/L$ turns per unit length has the value $B_z = \mu_0 nI$ everywhere inside the solenoid and $B_z = 0$ outside. This result can be shown by direct integration of the Biot-Savart law.

Figure 5–16 shows the magnetic field lines off the axis of the solenoid. We observe that away from the ends the field is nearly uniform inside the solenoid and is much smaller outside the solenoid.

## 5-9 MAGNETIC FIELD EXPRESSED AS THE CURL OF A VECTOR POTENTIAL

The vector operation $\nabla \times \mathbf{A}$ is called the *curl* of **A**. As discussed in Section 1–4, the definition of the cross product and the Cartesian form of the vector operator $\nabla$ gives the curl of **A** to be

$$\nabla \times \mathbf{A} = \begin{vmatrix} \hat{\mathbf{x}} & \hat{\mathbf{y}} & \hat{\mathbf{z}} \\ \dfrac{\partial}{\partial x} & \dfrac{\partial}{\partial y} & \dfrac{\partial}{\partial z} \\ A_x & A_y & A_z \end{vmatrix}$$

$$\equiv \hat{\mathbf{x}}\left(\frac{\partial A_z}{\partial y} - \frac{\partial A_y}{\partial z}\right) + \hat{\mathbf{y}}\left(\frac{\partial A_x}{\partial z} - \frac{\partial A_z}{\partial x}\right) \hat{\mathbf{z}}\left(\frac{\partial A_y}{\partial x} - \frac{\partial A_x}{\partial y}\right)$$

In cylindrical coordinates the curl of **A** is

$$\nabla \times \mathbf{A} = \left(\frac{1}{\imath}\frac{\partial A_z}{\partial \phi} - \frac{\partial A_\phi}{\partial z}\right)\hat{\boldsymbol{\imath}} + \left(\frac{\partial A_\imath}{\partial z} - \frac{\partial A_z}{\partial \imath}\right)\hat{\boldsymbol{\phi}}$$

$$+ \frac{1}{\imath}\left[\frac{\partial}{\partial \imath}(\imath A_\phi) - \frac{\partial A_\imath}{\partial \phi}\right]\hat{\mathbf{z}} \tag{5-53}$$

While in spherical coordinates the expression is

$$\nabla \times \mathbf{A} = \frac{1}{r \sin\theta}\left[\frac{\partial}{\partial\theta}(A_\phi \sin\theta) \cdot \frac{\partial A_\theta}{\partial\phi}\right]\hat{\mathbf{r}} + \frac{1}{r}\left[\frac{1}{\sin\theta}\frac{\partial A_r}{\partial\phi} - \frac{\partial(rA_\phi)}{\partial r}\right]\hat{\boldsymbol{\theta}} + \frac{1}{r}\left[\frac{\partial(rA_\theta)}{\partial r} - \frac{\partial A_r}{\partial\theta}\right]\hat{\boldsymbol{\phi}} \tag{5-54}$$

The Biot-Savart law for the magnetic field of a steady current distribution is

$$\mathbf{B}(\mathbf{r}) = \frac{\mu_0}{4\pi}\int \mathbf{J}(r') \times \frac{\hat{\mathbf{R}}}{R^2}\, dV'$$

Using the result of Equation (1–71) with $n = 1$,

$$\frac{\hat{\mathbf{R}}}{R^2} = -\nabla_{\mathbf{r}}\left(\frac{1}{R}\right)$$

we can write the Biot-Savart law as

$$\mathbf{B}(\mathbf{r}) = -\frac{\mu_0}{4\pi}\int \mathbf{J}(\mathbf{r}') \times \nabla_{\mathbf{r}}\left(\frac{1}{R}\right) dV' \tag{5-55}$$

Since $\mathbf{J}(\mathbf{r}')$ does not depend on $\mathbf{r}$, we can reverse the order of $\mathbf{J}(\mathbf{r}')$ and $\nabla_{\mathbf{r}}$ in the integrand, $\mathbf{J}(\mathbf{r}') \times \nabla_{\mathbf{r}} = -\nabla_{\mathbf{r}} \times \mathbf{J}(\mathbf{r}')$, giving

$$\mathbf{B}(\mathbf{r}) = \frac{\mu_0}{4\pi}\int \nabla_{\mathbf{r}} \times \left(\frac{\mathbf{J}}{R}\right) dV' \tag{5-56}$$

and then take the $\nabla_{\mathbf{r}}$ outside the integration to obtain the Biot-Savart law as

$$\mathbf{B}(\mathbf{r}) = \nabla_{\mathbf{r}} \times \left(\frac{\mu_0}{4\pi}\int \frac{\mathbf{J}}{R}\, dV'\right) \tag{5-57}$$

This relation has the form

$$\mathbf{B} = \nabla \times \mathbf{A} \tag{5-58}$$

where

**Vector Potential**

$$\mathbf{A}(\mathbf{r}) = \frac{\mu_0}{4\pi}\int \frac{\mathbf{J}(\mathbf{r}')}{R}\, dV' \tag{5-59}$$

The quantity $\mathbf{A}$ is called the vector potential.

Equation (5–58) represents $\mathbf{B}$ as the curl of the vector potential $\mathbf{A}$, analogous to the representation in Equation (1–67a) of the electrostatic field $\mathbf{E}$ as the gradient of the electrostatic potential $\Phi$.

An important implication of Equation (5–58) is that the divergence of **B** is zero:

$$\nabla \cdot \mathbf{B} = \nabla \cdot \nabla \times \mathbf{A} = \nabla \times \nabla \cdot \mathbf{A} = 0 \qquad \textbf{(5-60)}$$

Furthermore, the divergence theorem can be used to show that the integral of **B** through a closed surface is zero:

$$\oint \mathbf{B} \cdot d\mathbf{S} = \int \nabla \cdot \mathbf{B}\, dV = 0 \qquad \textbf{(5-61)}$$

Equation (5–58) does not completely define **A**, because adding to **A** a term $\nabla\chi(\mathbf{r})$ will not affect the value of **B**, since $\nabla \times \nabla\chi = 0$. Equation (5–59) is the simplest form for **A** that gives the Biot-Savart **B** field. For a surface current source **j** the expression for **A** is

$$\mathbf{A}(\mathbf{r}) = \frac{\mu_0}{4\pi} \int \frac{\mathbf{j}(\mathbf{r}')}{R}\, dS' \qquad \textbf{(5-62)}$$

where $dS'$ is an element of the surface in which the current flows. For a line current the result is

$$\mathbf{A}(\mathbf{r}) = \frac{\mu_0 I}{4\pi} \int \frac{d\mathbf{r}'}{R} \qquad \textbf{(5-63)}$$

Table 5–2 summarizes the correspondence between electrical and magnetic quantities and the relation to their charge and current sources.

As an example, we calculate the vector potential due to a straight wire. For an infinite straight wire, **A** will be independent of $z$ by symmetry, so we can evaluate it at $z = 0$. To avoid dealing with a divergent integral, we first evaluate the result for a finite wire of length $2L$, and later we consider the $L \to \infty$ limit. By Equation (5–63) and the geometry of Figure 5–4, the expression for the vector potential at $z = 0$ is

$$\mathbf{A} = \hat{\mathbf{z}} \frac{\mu_0 I}{4\pi} \int_{-L}^{+L} \frac{dz'}{\sqrt{\mathscr{r}^2 + z'^2}} = \hat{\mathbf{z}} \frac{\mu_0 I}{4\pi} \ln[z' + (\mathscr{r}^2 + z'^2)^{1/2}] \Big|_{-L}^{+L} \qquad \textbf{(5-64)}$$

or

$$\mathbf{A} = \hat{\mathbf{z}} \frac{\mu_0 I}{4\pi} \ln\left(\frac{L + \sqrt{L^2 + \mathscr{r}^2}}{-L + \sqrt{L^2 + \mathscr{r}^2}}\right) \qquad \textbf{(5-65)}$$

For $L$ much larger than $\mathscr{r}$ we make a series expansion:

$$\sqrt{L^2 + \mathscr{r}^2} = L + \frac{\mathscr{r}^2}{2L} + \cdots$$

$$\ln\left(1 + \frac{4L^2}{\mathscr{r}^2}\right) = \ln \frac{4L^2}{\mathscr{r}^2} + \ln\left(1 + \frac{\mathscr{r}^2}{4L^2}\right)$$

$$= \ln 4L^2 - \ln \mathscr{r}^2 + \frac{\mathscr{r}^2}{4L^2} + \cdots$$

**TABLE 5-2** Comparison of steady-state electrical and magnetic quantities

| Electrical | Magnetic |
|---|---|
| Charge, $q$ | Current, $q\mathbf{v}$ or $I\,d\mathbf{r}$ |
| Charge density, $\rho(\mathbf{r})$ | Current density, $\mathbf{J}(\mathbf{r})$ |
| Electric field definition: | Magnetic field definition: |
| $\mathbf{F} = q\mathbf{E} = \int \rho\mathbf{E}\,dV$ | $\mathbf{F} = q\mathbf{v} \times \mathbf{B} = \int \mathbf{J} \times \mathbf{B}\,dV$ |
| Coulomb's force between two charges: | Ampère's force between two currents: |
| $\mathbf{F} = \dfrac{qq_1}{4\pi\varepsilon_0}\dfrac{\hat{\mathbf{R}}}{R^2}$ | $\mathbf{F} = \dfrac{\mu_0}{4\pi}\iint \dfrac{(I_2\,d\mathbf{r}_2) \times [(I_1\,d\mathbf{r}_1) \times \hat{\mathbf{R}}]}{R^2}$ |
| Electric field due to charge density: | Magnetic field due to current density $\mathbf{J}$: |
| $\mathbf{E} = \dfrac{1}{4\pi\varepsilon_0}\int \rho\dfrac{\hat{\mathbf{R}}}{R^2}\,dV'$ | $\mathbf{B} = \dfrac{\mu_0}{4\pi}\int \mathbf{J} \times \dfrac{\hat{\mathbf{R}}}{R^2}\,dV'$ |
| Electrostatic potential, $\Phi$: | Vector potential, $\mathbf{A}$: |
| $\mathbf{E} = -\nabla\Phi$ | $\mathbf{B} = \nabla \times \mathbf{A}$ |
| $\Phi = \dfrac{1}{4\pi\varepsilon_0}\int \dfrac{\rho}{R}\,dV'$ | $\mathbf{A} = \dfrac{\mu_0}{4\pi}\int \dfrac{\mathbf{J}}{R}\,dV'$ |
| $\nabla^2\phi = -\dfrac{\rho}{\varepsilon_0}$ | $\nabla^2\mathbf{A} = -\mu_0\mathbf{J}$ |

In the limit $L \gg \imath$ the vector potential becomes

$$\mathbf{A} \approx \hat{\mathbf{z}}\frac{\mu_0 I}{4\pi}(\ln 4L^2 - \ln \imath^2) \tag{5-66}$$

The constant $\ln 4L^2$ term has no consequence for the **B** field, and so we can drop it. Hence the vector potential of an infinite-line current is

$$\mathbf{A} = -\hat{\mathbf{z}}\frac{\mu_0 I}{2\pi}\ln \imath \tag{5-67}$$

If we calculate **B** from it, using Equation (5–53) for the curl in cylindrical coordinates, we find

$$\begin{aligned}\mathbf{B} &= \nabla \times \mathbf{A} \\ &= -\frac{\partial A_z}{\partial \imath}\hat{\boldsymbol{\phi}} = \frac{\mu_0 I}{2\pi\imath}\hat{\boldsymbol{\phi}}\end{aligned}$$

which agrees with the result of Equation (5–29) from direct calculation by using the Biot-Savart law.

# 5-10 APPLICATION: MAGNETIC FIELD AND VECTOR POTENTIAL OF AN INFINITE CURRENT SHEET

An infinite, uniform plane current sheet is an instructive example in which the magnetic field and the vector potential can be easily evaluated. For a coordinate system we take $x$ and $y$ axes in the plane; the specific location of the origin is immaterial owing to the infinite extent of the sheet. For definiteness we take the surface current to flow in the $x$ direction, $\mathbf{j} = j\hat{\mathbf{x}}$, as illustrated in Figure 5–17. The line current along $\hat{\mathbf{x}}$ in a differential width $dy'$ is

$$dI = j\,dy' \qquad \textbf{(5-68)}$$

The corresponding magnetic field at coordinates (0, 0, $z$) due to this line current is

$$d\mathbf{B} = \frac{\mu_0}{2\pi}\frac{dI}{r}\hat{\boldsymbol{\phi}} \qquad \textbf{(5-69)}$$

from Equation (5–29), where $r = (z^2 + y'^2)^{1/2}$ and $\hat{\boldsymbol{\phi}}$ is an azimuthal unit vector in the $zy'$ plane around the line current, with

$$\sin\phi = \frac{z}{r} \qquad \textbf{(5-70)}$$

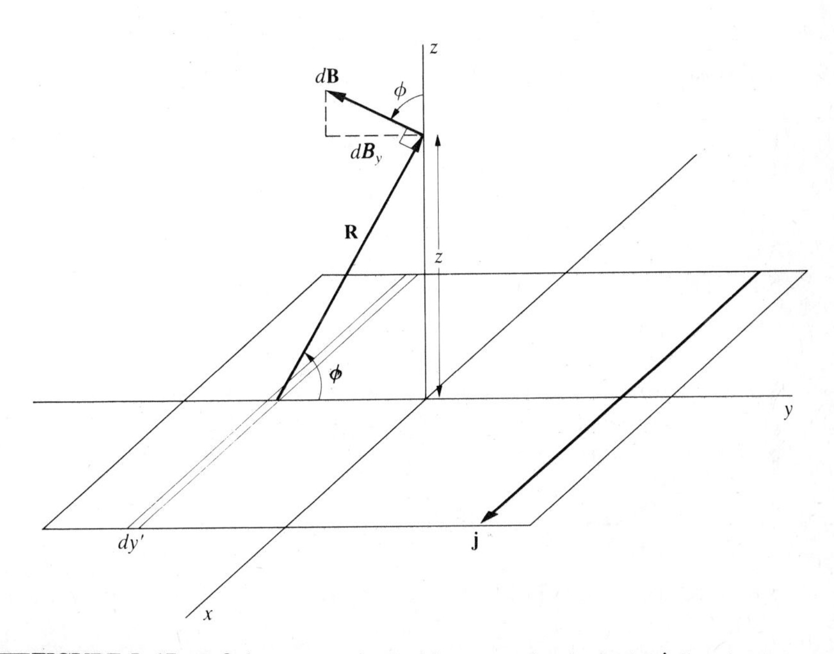


**FIGURE 5-17** Infinite current sheet with current density in the $\hat{\mathbf{x}}$ direction

The $dB_z$ components from $dI$ elements symmetrically placed around $y = 0$ cancel. The only nonzero component of $d\mathbf{B}$ is then $dB_y$, given by

$$dB_y = -dB \sin \phi = -dB \frac{z}{\imath} = -\frac{\mu_0 \, j \, dy'}{2\pi} \frac{z}{\imath} \frac{}{\imath} \tag{5-71}$$

Summing up the differential line currents by integration over $dy'$, we find the resulting magnetic field at $(0, 0, z)$ of the infinite current sheet:

$$B_y = -\frac{\mu_0}{2\pi} jz \int_{-\infty}^{+\infty} \frac{dy'}{z^2 + y'^2} = -\frac{\mu_0}{2\pi} j \tan^{-1}\left(\frac{y'}{z}\right)\Bigg|_{-\infty}^{+\infty} \tag{5-72}$$

Thus

$$B_y = \begin{cases} -\dfrac{\mu_0}{2} j, & \text{for } z > 0 \\ \dfrac{\mu_0}{2} j, & \text{for } z < 0 \end{cases} \tag{5-73}$$

$$B_z = 0 \qquad B_x = 0$$

In vector form this result can be written as

$$\mathbf{B} = \frac{\mu_0}{2} \mathbf{j} \times \hat{\mathbf{n}} \tag{5-74}$$

where $\hat{\mathbf{n}}$ is the normal to the plane pointing toward the observation point. The magnetic field of an infinite current sheet is constant, and its direction is defined by the right-hand rule, with $\hat{\mathbf{j}}$, $\hat{\mathbf{n}}$, and $\mathbf{B}$ as the three perpendicular vectors.

We turn now to the calculation of the vector potential for an infinite current sheet. It is defined by Equation (5–62):

$$\mathbf{A}(\mathbf{r}) = \frac{\mu_0}{4\pi} \int \frac{\mathbf{j}(\mathbf{r}')}{R} \, dS'$$

Since $\mathbf{j}(r') = j\hat{\mathbf{x}}$, the direction of $\mathbf{A}$ is also along $\hat{\mathbf{x}}$:

$$\mathbf{A}(\mathbf{r}) = A_x(\mathbf{r})\hat{\mathbf{x}} \tag{5-75}$$

We can again represent the current sheet as a superposition of line currents given by Equation (5–68). From Equation (5–67) the differential vector potential from an element $dI = j \, dy'$ is

$$d\mathbf{A} = -\hat{\mathbf{x}} \frac{\mu_0 (j \, dy')}{2\pi} \ln \imath$$

Summing over a band of width $2L$ centered at $y' = 0$, we find

$$\mathbf{A} = -\hat{\mathbf{x}} \frac{\mu_0 j}{2\pi} \int_{-L}^{+L} dy' \ln \sqrt{(y'^2) + (z)^2} \tag{5-76}$$

Integration yields

$$\mathbf{A} = -\hat{\mathbf{x}}\frac{\mu_0 j}{2\pi}\left(2L\ln\sqrt{L^2+z^2} - 2L + 2z\tan^{-1}\frac{L}{z}\right) \tag{5-77}$$

In the limit $L \gg |z|$ we make the series expansions

$$\sqrt{L^2+z^2} = L + \frac{z^2}{2L} + \cdots$$

$$\tan^{-1}\frac{L}{z} = \begin{cases} \dfrac{\pi}{2} - \dfrac{z}{L} + \cdots, & \text{for } z > 0 \\[2ex] -\dfrac{\pi}{2} - \dfrac{z}{L} + \cdots, & \text{for } z < 0 \end{cases}$$

We thereby obtain, to lowest order in $|z|/L$,

$$\mathbf{A} \approx -\hat{\mathbf{x}}\frac{\mu_0 j}{\pi}\left(L\ln L - L + \frac{\pi}{2}|z|\right) \tag{5-78}$$

We can drop the constant term $L \ln L - L$. Hence for the infinite sheet the exact expression for the vector potential is

$$A_x = \begin{cases} -\dfrac{\mu_0 j}{2}z, & \text{for } z > 0 \\[2ex] \dfrac{\mu_0 j}{2}z, & \text{for } z < 0 \end{cases} \tag{5-79}$$

The curl of Equation (5–79) reduces to

$$B_y = \frac{\partial A_x}{\partial z} \tag{5-80}$$

which reproduces the result of Equation (5–73).

## 5-11 STEADY CURRENT FLOW AND OHM'S LAW

In many materials, especially metals, the electric current density is directly proportional to the imposed electric field:

**Ohm's Law** $\mathbf{J} = g\mathbf{E}$ (5-81)

This relation is known as Ohm's law (due to Georg Simon Ohm in 1826). The constant of proportionality $g$ is called the *electric conductivity*; $g$ has units of amperes per (volt-meter). Many texts use $\sigma$ as the conductivity, but to avoid confusion with the surface charge density, we will use the symbol $g$. The conductivity depends on the material, the ambient conditions (temperature, pressure), and the state of the material (liquid, solid crystalline, concentration of crystal defects, etc.)

The form of Ohm's law in Equation (5–81) can be readily converted to the more familiar form $V = IR$ for a conductor. We consider a straight conducting wire of length $\ell$ and cross-sectional area $S$. The current density in the wire is related to the current $I$ by

$$J = \frac{I}{S} \tag{5-82}$$

and the constant electric field is related to the potential difference $V$ between the ends of the wire by

$$E = \frac{V}{\ell} \tag{5-83}$$

Substituting into $J = g\mathbf{E}$, we obtain

$$\frac{V}{I} = \left(\frac{\ell}{Sg}\right) \equiv R \tag{5-84}$$

where the resistance $R$ is a parameter that characterizes the conductor. The SI unit of resistance is

$$\text{ohm } (\Omega) = \frac{\text{volt}}{\text{ampere}}$$

and the conductivity is normally expressed in reciprocal units of ohm-meters, or $(\Omega\cdot\text{m})^{-1}$.

Values of $g$ for known materials vary widely, ranging from infinity for a superconductor, to $g \approx 10^8\ (\Omega\cdot\text{m})^{-1}$ for metals, to $g \approx 10^{-12}\ (\Omega\cdot\text{m})^{-1}$ for a poor conductor such as glass, to even smaller values for other materials. Poor conductors are called *dielectrics* or *insulators*. In this book the intermediate conductors, or *semiconductors*, and good conductors will be treated together and hereafter termed *conductors*. Table 5–3 lists typical conductivities of some representative materials.

From Equation (1–29) the work done by the microscopic electric field on a charge $q_i$ is

$$dW_i = q_i \mathbf{E}_\text{m}(\mathbf{r}_i) \cdot d\mathbf{r}_i \tag{5-85}$$

where $d\mathbf{r}_i$ is the particle displacement. For a system of charges the power transmitted to the charges is therefore

**TABLE 5-3** Conductivity $g$ for representative materials at room temperature

| Type | Material | $g(\Omega \cdot \text{m})^{-1}$ |
|---|---|---|
| Conductors | Silver | $6.1 \times 10^7$ |
| | Copper | $5.8 \times 10^7$ |
| | Gold | $4.1 \times 10^7$ |
| | Aluminum | $3.5 \times 10^7$ |
| | Iron | $1.0 \times 10^7$ |
| | Mercury | $1.0 \times 10^6$ |
| | Graphite | $1.0 \times 10^5$ |
| Semiconductors and ionic conductors | Seawater | $\approx 5$ |
| | Germanium | 2.2 |
| | Silicon | $1.6 \times 10^{-3}$ |
| | Pure water | $2 \times 10^{-4}$ |
| | Earth's crust | $\approx 10^{-4}$ |
| Insulators (dielectrics) | Wood | $\approx 10^{-10}$ |
| | Glass | $\approx 10^{-12}$ |
| | Petroleum | $\approx 10^{-14}$ |
| | Fused quartz | $\approx 10^{-18}$ |

$$P = \frac{dW}{dt} = \sum_i q_i \mathbf{E}_\text{m}(\mathbf{r}_i) \cdot \mathbf{v}_i = \int \mathbf{E}_\text{m}(\mathbf{r}) \cdot \mathbf{J}_\text{m}(\mathbf{r})\, dV \tag{5-86}$$

where $\mathbf{v}_i = d\mathbf{r}_i/dt$ is the particle velocity.

The *power* transmitted to the charges within a macroscopic volume $\Delta V$ can also be expressed in terms of macroscopic variables and correlations. Proceeding as for the macroscopic Lorentz force of Equation (5–20), we find

$$\Delta P = \mathbf{E} \cdot \mathbf{J}\, \Delta V + \int_{\Delta V} (\mathbf{E}_\text{m} - \mathbf{E}) \cdot (\mathbf{J}_\text{m} - \mathbf{J})\, dV \tag{5-87}$$

The second term in Equation (5–87) involves microscopic correlations. Consider, for example, the case of a steady conduction current in a wire. The power supplied to the conduction electrons by the battery must exactly balance the power absorbed in collisions of electrons with the ions, since otherwise the electrons would gain energy. The first term in Equation (5–87),

$$\Delta P = \mathbf{E} \cdot \mathbf{J}\, \Delta V \tag{5-88}$$

is the power delivered to the charges by the macroscopic electric field $\mathbf{E}$. This power is determined by macroscopic quantities alone and is therefore directly measurable.

For a uniform wire $\mathbf{J}\, dV' = I\, d\mathbf{r}'$ and $\int \mathbf{E} \cdot d\mathbf{r}' = -(V_2 - V_1) \equiv V$, where $V$ is the potential drop between the ends of the wire. Thus Equation (5–88) becomes

$$P = VI \tag{5-89}$$

From Ohm's law, $V = IR$, the power in this case is

$$P = I^2 R = \frac{V^2}{R} \tag{5-90}$$

Collisions of the conduction electrons with the ionic lattice of the conductor absorb the power input by the electric field, and the energy of the resulting ionic oscillations becomes heat energy.

The charge density $\rho(\mathbf{r}, t)$ and the current density $\mathbf{J}(\mathbf{r}, t)$ are related by the continuity equation of Equation (5–14),

$$\nabla \cdot \mathbf{J} + \frac{\partial \rho}{\partial t} = 0 \tag{5-91}$$

Suppose we consider steady current flow, meaning that at a given point in space $\mathbf{J}$ and $\rho$ are not functions of time. From Equation (5–91) we have

**Steady Current Flow**

$$\nabla \cdot \mathbf{J}(\mathbf{r}) = 0 \tag{5-92}$$

A static electric field can be written as the gradient of a scalar potential in the usual manner:

$$\mathbf{E} = -\nabla\Phi$$

If the conducting medium obeys Ohm's law, we have

$$\nabla \cdot [g(\mathbf{r})\nabla\Phi] = 0 \tag{5-93}$$

where $g$ can, in general, vary with position [i.e., $g = g(\mathbf{r})$]. In a region of constant conductivity we have simply

$$\nabla^2\Phi = 0 \tag{5-94}$$

So steady current flow in a region of uniform conductivity is governed by the Laplace equation.

As an example of a steady-current problem, we return to the concentric-sphere system of Figure 3–3. Suppose the space $a < r < b$ is a material of uniform conductivity $g$. The system is now a leaky capacitor. If the potentials $V_1$ and $V_2$ are maintained by batteries, a steady current will flow between the spheres. The electric field in the interior is still given by Equation (3–38), since the boundary conditions have not been changed and the potential satisfies Laplace's equation (i.e., there is no charge density in the interior region). The current through a spherical surface at radius $r$ is

$$I = J_r(4\pi r^2) = gE_r(4\pi r^2) = 4\pi abg\,\frac{V_1 - V_2}{b - a} \tag{5-95}$$

The current is independent of $r$; that is, the same charge passes through all spheres in a given time interval, as it must in this steady flow. The resistance of this leaky capacitor is

$$R = \frac{V_1 - V_2}{I} = \frac{b - a}{4\pi gab} \tag{5-96}$$

where $b > a$. The product $RC$ is given by

$$RC = \frac{\varepsilon_0}{g} \tag{5-97}$$

The result in Equation (5–97) holds for general geometries for capacitors in which the material between the surfaces has constant conductivity $g$ and dielectric constant $\varepsilon_0$. A similar result holds for a capacitor with a dielectric medium, with $\varepsilon_0$ replaced by $\varepsilon$.

We conclude this section with some practical considerations of conduction in real circuits. In Figure 5–18 a light bulb circuit is shown in which a 100-W light bulb is connected by 10 m of no. 14 copper wire to a 100-V, fixed-voltage source. In a practical circuit almost none of the voltage drop in the circuit occurs through the wires, so by Equation (5–89) the current is

$$I = \frac{P}{V} = \frac{100\ \text{W}}{100\ \text{V}} = 1\ \text{A} \tag{5-98}$$

The no. 14 wire has a diameter of about $1\frac{1}{2}$ mm and thus a cross-sectional area of $S \approx 2 \times 10^{-6}\ \text{m}^2$. The current is constant through the circuit, and hence by Ohm's law the electric field is also constant within the wire. Nu-

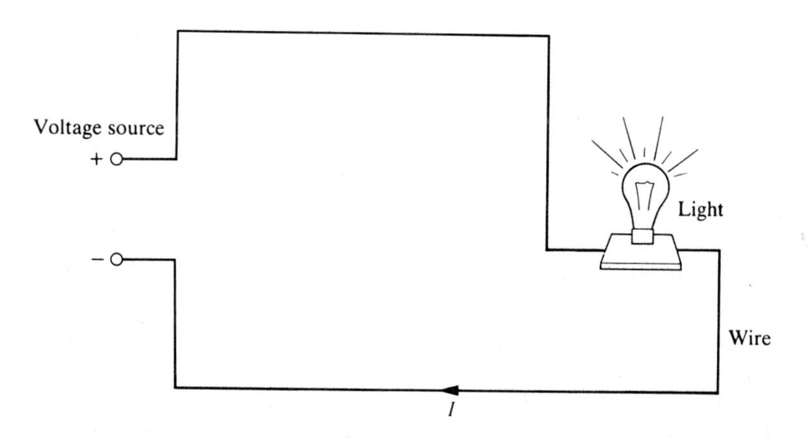


**FIGURE 5–18** Lighting circuit consisting of a voltage source $V$ and a light bulb

merically, since $g = 5.8 \times 10^6(\Omega\text{·m})^{-1}$ for copper,

$$E = \frac{J}{g} = \frac{1\ \text{A}/(2 \times 10^{-6}\ \text{m}^2)}{5.8 \times 10^6\ (\Omega\text{·m})^{-1}} = 0.007\ \text{V/m} \tag{5-99}$$

The resulting voltage drop in the wire is then $E\ell = (0.007)(10) = 0.07$ V, so indeed, as assumed, almost all the 100-V drop is across the light terminals.

By Equation (5–7) the current density is opposite to the electron flow and is given by

$$J = -n_- ev \tag{5-100}$$

For copper the number of conduction electrons per unit volume is $n_- \approx 8.5 \times 10^{28}\ \text{m}^{-3}$; the drift velocity $v$ is then

$$v = \frac{1/(2 \times 10^{-6})}{(1.6 \times 10^{-19})(8.5 \times 10^{28})} \approx 0.04\ \text{mm/s} \tag{5-101}$$

At this average velocity it would take an electron about 8 h to travel 1 m. Experience tells us that when the switch closes, the light turns on in a fraction of a second. When the switch is closed, current begins flowing in the whole circuit nearly simultaneously. Current flow in a wire behaves like an incompressible fluid, just as water seems to flow in a pipe as soon as the faucet is turned on. Actually, there is a delay equal to the pipe length divided by the speed of sound in water. In an electric circuit there is also a delay equal to the circuit length divided by the speed of electromagnetic waves in the circuit.

The electric field outside the wire may vary in a complex manner, but as long as it is electrostatic, the resulting potential drop along the wire exactly equals the source voltage. As we have pointed out, the electric field inside a wire carrying a steady current is constant; therefore

$$V = E\ell \tag{5-102}$$

holds for any uniform wire not necessarily straight, although the bending radius must be much larger than the wire diameter.

For this state of affairs to hold, there must be surface charges strategically placed so that their fields cancel the external field and result in a uniform internal field. These surface charges are of two types. Ordinary electrostatic induction charges appear in response to external fields to shield the conductor. These charges are independent of the current flow. Another type of surface charge can be thought of as "guiding charges" since they guide the current along, confining it to the wire no matter how many twists and turns there are. To estimate the magnitude of these guiding charges, we consider a wire with a right-hand turn, as depicted in Figure 5–19. For the current direction to change by 90°, there must be a surface charge distribution roughly as shown. By Gauss's law $\sigma = \varepsilon_0 E$, and the total surface charge

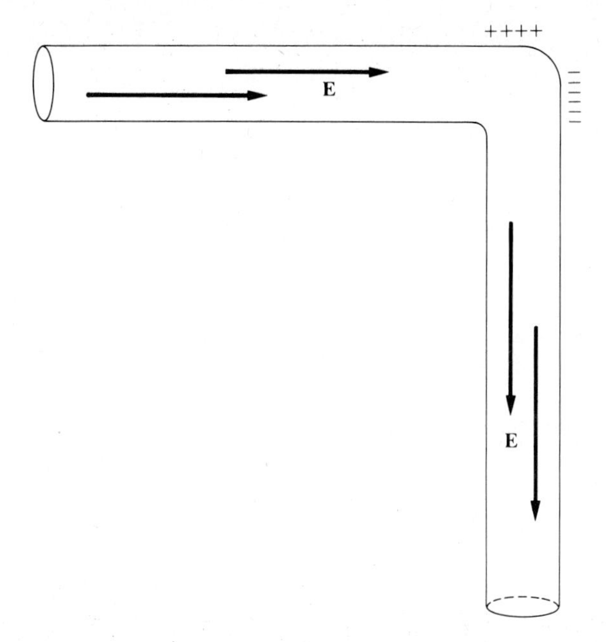


**FIGURE 5-19** Guiding charges on the surface of a conducting wire. These charges change the direction of the conduction electric field to follow the wire.

on one side of the bend is

$$q \approx \sigma S = \varepsilon_0 ES = \varepsilon_0 \frac{J}{g} S = \frac{\varepsilon_0 I}{g}$$

$$\approx \frac{(9 \times 10^{-12})I}{6 \times 10^7} \approx 1.5 \times 10^{-19} I \qquad \textbf{(5-103)}$$

The guiding charge is thus, on the average, very small, only amounting to about one electron for a current of 1 A.

## 5-12 ATMOSPHERIC ELECTRICITY

The earth's atmosphere is the arena for a wide range of electrical activity. At sea level air conductivity is about $3 \times 10^{-14}\,(\Omega \cdot \mathrm{m})^{-1}$; compared with air, the earth is a good conductor. Air is an excellent insulator, but given the size of the earth and the large atmospheric potential differences, significant currents can occur.

To develop a feeling for air conductivities, we consider an isolated conducting sphere of radius $a$ surrounded by a medium of conductivity $g$. The electric field just outside the surface of the sphere is $E_r = Q/(4\pi\varepsilon_0 a^2)$, and the resulting current density is $J_r = gE_r$. The total current flowing away from the sphere is then

$$I = -\frac{dQ}{dt} = gE_r(4\pi a^2) = \frac{g}{\varepsilon_0}Q \tag{5-104}$$

This differential equation for $Q$ has the solution

$$Q(t) = Q(0)e^{-t/(\varepsilon_0/g)} \tag{5-105}$$

The conductivity at sea level is $g \approx 3 \times 10^{-14}\,(\Omega\text{·m})^{-1}$. Thus the decay time constant $\varepsilon_0/g$ of the charge is

$$\frac{\varepsilon_0}{g} \approx \frac{9 \times 10^{-12}}{3 \times 10^{-14}} \approx 5 \text{ min} \tag{5-106}$$

In a time interval of order 5 min charge will leak from any conductor surrounded by air. This result does not depend on the particular geometry of the conductor (see Problem 5–23).

The conductivity of the atmosphere is largely due to the presence of ions that move in the electric field. These ions are formed by the earth's natural radioactivity and by the ionizing radiation of cosmic rays. Industrial pollution decreases air conductivity because the conducting ions attach to the pollutants, decreasing the mobility of the charge carriers.

In the early part of the twentieth century experimenters found that air conductivity increases with height. We now know that the conductivity is roughly given by

$$g(z) = (3.0 + 0.5z^2) \times 10^{-14}\,(\Omega\text{·m})^{-1} \tag{5-107}$$

where the height $z$ above sea level is given in kilometers. At $z = 50$ km the conductivity is 400 times its $z = 0$ value, and the conduction time constant drops to less than a second. The rise of conductivity was a mystery until the discovery of cosmic rays.

A second fact of atmospheric electricity is that near the earth's surface there is a fair-weather field of $E \approx 100$ V/m directed downward. Since the earth's conductivity $[g \approx 10^{-4}\,(\Omega\text{·m})^{-1}]$ is large compared with the atmosphere's, the earth can be considered an equipotential. The charge density induced on the earth's surface by the fair-weather field is

$$\sigma = \varepsilon_0 E \approx -(10^{-11})(10^2) = -10^{-9}\text{ C/m}^2 \tag{5-108}$$

This very small surface charge density becomes appreciable when we consider the charge of the entire earth's surface. The total charge is

$$Q_E = 4\pi R_E^2\sigma \approx -500{,}000\text{ C} \tag{5-109}$$

You may wonder why we are not electrocuted by this large fair-weather field.

The reason is that the low conductivity of the air means that very little current will flow.

Owing to the fair-weather electric field, a current flows downward to the earth. The current density evaluated at the earth's surface is

$$J = gE = (3 \times 10^{-14})(100) = 3 \times 10^{-12} \text{ A/m}^2 \tag{5-110}$$

Integrating this current density over the earth's surface, we find that the total downward current is

$$I = J(4\pi R_E^2) \approx 1500 \text{ A} \tag{5-111}$$

In the absence of a regenerative mechanism the fair-weather field would dissipate in 5 min. The regenerative mechanism is thunderstorm activity. On the average there are about two thousand thunderstorms in progress at any given time. Each storm generates a net upward current of about 1 A from the ground to the upper atmosphere; on a global scale this flow balances the total fair-weather current, as depicted in Figure 5–20. This upward current has been measured by flights above the top of the thunderstorm clouds, which can rise to over 15 km in height.

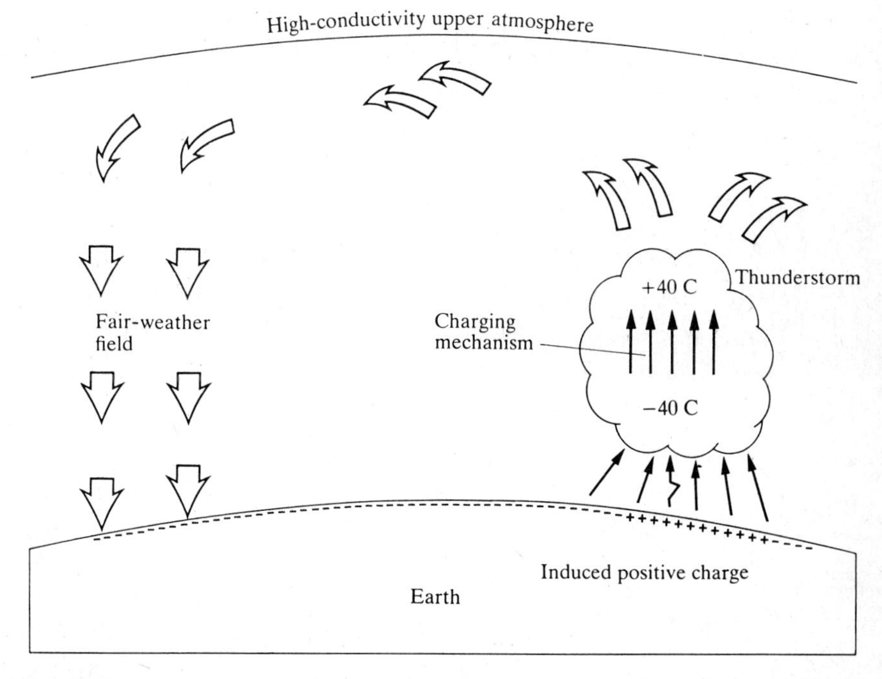


**FIGURE 5–20** Schematic drawing of atmospheric charging from thunderstorms, which is responsible for the fair-weather electric field around the earth

The upward current from the earth to the bottom of the cloud is caused by rather large upward electric fields of about 10 kV/m between the cloud and the ground. The bulk of this 1-A current flows quietly to the bottom of the thundercloud. A minor but very noticeable source of upward current is direct-cloud-to-ground sparks. A single lightning flash can deliver 10 C of electrons to the ground in 50 $\mu$s, but on the average an upward current of only 1/10 A is carried by lightning.

In a thundercloud positive and negative charges become separated, as illustrated in Figure 5–20, with typical charges of about 40 C at the top and $-40$ C at the bottom. The mechanism of charge separation presumably results because the smaller droplets of water or pieces of ice are positively charged. These smaller particles, which have a smaller rate of fall in still air, are carried upward in the thundercloud. The reason for the relation between size and charge is still not understood. It is the charge separation in a thundercloud that gives rise to the electric fields, which cause currents to flow from the ground to the bottom of the thundercloud and from the top of the thundercloud to the upper atmosphere.

# 5-13 CURL OF **B** FROM THE BIOT-SAVART LAW

For a time-independent current density $\mathbf{J}(\mathbf{r})$ the magnetic field is determined by the Biot-Savart law of Equation (5–27). In Equations (5–58) and (5–59) we have expressed the Biot-Savart result in terms of a vector potential $\mathbf{A}(\mathbf{r})$ as

$$\mathbf{B}(\mathbf{r}) = \nabla \times \mathbf{A}(\mathbf{r})$$

with

$$\mathbf{A}(\mathbf{r}) = \frac{\mu_0}{4\pi} \int \frac{\mathbf{J}(\mathbf{r}')}{R}\, dV'$$

From this expression for $\mathbf{B}$ we will now calculate the curl of $\mathbf{B}$; the result gives Ampère's law in differential form.

In terms of $\mathbf{A}$ the curl of $\mathbf{B}$ is

$$\begin{aligned} \nabla \times \mathbf{B} &= \nabla \times (\nabla \times \mathbf{A}) \\ &= \nabla(\nabla \cdot \mathbf{A}) - \nabla^2 \mathbf{A} \end{aligned} \tag{5-112}$$

where the second equation is the vector identity for the triple cross product. From the previous expression for $\mathbf{A}$ the $\nabla^2\mathbf{A}$ term in Equation (5–112) can be written as

$$\nabla_r^2 \mathbf{A} = \frac{\mu_0}{4\pi} \int \mathbf{J}(\mathbf{r}')\, \nabla_r^2 \left(\frac{1}{R}\right) dV' \tag{5-113}$$

We recall from Equations (1–55) and (1–71) that

$$\nabla_{\mathbf{r}}\left(\frac{1}{R}\right) = -\frac{\hat{\mathbf{R}}}{R^2}$$

and that

$$\nabla_{\mathbf{r}} \cdot \left(\frac{\hat{\mathbf{R}}}{R^2}\right) = 4\pi\delta^3(\mathbf{R})$$

from Equation (2–32). Hence the Laplacian of $1/R$ is

$$\nabla_{\mathbf{r}}^2\left(\frac{1}{R}\right) = -4\pi\delta^3(\mathbf{r} - \mathbf{r}') \tag{5-114}$$

Substituting this result into Equation (5–113), we obtain

$$\nabla_{\mathbf{r}}^2\mathbf{A}(\mathbf{r}) = -\mu_0\mathbf{J}(\mathbf{r}) \tag{5-115}$$

This relation is the analog for the vector potential of Poisson's equation for the scalar potentials.

We turn next to the $\nabla(\nabla \cdot \mathbf{A})$ term in Equation (5–112). The divergence of $\mathbf{A}$ is given by

$$\nabla \cdot \mathbf{A} = \frac{\mu_0}{4\pi}\int \mathbf{J}(\mathbf{r}') \cdot \nabla_{\mathbf{r}}\left(\frac{1}{R}\right) dV' = -\frac{\mu_0}{4\pi}\int \mathbf{J}(\mathbf{r}') \cdot \nabla_{\mathbf{r}'}\left(\frac{1}{R}\right) dV' \tag{5-116}$$

With a vector identity this expression becomes

$$\nabla \cdot \mathbf{A} = \frac{\mu_0}{4\pi}\int \frac{\nabla_{\mathbf{r}'} \cdot \mathbf{J}(\mathbf{r}')}{R}\, dV' - \frac{\mu_0}{4\pi}\int \nabla_{\mathbf{r}'} \cdot \left[\frac{\mathbf{J}(\mathbf{r}')}{R}\right] dV' \tag{5-117}$$

The second term in Equation (5–117) can be written as a surface integral, via the divergence theorem, over a surface at infinity. Assuming that the current $\mathbf{J}$ is confined to a finite region of space, this surface integral vanishes, and we are left with

$$\nabla \cdot \mathbf{A} = \frac{\mu_0}{4\pi}\int \frac{\nabla_{\mathbf{r}'} \cdot \mathbf{J}(\mathbf{r}')}{R}\, dV' \tag{5-118}$$

For steady currents $\nabla_{\mathbf{r}'} \cdot \mathbf{J}(\mathbf{r}') = 0$ from the continuity equation, Equation (5–92), and hence

$$\nabla \cdot \mathbf{A} = 0 \tag{5-119}$$

Combining the results of Equations (5–112), (5–115), and (5–119), we have

**Ampère's Law (Differential Form)**

$$\nabla \times \mathbf{B} = \mu_0 \mathbf{J} \tag{5-120}$$

which is valid for steady currents. This equation is the differential form of Ampère's law. After proving the curl theorem of vector calculus in the next section, we will be able to cast this result in an integral form.

## 5-14 CURL THEOREM

The curl theorem of vector calculus, also called Stokes's theorem, relates a line integral about a closed circuit to a surface integral over a region bounded by the circuit. The statement of the curl theorem is

**Curl Theorem**

$$\oint_C \mathbf{E} \cdot d\mathbf{r} = \int_S (\nabla \times \mathbf{E}) \cdot d\mathbf{S} \tag{5-121}$$

where $\mathbf{E}$ is an arbitrary vector function and $S$ is any surface capping the circuit $C$. The unit normal $\hat{\mathbf{n}}$ to the surface is defined in the sense of the right-hand rule with respect to the direction of the line integral around $C$; see Figure 5–21. We proceed now with a proof.

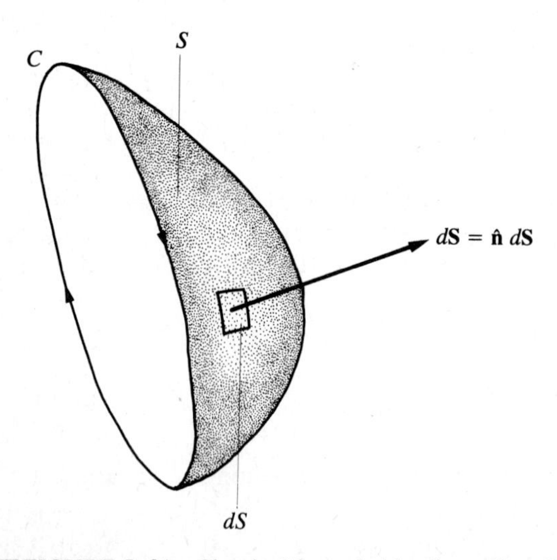


**FIGURE 5-21** Circuit $C$ bounded by the surface $S$

We express the surface integral on the right-hand side of Equation (5–121) in Cartesian components:

$$\int (\nabla \times \mathbf{E}) \cdot d\mathbf{S} = \int \left[ \left( \frac{\partial E_z}{\partial y} - \frac{\partial E_y}{\partial z} \right) dS_x + \left( \frac{\partial E_x}{\partial z} - \frac{\partial E_z}{\partial x} \right) dS_y + \left( \frac{\partial E_y}{\partial x} - \frac{\partial E_x}{\partial y} \right) dS_z \right] \tag{5-122}$$

We first consider the terms involving $E_x$:

$$I_x \equiv \int \left( \frac{\partial E_x}{\partial z} dS_y - \frac{\partial E_x}{\partial y} dS_z \right) \tag{5-123}$$

For definiteness we consider the surface $S$ and the circuit $C$ illustrated in Figure 5–22. For this surface the area elements are

$$dS_y = -dx\, dz \qquad dS_z = dx\, dy \tag{5-124}$$

The minus sign in $dS_y$ is due to the fact that $\hat{\mathbf{y}} \cdot \hat{\mathbf{n}}$ is negative for $dx > 0$, $dz > 0$, and is positive for $dx > 0$, $dz < 0$. Inserting Equation (5–124) in Equation (5–123), we have

$$I_x = -\int dx \int \left( \frac{\partial E_x}{\partial z} dz + \frac{\partial E_x}{\partial y} dy \right) \tag{5-125}$$

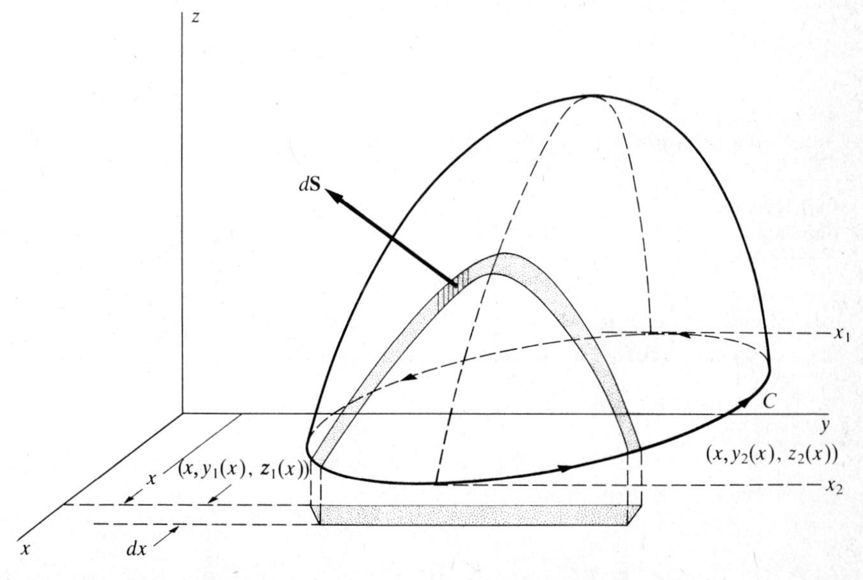


**FIGURE 5–22** Surface $S$ and curve $C$ used in the proof of the curl theorem. The shaded element illustrates the change in direction of $d\mathbf{S}$ as the $x$ coordinate varies.

In the inside integral $x$ is fixed. The integrand is the total differential at fixed $x$:

$$dE_x(\text{fixed } x) = \frac{\partial E_x}{\partial z}\,dz + \frac{\partial E_x}{\partial y}\,dy \tag{5-126}$$

The integral of the total differential $dE_x$ is $E_x$ evaluated at the integration limits; hence

$$I_x = -\int_{x_1}^{x_2} [E_x(x, y_2, z_2) - E_x(x, y_1, z_1)]\,dx \tag{5-127}$$

where the limits $y_1, z_1$ and $y_2, z_2$, which lie on the curve $C$, depend on the value of $x$. The integral in Equation (5–127) can be rewritten as

$$I_x = \int_{x_2}^{x_1} E_x(x, y_2, z_2)\,dx + \int_{x_1}^{x_2} E_x(x, y_1, z_1)\,dx \tag{5-128}$$

We recognize this expression as the line integral of $E_x\,dx$ around the bounding curve $C$:

$$I_x = \oint_C E_x\,dx \tag{5-129}$$

Similar considerations for the other components of **E** complete the proof of the curl theorem of Equation (5–121).

A corollary of the curl theorem is obtained by the substitution of $\mathbf{E} = \mathbf{c}f(\mathbf{r})$ into Equation (5–121), where **c** is an arbitrary constant vector and $f(\mathbf{r})$ is a scalar function. Using the identity

$$\nabla \times (\mathbf{c}f) = -\mathbf{c} \times \nabla f \tag{5-130}$$

we obtain

**Curl Theorem Corollary**

$$\oint f\,d\mathbf{r} = \int d\mathbf{S} \times \nabla f \tag{5-131}$$

As a simple application of the curl theorem, we now show that electric fields for which $\nabla \times \mathbf{E} = 0$ have a unique potential (see Section 1–4). By the theorem a vanishing curl implies that

$$\oint \mathbf{E} \cdot d\mathbf{r} = 0 \tag{5-132}$$

For one path (I) from point $A$ to point $B$ and a different path (II) from $B$ back to $A$, Equation (5–132) yields

$$\underset{\text{path I}}{\int_A^B \mathbf{E} \cdot d\mathbf{r}} = \underset{\text{path II}}{\int_A^B \mathbf{E} \cdot d\mathbf{r}} \tag{5-133}$$

From $\mathbf{E} \cdot d\mathbf{r} = -d\Phi$, we conclude that the potential difference is independent of path, and therefore a unique potential exists up to an additive constant. So we complete the proof that $\nabla \times \mathbf{E} = 0$ is a necessary and sufficient condition for the existence of a unique potential.

## 5-15 AMPÈRE'S CIRCUITAL LAW

The differential form of Ampère's law of Equation (5–120) can be cast into a useful integral form by use of the curl theorem. In the curl theorem, Equation (5–121), we take the vector field **E** to be the **B** field, whose curl is $\nabla \times \mathbf{B} = \mu_0 \mathbf{J}$. So the theorem reads

**Ampère's Circuital Law**

$$\oint_C \mathbf{B} \cdot d\mathbf{r} = \int_S \mu_0 \mathbf{J} \cdot d\mathbf{S} = \mu_0 I \qquad \textbf{(5–134)}$$

where $I$ is the (steady) current flowing through a surface bounded by the circuit $C$. In symmetrical situations Ampère's circuital law provides a convenient means of calculating **B** when the tangential component of **B** along the path of integration is a constant. This use parallels the use of Gauss's law in electrostatics in computing **E** when the normal component of **E** is constant.

## 5-16 APPLICATION: MAGNETIC FIELD INSIDE A CURRENT-CARRYING WIRE

As a simple application of Ampère's circuital law, we consider steady current flow through a long uniform straight wire. By Ohm's law

$$\mathbf{E} = \frac{\mathbf{J}}{g}$$

holds, where **E** is the electrostatic field in the wire. For a uniform ideal wire $g$ is constant inside the wire. From Equation (5–94) the electric potential, from which **E** can be derived, satisfies Laplace's equation. In cylindrical coordinates Laplace's equation for a potential $\Phi$ that is symmetrical around the $z$ axis is

$$\frac{1}{r}\frac{\partial}{\partial r}\left(r\frac{\partial \Phi}{\partial r}\right) + \frac{\partial^2 \Phi}{\partial z^2} = 0 \qquad \textbf{(5–135)}$$

We try a separation of variables,

$$\Phi(r, z) = R(r)Z(z) \qquad \textbf{(5–136)}$$

obtaining

$$\frac{1}{\mathscr{r}}\frac{1}{R}\frac{d}{d\mathscr{r}}\left(\mathscr{r}\frac{dR}{d\mathscr{r}}\right) = -\frac{1}{Z}\frac{d^2Z}{dz^2} = \text{constant} \tag{5-137}$$

The separation constant must be zero to realize a uniform electric field in the $z$ direction. With a zero value for the separation constant the solutions are

$$R(\mathscr{r}) = A + B \ln \mathscr{r} \qquad Z(z) = C + Dz \tag{5-138}$$

Since the current is confined to the wire, the radial component of **J** must vanish at $\mathscr{r} = a$. Using $J_{\mathscr{r}} = -g(\partial\Phi/\partial\mathscr{r})$, we conclude that the parameter $B$ in Equation (5–138) vanishes. The final potential solution is then

$$\Phi = C + Dz \tag{5-139}$$

where we have set $A = 1$ without loss of generality. This solution yields

$$\mathbf{E} = -D\hat{\mathbf{z}} \tag{5-140}$$

Since $\mathbf{J} = g\mathbf{E}$, this relation implies that **J** is uniform over the cross section of wire, since $g$ and **E** are uniform. Having established this result, we can now calculate the **B** field inside and outside the wire by use of Ampère's circuital law.

The current $I$ along the wire gives rise to a magnetic field **B**, which by symmetry arguments is azimuthal in direction and independent of $\phi$ and $z$. To evaluate the line integral of Ampère's circuital law, Equation (5–134), we choose a circular path around the wire of radius $\mathscr{r}$, as illustrated in Figure 5–23. We obtain

$$2\pi\mathscr{r}B_\phi = \begin{cases} \mu_0 I, & \text{for } \mathscr{r} > a \\ \mu_0 I\left(\dfrac{\pi\mathscr{r}^2}{\pi a^2}\right), & \text{for } \mathscr{r} < a \end{cases} \tag{5-141}$$

where $a$ is the radius of the wire. Thus $B_\phi$ is given by

$$B_\phi = \begin{cases} \dfrac{\mu_0 I}{2\pi\mathscr{r}}, & \text{for } \mathscr{r} > a \\ \dfrac{\mu_0 I\mathscr{r}}{2\pi a^2}, & \text{for } \mathscr{r} < a \end{cases} \tag{5-142}$$

A sketch of the magnitude of $B_\phi$ is given in Figure 5–24. The result that **J** is constant over the conductor cross section holds only for steady currents. For high-frequency alternating currents the current mostly flows near the surface of the conductor.

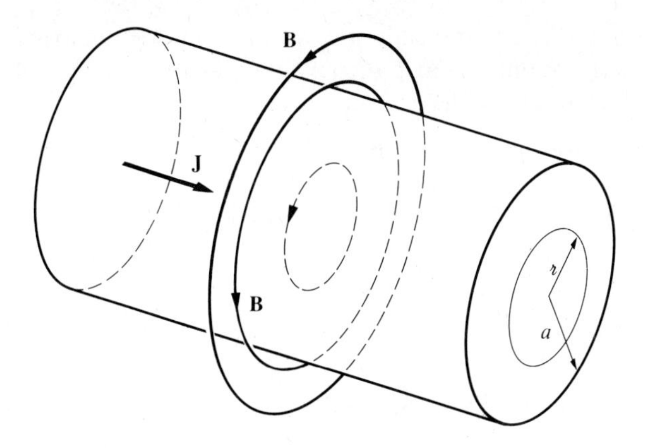


**FIGURE 5-23** Magnetic field due to a current-carrying wire of radius $a$, determined by Ampère's circuital law

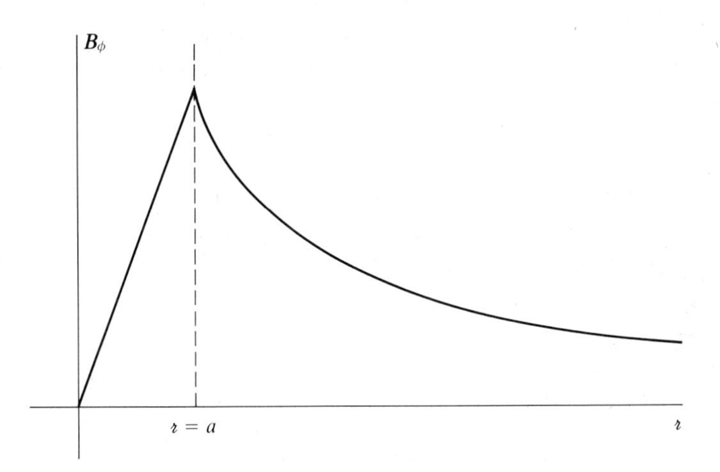


**FIGURE 5-24** Magnetic field of a current-carrying wire of radius $a$

## 5-17 APPLICATION: MAGNETIC FIELD OF AN INFINITE SOLENOID

In Section 5–8 the magnetic field on the axis of a finite solenoid was calculated from the Biot-Savart law. Here we use Ampère's circuital law to determine the magnetic field of an infinite solenoid, both inside and outside.

We consider a solenoid of $n$ turns per unit length with layers of helical windings of opposite pitch, so the net current flow is azimuthal around a cylinder of radius $a$. From the Biot-Savart law $B_\phi = 0$, since the current elements are azimuthal; also, $B_r = 0$, since current elements symmetrical about

$z = 0$ have opposing $B_{\mathscr{r}}$. The translational symmetry along the axis and the rotational symmetry about the axis of the solenoid lead us to look for a solution in which $B_z$ does not depend on $z$ or $\phi$:

$$\mathbf{B}(\mathscr{r}, z, \phi) = B_z(\mathscr{r})\hat{\mathbf{z}} \tag{5-143}$$

Next, we apply Ampère's law to a rectangular circuit that is outside the solenoid (circuit 1 in Figure 5–25). Since no current is enclosed,

$$\int_{C_1} \mathbf{B} \cdot d\mathbf{r} = 0 \tag{5-144}$$

From Equations (5–143) and (5–144) we get

$$B_z(\mathscr{r}_2) - B_z(\mathscr{r}_1) = 0 \tag{5-145}$$

Hence $B_z$ is a constant outside the solenoid. A similar argument for a circuit totally inside the solenoid (circuit 2 in Figure 5–25) leads to the conclusion that $B_z$ is also independent of $\mathscr{r}$ inside the solenoid.

To find the constant value of $B_z$ outside, we consider a plane perpendicular to the solenoid axis. The flux that passes through the plane on the inside of the solenoid must return on the outside. Since the outside flux is the product of the constant outside value of $B_z$ times the outside area, which is infinite, the value of $B_z$ outside must be zero.

To determine the constant value of $B_z$ inside the solenoid, we apply Ampère's law to a rectangular circuit intersecting the sides of the solenoid (circuit 3 of Figure 5–25). Since $B_z = 0$ on the outer side and $B_{\mathscr{r}} = 0$ on the

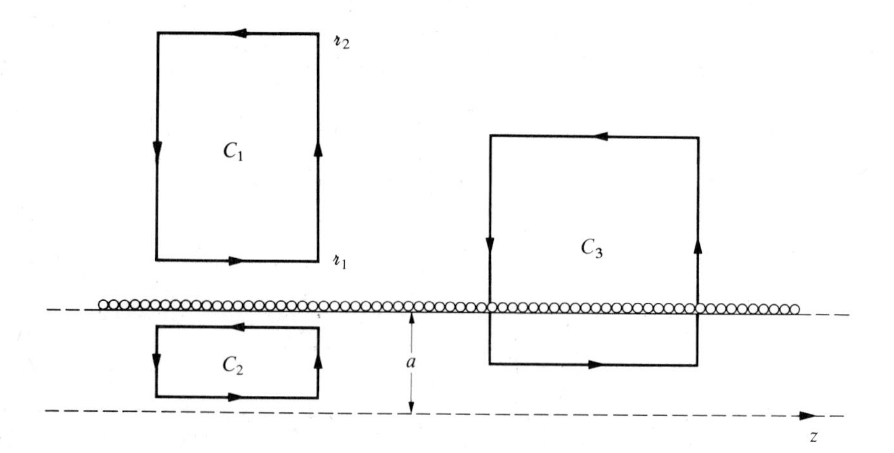


**FIGURE 5–25** Three circuits used to determine the magnetic field of the solenoid from Ampère's circuital law

perpendicular sides, we find

$$B_z(\text{inside})\ell = \mu_0 n \ell I \tag{5-146}$$

Hence the magnetic field of an infinite solenoid is

$$\mathbf{B} = \begin{cases} \mu_0 n I \hat{\mathbf{z}}, & \text{for } \imath < a \\ 0, & \text{for } \imath > a \end{cases} \tag{5-147}$$

This result agrees with Equation (5–52) for the magnetic field on the axis of a long solenoid.

To find the vector potential that gives the **B** field of the solenoid, we consider the magnetic flux through an open surface $S$ (i.e., a surface with an edge),

$$\mathscr{F} = \int_S \mathbf{B} \cdot d\mathbf{S} = \int_S \nabla \times \mathbf{A} \cdot d\mathbf{S} \tag{5-148}$$

This relation can be rewritten, via the curl theorem of Equation (5–121), as

$$\mathscr{F} = \oint_C \mathbf{A} \cdot d\mathbf{r} \tag{5-149}$$

where $C$ is the circuit bounding $S$. We specialize to a circular surface of radius $\imath$ perpendicular to the solenoid axis, as illustrated in Figure 5–26. Equation (5–149) yields

$$\mathscr{F} = 2\pi \imath A_\phi \tag{5-150}$$

which we equate to

$$\mathscr{F} = \begin{cases} \pi \imath^2 B, & \text{for } \imath < a \\ \pi a^2 B, & \text{for } \imath > a \end{cases} \tag{5-151}$$

We obtain

$$A_\phi = \begin{cases} \dfrac{\mu_0 n I \imath}{2}, & \text{for } \imath < a \\ \\ \dfrac{\mu_0 n I a^2}{2\imath}, & \text{for } \imath > a \end{cases} \tag{5-152}$$

as the vector potential of a long solenoid. The discerning student will observe the similarity of this calculation to that of the magnetic field of a long, straight uniform wire. The two calculations employ $\nabla \times \mathbf{A} = \mathbf{B}$ and $\nabla \times \mathbf{B} = \mu_0 \mathbf{J}$, respectively.

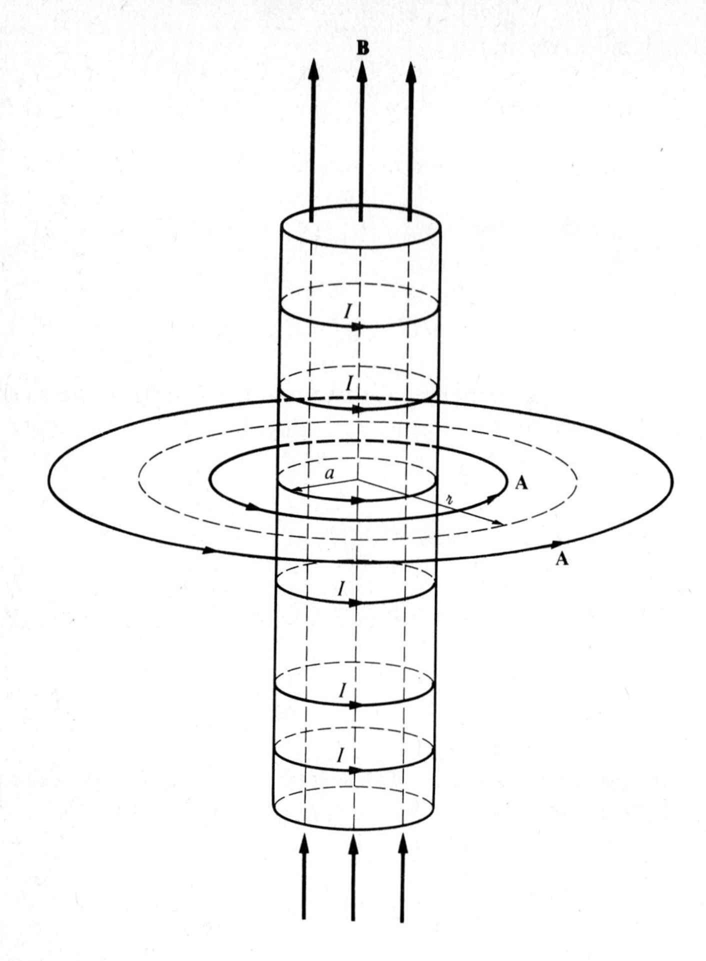


**FIGURE 5-26** Circular loop of radius $\mathscr{r}$. The line integral of the vector potential **A** around the loop gives the magnetic flux through the surface bounded by the loop.

## SUMMARY

### Important Concepts and Equations

Current density (**J**)

*The current density* **J** *is the current per unit area perpendicular to the direction of current flow.*

$$\nabla \cdot \mathbf{J} + \frac{\partial \rho}{\partial t} = 0 \qquad \text{(continuity equation)}$$

*This equation is a consequence of charge conservation.*

$$I = \frac{dq}{dt} = \int \mathbf{J} \cdot d\mathbf{S}$$

*The current $I$ (the rate of charge flow $dq/dt$) through a surface is equal to the normal component of current density* **J** *integrated over the surface.*

$$d\mathbf{F} = (\rho\mathbf{E} + \mathbf{J} \times \mathbf{B})\, dV \qquad \text{(Lorentz force, volume distributions)}$$

*The differential form of the Lorentz force for a volume charge density $\rho$ and current density* **J**.

$$d\mathbf{F} = (\sigma\mathbf{E} + \mathbf{j} \times \mathbf{B})\, dS \qquad \text{(Lorentz force, surface distributions)}$$

*The differential form of the Lorentz force for a surface charge density $\sigma$ and a surface current density* **j**.

$$d\mathbf{F} = \lambda\mathbf{E}\, dr + I\, d\mathbf{r} \times \mathbf{B} \qquad \text{(Lorentz force, line distributions)}$$

*The differential form of the Lorentz force for a line charge density $\lambda$ and a current $I$; $I$ is the charge per unit time passing a point on the line.*

$$\mathbf{B} = \frac{\mu_0}{4\pi} \int \mathbf{J}(\mathbf{r}') \times \frac{\hat{\mathbf{R}}}{R^2}\, dV' \qquad \text{(Biot-Savart law)}$$

*The magnetic field* **B** *produced from a steady current distribution; $\mu_0$ is the permeability of empty space, and $\mathbf{R} = \mathbf{r} - \mathbf{r}'$. For surface currents $\mathbf{J}\, dV'$ is replaced by $\mathbf{j}\, dS$. For line currents $\mathbf{J}\, dV'$ is replaced by $I\, d\mathbf{r}$.*

$$\mathbf{B} = \frac{\mu_0 I}{2\pi \imath}\, \hat{\phi}$$

*The magnetic field of a long straight wire is directed around the wire and falls off inversely with the distance from the wire.*

$$\mathbf{F}_{21} = -\frac{\mu_0 I_1 I_2}{2\pi \imath}\, \hat{\imath}$$

*The force between two long parallel wires carrying currents $I_1$ and $I_2$, separated by a distance $\imath$. If the currents are in the same direction, the wires attract each other.*

$$\mathbf{F}_{21} = \mu_0 \frac{I_1 I_2}{4\pi} \oint_{C_2} \oint_{C_1} \frac{d\mathbf{r}_2 \times (d\mathbf{r}_1 \times \hat{\mathbf{R}})}{R^2} \qquad \text{(Ampère's force law)}$$

*The general expression for the force between two closed circuits $C_1$ and $C_2$ carrying steady currents $I_1$ and $I_2$, where $\mathbf{R} = \mathbf{r}_2 - \mathbf{r}_1$.*

Helmholtz coil

*A Helmholtz coil is two circular coils with the same radius $a$ and a common axis separated by a distance $\ell = a$; they give an approximately uniform $B$ field between the coils.*

$$\mathbf{A} = \frac{\mu_0}{4\pi}\int \frac{\mathbf{J}(\mathbf{r}')}{R}\,dV' \qquad \text{(vector potential)}$$

*The vector potential* $\mathbf{A}$*;* $\mathbf{J}(\mathbf{r}')$ *is the current density,* $R$ *is the distance from the source point to the field point, and the integral is over the volume of the source.*

$$\mathbf{B} = \nabla \times \mathbf{A}$$

*The magnetic field* $\mathbf{B}$ *is given by the curl of the vector potential* $\mathbf{A}$*.*

$$\mathbf{J} = g\mathbf{E} \qquad \text{(Ohm's law)}$$

*The current density* $\mathbf{J}$ *is given by the conductivity* $g$ *multiplied by the electric field* $\mathbf{E}$*.*

$$R \equiv \frac{\ell}{Sg}$$

*The resistance of a uniform conductor* $R$ *is directly proportional to the length* $\ell$ *and inversely proportional to the cross-sectional area* $S$ *and the conductivity* $g$*. Ohm's law is then* $V = IR$*.*

$$P = \int \mathbf{E}\cdot\mathbf{J}\,dV$$

*The power* $P$ *delivered to charges is given by the integral of the dot product of the electric field* $\mathbf{E}$ *and the current density* $\mathbf{J}$*. For a uniform wire* $P = VI$*, which can be expressed by using Ohm's law as* $P = I^2R$*.*

$$\nabla\cdot\mathbf{J} = 0 \qquad \text{(steady current flow)}$$

*Steady current flow means that at a given point in space* $\mathbf{J}$ *or* $\rho$ *is not a function of time. From the equation of continuity* $\nabla\cdot\mathbf{J} = 0$ *for steady current flow. If the conducting medium obeys Ohm's law, then* $\nabla\cdot[g(\mathbf{r})\,\nabla\Phi] = 0$*, where the conductivity* $g$ *may be a function of the radius. For constant conductivity* $\nabla^2\Phi = 0$*.*

$$\nabla\times\mathbf{B} = \mu_0\mathbf{J} \qquad \text{(Ampère's law, differential form)}$$

*The curl of the magnetic field* $\nabla\times\mathbf{B}$ *is equal to the permeability of free space* $\mu_0$ *times the current density* $\mathbf{J}$*.*

$$\oint_C \mathbf{E}\cdot d\mathbf{r} = \int_S (\nabla\times\mathbf{E})\cdot d\mathbf{S} \qquad \text{(curl theorem)}$$

*The line integral of a vector field* $\mathbf{E}$ *around a closed circuit* $C$ *is equal to the surface integral of the curl of the vector field* $\nabla\times\mathbf{E}$ *over any surface* $S$ *capping the circuit* $C$*.*

$$\oint_C f\,d\mathbf{r} = \int_S d\mathbf{S} \times \nabla f \qquad \text{(curl theorem corollary)}$$

*The integral of a scalar quantity around a closed circuit is equal to the integral of the surface element crossed into the gradient of the scalar quantity, where the surface caps the circuit.*

$$\oint_C \mathbf{B} \cdot d\mathbf{r} = \mu_0 I \qquad \text{(Ampère's circuit law, steady current)}$$

*The line integral of the magnetic field* **B** *around a closed circuit is directly proportional to the current I passing through the surface bounded by the circuit.*

$$\mathbf{B} = \begin{cases} \mu_0 n I \hat{\mathbf{z}}, & \text{inside} \\ 0, & \text{outside} \end{cases}$$

*The magnetic field of an infinitely long solenoid along the z axis is constant inside the solenoid and zero outside. The field is proportional to n, the number of turns per unit length, and the current I.*

# PROBLEMS

## Section 5-1: Magnetic Forces and the Continuity Equation

**5-1** For most purposes it does not matter whether the charge carriers giving rise to conduction are positive or negative. The Hall effect gives striking experimental evidence not only of the sign but also of the carrier density. When a conductor is placed in a magnetic field perpendicular to the direction of current flow, a voltage is developed in a direction perpendicular to both the current and the magnetic field. This voltage is the Hall voltage. For the strip conductor shown in Figure 5–27, find an expression for $ne$ (where $n$ is the volume carrier density) in terms of the applied magnetic field, the Hall voltage, the current, and the thickness of the strip. Discuss the two cases of positive or negative carriers. Compute numerically the Hall voltage for a copper strip with $B = 1$ T, $I = 1$ A, $n = 8.5 \times 10^{28}$ m$^{-3}$, and a strip thickness of 1 mm.

## Section 5-2: Biot-Savart Law

**5-2** A nonconducting sphere of radius $a$ has a uniform volume charge density with total charge $Q$. The sphere rotates about an axis through its center with constant angular velocity $\omega$. Find the magnetic field at the center of the sphere.

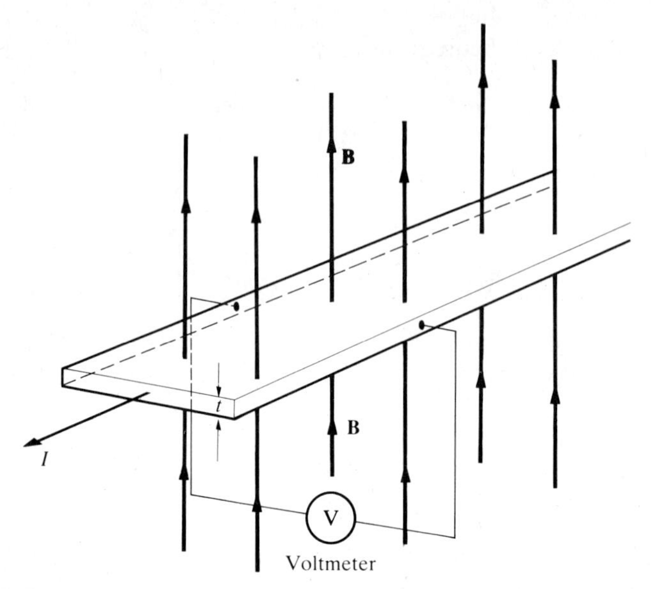


**FIGURE 5-27** Conducting strip and Hall voltage

## Section 5-3: Magnetic Field of a Straight Wire

**5-3** A particle with charge $q$ is traveling with velocity $\mathbf{v}$ parallel to a wire at a distance $\imath$. The wire has a uniform linear charge distribution $\lambda$ per unit length and also carries a current $I$. For what velocity $\mathbf{v}$ will the force on the particle vanish?

**5-4** The angle the magnetic field of the earth makes with the horizontal is 60°, and the magnitude of the field is $5 \times 10^{-5}$ T. Find the force on a 3-m-long wire carrying 10 A in the following positions.

(i) The wire is vertical.

(ii) The wire is horizontal and parallel to the horizontal component of the earth's field.

(iii) The wire is horizontal and perpendicular to the horizontal component of the earth's field.

**5-5** From Ampère's law for the force between two closed circuits, show that $\mathbf{F}_{21} = -\mathbf{F}_{12}$, that is, that Newton's third law holds. Find an alternate form of Ampère's force law that explicitly exhibits this property.

**5-6** An infinitely long, cylindrical conductor has an off-center cylindrical hole cut in it, as shown in Figure 5-28. The remaining portion of the conductor carries a uniform current density $J$ into the plane of the paper. Find the force exerted on this conductor by a parallel current filament a distance $D$ away.

**5-7** Consider three straight, long coplanar wires of negligible radius equally spaced a distance $d$ apart, each carrying a current $I$ in the same direction.

(i) Calculate the location of the two zeros in the magnetic field.

(ii) Sketch the magnetic field line pattern.

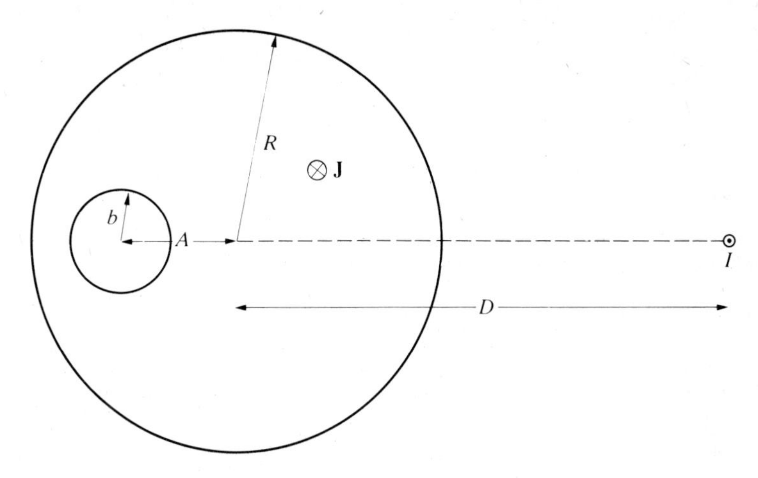


**FIGURE 5-28** Infinitely long current-carrying cylindrical conductor with a cylindrical cavity in the vicinity of a parallel line current

(iii) The middle wire is rigidly displaced a very small distance ($\ll d$) toward one of the wires while the outer wires are held fixed. Describe qualitatively the subsequent motion of the middle wire.

**5-8** A current $I$ flows in along the positive part of the $x$ axis and out along the positive part of the $y$ axis. What is the magnetic field (all three components) along the positive part of the $z$ axis?

## Section 5-5: Magnetic Fields in the Laboratory and Nature

**5-9** The Faraday disk dynamo was discussed in Section 5–5. A uniform magnetic field $\mathbf{B}$ passes through the disk parallel to its axis.

(i) Find the potential difference between the center and the rim of the disk in terms of the field $\mathbf{B}$, the rim radius $a$, and the disk angular velocity $\omega$.

(ii) The resistance of the circuit is $R$. Find the torque required to maintain a constant angular velocity.

**5-10** Suppose that a star of radius $10^6$ km and a uniform internal magnetic field of $10^{-2}$ T undergoes gravitational collapse to a neutron star of radius 5 km. Assume that the magnetic field remains uniform and that the magnetic flux going through the star remains constant owing to the high conductivity of the stellar interior. Find the magnitude of the magnetic field in the neutron star.

## Section 5-6: Magnetic Field on the Axis of a Current Loop

**5-11** A direct current $I$ through a wire formed into a circular loop makes a magnetic field $\mathbf{B}_1$ at the center of the loop. If the same length of wire is wound in a smaller loop of $N$ turns, what is the magnetic field $\mathbf{B}_N$ at the center of the loop in terms of $N$ and $\mathbf{B}_1$ if the current is kept the same? Neglect the cross-sectional area of the wire.

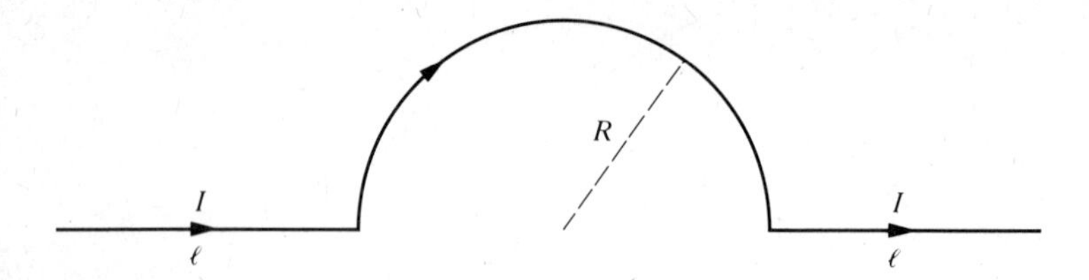


**FIGURE 5-29** Line current with a semicircular segment

**5-12** A current $I$ flows through a wire bent into a semicircular arc of radius $R$ and two straight sections of length $\ell$, as shown in Figure 5-29. Using the Biot-Savart law, find the following magnetic fields at the center of the semicircle.

(i) Resulting from the semicircular segment.

(ii) Resulting from the two straight-line segments.

## Section 5-7: Helmholtz Coil

**5-13** In Section 5-7 we compute the magnetic field of a Helmholtz coil on the axis nearly equidistant between the coils ($z \approx 0$). The result can be written as

$$B_z(z) = B_z(0)[1 - \beta(z/a)^4 + \cdots]$$

where $B_z(0)$ was explicitly calculated. Continue the procedure described in the text to show that $\beta = 144/125$.

## Section 5-8: Solenoid

**5-14** Find the magnetic field on the axis of a long solenoid with $n$ turns per unit length and square cross section of side dimension $a$.

**5-15** Use the Biot-Savart law to show that the magnetic field of an infinitely long solenoid of radius $a$, with $n$ turns per unit length, each carrying current $I$, is given at an arbitrary point $(\imath, z, \phi)$ by

$$\mathbf{B} = \hat{\mathbf{z}}\frac{\mu_0 n I a}{2\pi}\int_0^{2\pi} \frac{(a - \imath\cos\phi')\,d\phi'}{(a^2 + \imath^2 - 2a\imath\cos\phi')}$$

Then show that integration gives

$$\mathbf{B} = \begin{cases} \hat{\mathbf{z}}\mu_0 n I, & \text{for } \imath < a \\ 0, & \text{for } \imath > a \end{cases}$$

## Section 5-9: Vector Potential and Divergence of B

**5-16** Calculate the magnetic fields associated with the following vector potentials.

(i) $\mathbf{A}(\mathbf{r}) = \frac{1}{c}\mathbf{c} \times \mathbf{r}$ (ii) $\mathbf{A}(\mathbf{r}) = c\mathbf{r}$

(iii) $\mathbf{A}(\mathbf{r}) = \dfrac{\mathbf{c} \times \mathbf{r}}{r^3}$ for $r \neq 0$ (iv) $\mathbf{A}(\mathbf{r}) = -\hat{\mathbf{z}}\dfrac{\mu_0 I}{2\pi} \ln\left(\dfrac{\imath}{\ell}\right)$ for $\imath \neq 0$

where $\mathbf{c}$ is a constant vector.

**5-17** A cylindrically symmetric magnetic field in a current-free region has a $z$ component given by $B_z = Cz$. Find the other components, using $\nabla \cdot \mathbf{B} = 0$.

**5-18** For a circular wire loop of radius $a$ carrying current $I$, calculate the magnetic field on the axis of the loop, using the Biot-Savart law and symmetry. Then, use $\nabla \cdot \mathbf{B} = 0$ to find the cylindrical component $B_\imath(\imath, z)$ near the axis.

**5-19** A wire loop of radius $a_1$ carrying current $I_1$ is coaxial with a large loop of radius $a_2 \gg a_1$ carrying current $I_2$. The planes of the loops are separated by a distance $d$. Find the force on the smaller loop. (*Hint*: Use the result of the previous problem.)

**5-20** The magnetic field on the axis of a Helmholtz coil was found in Section 5-7 and Problem 5-13 to be $B_z(z) \approx B_z(0)[1 - \beta(z/a)^4]$ for small $z$ (i.e., near the midpoint between the coils). Use this result to find the total magnetic field for small $z$ and $\imath$. [*Hint:* In a current-free region $\nabla \times \mathbf{B} = 0$ and hence $\mathbf{B} = \nabla\Phi$ for some scalar function $\Phi$. The divergence condition $\nabla \cdot \mathbf{B} = 0$ implies $\Phi$ satisfies the Laplace equation. Look for solutions of the form $r^n P_n(\cos\theta)$, which give the correct $B_z(z)$.]

## Section 5-10 : Field of a Current Sheet

**5-21** Two parallel sheets of metal, each 0.1 m in width and infinitely long (in a direction perpendicular to the paper), each carry a current of 100 A. The current in the left sheet is out of the paper, and the current in the right sheet is into the paper. The sheets are separated by 1 mm.

(i) Calculate (ignoring fringing) the **B** field between the two sheets and immediately to the left of the left sheet.

(ii) Calculate the force per unit area on the left sheet.

**5-22** An infinitely long, nonconducting conveyor belt with a uniform charge density $\sigma$ moves with a constant velocity $\mathbf{v} = v\hat{\mathbf{z}}$ along its length.

(i) What is the surface current density $\mathbf{j}$?

(ii) Relate the vector potential $\mathbf{A}(\mathbf{r})$ of the moving belt at any fixed position $\mathbf{r}$ above the belt to the electrostatic potential $\Phi(\mathbf{r})$ of the belt at rest. (*Hint:* You can use integral expressions for $\mathbf{A}$ and $\Phi$ without evaluation of the integrals.)

(iii) Using the relation between $\mathbf{A}$ and $\Phi$ deduced in part (ii), relate the magnetic field $\mathbf{B}(\mathbf{r})$ due to the moving belt to the electric field $\mathbf{E}(\mathbf{r})$ of the static belt.

## Section 5-11: Current Flow and Ohm's Law

**5-23** At time $t = 0$ the charge distribution in a large conductor is $\rho(\mathbf{r}, 0)$. Obtain $\rho(\mathbf{r}, t)$ for subsequent times, and find an expression for the time constant describing the approach to the equilibrium condition. Find this time constant for copper and for air. [For copper $g = 5.8 \times 10^7\,(\Omega \cdot \mathrm{m})^{-1}$, and for air $g = 3 \times 10^{-14}\,(\Omega \cdot \mathrm{m})^{-1}$. *Hint:* Use the equation of continuity.]

**5-24** The space between concentric cylindrical conductors of radii $a$ and $b$ is filled with a conducting material of conductivity $g$. Determine the electric current per unit length and the resistance per unit length, $R = \Delta V/I$, where $\Delta V$ is the potential difference between the conductors.

**5-25** A spherical conducting electrode of radius $a$ is surrounded by a medium obeying

Ohm's law and whose conductivity $g(r)$ is a function of distance from the center of the electrode. A steady current flows from the electrode.

(i) Derive a relation between the electric field, the conductivity $g(r)$, and the total current $I$.

(ii) What should $g(r)$ be so that the charge density in the medium is (a) zero and (b) constant?

**5-26** A metal sphere of radius $a$ is surrounded by a concentric metal sphere of inner radius $b$, where $b > a$. The space between the spheres is filled with a material whose electric conductivity $g$ varies with the electric field strength $E$ according to the relation $g = kE$, where $k$ is a constant. A potential difference $V$ is maintained between the two spheres. What is the current between the spheres?

**5-27** The potential on the surface of a homogeneous, isotropic sphere of conductivity $g$ is $B_0 \cos\theta$ ($\theta$ is the usual polar angle measured with respect to an axis through the center of the sphere). Find the current density $\mathbf{J}$ at all points inside the sphere.

**5-28** The capacitance of the region between two isolated conductors separated by a dielectric of permittivity $\varepsilon$ is $C$. The region is filled with a material of conductivity $g$, and the resistance between the conductors is $R$. Show that $RC = \varepsilon/g$.

## Section 5-12: Atmospheric Electricity

**5-29** From the air conductivity given in Equation (5–107) and the fair-weather field $E = 100$ V/m at the earth's surface, find the potential difference between the surface and an altitude where the conductivity is much larger than the surface value. You may assume the earth is locally flat.

**5-30** Using the air conductivity of Equation (5–107), find the volume space charge density. Integrate to find the total space charge in a 1-m$^2$ vertical prism, and compare this result with the surface charge density. Assume the earth is locally flat for this calculation.

## Section 5-14: Curl Theorem

**5-31** With the divergence theorem corollary of Equation (2–15), the curl can be defined as

$$\text{curl } \mathbf{E} = \lim_{\Delta V \to 0} \frac{\oint d\mathbf{S} \times \mathbf{E}}{\Delta V}$$

Use this definition to find the curl $\mathbf{E}$ in spherical coordinates (in analogy to Section 2–2).

**5-32** By expansion of $\nabla \times \mathbf{E}$ in Cartesian coordinates, find the cylindrical expression for curl $\mathbf{E}$ by the method of Section 2–5C.

## Section 5-15: Ampère's Law

**5-33** A long coaxial cable consists of two concentric conductors. The inner conductor has radius $a$. The outer conductor has inner radius $b$ and outer radius $c$. The conductors carry equal and opposite steady currents $I$. Find the magnetic field $\mathbf{B}$ in the four regions $\imath < a$, $a < \imath < b$, $b < \imath < c$, and $\imath > c$.

**5-34** A steady uniform surface current flows on the surface of a hollow conical conductor from the apex of the cone to its base. The total current $I$ has no azimuthal component about the axis of the cone.

(i) Find the azimuthal magnetic field inside and outside the cone.

(ii) A proton beam strikes a fixed target located just outside the apex of the cone and charged pions are produced, which enter the cone. The pions can pass freely through the surface of the cone. Does the magnetic field due to the device act to focus or defocus the pions of positive charge? This apparatus is known as a neutrino horn since collimation of the charged pions serves to focus the neutrinos that originate in pion decay.

**5-35** Use Ampère's law to find the magnetic field due to a uniform infinite current sheet with constant surface current density $\mathbf{j}$.

**5-36** Use Equation (5–149) to find the vector potential of an infinitely long straight wire of radius $a$, uniform finite conductivity, and carrying a current $I$ surrounded by a thin cylindrical shell of radius $b$ (coaxial to the wire) carrying current $-I$.

## Section 5-17: Solenoid and Toroid

**5-37** A toroid with radius $R$ and circular cross section $S$ is wound uniformly with $n$ turns of wire carrying current $I$.

(i) Calculate the magnetic field in the toroid.

(ii) In what direction is the magnetic force on the winding?

# Electric Fields from Time-Varying Magnetic Fields

## CHAPTER CONTENTS

After the dramatic discovery in 1820 that an electric field, by producing a current in a wire, sets up a magnetic field, scientists widely suspected that a converse effect might exist in which a magnetic field causes an electric effect. Searches for this effect in a steady state did not succeed. Finally, in 1831 Michael Faraday in England and Joseph Henry in the United States observed independently that when the magnetic field through a conducting circuit changes, a current is induced in the circuit. This induction effect became known as Faraday's law, since Henry's work was not widely known at the time. However, Henry eventually received recognition, and the unit of inductance bears his name.

In Faraday's experiment two coils of wire were independently wound around an iron ring; see Figure 6–1. When a current from a battery was started through the first coil by closing the switch in that circuit, a pulse of current occurred in the second coil. There was again a pulse of current in the second coil, but in the opposite direction, when the current in the first coil was stopped by opening its circuit. Faraday carried out an extensive series of experiments on induced currents. His results on the directions of the induced currents were summarized by Friedrich Emil Lenz in 1833 in a simple form known as Lenz's rule.

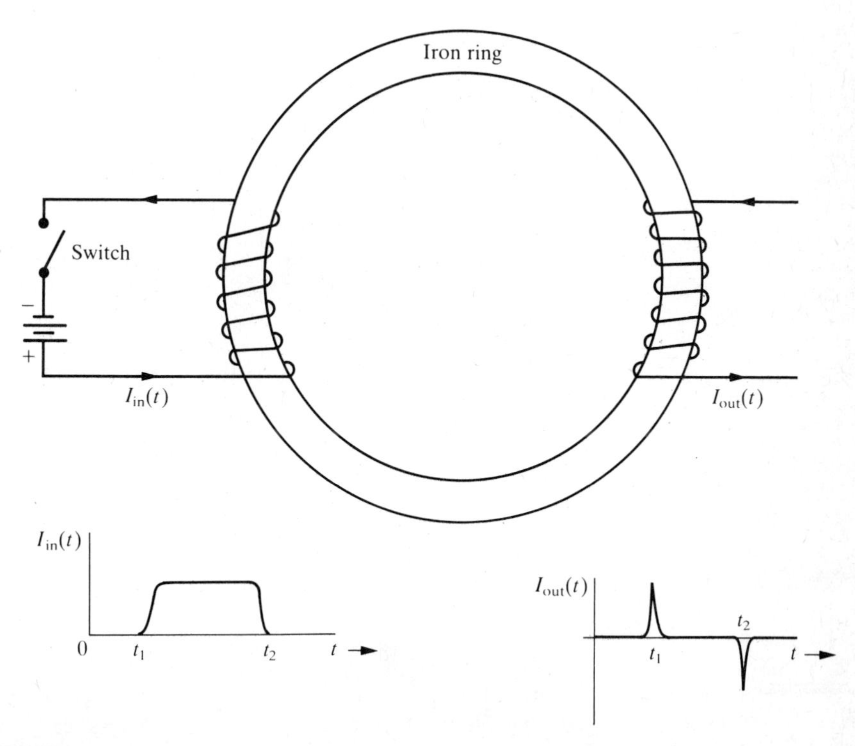


**FIGURE 6–1** Iron ring used by Faraday to detect currents induced by changing magnetic fields

Induction machines to make currents independently of batteries soon followed Faraday's discovery; they produced the varying magnetic field through a circuit by using permanent magnets moving relative to the circuit. However, the large-scale development of motors and generators did not come until after 1867, when Werner von Siemens replaced the permanent magnets by electromagnets powered by currents.

## 6-1 ELECTROMOTIVE FORCE

An electric battery, such as used in a flashlight or an automobile, will cause a current to flow along a conducting path from one of its electrodes to the other, because chemical reactions in the battery make the potentials of the battery electrodes different. When no external conducting path connects the electrodes, this potential difference is called the *open-circuit voltage* $V_0$ of the battery. When the battery is part of a closed circuit of total resistance $R$, in a steady state a current $I$ flows through the circuit such that the $IR$ voltage drop balances the open-circuit voltage $V_0$. (The voltage difference across the battery's electrodes is then $V_0 - R_{\text{int}}I$, where the internal resistance $R_{\text{int}}$ is the part of $R$ that is inside the battery.) The common term for a voltage difference across part (or all) of a circuit due to a particular mechanism is *electromotive force* (emf, pronounced "ee–em–ef"). In this example the

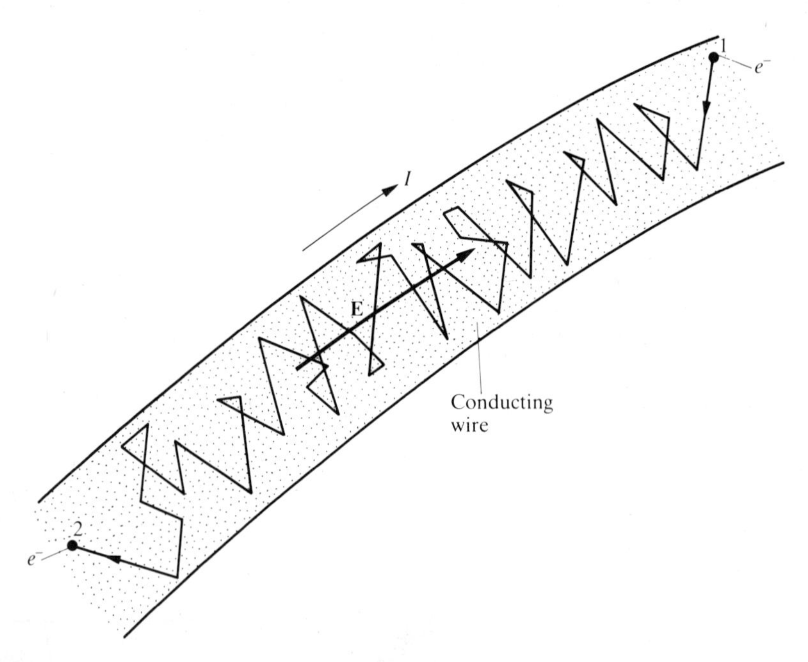


**FIGURE 6-2** Schematic of an electron moving in a conducting wire under the influence of a macroscopic electric field **E**. The electron moves erratically owing to collisions with ions. The average ionic electric field acting on the electron is exactly the negative of the macroscopic field.

chemical emf of the circuit is $V_0$, and the resistive emf (or *back emf*) is $-IR$. One speaks of the chemical emf in this circuit as the "driving" or "source" emf; after the circuit is closed, the current $I$ rises until the resistive back emf balances the chemical driving emf. The analogy of this system to a mechanical system in which a driving force produces motion is the historical reason for the use of the word *force* in emf. As stated above, emf is a voltage, that is, energy per unit charge.

A more precise definition of an emf is that it is the work done for each unit of charge transferred around the circuit. In a normal circuit the total electron kinetic energy is negligible compared with other circuit energies. Then by the work-energy theorem (see Section 1-4) the total work must vanish. The sum of all the emfs of a circuit is thus zero (e.g., $V_0 - IR = 0$ in the example just given).

There is only one emf that is given by the macroscopic electromagnetic fields, namely, the inductive emf of Faraday's law, which is the subject of the bulk of this chapter. All other emfs arise on the microscopic level. Chemical and resistive emfs arise because the field acting on each charge carrier in the circuit is different from the macroscopic (spatially averaged) field. To see how this emf arises, we consider an individual charge as it moves a macroscopic distance between points 1 and 2, as shown in Figure 6–2. Since, on the average, the charge carrier does not accelerate, the work done per unit charge over a macroscopic distance by the microscopic electric field $\mathbf{E}_\mathrm{m}$ must vanish:

$$\int_{\mathbf{r}_1}^{\mathbf{r}_2} \mathbf{E}_\mathrm{m}(\mathbf{r}, t) \cdot d\mathbf{r} = 0 \tag{6–1}$$

Since here $\mathbf{r}$ is a function of time, the integral must be evaluated as

$$\int_{t_1}^{t_2} \mathbf{E}_\mathrm{m}(\mathbf{r}(t), t) \cdot \dot{\mathbf{r}}(t)\, dt = 0 \tag{6–2}$$

We separate the microscopic field $\mathbf{E}_\mathrm{m}$ into the macroscopic part $\mathbf{E}$ and a fluctuating part. The emf of microscopic origin is defined as the work done per unit charge by the electric field difference $\mathbf{E}_\mathrm{m} - \mathbf{E}$:

$$\mathscr{E} \equiv \int_1^2 (\mathbf{E}_\mathrm{m} - \mathbf{E}) \cdot d\mathbf{r} = -\int_1^2 \mathbf{E} \cdot d\mathbf{r} = \Phi(2) - \Phi(1) \tag{6–3}$$

where 1 and 2 denote $\mathbf{r}_1$ and $\mathbf{r}_2$, respectively. The first integral must be suitably averaged over charges if there is more than one kind of charge carrier. In Equation (6–3) we have made use of Equation (6–1). The right-hand side of Equation (6–3) is just the macroscopic potential change. Applying Equation (6–3) to a closed circuit, we see that the total emf around the circuit vanishes, since $\Phi(2) = \Phi(1)$.

## 6-1A Resistive emf

If the two points in Figure 6–2 are in a conducting wire of cross section $S$ and conductivity $g$ and are separated by a distance $\ell$, we use Ohm's law in Equation (6–3) to obtain the *resistve emf*:

$$\mathscr{E}_{\mathrm{R}} = -\int_1^2 \mathbf{E} \cdot d\mathbf{r} = -\frac{1}{g}\int_1^2 \mathbf{J} \cdot d\mathbf{r}$$

or

$$\mathscr{E}_{\mathrm{R}} = -\left(\frac{\ell}{gS}\right) I = -RI \tag{6-4}$$

where $R$ is the resistance of the wire segment. To summarize the above derivation, the resistive emf is

$$\mathscr{E}_{\mathrm{R}} = -RI = \int_1^2 (\mathbf{E}_{\mathrm{m}} - \mathbf{E}) \cdot d\mathbf{r} = \Phi(2) - \Phi(1) \tag{6-5}$$

and the potential at point 1 exceeds the potential at point 2 by $RI$. The integrand in this case represents fluctuations around the macroscopic field due to collisions with the ions of the metallic crystal lattice.

## 6-1B Battery emf

The chemical battery was developed by Alessandro Volta in 1796. This invention was of great significance because it made possible the early-nineteenth-century discoveries of the relation between magnetic fields and currents.

To gain some feeling for the generation of an emf by chemical reactions, consider the energy-releasing reaction

$$\mathrm{Zn} + \mathrm{Cu}^{++} \rightarrow \mathrm{Zn}^{++} + 2e^- + \mathrm{Cu}^{++} \rightarrow \mathrm{Zn}^{++} + \mathrm{Cu}$$

If this reaction goes on in a homogeneous solution, the energy released appears as heat. But if the Zn and $\mathrm{Cu}^{++}$ are at different locations, the reaction can proceed only by transfer of the electrons between the locations, Thus a current of electrons from the Zn location to the Cu location results, and energy can be extracted from this current. Of course, in a steady state the charge density is constant, so there must be a balancing current of ions; this ion current will locally cancel the electron current unless the two currents are spatially separated.

A simple way to utilize this chemical reaction to produce an emf in a circuit is to attach one end of a conducting wire to a zinc electrode immersed in an aqueous solution of $\mathrm{ZnSO_4}$ and the other end to a copper electrode immersed in an aqueous solution of $\mathrm{CuSO_4}$, as shown in Figure 6–3. So that

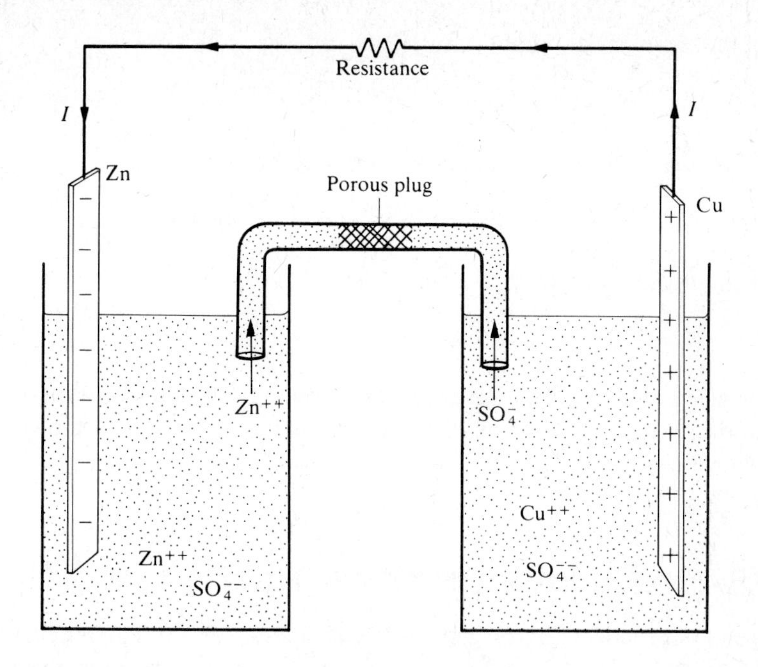


**FIGURE 6–3** Simple $ZnSO_4$ and $CuSO_4$ battery, which is a chemical source of emf

ionic conduction occurs, the two solutions are connected by a tube that has a porous plug (e.g., cotton) to minimize gross convection. The zinc goes into solution as $Zn^{++}$ and releases electrons to the Zn electrode. The $Cu^{++}$ plates out onto the other electrode by taking up electrons that travel through the wire circuit. The $Zn^{++}$ and $SO_4^{--}$ ions migrate through the plug as shown. The electronic and ionic currents are spatially separated because ions do not move freely in the metallic wire, and electrons do not move into the solution.

## 6-1C Induction emf

For an induced emf the forces acting on the charge carriers are macroscopic, and we can directly define the emf as the work done by the macroscopic fields on a unit charge moving around the circuit. When a magnetic field is present, the magnetic part of the Lorentz force can also contribute, giving

$$\mathscr{E} = \oint (\mathbf{E} + \mathbf{v}_e \times \mathbf{B}) \cdot d\mathbf{r} \qquad \textbf{(6–6)}$$

where $d\mathbf{r}$ is a coordinate differential along the circuit path at position $\mathbf{r}$. If $\mathbf{B}$ is time varying, then, as we will see, $\oint \mathbf{E} \cdot d\mathbf{r}$ is nonzero, so we need to retain both the $\mathbf{E}$ and the $\mathbf{v}_e \times \mathbf{B}$ terms here. The average velocity of the electrons $\mathbf{v}_e$ is due both to the motion of the wire $\mathbf{v}$ and to the drift velocity

$\mathbf{v}_d$ of the electrons along the wire:

$$\mathbf{v}_e = \mathbf{v} + \mathbf{v}_d \tag{6-7}$$

Since the drift velocity $\mathbf{v}_d$ is parallel to the wire element $d\mathbf{r}$,

$$\mathbf{v}_d \times \mathbf{B} \cdot d\mathbf{r} = -\mathbf{B} \times \mathbf{v}_d \cdot d\mathbf{r} = -\mathbf{B} \cdot \mathbf{v}_d \times d\mathbf{r} = 0$$

The emf of Equation (6–6) becomes

$$\mathscr{E} = \oint (\mathbf{E} + \mathbf{v} \times \mathbf{B}) \cdot d\mathbf{r} \tag{6-8}$$

Thus when an element of the circuit has a velocity component transverse to itself (i.e., $\mathbf{v} \times d\mathbf{r} \neq 0$ above), the magnetic field contributes to the work in moving a charge around the circuit.

# 6-2 FARADAY'S LAW OF INDUCTION

The emf of practical electric generators originates from a changing magnetic flux linking a circuit. The quantitative measure of this emf is expressed by Faraday's empirical law of induction:

**Faraday's Law**

$$\mathscr{E} = -\frac{d\mathscr{F}}{dt} \tag{6-9}$$

where $\mathscr{F}$ is the magnetic flux through a surface that caps the circuit and is given by

$$\mathscr{F} = \int \mathbf{B} \cdot d\mathbf{S} \tag{6-10}$$

Here the direction of $d\mathbf{S}$ is given by the right-hand rule in terms of the sense chosen for the line integral; see Figure 6–4. From Equations (6–8) and (6–10) the explicit expression of Faraday's law is

$$\begin{aligned} \mathscr{E} = \oint (\mathbf{E} + \mathbf{v} \times \mathbf{B}) \cdot d\mathbf{r} &= -\frac{d}{dt}\left(\int_S d\mathbf{S}\right) \cdot \mathbf{B} - \int_S d\mathbf{S} \cdot \frac{\partial \mathbf{B}}{\partial t} \\ &\equiv -\frac{d\mathscr{F}^{\dot{C}}}{dt} - \frac{d\mathscr{F}^{\dot{B}}}{dt} \end{aligned} \tag{6-11}$$

The term $d\mathscr{F}^{\dot{C}}/dt$ is the time rate of change of the flux due to the motion of the circuit, and the term $d\mathscr{F}^{\dot{B}}/dt$ is the time rate of change of the flux due to the time dependence of the field. Notice that $d\mathscr{F}^{\dot{C}}/dt$ is what one would get if the field was time-independent, and $d\mathscr{F}^{\dot{B}}/dt$ is what one would get if the circuit was fixed.

To develop some insight about Faraday's law, we consider the situa-

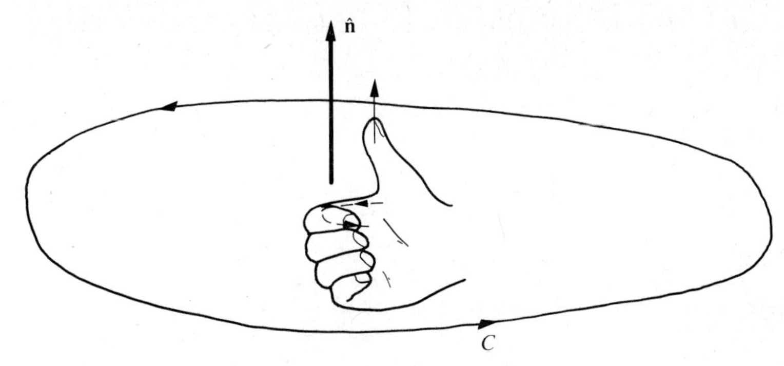


**FIGURE 6–4** Direction of the unit normal vector for a surface as determined by the right-hand rule. With the fingers curled in the direction of the path bounding the surface, the thumb gives the direction of the normal vector.

tion where **B** is time-independent and the circuit moves. In time $dt$ a point **r** on the circuit moves a distance $\mathbf{v}\,dt$, where **v** is the velocity of the circuit, which may vary with the position on the circuit (i.e., the circuit may rotate or even distort). The differential area swept out by a differential circuit element $d\mathbf{r}$ is

$$\hat{\mathbf{n}}\,dS = (\mathbf{v}\,dt) \times d\mathbf{r} \tag{6–12}$$

as illustrated in Figure 6–5. The corresponding change in flux is

$$d\dot{\mathscr{F}}^{C} = \mathbf{B} \cdot \hat{\mathbf{n}}\,dS = \mathbf{B} \cdot \mathbf{v} \times d\mathbf{r}\,dt \tag{6–13}$$

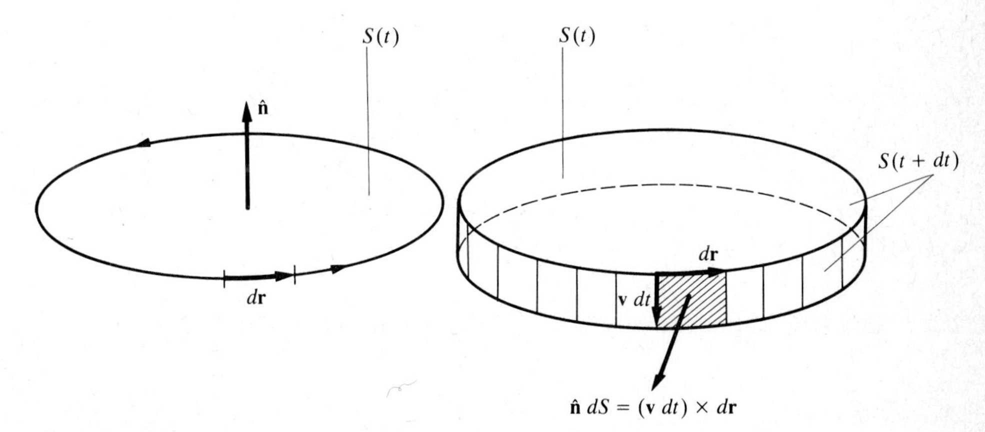


**FIGURE 6–5** Differential surface area swept out by a closed circuit in motion in a magnetic field

For the complete circuit the time rate of change of magnetic flux is

$$\frac{d\mathscr{F}^{\dot{C}}}{dt} = \oint \mathbf{B} \cdot \mathbf{v} \times d\mathbf{r} \tag{6-14}$$

By interchange of dot and cross products,

$$\mathbf{B} \cdot \mathbf{v} \times d\mathbf{r} = \mathbf{B} \times \mathbf{v} \cdot d\mathbf{r} = -\mathbf{v} \times \mathbf{B} \cdot d\mathbf{r} \tag{6-15}$$

we obtain, from Equations (6–14) and (6–15), the result

$$-\frac{d\mathscr{F}^{\dot{C}}}{dt} = \oint \mathbf{v} \times \mathbf{B} \cdot d\mathbf{r} \tag{6-16}$$

Thus the emf due to the motion of the circuit can be interpreted as the work done by the magnetic force per unit charge due to the motion of the circuit.

We can make an intuitive argument based on relativity of motion that an emf should also result from the time variation of a general **B** field. Consider the following simple example. A stationary permanent magnet is a time-independent source of **B**. If a rigid circuit is moved with velocity **v** away from the magnet, as in Figure 6–6(a), an emf is induced, which we have explained in terms of the magnetic Lorentz force law. If, in contrast, the circuit is held fixed and the magnet is moved with velocity $-v$ away from the circuit, as in Figure 6–6(b), we should expect to obtain the same physical effect—an emf in the circuit. It is this contribution that appears in the $d\mathscr{F}^{\dot{B}}/dt$ term in Faraday's law of Equation (6–11).

The emf associated with a time-varying magnetic field is the basic new content of Faraday's law, beyond the Lorentz force law. Subtracting the contribution of the moving circuit in Equation (6–16) from Equation (6–11), we arrive at

$$\oint \mathbf{E} \cdot d\mathbf{r} = -\int \frac{\partial \mathbf{B}}{\partial t} \cdot d\mathbf{S} \tag{6-17}$$

as an integral relation between **E** and **B** fields implied by Faraday's law. Using the curl theorem to express the line integral as a surface integral, we get

$$\int \left( \nabla \times \mathbf{E} + \frac{\partial \mathbf{B}}{\partial t} \right) \cdot d\mathbf{S} = 0 \tag{6-18}$$

Since the surface of integration is arbitrary, the integrand must vanish:

**Faraday's Law (Differential Form)**

$$\nabla \times \mathbf{E} = -\frac{\partial \mathbf{B}}{\partial t} \tag{6-19}$$

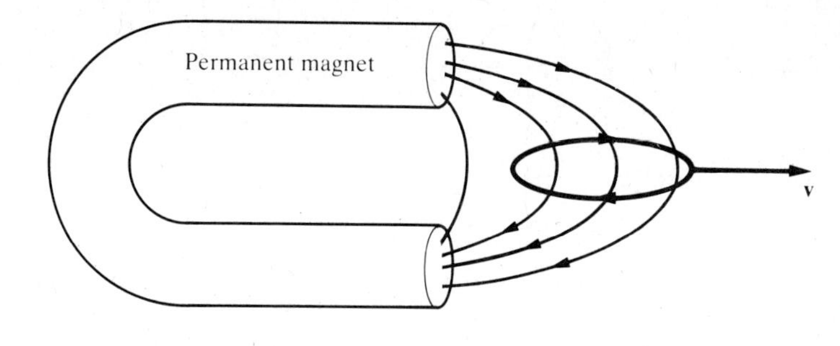


(a) Moving circuit

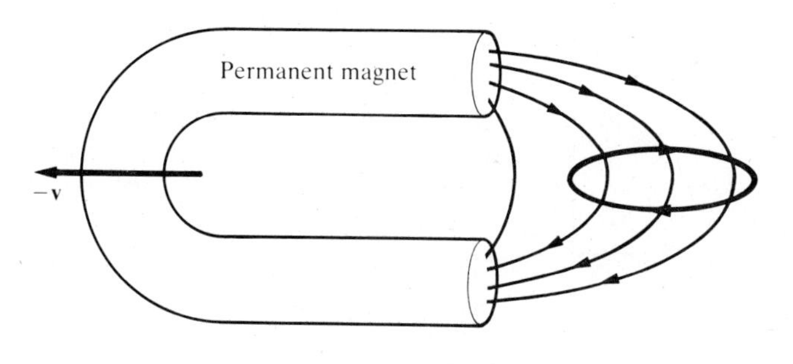


(b) Moving magnet

**FIGURE 6–6** Induced emf. An emf can be induced by either (a) a moving circuit in a static magnetic field or (b) a stationary circuit in a time-dependent magnetic field.

A changing magnetic field thereby implies a nonconservative electric field.

An induced emf in a conducting circuit causes a current to flow, which in turn produces an additional magnetic field. This **B** field produced by the induced current always opposes the change in flux through the circuit, as demonstrated in Figure 6–7 for a fixed circuit and a time-varying **B** field. In Figure 6–7(a) the magnetic field is increasing in the direction of the normal $\hat{\mathbf{n}}$ to the surface, so the flux is increasing also. Faraday's flux law then requires a negative emf (i.e., the induced current flows in the negative sense) around the circuit. By the right-hand rule the magnetic field due to the induced current produces a negative flux and so opposes the change in flux. A similar result follows for a decreasing magnetic flux, as illustrated in Figure 6–7(b). For the case of flux changes due to circuit motion, analogous conclusions can be drawn. This observation that *the sense of the induced emf opposes the change of flux through the circuit* is known as *Lenz's rule.*

In the calculation of the flux through a circuit that is not a single loop, the normal component of **B** must be integrated over the entire surface of all loops. A circuit of two loops is illustrated in Figure 6–8. For example, in a coil of $N$ closely wound loops, the flux through the coil is approximately

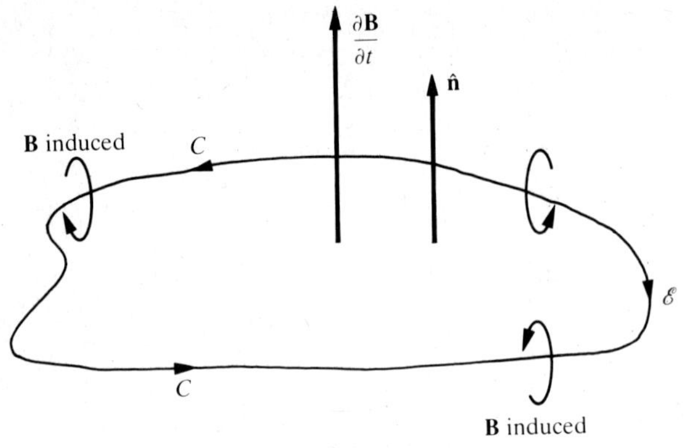


(a) Increasing flux

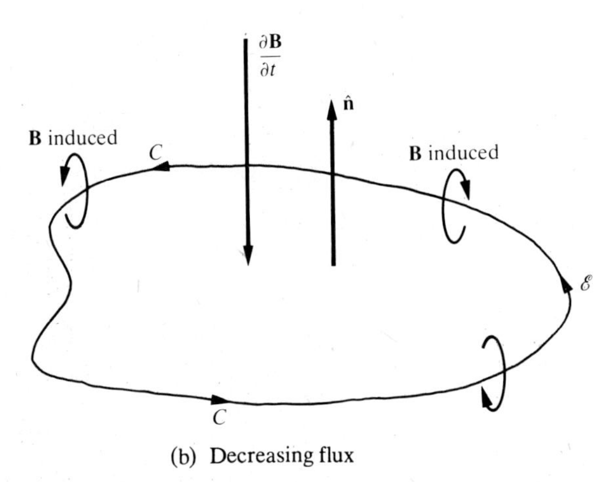


(b) Decreasing flux

**FIGURE 6-7** Induced magnetic field, which always opposes the change in flux through a circuit (Lenz's rule)

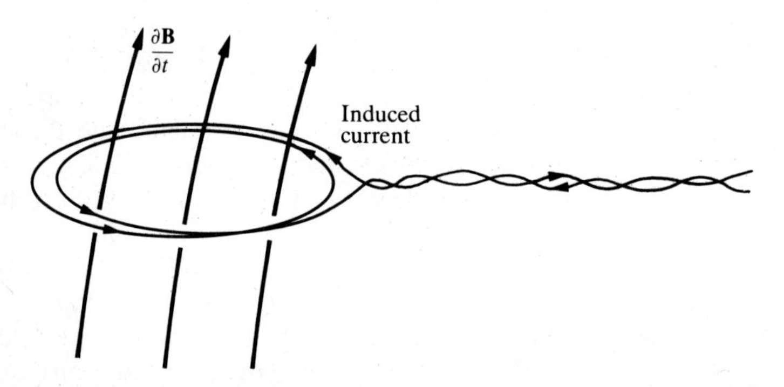


**FIGURE 6-8** Circuit with two loops, which a time-dependent magnetic field threads. The same magnetic field passes through each loop.

$N$ times the flux through one loop of the coil, and the emf is given by

$$\mathscr{E} \approx -N \frac{d}{dt} \int_{1\ \text{loop}} \mathbf{B} \cdot d\mathbf{S} \tag{6-20}$$

# 6-3 FLIP COIL: SIMPLE EXAMPLE OF FARADAY'S LAW

A flip coil is a common instrument for measuring a magnetic field. Figure 6–9 illustrates a circular flip coil of radius $a$ and $N$ turns, which is part of a circuit of resistance $R$. The leads of the coil are connected to a galvanometer, which measures the net charge flow. The flip coil provides a concrete example of the relationship between Faraday's law and the Lorentz force, as we now demonstrate for the case of measuring a uniform magnetic field.

## 6-3A Charge Flow Calculation from Faraday's Law

The flip coil is placed in the field with the plane of the coil normal to **B**, whose direction is assumed to be known for our present purposes, though the direction of **B** can be determined as well by using the flip coil. The coil is now quickly turned through 180°, reversing the flux through the coil. The emf across the galvanometer caused by the rotation of the coil is

$$\mathscr{E} = -\frac{d\mathscr{F}}{dt}$$

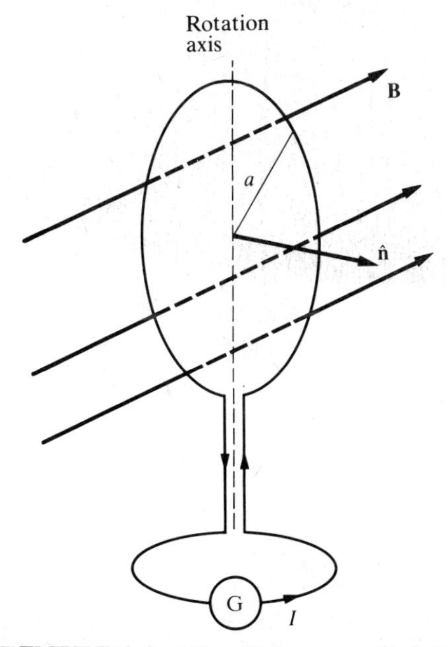


**FIGURE 6–9** Flip coil in a magnetic field connected to a galvanometer (G)

by Faraday's law. This emf causes a current to flow in the circuit. By Equation (6–5)

$$\mathscr{E} = R\frac{dq}{dt}$$

where $q(t)$ is the net charge that has passed through the galvanometer at time $t$. Equating these two expressions for the emf, we obtain

$$d\mathscr{F} = -R\,dq \tag{6–21}$$

Integrating, we find

$$\mathscr{F}(t) - \mathscr{F}(0) = -R[q(t) - q(0)] \tag{6–22}$$

At time $t = 0$, $q(0) = 0$ and $\mathscr{F}(0) = N\pi a^2 B$. After a sufficiently long time the charge flow ceases, and $\mathscr{F}(\infty) = -\mathscr{F}(0)$, $q(\infty) = Q$. Hence the magnetic field strength is determined by

$$B = \frac{R}{2N\pi a^2}Q \tag{6–23}$$

where the total charge $Q$ that passes through the circuit is measured by the galvanometer. We could equally well measure $B$ by pulling the flip coil out of the magnetic field region or by turning off the generator of the field. In each circumstance the initial flux would be the same as discussed here, the final flux would be zero, and $Q$ would be half that measured by flipping the coil.

## 6-3B Charge Flow Calculation from the Lorentz Force Law

We redo the preceding calculation of $B$ by using the Lorentz force law expression for the emf,

$$\mathscr{E} = N\oint \mathbf{v} \times \mathbf{B} \cdot d\mathbf{r} \tag{6–24}$$

where the integral is once around the circular loop. Using the geometry of Figure 6–10, with $\phi$ the angle between the $y$ axis and the normal $\hat{\mathbf{n}}$ to the coil, we have

$$\mathbf{v} = a \sin\theta \frac{d\phi}{dt}\hat{\mathbf{n}} \tag{6–25}$$

and

$$\mathbf{B} = B\hat{\mathbf{y}} \qquad d\mathbf{r} = -a\,d\theta\hat{\boldsymbol{\theta}}$$

$$\mathbf{v} \times \mathbf{B} = vB\sin\phi\,\hat{\mathbf{z}}$$

$$\mathbf{v} \times \mathbf{B} \cdot d\mathbf{r} = vBa\sin\phi\sin\theta\,d\theta$$

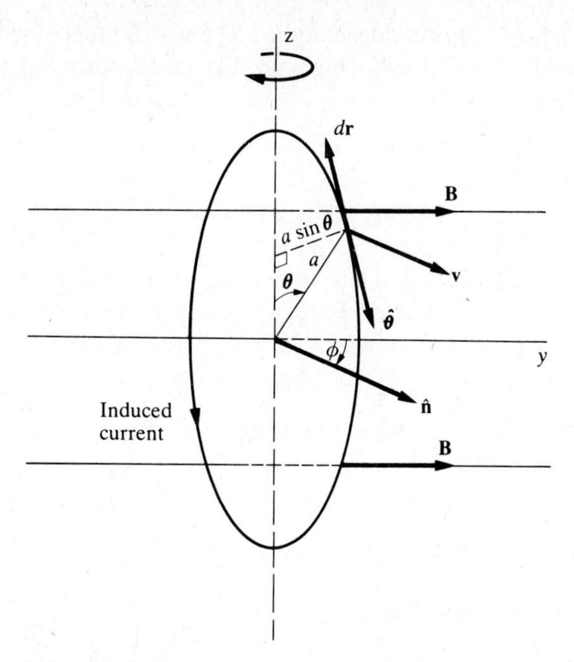


**FIGURE 6–10** Geometry for calculating the charge flow in a flip coil, using the Lorentz force law

Equation (6–24) becomes

$$\mathscr{E} = Na^2 B \sin\phi \frac{d\phi}{dt} \int_0^{2\pi} \sin^2\theta \, d\theta = N\pi a^2 B \sin\phi \frac{d\phi}{dt} \tag{6–26}$$

Equating this expression for $\mathscr{E}$ to $\mathscr{E} = R\,(dq/dt)$, we obtain

$$dq = \frac{N\pi a^2 B}{R} \sin\phi \, d\phi \tag{6–27}$$

For a 180° flip in $\phi$ Equation (6–27) yields

$$q(\infty) - q(0) = \frac{N\pi a^2 B}{R} \int_0^{\pi} \sin\phi \, d\phi$$

$$Q = \frac{2N\pi a^2 B}{R} \tag{6–28}$$

This result is the same as that obtained via Faraday's law in Equation (6–23).

In the preceding discussions of the flip coil we have disregarded magnetic flux due to the induced current. As long as we are only interested in the total charge that passes through the coil (not in $q$ as a function of time), this treatment is justified, as we will now establish.

A current increment $dI$ induced in the coil gives rise to an increment in the magnetic field. The induced magnetic flux is proportional to this induced current. Thus we write

$$d\mathscr{F}_{\text{ind}} = L\,dI \tag{6-29}$$

where $L$ is a positive proportionality constant known as the *self-inductance* of the coil. The positivity of $L$ can be directly established from the Biot-Savart law result for the **B** field of a current flowing in a circular ring.

The magnetic flux that enters Faraday's law is a sum of the flux due to the external field $\mathscr{F}_{\text{ext}}$ and the induced flux $\mathscr{F}_{\text{ind}}$. Thus the exact statement of Faraday's law in Equation (6–21) for the flip coil is

$$d\mathscr{F} = d\mathscr{F}_{\text{ext}} + d\mathscr{F}_{\text{ind}} = d\mathscr{F}_{\text{ext}} + L\,dI \tag{6-30}$$

where

$$d\mathscr{F} = -\mathscr{E}\,dt = -R\,dq \tag{6-31}$$

Integration over all time yields [since $I(\infty) = I(0) = 0$]

$$\mathscr{F}_{\text{ext}}(\infty) - \mathscr{F}_{\text{ext}}(0) = -R[q(\infty) - q(0)]$$

as previously found in Equation (6–22). The inductance term vanishes since the initial and final currents are zero. However, the time dependence of the induced current does depend on the inductance term.

## 6-3C Alternating-Current Generator

The rotation of a flip coil about a diameter illustrates the principle of an alternating-current generator. Combining Equations (6–30) and (6–31), we obtain

$$L\frac{dI}{dt} + RI = -\frac{d\mathscr{F}_{\text{ext}}}{dt} \tag{6-32}$$

For uniform rotation with angular frequency $\omega$,

$$\mathscr{F}_{\text{ext}}(t) = \mathscr{F}(0)\cos\omega t \tag{6-33}$$

the generator equation is

$$L\frac{dI}{dt} + RI = \omega\mathscr{F}(0)\sin\omega t \tag{6-34}$$

The transient solution to the homogeneous equation

$$I(t) = I(0)e^{-(R/L)t} \tag{6-35}$$

damps exponentially to zero with increasing time. The steady-state (or particular) solution is of the form

$$I(t) = A\sin(\omega t - \delta) \tag{6-36}$$

By substitution of Equation (6–36) into Equation (6–34), we find

$$\tan\delta = \frac{\omega L}{R} \qquad A = \frac{\omega\mathscr{F}(0)}{\sqrt{(\omega L)^2 + R^2}} \tag{6-37}$$

The induced steady-state current lags the emf by the angle $\delta$.

The induced current through the flip coil experiences a Lorentz force due to the external **B** field. This force results in a torque that opposes the rotation. So that the rotation of the flip coil generator is maintained, an external torque about the rotation axis is required. Thus mechanical energy is converted to electric energy via the coil motion in the magnetic field.

### 6-3D Induced Currents in Superconductors

The electric resistance of some metals is exactly zero below a certain critical temperature. This property is known as *superconductivity*. The critical temperature is characteristic of the material; for known superconductors the critical temperatures are in the range 0°–20°K. Since the resistance is zero, you might think that an infinite current would be induced in a superconducting coil if the magnetic flux linking the coil were changed. Since this result is physically impossible, a finite current is induced in the coil, whose magnetic field exactly compensates for changes in the external field. Thus the total flux linking a superconducting ring never changes. When a superconducting coil with zero current is moved into a magnetic field, a current flow in the coil is induced; this current is maintained as long as the coil is in the field region.

We can induce a permanent circulating current in a superconducting ring by starting with the coil in the magnetic field while it is a normal conductor. The coil is then cooled down to below the superconducting temperature. When the coil is removed from the magnetic field region, a permanent current will be generated, which maintains the flux through the coil. Such induced currents can be easily measured by their associated magnetic field. These superconducting currents flow with no apparent diminution, some having been observed to continue flowing for years.

## 6-4 MAGNETIC POTENTIAL ENERGY

In the assembly of a charge distribution, work is done, which results in potential energy stored in the electric field. When a current flow is established, work must also be done, which results in potential energy stored in the magnetic field. In this section we derive the expression for the energy stored in a magnetic field.

As we found in Section 5–11, the power delivered to a system's charges by an electric field is

$$P = \int \mathbf{E} \cdot \mathbf{J} \, dV \tag{6-38}$$

For current flow in a resistive conductor this power appears as an $I^2R$ heat loss. One possible source of the electric field in Equation (6–38) that supplies the power to the charges is a time-varying magnetic field [see Equation (6–19)]. In that case the origin of the heat energy would be a decrease in the energy stored in the magnetic field. To derive the power transferred from the magnetic field to the charges, we replace $\mathbf{J}$ in Equation (6–38) by $(\nabla \times \mathbf{B})/\mu_0$:

$$P = \frac{1}{\mu_0} \int \mathbf{E} \cdot (\nabla \times \mathbf{B}) \, dV \tag{6-39}$$

Using the vector identity

$$\nabla \cdot (\mathbf{E} \times \mathbf{B}) = \mathbf{E} \cdot \nabla \times \mathbf{B} - \mathbf{B} \cdot (\nabla \times \mathbf{E}) \tag{6-40}$$

and Faraday's law in differential form, $\nabla \times \mathbf{E} = -\partial \mathbf{B}/\partial t$, we find

$$P = \frac{1}{\mu_0} \int \nabla \cdot (\mathbf{E} \times \mathbf{B}) \, dV - \frac{1}{\mu_0} \int \mathbf{B} \cdot \frac{\partial \mathbf{B}}{\partial t} \, dV \tag{6-41}$$

The first integral can be converted to a surface integral by using the divergence theorem. In the second integral we write

$$\mathbf{B} \cdot \frac{\partial \mathbf{B}}{\partial t} = \frac{1}{2} \frac{\partial}{\partial t} (B^2)$$

Thus

$$P = \frac{1}{\mu_0} \int \mathbf{E} \times \mathbf{B} \cdot d\mathbf{S} - \int \frac{\partial}{\partial t} \left( \frac{B^2}{2\mu_0} \right) dV \tag{6-42}$$

For a bounding surface at infinity the surface integral vanishes if the charges and currents are localized in a finite region of space (i.e., in that case $|\mathbf{E} \times \mathbf{B}| \approx 1/r^4$). Then the power is given by the volume integral over all space:

$$P = -\frac{d}{dt} \int_{\text{all space}} \frac{B^2}{2\mu_0} \, dV \tag{6-43}$$

This equation expresses conservation of energy flow. The energy input to the charges $P\,dt$ equals the decrease $dU$ in the potential energy:

**Magnetostatic Potential Energy**

$$U = \frac{1}{2\mu_0} \int_{\text{all space}} B^2 \, dV \tag{6-44}$$

We conclude that Equation (6–44) is the potential energy stored in the magnetic field. This relation is the magnetic analog of Equation (2–100) for electric potential energy, $U = \frac{1}{2}\varepsilon_0 \int E^2 \, dV$.

Equation (6–42) is the corresponding expression of energy conservation for a finite volume. The quantity $(1/\mu_0)\mathbf{E} \times \mathbf{B}$ is known as *Poynting's vector* and represents the power flow per unit area across the bounding surface. Equation (6–42) says that the power delivered to electric charges within a fixed volume equals the power flowing into the volume minus the rate in which the magnetic field energy is increasing. The general energy conservation theorem, known as Poynting's theorem, will be discussed further in Chapter 8.

We will next obtain alternative expressions for the magnetic energy in terms of the source current **J** and the vector potential **A**. We start with the identity

$$\nabla \cdot (\mathbf{A} \times \mathbf{B}) = \mathbf{B} \cdot \nabla \times \mathbf{A} - \mathbf{A} \cdot \nabla \times \mathbf{B}$$

and use $\mathbf{B} = \nabla \times \mathbf{A}$ and $\nabla \times \mathbf{B} = \mu_0 \mathbf{J}$ to obtain

$$\nabla \cdot (\mathbf{A} \times \mathbf{B}) = B^2 - \mu_0 \mathbf{A} \cdot \mathbf{J} \tag{6-45}$$

Then Equation (6–44) can be rewritten by use of the divergence theorem as

$$U = \frac{1}{2} \int_{\text{all space}} \mathbf{J} \cdot \mathbf{A} \, dV + \frac{1}{2\mu_0} \int_{\text{surface at } \infty} \mathbf{A} \times \mathbf{B} \cdot d\mathbf{S} \tag{6-46}$$

From arguments similar to those used in deriving Equation (6–44), the surface integral vanishes and

$$U = \frac{1}{2} \int_{\text{all space}} \mathbf{J} \cdot \mathbf{A} \, dV \tag{6-47}$$

This relation is the analog of Equation (2–93) for electric potential energy, $U = \frac{1}{2} \int \rho \Phi \, dV$. Another form for the magnetic energy is obtained from Equation (6–47) by expressing **A** in terms of its source currents:

$$\mathbf{A} = \frac{\mu_0}{4\pi} \int \frac{\mathbf{J} \, dV}{R}$$

This expression leads to the formula

$$U = \frac{\mu_0}{8\pi} \iint \frac{\mathbf{J}(\mathbf{r}) \cdot \mathbf{J}(\mathbf{r}') \, dV \, dV'}{R} \tag{6-48}$$

For the interaction energy of two filamentary circuits we replace $\mathbf{J} \, dV$ by

$I\,d\mathbf{r}$ in Equation (6–48), to obtain

$$U = \frac{\mu_0}{8\pi} I_1 I_2 \oint_{C_1} \oint_{C_2} \frac{d\mathbf{r}_1 \cdot d\mathbf{r}_2}{|\mathbf{r}_1 - \mathbf{r}_2|} \tag{6-49}$$

The self-energy of a filamentary circuit is infinite, as was the electrostatic self-energy of a point charge.

# 6-5 INDUCTANCE

Induced currents caused by time-varying magnetic fields are of particular importance in electric circuits. In this section we express the magnetic energy of a collection of current-carrying, closed-circuit loops in terms of the currents and of geometric factors known as coefficients of inductance. The physics of interacting circuits, such as mutual forces, torques, and induced emfs, can be described in terms of the coefficients of inductance.

By Equation (5–149) the magnetic flux through a circuit $C_i$ due to a current in a different circuit $C_j$ is

$$\mathscr{F}_i^j = \oint_{C_i} \mathbf{A}_j(\mathbf{r}_i) \cdot d\mathbf{r}_i \tag{6-50}$$

where the vector potential is expressed as

$$\mathbf{A}_j(\mathbf{r}_i) = \frac{\mu_0}{4\pi} I_j \oint_{C_j} \frac{d\mathbf{r}_j}{|\mathbf{r}_j - \mathbf{r}_i|} \tag{6-51}$$

according to Equation (5–63), with $\mathbf{J}\,dV = I\,d\mathbf{r}$. Thus $\mathscr{F}_i^j$ has the form

$$\mathscr{F}_i^j = L_{ij} I_j \tag{6-52}$$

where

$$L_{ij} = \frac{\mu_0}{4\pi} \oint_{C_i} \oint_{C_j} \frac{d\mathbf{r}_j \cdot d\mathbf{r}_i}{|\mathbf{r}_i - \mathbf{r}_j|} \tag{6-53}$$

is known as the *mutual inductance*. The value of $L_{ij}$ is determined by the geometry of the circuits. The flux linking $C_j$ due to $C_i$ is

$$\mathscr{F}_j^i = L_{ji} I_i \tag{6-54}$$

From Equation (6–53) we can deduce that

$$L_{ij} = L_{ji}$$

Hence a current $I_i$ in circuit $C_i$ gives the same flux through $C_j$ as an equal current in $C_j$ gives through $C_i$.

A circuit is also linked by flux associated with its own current. The flux in this case is

$$\mathscr{F}_i^i = L_{ii} I_i \tag{6-55}$$

where $L_{ii}$ is a positive constant known as the *coefficient of self-inductance.* Unlike the case of mutual inductance, the self-inductance depends strongly on the diameter of the wire and is infinite for zero diameter. Since Equation (6–53) is based on zero diameter, it cannot be used in calculating the self-inductance of a wire. To determine the self-inductance, one must use the expression for **A** in terms of the volume current density **J** in Equation (6–50).

In general, the *total flux* linking the $i$th circuit of a collection of $N$ circuits is

$$\mathscr{F}_i = \sum_{j=1}^{N} L_{ij} I_j \tag{6-56}$$

For currents confined to a system of closed circuits, the expression for the magnetic energy in Equation (6–47) becomes

$$U = \frac{1}{2} \sum_i I_i \oint \mathbf{A}(\mathbf{r}_i) \cdot d\mathbf{r}_i \tag{6-57}$$

where $\mathbf{A}(\mathbf{r}_i)$ is the total vector potential at position $\mathbf{r}_i$ due to the currents in all circuits. With Equation (6–50), Equation (6–57) can be written as

$$U = \frac{1}{2} \sum_i I_i \mathscr{F}_i \tag{6-58}$$

Then substitution of Equation (6–56) gives

$$U = \frac{1}{2} \sum_i \sum_j L_{ij} I_i I_j \tag{6-59}$$

For a single circuit

$$\mathscr{F} = LI \qquad U = \frac{1}{2} LI^2 \tag{6-60}$$

For two circuits

$$\begin{aligned} \mathscr{F}_1 &= L_1 I_1 + M I_2 \qquad \mathscr{F}_2 = L_2 I_2 + M I_1 \\ U &= \frac{1}{2} L_1 I_1^2 + \frac{1}{2} L_2 I_2^2 + M I_1 I_2 \end{aligned} \tag{6-61}$$

where $M$ is the mutual inductance, $M \equiv L_{12} = L_{21}$.

The unit of inductance is the henry (H), named after Joseph Henry, codiscoverer of magnetic induction:

$$1 \frac{\text{joule}}{(\text{ampere})^2} = 1 \text{ henry}$$

Coils (i.e., inductors) used in most electronics applications are much smaller than 1 H.

As an example of self-inductance, we consider a very long solenoid. Using the magnetic field $B_z$ from Equation (5–147), we can express the flux per unit length in the solenoid as

$$\mathscr{F} = (\pi a^2 n)(\mu_0 n I) = LI \tag{6-62}$$

where $a$ is the radius and $n$ is the number of turns per meter. For a solenoid of radius 1 cm having $10^4$ turns per meter, the self-inductance per meter is

$$\begin{aligned} L = \pi a^2 \mu_0 n^2 &= \pi(0.01)^2(4\pi \times 10^{-7})(10^4)^2 \\ &\approx 4 \times 10^{-2} \text{ H/m} \end{aligned} \tag{6-63}$$

As we will see in the next chapter, an iron core in the solenoid will increase the inductance about ten thousand times.

As an example of mutual inductance, we consider a finite solenoid of $N$ turns (circuit 2) wound around an infinite solenoid, with radius $a$ and $n$ turns per unit length, which carries current $I_1$ (circuit 1). The magnetic flux due to the infinite solenoid that links the finite solenoid is

$$\mathscr{F}_2^1 = (\mu_0 n I_1)(N\pi a^2) \tag{6-64}$$

Since by definition

$$\mathscr{F}_2^1 = M I_1 \tag{6-65}$$

we deduce that the mutual inductance is

$$M = \mu_0 n N \pi a^2 \tag{6-66}$$

This result does not depend on the outside coil's radius, so the geometry of the outside winding is irrelevant. A direct calculation of the flux linking the infinite solenoid due to the outside coil would be much more difficult, but the mutual inductance result must be the same.

In many cases a simpler procedure is to determine inductances by calculating the magnetic energy and comparing the result with Equation (6–59). As an example, we find the self-inductance of a coaxial conductor. This conductor system is shown schematically in Figure 6–11. A current passes down a central conducting tube of radius $a$ of negligible thickness and returns in a concentric tube of radius $b$, also of negligible thickness.

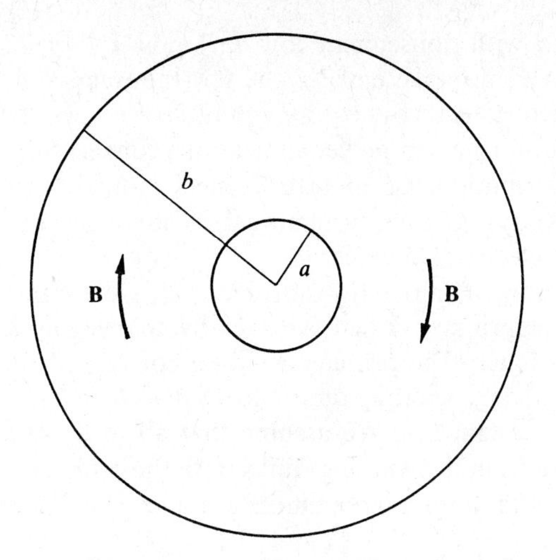


**FIGURE 6-11** Cross section of a coaxial cable

From the results in Section 5–16 the magnetic field of a coaxial cable is

$$\mathbf{B} = \begin{cases} 0, & \text{for } r < a \\ \dfrac{\mu_0 I}{2\pi r}\hat{\phi}, & \text{for } a < r < b \\ 0, & \text{for } r > b \end{cases} \tag{6-67}$$

The magnetic field vanishes for $r > b$ since there is no net current enclosed in the circuit integral of Ampère's law. The magnetic field energy per meter of cable is consequently

$$U = \frac{1}{2\mu_0}\left[\frac{\mu_0^2 I^2}{4\pi^2}\int_a^b \frac{(2\pi r\, dr)}{r^2}\right] = \frac{\mu_0 I^2}{4\pi}\int_a^b \frac{dr}{r} = \frac{\mu_0 I^2}{4\pi}\ln\frac{b}{a} \tag{6-68}$$

By comparison with $U = \frac{1}{2}LI^2$, we conclude that the coefficient of self-inductance per unit length of cable is

$$L = \frac{\mu_0}{2\pi}\ln\frac{b}{a} \tag{6-69}$$

This inductance could alternatively be calculated by finding the flux linkage between inner and outer conductors.

# 6-6 APPLICATION: VOLTAGE TRANSFORMER

The widespread use of alternating currents (AC) in preference to direct currents is due to the ease by which the power can be transformed to high voltage and low current, with consequent low $I^2R$ power loss in transmission. Electric generators produce emfs in the few-thousand-volt range. This power is transmitted across country at voltages of several hundred thousand volts. Then transformers at power substations convert the voltage to 1500 V for urban distribution. Local transformers reduce the voltage further to $\pm 110$ V. The AC transformer system was first advocated by Nikola Tesla, a Serbo-American electrical engineer.

We simplify the nature of a real transformer to the ideal transformer shown in Figure 6–12. The primary circuit is driven by an emf source $\mathscr{E}_P(t)$, which has negligible resistance. The primary winding consists of $N_P$ turns on an iron torus. The secondary winding on the torus has $N_S$ turns, and the secondary circuit has a resistance $R$. We assume that all of the magnetic flux produced by the currents in the windings links both the primary and the secondary coils. The self- and mutual inductances are then given by

$$L_P = L_0 N_P^2 \qquad L_S = L_0 N_S^2$$
$$M = L_0 N_P N_S \tag{6-70}$$

where $L_0$ is the inductance per turn. Thus no magnetic flux loss implies that

$$M^2 = L_P L_S \tag{6-71}$$

which is the maximum value of the mutual inductance (see Problem 6–34).

The sum of the emf drops around each circuit must vanish, giving the equations

$$\begin{aligned} \mathscr{E}_P(t) &= L_P \frac{dI_P}{dt} - M \frac{dI_S}{dt} \\ 0 &= L_S \frac{dI_S}{dt} - M \frac{dI_P}{dt} + RI_S \end{aligned} \tag{6-72}$$

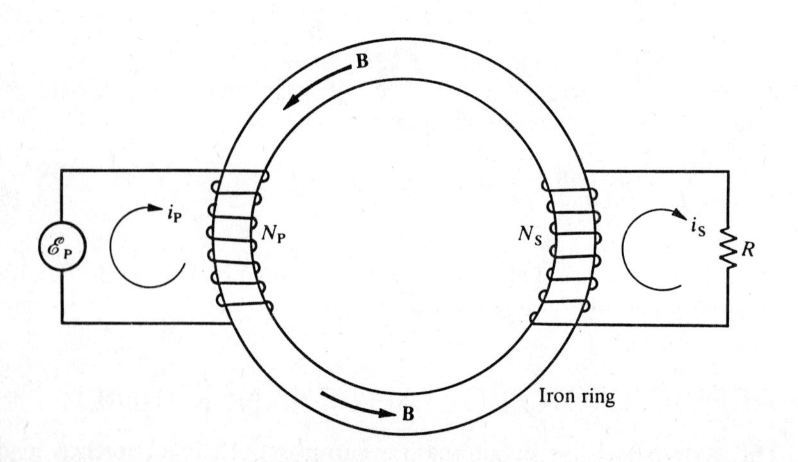


**FIGURE 6–12** Transformer consisting of primary and secondary windings on a common iron ring

The mutual inductance term would appear with the opposite signs in Equations (6–72) if one of the coils were wound with the other handedness. The correct sign can be ascertained by appealing to Lenz's rule.

For a sinusoidal emf we write, in complex notation,

$$\mathscr{E}_P(t) = \mathscr{E}_P e^{-i\omega t} \tag{6–73}$$

The steady-state (particular) solution to Equations (6–72) is achieved by taking

$$I_P(t) = I_P e^{-i\omega t} \qquad I_S(t) = I_S e^{-i\omega t} \tag{6–74}$$

Substitution of Equations (6–73) and (6–74) into Equations (6–72) yields

$$\mathscr{E}_P = -i\omega L_P I_P + i\omega M I_S \tag{6–75a}$$

$$0 = -i\omega L_S I_S + i\omega M I_P + R I_S \tag{6–75b}$$

Using the second equation, we eliminate $I_P$ in Equation (6–75a). From the no–flux loss condition of Equation (6–70), we then obtain

$$\mathscr{E}_P = \frac{L_P}{M}\mathscr{E}_S \tag{6–76}$$

where $\mathscr{E}_S = RI_S$ is the voltage across the secondary winding. Using Equation (6–70), we obtain the ideal voltage transformation:

$$\frac{\mathscr{E}_S}{\mathscr{E}_P} = \frac{N_S}{N_P} \tag{6–77}$$

Thus if the number of secondary windings exceeds the number of primary windings, the secondary voltage will be higher. Independent of the details of the primary and secondary circuit elements, Equation (6–77) must hold, since the same flux links both circuits and hence the emfs must be in the same ratio as the number of windings.

By Equation (6–75b) the current ratio is

$$\frac{I_P}{I_S} = \frac{i\omega L_S - R}{i\omega M} = \frac{L_S}{M} + \frac{iR}{\omega M} \tag{6–78}$$

If the secondary resistance is large, Equation (6–78) shows that the magnitude of the secondary current is small. For a small resistance in the secondary circuit, $R \ll \omega M$, we obtain

$$\frac{I_S}{I_P} = \frac{N_P}{N_S} \tag{6–79}$$

The relationships in Equations (6–77) and (6–79) are the properties of an *ideal transformer*. Combining Equations (6–77) and (6–79), we see that the product of emf and current is the same for the primary and secondary circuits of an ideal transformer:

$$\frac{\mathscr{E}_S I_S}{\mathscr{E}_P I_P} = \frac{N_S}{N_P}\frac{N_P}{N_S} = 1 \tag{6-80}$$

By use of a transformer the 110-V line voltage can be changed to 10 V for the transistors in an FM tuner or to 10,000 V for an X-ray tube.

There is another useful role the transformer plays, that of impedance matching. As an example of impedance matching, we consider the familiar situation of connecting a 75-Ω coaxial TV antenna cable to a 300-Ω, flat, two-wire lead-in. If a direct connection were made, a fraction of the antenna signal would be reflected at the junction, reducing the received power and possibly introducing spurious reflections (ghost images). The situation is analogous to two different connected strings under tension. Transverse waves on the string will be partially reflected at the connection unless the wave velocity is the same in each string. For an electric circuit the corresponding quantity is the impedance $Z$,

$$Z \equiv \frac{\mathscr{E}}{I} \tag{6-81}$$

By Equations (6–77) and (6–79) the primary and secondary impedances are related by

$$\frac{Z_S}{Z_P} = \left(\frac{N_S}{N_P}\right)^2 \tag{6-82}$$

By Equation (6–82) we observe that as far as the impedance seen by terminals $ab$ of Figure 6–13 are concerned, we can replace the whole circuit to the right of $ab$ by a single effective impedance,

$$Z_P = \left(\frac{N_P}{N_S}\right)^2 Z_S \tag{6-83}$$

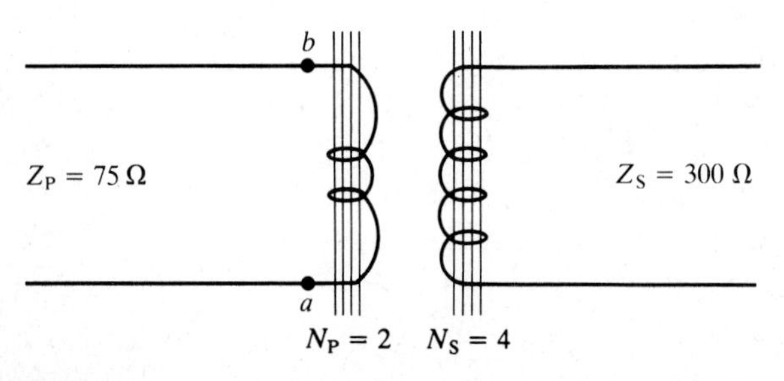


**FIGURE 6–13** Two cables having apparent impedances $Z_P$ and $Z_S$ matched by a transformer to eliminate reflections. The vertical lines represent segments of a closed iron core.

For our case we want $Z_P = 75\ \Omega$ and $Z_S = 300\ \Omega$ so that the ratio of winding is

$$\frac{N_P}{N_S} = \frac{1}{2} \tag{6-84}$$

By use of this impedance-matching device the secondary circuit "looks" like a continuation of the primary circuit, so no reflections will occur.

## 6-7 FORCES BETWEEN CIRCUITS

Current-carrying circuits generally exert forces on each other. A particularly simple example is the force between two parallel conducting wires, discussed in Section 5–3. The straight wires can be viewed as closed circuits with return currents at infinity. In the parallel-wire case we calculated the force between the wires by using the Lorentz force. In this section we derive a general expression for the forces and torques between two conducting circuits that carry constant currents.

By Equation (5–24) the total magnetic force on a circuit 1 due to a circuit 2 is

$$\mathbf{F} = I_1 \oint d\mathbf{r}_1 \times \mathbf{B}(\mathbf{r}_1) \tag{6-85}$$

where $\mathbf{B}(\mathbf{r}_1)$ is the magnetic field due to circuit 2 at a position $\mathbf{r}_1$ on circuit 1. As this force moves circuit 1 relative to circuit 2, the flux through circuit 1 changes. From Equation (6–14) the time rate of change of this flux is

$$\frac{d\mathscr{F}_1^2}{dt} = \oint \mathbf{B}(\mathbf{r}_1) \cdot \mathbf{v} \times d\mathbf{r}_1 \tag{6-86}$$

Rearranging the dot and cross products and noting that for translational motion $\mathbf{v}$ is the same for each element of circuit 1, we obtain

$$\frac{d\mathscr{F}_1^2}{dt} = \mathbf{v} \cdot \oint d\mathbf{r}_1 \times \mathbf{B}(\mathbf{r}_1) \tag{6-87}$$

Substituting for the integral the result of Equation (6–85), we have

$$\frac{d\mathscr{F}_1^2}{dt} = \frac{\mathbf{v} \cdot \mathbf{F}}{I_1} \tag{6-88}$$

Since only the mutual inductance $M$ changes with time, the time rate of change of the flux $\mathscr{F}_1^2 = I_2 M$ due to circuit motion is

$$\begin{aligned} \frac{d\mathscr{F}_1^2}{dt} &= I_2 \frac{dM}{dt} = I_2 \left( \frac{\partial M}{\partial x} \frac{dx}{dt} + \frac{\partial M}{\partial y} \frac{dy}{dt} + \frac{\partial M}{\partial z} \frac{dz}{dt} \right) \\ &= I_2 (\nabla M) \cdot \mathbf{v} \end{aligned} \tag{6-89}$$

where $(x, y, z)$ are the coordinates of circuit 1 relative to circuit 2. Equating this expression to the preceding result for $d\mathscr{F}_1^2/dt$, we obtain the *force law*

$$\mathbf{F} = I_1 I_2 \nabla M \tag{6-90}$$

Since the mutual inductance generally decreases as two circuits are separated, the force between two circuits with currents in the same sense is attractive [the sense of $I_1$ and $I_2$ are the same if both give a flux contribution of the same sign in circuit 1 (or circuit 2)]. This conclusion agrees with the force calculation in Section 5–3 for the force between parallel, current-carrying wires.

We can similarly determine the *torque* exerted on circuit 1 due to the magnetic field of circuit 2. To find this torque, we take $\mathbf{v} = \boldsymbol{\omega} \times \mathbf{r}_1$, where $\boldsymbol{\omega}$ is the angular velocity of rotation of circuit 1. Equation (6–86) becomes

$$\begin{aligned}\frac{d\mathscr{F}_1^2}{dt} &= \oint \mathbf{B}\cdot(\boldsymbol{\omega} \times \mathbf{r}_1) \times d\mathbf{r}_1 = \oint (\boldsymbol{\omega} \times \mathbf{r}_1)\cdot(d\mathbf{r}_1 \times \mathbf{B}) \\ &= \boldsymbol{\omega}\cdot\oint \mathbf{r}_1 \times (d\mathbf{r}_1 \times \mathbf{B})\end{aligned} \tag{6-91}$$

by interchange of dot and cross products. Identifying the factor $d\mathbf{r}_1 \times \mathbf{B}(\mathbf{r}_1)$ with $d\mathbf{F}/I_1$ from Equation (6–87), we obtain

$$\frac{d\mathscr{F}_1^2}{dt} = \frac{\boldsymbol{\omega}\cdot\mathbf{N}}{I_1} \tag{6-92}$$

where $\mathbf{N}$ is the torque exerted on the circuit 1 by the magnetic field of circuit 2.

For a planar circuit 1 of directed area $\mathbf{S}$ linked by a constant magnetic field $\mathbf{B}$ from circuit 2, the time rate of flux change for a rotation is

$$\frac{d\mathscr{F}_1^2}{dt} = \frac{d}{dt}(\mathbf{B}\cdot\mathbf{S}) = \frac{d\mathbf{S}}{dt}\cdot\mathbf{B}$$

Under rotations $d\mathbf{S}/dt = \boldsymbol{\omega} \times \mathbf{S}$, so

$$\frac{d\mathscr{F}_1^2}{dt} = \boldsymbol{\omega} \times \mathbf{S}\cdot\mathbf{B} = \boldsymbol{\omega}\cdot(\mathbf{S} \times \mathbf{B}) \tag{6-93}$$

Comparing Equations (6–92) and (6–93), we see that the torque is given by

$$\mathbf{N} = I_1 \mathbf{S} \times \mathbf{B} \tag{6-94}$$

This torque acts to align the normal $\hat{\mathbf{S}}$ of circuit 1 with the (fixed) direction of the magnetic field $\mathbf{B}$ due to circuit 2; see Figure 6–14. In Chapter 7 we will show that the quantity $I_1\mathbf{S}$ is the magnetic dipole moment for a planar circuit and that, in general, the torque on a magnetic dipole is given by the cross product of the magnetic dipole moment and the imposed $\mathbf{B}$ field.

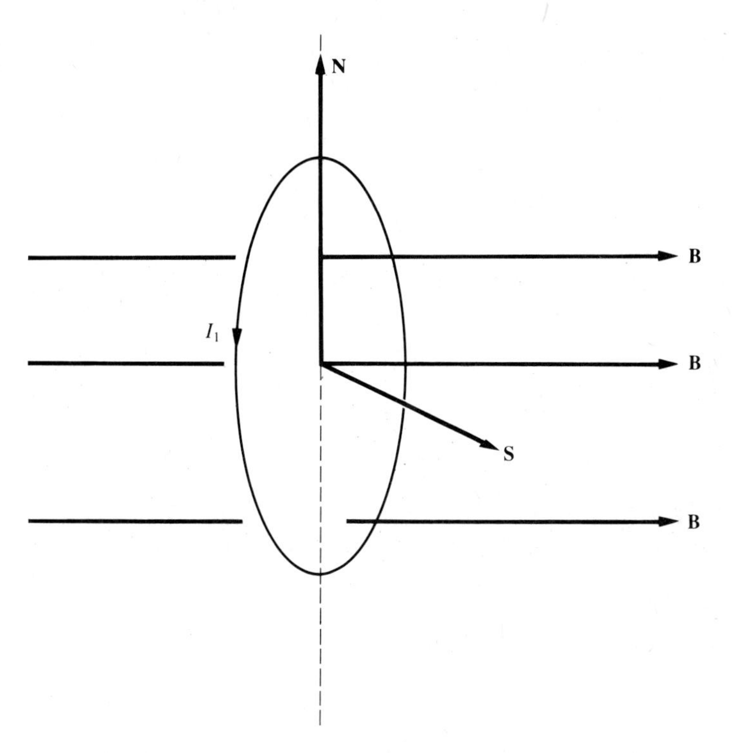


**FIGURE 6-14** Torque on a current-carrying loop in a static magnetic field. The torque acts so as to align the normal of the loop with the magnetic field.

## SUMMARY

### Important Concepts and Equations

Electromotive force (emf)

*The emf is the work done by a particular mechanism for each unit of charge transferred around a circuit. Batteries or generators are emf sources; resistances or inductors produce back emfs. The sum of emfs of a circuit vanishes.*

$\mathscr{E}_{\mathrm{R}} = RI$ (resistive emf)

*Resistive emf is the work per unit charge done by the ionic fields in a conductor on the charge carriers (electrons) as they move along the conductor.*

$$\mathscr{E} = \oint (\mathbf{E} + \mathbf{v} \times \mathbf{B}) \cdot d\mathbf{r}$$

*The emf $\mathscr{E}$ is the integral of macroscopic force per unit charge around the closed circuit at some instant of time; $d\mathbf{r}$ is a coordinate differential along the circuit path at position $\mathbf{r}$, and $\mathbf{v}$ is the velocity of the circuit element $d\mathbf{r}$.*

$$\mathscr{F} = \int_S \mathbf{B} \cdot d\mathbf{S}$$

*The magnetic flux $\mathscr{F}$ through a surface is given by the surface integral of the magnetic field* **B** *dotted into the surface element d***S***.*

$$\mathscr{E} = -\frac{d\mathscr{F}}{dt} \qquad \text{(Faraday's law)}$$

*The emf $\mathscr{E}$ induced in a circuit is given by the negative of the derivative of the magnetic flux $\mathscr{F}$ with respect to time. The magnetic flux through a circuit can change owing to motion of the circuit $d\mathscr{F}^{\dot{C}}/dt$ or a time-varying magnetic field $d\mathscr{F}^{\dot{B}}/dt$ or both.*

$$\nabla \times \mathbf{E} = -\frac{\partial \mathbf{B}}{\partial t} \qquad \text{(Faraday's law, differential form)}$$

*The curl of the electric field $\nabla \times \mathbf{E}$ is equal to minus the time rate of change of* **B**. *A changing magnetic field implies a nonconservative electric field.*

Lenz's rule

*The sense of the induced emf produces a* **B** *field that opposes the change of flux through the circuit.*

$$U = \frac{1}{2\mu_0} \int_{\text{all space}} B^2 \, dV$$

*The potential energy U stored in a magnetic field* **B**; *the integration must be done over all space.*

$$U = \frac{1}{2} \int_{\text{all space}} \mathbf{J} \cdot \mathbf{A} \, dV \qquad U = \frac{\mu_0}{8\pi} \iint \frac{\mathbf{J}(\mathbf{r}) \cdot \mathbf{J}(\mathbf{r}') \, dV \, dV'}{R}$$

*The potential energy U stored in a magnetic field in terms of the vector potential* **A** *and current density* **J**; *the integration is over all space. From the expression for the vector potential, U can be written in terms of the currents.*

$$L_{ij} = \frac{\mu_0}{4\pi} \oint_{C_i} \oint_{C_j} \frac{d\mathbf{r}_i \cdot d\mathbf{r}_j}{|\mathbf{r}_i - \mathbf{r}_j|}$$

*The mutual inductance $L_{ij}$ between circuits $C_i$ and $C_j$. The value of $L_{ij}$ is determined by the geometry of the circuits and is symmetric in the indices: $L_{ij} = L_{ji}$.*

$$\mathscr{F}_i = \sum_{j=1}^{N} L_{ij} I_j$$

*The total flux $\mathscr{F}_i$ linking the ith circuit of a collection of N circuits, where $I_j$ is the current of the jth circuit and $L_{ij}$ is the mutual inductance. The $L_{ii}$ term is the self-inductance; it gives the flux linking the circuit due to its own current.*

$$U = \frac{1}{2}\sum_i \sum_j L_{ij} I_i I_j$$

*The magnetic energy $U$ for currents confined to a closed system of circuits, where $L_{ij}$ is the mutual inductance and $I_i$ and $I_j$ are the currents in the circuits.*

$$\mathbf{F} = I_1 I_2 \nabla M$$

*The force between two circuits carrying currents $I_1$ and $I_2$, respectively; $\nabla M$ is the gradient of the mutual inductance. The force between circuits with currents in the same sense is attractive, while those with different senses repel each other. [The senses of $I_1$ and $I_2$ are the same if both give a flux contribution of the same sign in circuit 1 (or circuit 2).]*

$$\mathbf{N} = I\mathbf{S} \times \mathbf{B}$$

*The torque* **N** *on a circuit carrying current $I$ with directed area* **S** *in a magnetic field* **B**.

## PROBLEMS

### Section 6-2: Faraday's Law of Induction

**6-1** A thin metal rod of length $L$ and zero resistance falls in a horizontal orientation through a constant magnetic field **B** that is horizontal and perpendicular to the rod. At what rate does the potential difference between the ends of the rod increase?

**6-2** Refer to Figure 6–15. What is the direction of the current in the resistor $r$ (from $A$ to $B$ or from $B$ to $A$) when the following operations are performed? In each case, give a brief explanation of your reasoning.

(i) The switch S is closed.
(ii) Coil 2 is moved closer to coil 1 with the switch closed.
(iii) The resistance $R$ is decreased with the switch closed.

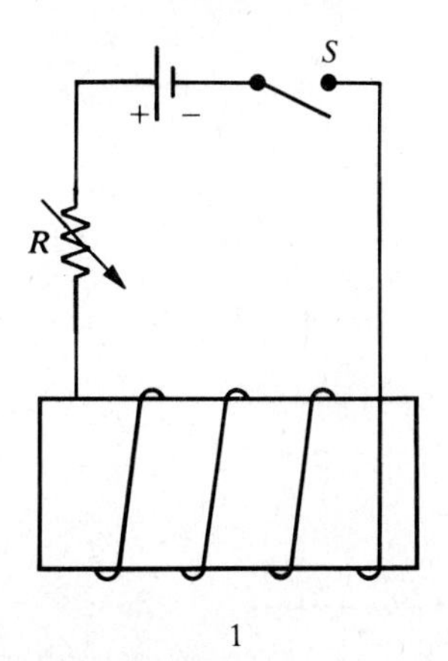


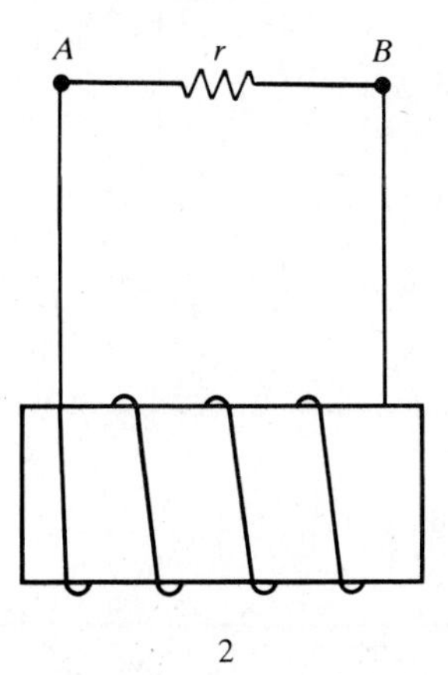


**FIGURE 6–15** Two current-carrying coils

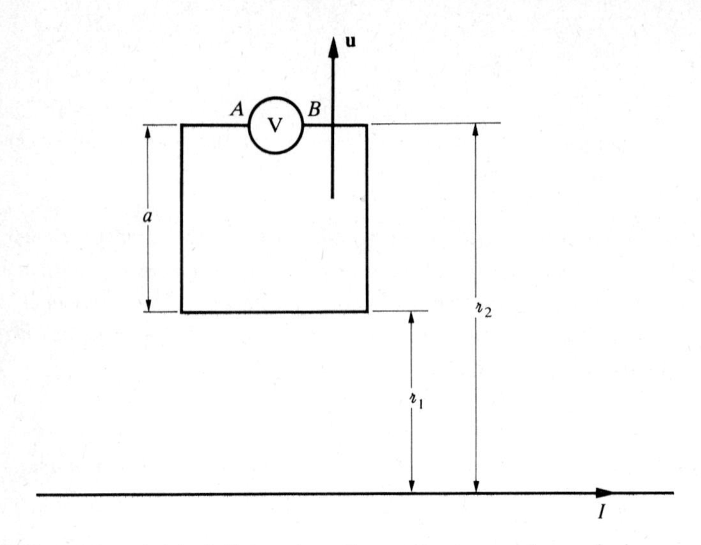


**FIGURE 6-16** Infinite wire adjacent to a square loop of wire

**6-3** An infinite wire carries a current $I$ in the $+z$ direction. A square loop of wire is connected to a voltmeter and moved with velocity **u** radially away from the wire. In Figure 6-16, which terminal of the voltmeter is positive? Calculate the reading on the voltmeter in terms of the distances $r_1$, $r_2$, and $a$.

**6-4** Two infinite parallel wires separated by a distance $d$ carry equal currents $I$ in opposite directions, with $I$ increasing at a rate $dI/dt$ sufficiently small that Ampère's law still applies. A square loop of wire of length $d$ on a side lies in the plane of the wires, at a distance $d$ from one of the parallel wires, as illustrated in Figure 6-17. Calculate the

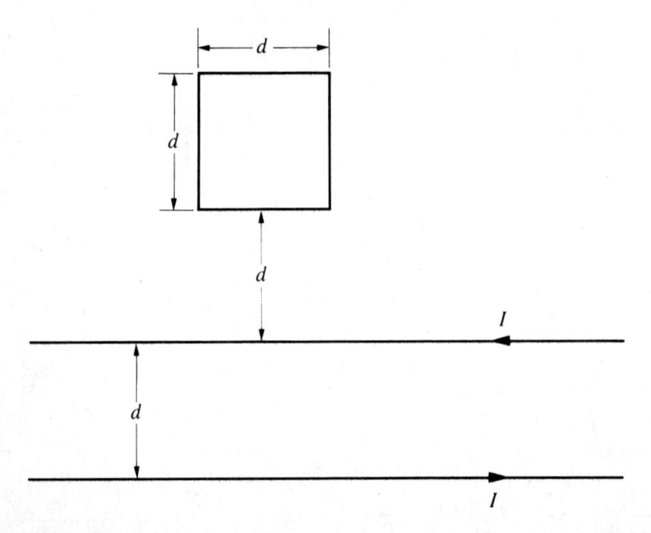


**FIGURE 6-17** Two infinite parallel wires adjacent to a square loop circuit

magnetic flux through the square loop, neglecting the field due to the induced current. Using Faraday's law, find the emf induced in the square loop. Is the induced current clockwise or counterclockwise? Justify your answer.

**6-5** In the circuit shown in Figure 6–18, a conducting rod of mass $m$ and length $L$ is acted upon by a constant force $F$. The rod, which starts from rest, moves through a region of uniform magnetic field on conducting rails. Assume there is no friction between the rod and the rails, and ignore the field due to the induced current.

(i) Calculate the velocity of the rod as a function of time.

(ii) Find the magnitude and the direction of the current through the resistor $R$.

**6-6** Two semi-infinite parallel rails terminated by a generator G are oriented in a uniform magnetic field, as shown in Figure 6–19. A rod of mass $m$ is in contact with the rails, separated by a distance $L$, which are frictionless. The generator produces a current in the direction shown given by

$$i(t) = \begin{cases} 0, & \text{for } t < 0 \\ bt, & \text{for } 0 < t < T \\ 0, & \text{for } t > T \end{cases}$$

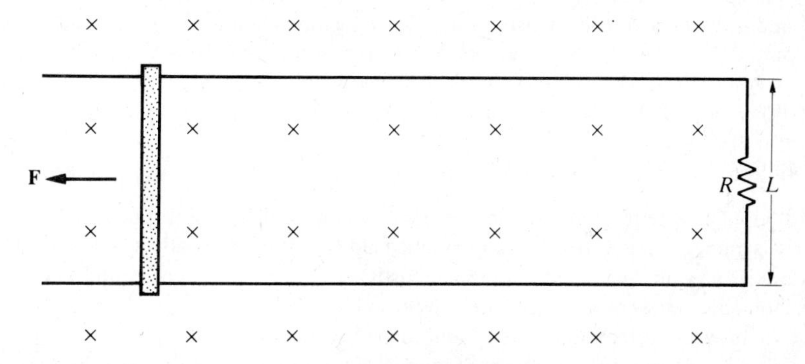


**FIGURE 6-18** Conducting rod of mass $m$ moving on conducting rails in a magnetic field under the action of a constant applied force **F**

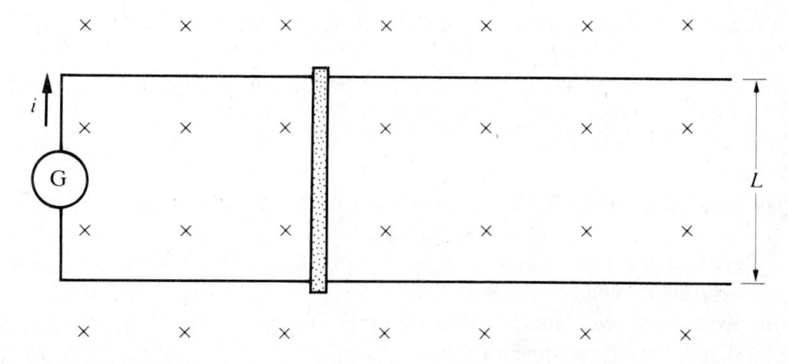


**FIGURE 6-19** Two semi-infinite parallel rails connected by a sliding rod and a current generator, in a magnetic field

If the rod is initially at rest, find the acceleration and the velocity of the rod and the voltage produced by the generator. Assume the magnetic field produced by both the current $i(t)$ and the induced current can be ignored.

**6-7** Consider a circular loop of wire of initial radius $r_0 = 0.1$ m and resistance $1\ \Omega$ (the resistance of the loop is kept constant for this problem). A variable-voltage battery is initially set for 1 V, and so initially the current through the wire loop is 1 A. An external magnetic field **B** of 1 T is applied normal to the loop. At time $t = 0$ the radius of the loop begins to increase at a constant rate $v = 1$ m/s, so $r = r_0 + vt$ for times greater than $t = 0$. Neglecting the magnetic field due to the current in the circuit, what voltage must be supplied by the battery to maintain a constant current of 1 A in the wire loop? What is the value of the voltage when $r = 10r_0$?

**6-8** A uniform magnetic field (into the paper) through a circular area 0.5 m in radius increases at the rate of 2 T/s. Assume no magnetic field *outside* this area.

(i) Find the emf induced in a single loop of wire ($r = 1.0$ m) concentric with the area. Label which lead will be positive.

(ii) Find the magnitude and the direction of the electric field at point $P$ a distance of 1.5 m from the center of the magnetic field.

**6-9** A magnetic field $B(\imath, t)$ is axially symmetric about the $z$ axis within a cylinder of radius $R$ and is increasing at the constant rate $\dot{B}(\imath)$, which points into the page. Use the approximation that $\dot{B}$ is constant (independent of $\imath$) in some region near the $z$ axis.

(i) In terms of $\dot{B}(\imath)$, what is the electric field within the cylinder?

(ii) What is the emf $\mathscr{E}$ around a square path of sides $\ell$ lying within the cylinder? Specify direction.

(iii) Calculate the potential difference between two adjacent corners in terms of $\mathscr{E}$.

**6-10** A circular wire loop of resistance $R$, mass $M$, and radius $a$ is dropped from $z = 0$ with its symmetry axis vertical in a magnetic field that is cylindrically symmetric about the $z$ axis and whose vertical ($z$) component is $B_z = Cz$, where $C$ is a constant. The axis of the loop falls along the symmetry axis of the field.

(i) In which direction will the induced current flow in the loop?

(ii) Find the current flowing as a function of the loop velocity. Neglect the contribution to the flux due to the field of the induced current.

(iii) Determine the forces acting on the loop from the axial and radial **B** components, and find the velocity of the loop after it has fallen a long time.

**6-11** A slender metal bar $AC$ of length $\ell$ and mass $m$ is pivoted at $A$; see Figure 6–20. The bar oscillates in a vertical plane with simple angular harmonic motion such that at any instant of time $t$ the angle $\theta$ between the vertical and the axis of the bar is given by $\theta = \theta_0 \sin(\omega t)$, where $\theta_0$ and $\omega$ are constants. The bar oscillates in a uniform magnetic field, where the field is everywhere perpendicular to the bar and directed into the plane of the diagram.

(i) Write the magnitude of $d\mathscr{E}$, the emf induced in an element of length $dr$, at an instant when $v$, the linear velocity of the element, is directed as shown in the figure. The element $dr$ is located at a distance of $r$ from $A$. Make the necessary substitutions to express $d\mathscr{E}$ explicitly in terms $r$ and $t$.

(ii) Evaluate the emf induced in the bar at the instant considered in part (i). Specify the sign of the charge appearing at end $C$.

(iii) For what value of $\theta$ is the induced emf a maximum? What is the maximum value?

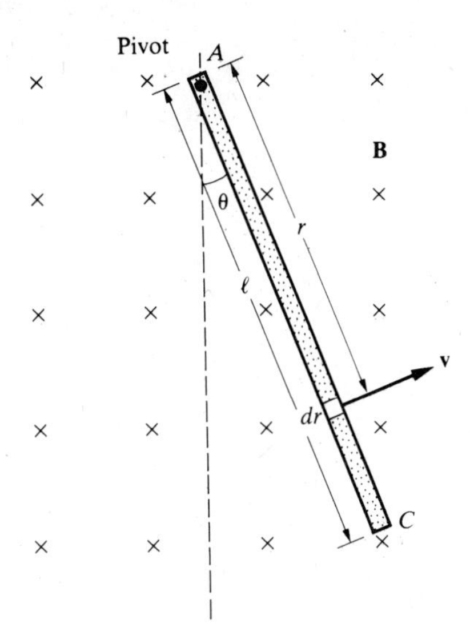


FIGURE 6-20 Metal bar pivoted at $A$ that swings in a magnetic field

**6-12** A plastic (nonconducting) wheel with moment of inertia $I$ has $n$ positive electric charges $q$ embedded in it at a radius $a$; see Figure 6–21. A uniform magnetic field **B** is now turned on, parallel to the axis of the wheel. Indicate the direction of rotation, and calculate the angular velocity.

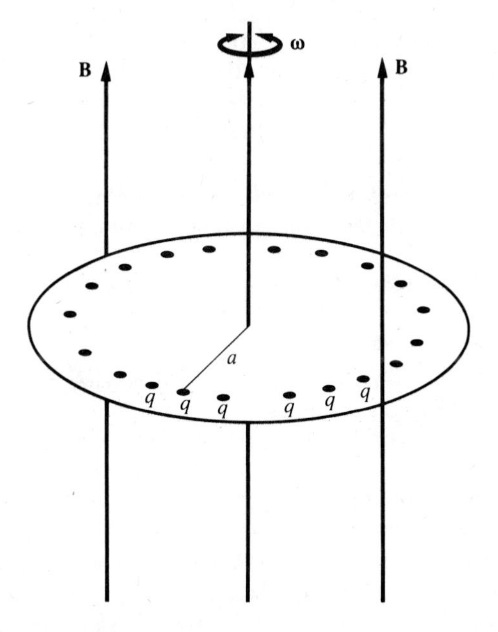


FIGURE 6-21 Plastic nonconducting wheel, with implanted charges, that rotates when the magnetic field changes

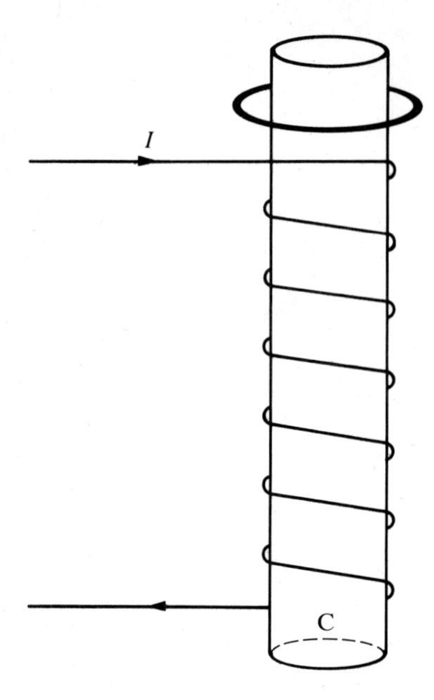


**FIGURE 6-22** Jumping ring

**6-13** A popular demonstration apparatus (see Figure 6-22) includes a coil wound around a core and an aluminum ring that fits loosely around the core. When a current $I$ is started in the direction shown in the figure, the ring jumps up off the core.

(i) What force throws the ring up? (Show a clear diagram of the forces involved.)

(ii) Will the ring also be thrown upward if the direction of the current change is reversed? Give a reason for your answer.

**6-14** A circular wire loop of radius $a$ is located in the $xy$ plane with its center at the origin. A point charge $q$ moves in the $xy$ plane along the line $y = D$ with constant nonrelativistic velocity $v$.

(i) If the loop is open, what emf is induced in it as a function of the $x$ coordinate of the charge? Assume that $D \gg a$.

(ii) Find the positions of the charge at which the emf is maximum.

**6-15** Homopolar generator: Consider a copper disk of radius $b$ that is mechanically rotated with angular velocity $\omega$ in a uniform magnetic field, as shown in Figure 6-23.

(i) When the switch S is open, show that the emf generated is $\mathscr{E} = \frac{1}{2}\omega B b^2$.

(ii) When the switch S is closed, a current flows through the circuit (assume that the resistance can be lumped into $R$). A torque is now required to maintain the angular velocity. Find this torque and the associated power required to run the generator.

**6-16** Homopolar motor: Use the apparatus depicted in Figure 6-23 with the following change: A battery of emf $V$ is now inserted into the circuit.

(i) Compute the locked-rotor torque ($\omega = 0$).

(ii) As the disk starts turning, a back emf is generated. Find the torque as a function of angular velocity. Make a rough plot of the torque-speed curve and the power output as a function of $\omega$.

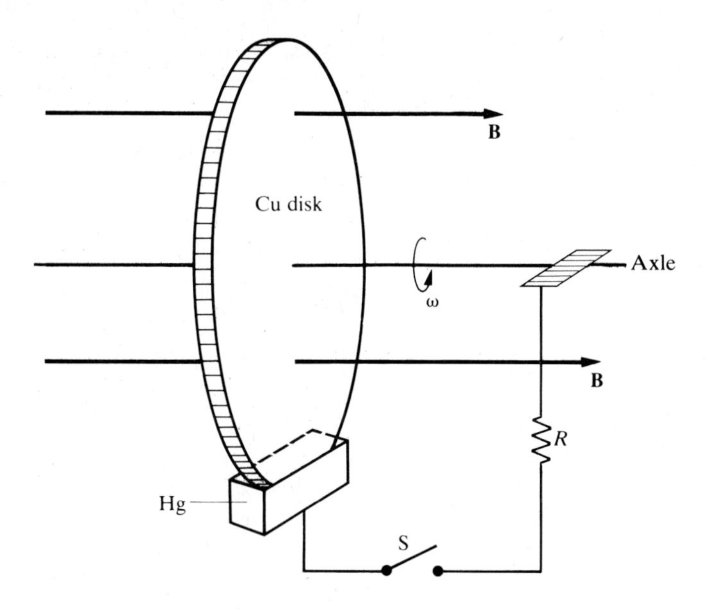


**FIGURE 6–23** Homopolar generator

**6–17** A spherically symmetric electric field due to a charge distribution is given by

$$\mathbf{E}(\mathbf{r}, t) = Ae^{-\alpha rt}\frac{\hat{\mathbf{r}}}{r^2}$$

where $A$ and $\alpha$ are constants. Find the charge density $\rho(\mathbf{r}, t)$. Verify your answer in the limit $\alpha \to 0$. For a radial current density $\mathbf{J} = J(r)\hat{\mathbf{r}}$, find the magnetic field $\mathbf{B}$.

**6–18** In a betatron accelerator an electron moves in a circular orbit of constant radius $\imath_0$ in the $z = 0$ plane. The plane of the orbit is perpendicular to the axis of a magnetic field $\mathbf{B}(\imath, z, t)$; the field is axially symmetric around the $z$ axis and symmetric about the $z = 0$ plane. The average magnetic field $B_{avg}(z, t)$ linking the orbit is

$$B_{avg}(z, t) = \frac{\int \mathbf{B}(\imath, z, t)\cdot d\mathbf{S}}{\pi \imath_0^2}$$

(i) In terms of $B_{avg}(z, t)$, calculate the electric field $E_\phi$ acting on the electron. (Use Faraday's law in integral form.)

(ii) The electron is accelerated in the direction tangent to its circular orbit by $E_\phi$. What is the corresponding time rate of change of momentum ($\dot{p}$)?

(iii) Find the circular-orbit relation between momentum $p$, the radius $\imath_0$, and the magnetic field $B_z(\imath = \imath_0, z = 0, t)$.

(iv) Using the results of parts (ii) and (iii), find a relation between $B_{avg}(z = 0, t)$ and $B(\imath = \imath_0, z = 0, t)$ that is necessary for circular motion at radius $\imath_0$ at all $t$.

**6–19** A very long air-core solenoid of radius $b$ has $n$ turns per meter and carries a current $i = i_0 \sin \omega t$. Give an expression for the magnetic field $B$ inside the solenoid and expressions for the electric field $\mathbf{E}$ inside and outside the solenoid as a function of time. (Assume that $\mathbf{B}$ is zero outside the solenoid.)

**6-20** An infinite solenoid of circular cross section has a slowly time-varying current $I(t)$ in its winding. Calculate the velocity that a charged particle a distance $\imath$ from the solenoid axis must have for the Lorentz force to vanish.

**6-21** A spatially constant but time-varying magnetic field is normally incident on a circular conducting disk of radius $a$, thickness $t$, and conductivity $g$.
(i) Find the current density and the total current induced in the disk. Currents of this type are known as *eddy currents*.
(ii) Find the joule heating power.

**6-22** A piece of wire of mass $m$ and electric resistance $R$ forms a circular ring of area $A$. It is suspended by an insulated, massless string of length $\ell$ to form a frictionless pendulum in the presence of a uniform, vertical magnetic field $B$. The normal to the ring is parallel to its velocity. Calculate the time required for the energy of a small oscillation of the pendulum to decay to $1/e$ of its initial value if the energy dissipated per cycle is small compared with the initial energy.

**6-23** A torsion pendulum consists of a solid disk of radius $a$ and thickness $t$, composed of metal with electric conductivity $g$. The pendulum turns in a magnetic field $B$ perpendicular to the disk. Calculate the damping torque of the pendulum, assuming that charge accumulation at the edge of the disk is negligible and that the uniform magnetic field is not affected by the induced currents.

## Section 6-3B: Flip Coil

**6-24** A rectangular coil with resistance $R$ and dimensions $a$ and $b$ moves with constant velocity $v$ into a uniform magnetic field, as shown in Figure 6-24. Find the force on the coil as the coil enters the field.

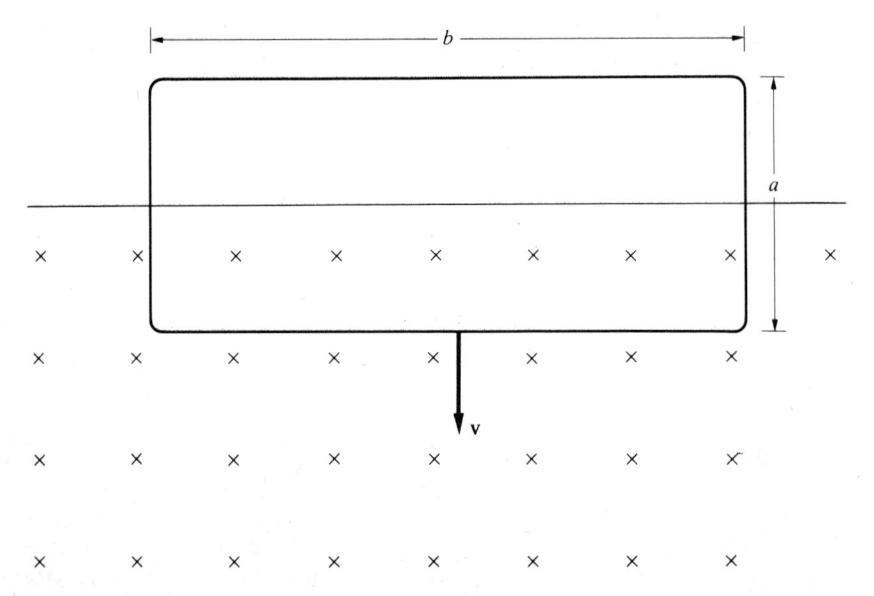


**FIGURE 6-24** Coil entering a magnetic field

## Section 6-3C: Alternating-Current Generator

**6-25** A flip coil of resistance $R$ and self-inductance $L$ is rotated at constant angular velocity $\omega$ from maximum to minimum magnetic flux. Compute the current flowing in the coil as a function of time. By explicit integration, verify that the total charge flow is $2\mathscr{F}(0)/R$. Sketch the current as a function of time for $\omega = L = R = 1$ in SI units.

## Section 6-3D: Induced Currents in Superconductors

**6-26** A superconducting flip coil circuit has self-inductance $L$ but zero resistance. Show, by integrating the circuit equation, that the total flux always remains constant.

## Section 6-4: Magnetic Energy

**6-27** A cylindrical thin shell of electric charge has length $\ell$ and radius $a$, where $\ell \gg a$. The surface charge density on the shell is $\sigma$. The shell rotates about its axis with an angular velocity $\omega$ that increases slowly with time as $\omega = kt$, where $k$ is a positive constant and $t \geqslant 0$. Neglect fringing effects.

(i) Determine the magnitude and the direction of the magnetic field $\mathbf{B}$ inside the cylinder, that is, for $r < a$.

(ii) Determine the magnitude and the direction of the electric field $\mathbf{E}$ inside the cylinder.

(iii) Determine the total electric field energy $U_E$ inside the cylinder and the total magnetic field energy $U_B$ inside the cylinder.

**6-28** A uniform magnetic field $\mathbf{B}_0$ is directed into the page in a vacuum region surrounded by a metal box with cross section as shown in Figure 6–25. The box has sides of length $a$ and is infinite in extent. The metal is a perfect conductor and hence can have no $\mathbf{E}$ or $\mathbf{B}$ fields in it.

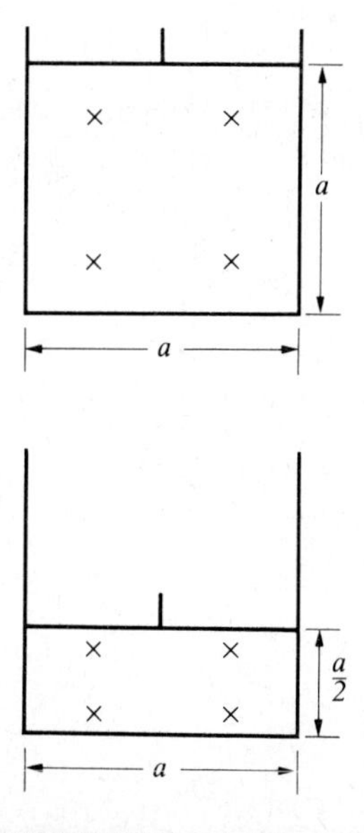


**FIGURE 6-25** Magnetic field compression in a metal box

(i) Suppose the top plane is moved downward to a distance $a/2$ as shown in the figure. Calculate the flux inside the box.

(ii) Find the work needed per unit length to perform the operation in part (i).

## Section 6-5: Inductance

**6-29** Two coils of self-inductances $L_1$ and $L_2$ connected in series in a circuit have a mutual inductance $M$. What is the resultant inductance of the circuit? If the wires to one coil are reversed, what is the resultant inductance?

**6-30** Show that the inductance per unit length of a long wire of radius $a$ carrying a uniform current due to the interior field alone is $\mu_0/8\pi$. (*Hint*: First compute the interior magnetic energy of the wire.)

**6-31** Compute an approximate value for the self-inductance of a wire of radius $a$ bent into the shape of a rectangle with sides $b$ and $c$. First, find the interior inductance by using the result of Problem 6–30, and then compute the flux linkage exterior to the wire.

**6-32** Two long solenoids each with length $\ell$, with $N$ turns, and with radii $r_1$ and $r_2$, respectively, are positioned coaxially. The larger solenoid is connected to a voltage source $V_0 \cos \omega t$. Calculate the self-inductance of the primary and secondary circuits and the mutual inductance. Write the differential equations that determine the currents in the primary and secondary circuits if the secondary is in series with a resistance $R$.

**6-33** A solenoid has an air core and has a length of $\frac{1}{2}$ m, a cross section of 1 $\text{cm}^2$, and 1000 turns. Neglecting end effects, what is the self-inductance? A secondary winding wrapped around the solenoid has 100 turns. What is the mutual inductance? A constant current of 1 A flows in the secondary, and the solenoid is connected to a load of $10^3\ \Omega$. The constant current is suddenly stopped. How much charge flows through the resistance?

**6-34** Show that for two magnetic circuits $M \leqslant \sqrt{L_1 L_2}$. (*Hint*: The magnetic energy cannot be negative.)

## Section 6-6: Transformers

**6-35** For the circuit shown in Figure 6–26, $L_A = L_B = M_{AB} = 3$ Hz. At time $t = 0$ the switch S is closed.

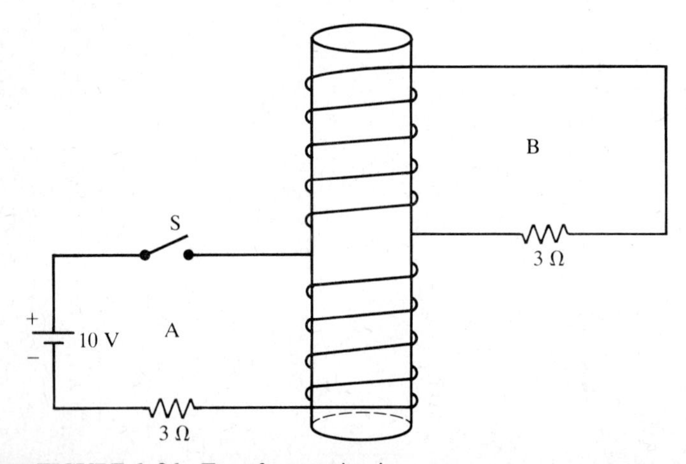


**FIGURE 6-26** Transformer circuit

(i) Which way (clockwise or counterclockwise) will current start to flow in circuits A and B?
(ii) What are the eventual currents (magnitude and direction) in circuits A and B?
(iii) Write the equations for emf drops for both circuits. Solve these equations by using the above initial conditions, and verify that the answers to part (ii) are correct.

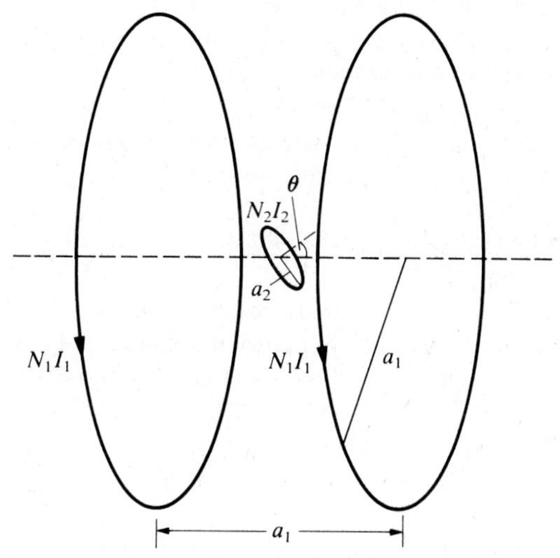


**FIGURE 6-27** Helmholtz coil arrangement

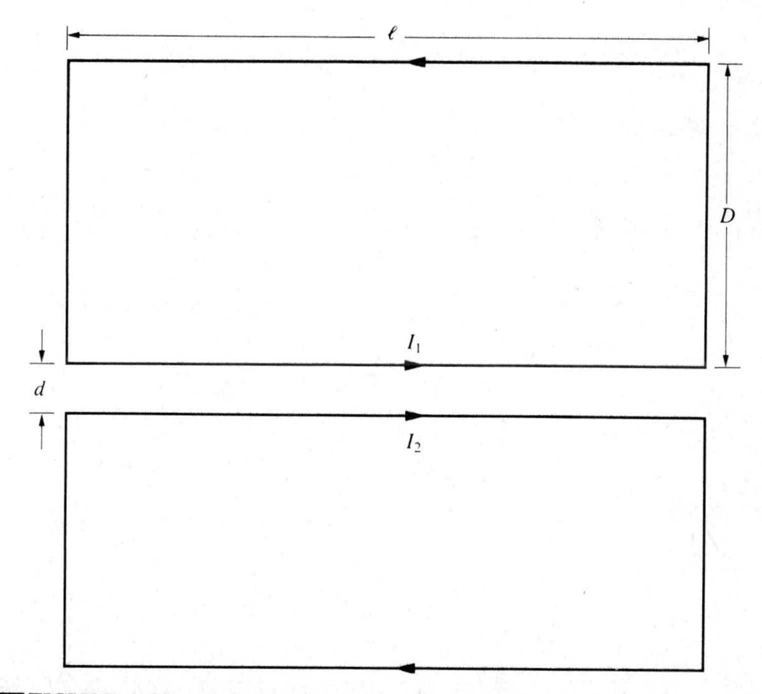


**FIGURE 6-28** Coplanar rectangular wire circuits

## Section 6-7: Forces and Torques on Current-Carrying Circuits

**6-36** A long nonconducting cylinder of radius $a$ carries a surface charge per unit length $\lambda$. It is wound with $n$ turns per meter of wire carrying a current $I$. This current gives rise to a uniform magnetic field $B$ inside the cylinder. The current $I$ is now turned off, which gives rise to an induced electric field (by Faraday's law) that exerts a torque $\mathbf{N} = \mathbf{r} \times \mathbf{F}$ on the cylinder. Find the magnitude and the direction of this torque per unit length in terms of $dI/dt$. (Neglect the fields due to the induced current.)

**6-37** A Helmholtz coil arrangement (see Figure 6–27) has radius $a_1$ and $N_1$ turns carrying current $I_1$. A small coil of radius $a_2 \ll a_1$ is located on the axis halfway between the Helmholtz coils. The small coil has $N_2$ turns, each carrying current $I_2$, and its axis is inclined at an angle $\theta$ to the Helmholtz axis.
(i) Find the mutual inductance between the Helmholtz coil and the small coil.
(ii) Find the torque exerted on the small coil.

**6-38** Two coplanar rectangular wire circuits have dimensions $D, \ell \gg d$, where $d$ is the closest separation, as indicated in Figure 6–28. Find the mutual inductance, and use this result to calculate the force between the circuits. Compare your result with the force between long parallel wires.

# 7

# Magnetic Fields in Matter

## CHAPTER CONTENTS

We examined the effects of an electric field on ordinary matter in Chapter 4. There we found that the external electric field induced an electric dipole volume density that contributed in turn to the electric field. In this chapter we will investigate similar effects involving a magnetic field. The external magnetic field induces magnetic moments in the matter that add to (vectorially, of course) the external field. For certain materials (ferromagnetic) the magnetic field due to the induced magnetism of the material can be many thousands of times larger than the original imposed field. The macroscopically observable magnetic property is the magnetic dipole moment density, the magnetization. Precisely (in parallel to the definition of electric polarization), the magnetization is the magnetic dipole moment per unit volume; it is a macroscopic quantity and a vector.

After constructing the macroscopic theory, we qualitatively examine the microscopic causes of the various kinds of magnetism. Diamagnetism arises from induced atomic currents. Paramagnetism is due to the partial alignment of intrinsic magnetic dipoles with the imposed field. Ferromagnetism also involves alignment of intrinsic magnetic dipoles and occurs when (nonmagnetic) atom-atom forces align neighboring intrinsic magnetic dipoles parallel to one another.

We begin our discussion of magnetism with the magnetic moment of a current and later generalize to include intrinsic magnetic moments.

## 7-1 MAGNETIC MULTIPOLE EXPANSION

If a steady current distribution $\mathbf{J}(\mathbf{r}')$ is known, then we can calculate the resulting magnetic field from the vector potential

$$\mathbf{A}(\mathbf{r}) = \frac{\mu_0}{4\pi} \int \frac{\mathbf{J}(\mathbf{r}')}{|\mathbf{r} - \mathbf{r}'|} \, dV'$$

When the currents are confined to a region that is far from the field point $\mathbf{r}$, many of the details of the current distribution become unimportant, and its first nonvanishing multipole moment suffices to describe the distant field. To show this result, we choose the origin of the coordinate system to be near the center of the region containing the current; see Figure 7–1. When $|\mathbf{r}| \gg |\mathbf{r}'|$, we can use Equation (4–5) and expand the denominator of the integrand:

$$\frac{1}{|\mathbf{r} - \mathbf{r}'|} \approx \frac{1}{r} + \frac{\mathbf{r} \cdot \mathbf{r}'}{r^3} + \cdots \tag{7-1}$$

Thus the first and second terms in the expansion of the vector potential are

$$\mathbf{A}(\mathbf{r}) = \frac{\mu_0}{4\pi r} \int \mathbf{J}(\mathbf{r}') \, dV' + \frac{\mu_0}{4\pi r^3} \int \mathbf{r} \cdot \mathbf{r}' \mathbf{J}(\mathbf{r}') \, dV' + \cdots \tag{7-2}$$

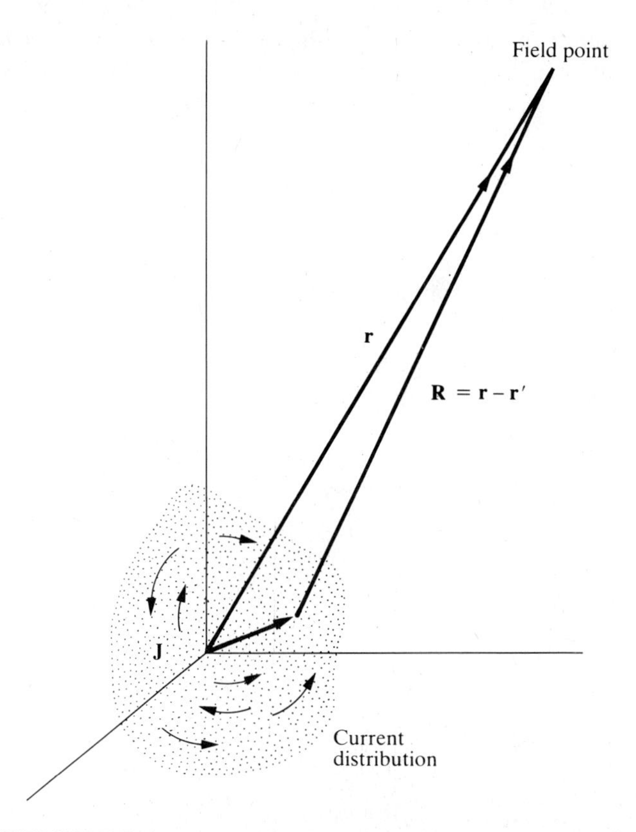


**FIGURE 7–1** Current distribution localized in a region around the origin

These terms are called the *monopole* and *dipole terms*, respectively; the next higher-order term in the expansion of **A** is called the magnetic *quadrupole*. This magnetic multipole expansion is more difficult to deal with than the electrostatic case because **A** is a vector quantity. We will first derive a useful relation to help us through these difficulties.

We consider the product $f\mathbf{J}$, where $f(\mathbf{r}')$ is an arbitrary function. Applying the divergence theorem for a bounding surface that completely encloses the current distribution, we find

$$\int \nabla \cdot (f\mathbf{J})\, dV' = \oint f\mathbf{J} \cdot d\mathbf{S} = 0 \tag{7–3a}$$

where $\nabla$ acts on the source coordinates $x'$, $y'$, and $z'$. Using the identity $\nabla \cdot (f\mathbf{J}) = (\nabla f) \cdot \mathbf{J} + f\, \nabla \cdot \mathbf{J}$ and the fact that $\nabla \cdot \mathbf{J} = 0$ for a steady current distribution, Equation (7–3a) gives

$$\int (\nabla f) \cdot \mathbf{J}\, dV' = 0 \tag{7–3b}$$

If we make the choice $f = x'_i$ (the $i$th component of the vector $\mathbf{r}'$), $\nabla f = \hat{\mathbf{x}}_i$ and we have

$$\int J_i dV' = 0 \tag{7-4}$$

so that $\int \mathbf{J}\, dV' = 0$. Thus the first term in the expansion, Equation (7–2), vanishes. This result is a consequence of the absence of magnetic charge.

The $i$th component of the dipole term in Equation (7–2) is

$$A_i = \frac{\mu_0}{4\pi r^3} \sum_j x_j \int x'_j J_i\, dV' \tag{7-5}$$

To put this expression into a more useful form, we separate the dyad $x'_j J_i$ in the integrand into symmetric and antisymmetric parts:

$$A_i = \frac{\mu_0}{4\pi r^3} \sum_j x_j \int \left[ \frac{1}{2}(x'_j J_i + x'_i J_j) + \frac{1}{2}(x'_j J_i - x'_i J_j) \right] dV' \tag{7-6}$$

Using Equation (7–3b) with $f = x'_i x'_j$, we find

$$\int (x'_j \hat{\mathbf{x}}_i + x'_i \hat{\mathbf{x}}_j) \cdot \mathbf{J}\, dV' = 0 \tag{7-7}$$

and thus the symmetric term in the integrand of Equation (7–6) vanishes, leaving

$$A_i = \frac{\mu_0}{4\pi r^3} \sum_j x_j \int \frac{1}{2}(x'_j J_i - x'_i J_j)\, dV' \tag{7-8}$$

The integrand can now be reexpressed in terms of vectors as

$$\sum_j x_j (x'_j J_i - x'_i J_j) = [\mathbf{J}(\mathbf{r} \cdot \mathbf{r}') - \mathbf{r}'(\mathbf{J} \cdot \mathbf{r})]_i = [(\mathbf{r}' \times \mathbf{J}) \times \mathbf{r}]_i \tag{7-9}$$

Then from Equations (7–8) and (7–9) the dipole term in the vector potential has the form

**Dipole Vector Potential**

$$\mathbf{A} = \frac{\mu_0}{4\pi} \left( \frac{\mathbf{m} \times \mathbf{r}}{r^3} \right) \tag{7-10}$$

where $\mathbf{m}$ is the magnetic dipole moment:

**Magnetic Dipole Moment**

$$\mathbf{m} = \frac{1}{2} \int (\mathbf{r}' \times \mathbf{J})\, dV' \tag{7-11}$$

From Equation (7–10) the magnetic dipole field is

$$\mathbf{B} = \frac{\mu_0}{4\pi} \nabla \times \left[\frac{1}{r^3}(\mathbf{m} \times \mathbf{r})\right] = \frac{\mu_0}{4\pi}\left[\frac{1}{r^3} \nabla \times (\mathbf{m} \times \mathbf{r}) - \frac{3\mathbf{r}}{r^5} \times (\mathbf{m} \times \mathbf{r})\right] \tag{7-12}$$

where we have used $\nabla \cdot \mathbf{r} = 3$. Since $\nabla \times (\mathbf{m} \times \mathbf{r}) = 2\mathbf{m}$, by expanding the vector triple product, we have

**Magnetic Dipole Field**

$$\mathbf{B} = \frac{\mu_0}{4\pi}\left[\frac{3\mathbf{r}(\mathbf{m} \cdot \mathbf{r}) - r^2\mathbf{m}}{r^5}\right] \tag{7-13}$$

We note that according to this formula, the magnetic dipole field has the same form as the electric dipole field of Equation (4–17), whose field lines were shown in Figure 4–4. But at $\mathbf{r} = 0$ the magnetic and electric dipoles differ; there is a delta function contribution at $\mathbf{r} = 0$, which can be deduced by using the divergence theorem corollary of Equation (2–15). The complete formula for the magnetic dipole field is

$$\mathbf{B} = \frac{\mu_0}{4\pi}\left[\frac{3\mathbf{r}(\mathbf{r} \cdot \mathbf{m}) - r^2\mathbf{m}}{r^5} + \frac{8\pi}{3}\mathbf{m}\delta^3(\mathbf{r})\right] \tag{7-14}$$

With the $z$ axis along $\mathbf{m}$ the spherical components of the magnetic field in Equation (7–13) are

$$B_r = \frac{\mu_0 m}{4\pi} \frac{2\cos\theta}{r^3} \qquad B_\theta = \frac{\mu_0 m}{4\pi} \frac{\sin\theta}{r^3} \tag{7-15}$$

To a good approximation, the field outside the earth is a magnetic dipole of dipole moment

$$|\mathbf{m}| = 8.2 \times 10^{22}\ \mathrm{A \cdot m^2}$$

inclined at an angle of 168.5° from the rotation axis. The magnetic field outside the earth is thus given by Equation (7–15). The magnetic north pole of the earth is defined as the point on the earth at which the magnetic field points vertically downward. This definition is opposite to the definition of the north pole on a compass magnet, which is the reason that the north pole of a compass points toward the magnetic north pole of the earth. The strength $|\mathbf{B}|$ of the earth's magnetic field is about 0.3 G at the equator and about 0.6 G in the polar region. The earth's magnetic field lines are shown in Figure 7–2. The dynamo mechanism producing these fields was briefly discussed in Section 5–5. Of the planets, Jupiter has the largest magnetic field, with moment $|\mathbf{m}| = 1.4 \times 10^{27}\ \mathrm{A \cdot m^2}$.

The magnetic moment of a steady current distribution does not depend on the origin of the coordinate system in which it is calculated. To

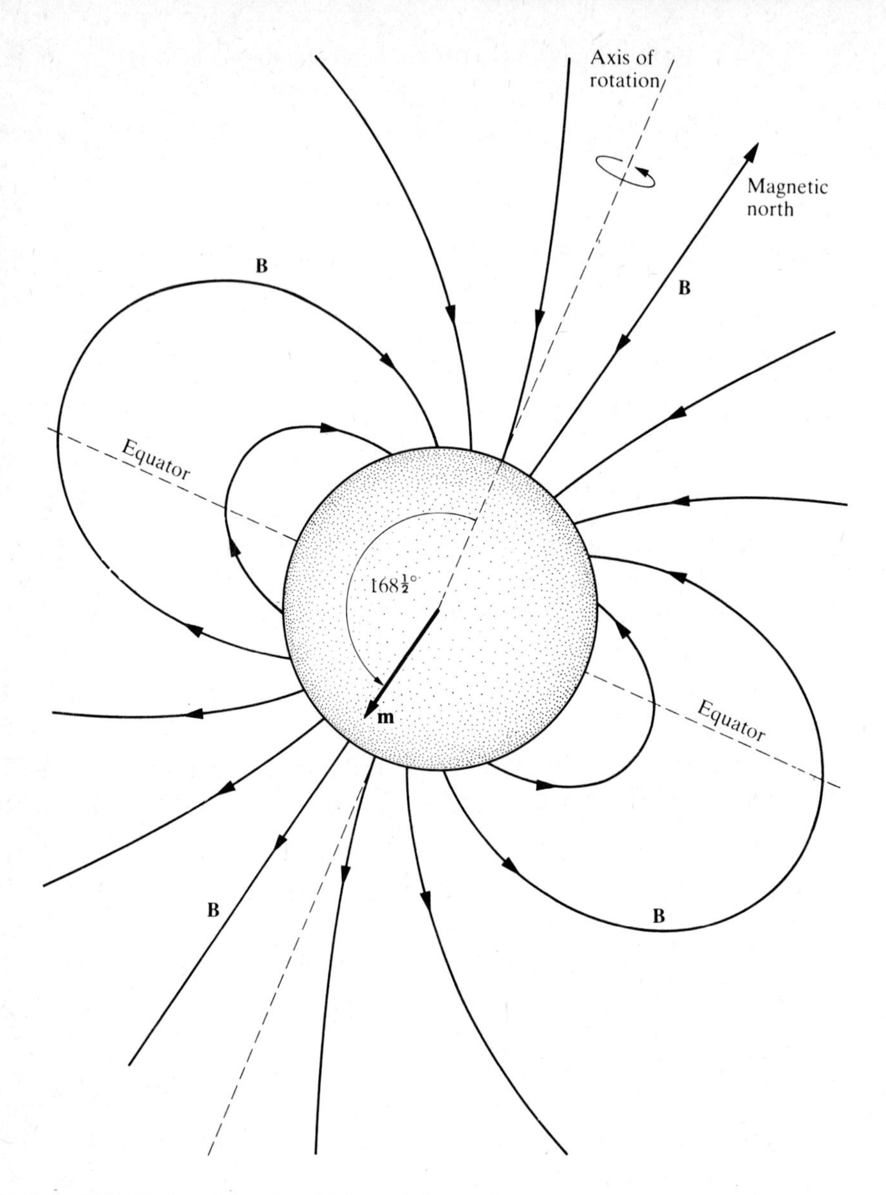


FIGURE 7-2 Magnetic field lines of the earth

show this result, we make the translation $\mathbf{r}' = \mathbf{r}'' + \mathbf{c}$ in Equation (7–11) for the magnetic moment:

$$\mathbf{m} = \frac{1}{2}\int \mathbf{r}' \times \mathbf{J}(\mathbf{r}')\, dV'$$

$$= \frac{1}{2}\int \mathbf{r}'' \times \mathbf{J}(r'')\, dV'' + \frac{1}{2}\mathbf{c} \times \int \mathbf{J}(\mathbf{r}')\, dV' \tag{7-16}$$

Since $\int \mathbf{J}(\mathbf{r}')\, dV' = 0$ for steady currents, $\mathbf{m}$ is unchanged by the translation. We will use this result later when we compute the total magnetic moment of an assembly of induced molecular magnetic moments.

For a planar circuit the magnetic moment has a simple form. Replacing $\mathbf{J}\, dV'$ by $I\, d\mathbf{r}'$ in Equation (7–11), we have

$$\mathbf{m} = I \oint \frac{1}{2} \mathbf{r}' \times d\mathbf{r}' \qquad \textbf{(7-17)}$$

The area of a parallelogram with sides $\mathbf{r}'$ and $\mathbf{r}' + d\mathbf{r}'$ is $\mathbf{r}' \times (\mathbf{r}' + d\mathbf{r})$, which is twice the area element swept out by the vector $\mathbf{r}'$ in moving to $\mathbf{r}' + d\mathbf{r}'$:

$$d\mathbf{S} = \hat{\mathbf{n}}\, dS = \frac{1}{2} \mathbf{r}' \times d\mathbf{r}' \qquad \textbf{(7-18)}$$

Here $\hat{\mathbf{n}}$ is the unit normal to the circuit, as shown in Figure 7–3. Thus for any planar circuit the magnetic moment is as follows:

**Magnetic Moment of Planar Circuit**

$$\mathbf{m} = IS\hat{\mathbf{n}} \qquad \textbf{(7-19)}$$

where $S$ is the area of the circuit.

Moving charges can also generate a magnetic moment. We begin with the current density of a system of moving charges from Equation (5–4):

$$\mathbf{J} = \sum_i q_i \mathbf{v}_i \delta^3[\mathbf{r}' - \mathbf{r}_i(t)]$$

Substituting this current density into the magnetic moment definition of

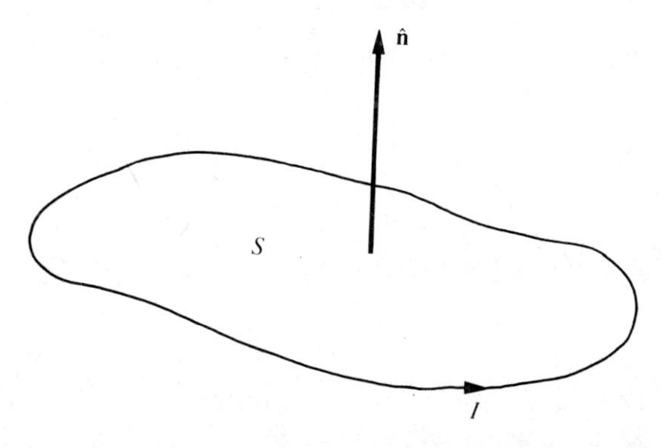


**FIGURE 7–3** Magnetic moment for a planar circuit. It is in the normal direction $\hat{\mathbf{n}}$ and has magnitude of current $I$ multiplied by the area of the circuit $S$.

Equation (7–11) yields

$$\mathbf{m} = \frac{1}{2}\int \mathbf{r}' \times \mathbf{J}\, dV' = \frac{1}{2}\sum_i q_i \int \mathbf{r}' \times \mathbf{v}_i \delta^3[\mathbf{r}' - \mathbf{r}_i(t)]\, dV'$$

$$= \frac{1}{2}\sum_i q_i \mathbf{r}_i \times \mathbf{v}_i \tag{7-20}$$

For a single particle moving in a circle in the $xy$ plane, this relation simplifies to

$$\mathbf{m} = \frac{q}{2} r v_\phi \hat{\mathbf{z}} \tag{7-21}$$

In terms of the angular velocity $\omega = v_\phi / r$, the magnetic moment can be expressed as

$$\mathbf{m} = \left(\frac{q}{2\pi/\omega}\right)(\pi r^2)\hat{\mathbf{z}} \tag{7-22}$$

Since $\tau = 2\pi/\omega$ is the period of the motion, we recognize that $\mathbf{m}$ is the product of the current $I = q/\tau$ of the circulating charge and the area $S = \pi r^2$ of the orbit:

$$\mathbf{m} = IS\hat{\mathbf{z}} \tag{7-23}$$

which is the same as the circuit result of Equation (7–19). Using $r = v_\phi/\omega$, we can also express the magnetic moment as

$$\mathbf{m} = \frac{q v_\phi^2}{2\omega}\hat{\mathbf{z}} \tag{7-24}$$

# 7-2 TORQUE ON A MAGNETIC DIPOLE

In Section 6–7 we considered the torque on a planar circuit due to a constant magnetic field. Our result there was

$$\mathbf{N} = IS\hat{\mathbf{n}} \times \mathbf{B}$$

where $S$ was the area of the circuit and $\hat{\mathbf{n}}$ was the unit normal to the circuit. In terms of the magnetic moment expression of Equation (7–19), the torque is

$$\mathbf{N} = \mathbf{m} \times \mathbf{B} \tag{7-25}$$

This expression for the torque also holds for an arbitrary current distribution in a constant magnetic field, as we will now show.

Since the magnetic part of the Lorentz force on a volume element is $d\mathbf{F} = \mathbf{J} \times \mathbf{B}\, dV'$, the torque on a current distribution is

$$\mathbf{N} = \int \mathbf{r}' \times (\mathbf{J} \times \mathbf{B})\, dV' \tag{7-26}$$

We expand the triple cross product of the integrand; in component form we have

$$N_i = \int [J_i(\mathbf{r}' \cdot \mathbf{B}) - B_i(\mathbf{r}' \cdot \mathbf{J})]\, dV' = \sum_j \int (J_i x'_j B_j - B_i x'_j J_j)\, dV' \tag{7-27}$$

To show that the second term in Equation (7–27) vanishes, we use Equation (7–3b) with $f = \frac{1}{2}(x'_j)^2 B_i$; then $\nabla f = \sum_j x'_j \hat{\mathbf{x}}_j B_i$ for constant $\mathbf{B}$, and Equation (7–3b) yields

$$\sum_j \int B_i x'_j J_j\, dV' = 0 \tag{7-28}$$

Thus the torque is given by

$$\mathbf{N} = \int (\mathbf{B} \cdot \mathbf{r}')\mathbf{J}\, dV' \tag{7-29}$$

From a comparison of the dipole terms in Equations (7–2) and (7–10), we see that

$$\int (\mathbf{r} \cdot \mathbf{r}')\mathbf{J}\, dV' = \mathbf{m} \times \mathbf{r} \tag{7-30}$$

This integral has the same structure as Equation (7–29) except that the constant vector is $\mathbf{r}$ instead of $\mathbf{B}$. Hence we conclude that the torque on a general current distribution due to a constant magnetic field is as follows:

**Torque on a Magnetic Moment**

$$\mathbf{N} = \mathbf{m} \times \mathbf{B} \tag{7-31}$$

This expression is analogous to that of the torque $\mathbf{N} = \mathbf{p} \times \mathbf{E}$ on an electric dipole in an $\mathbf{E}$ field.

Starting from the torque $\mathbf{N} = \mathbf{p} \times \mathbf{E}$ on an electric dipole, we showed in Section 4–1 that the potential energy of an electric dipole in an electric field was $U = -\mathbf{p} \cdot \mathbf{E}$. By a similar argument, the potential energy of a magnetic dipole in a magnetic field is as follows:

**Potential Energy of a Magnetic Moment**

$$U(\mathbf{r}) = -\mathbf{m} \cdot \mathbf{B}(\mathbf{r}) \tag{7-32}$$

The force on $\mathbf{m}$ due to the external field $\mathbf{B}$ is

$$\mathbf{F} = -\nabla U(\mathbf{r}) = \nabla[\mathbf{m} \cdot \mathbf{B}(\mathbf{r})] \tag{7-33}$$

The *energy of interaction* between two magnetic dipoles, $\mathbf{m}_1$ at $\mathbf{r} = 0$ and $\mathbf{m}_2$ at $\mathbf{r}$, is obtained from Equations (7–13) and (7–32):

**Dipole-Dipole Interaction Energy**

$$U(\mathbf{r}) = -\frac{\mu_0}{4\pi}\left[\frac{3(\mathbf{m}_1 \cdot \mathbf{r})(\mathbf{m}_2 \cdot \mathbf{r}) - r^2\mathbf{m}_1 \cdot \mathbf{m}_2}{r^5}\right] \tag{7-34}$$

The force on one magnetic dipole due to another magnetic dipole can be obtained from Equation (7–33). It follows directly from Equation (7–34) that two dipoles have an attractive force between them if they are aligned end to end and a repulsive force if they are oppositely aligned end to end, as shown in Figure 7–4. This result is the explanation of the familiar result that like magnetic poles repel and unlike attract.

| Dipole configuration | Energy, $U = \frac{\mu_0}{4\pi}\frac{m_1 m_2}{r^3}\lambda$ | Force, $\mathbf{F} = -\nabla_{\mathbf{r}} U$ |
|---|---|---|
| ⇨ ⇨ | $\lambda = -2$ | Attractive |
| ⇨ ⇦ | $\lambda = 4$ | Repulsive |
| ⇧ ⇧ | $\lambda = 1$ | Repulsive |
| ⇧ ⇩ | $\lambda = -1$ | Attractive |

**FIGURE 7-4** Dipole-dipole magnetic energy. The magnetic energy is evaluated from Equation (7–34) for four simple configurations. Two are attractive and two are repulsive.

# 7-3 MAGNETIC MIRRORS

The concept of the magnetic moment is useful in the description of the motion of a charged particle in a magnetic field. We consider a particle of charge $q$ and mass $M$ moving in a uniform magnetic field. The equation of motion is

$$M \frac{d\mathbf{v}}{dt} = q\mathbf{v} \times \mathbf{B} \tag{7-35}$$

As can be verified by substitution, the solution for $\mathbf{v}$ is the sum of a constant component $\mathbf{v}_0$ parallel to $\mathbf{B}$ and a rotating component $d\mathbf{v}_\perp/dt = \boldsymbol{\omega}_c \times \mathbf{r}$ perpendicular to $\mathbf{B}$, with angular velocity as follows:

**Cyclotron Frequency**

$$\boldsymbol{\omega}_c = -\frac{q}{M}\mathbf{B} \tag{7-36}$$

which is called the cyclotron angular frequency. The orbit is a helix with axis parallel to $\mathbf{B}$.

If we take the $z$ axis to be in the direction of $\mathbf{B}$, the orbit of the particle, projected onto the $xy$ plane, is a circle. Hence the $z$ component of the orbital magnetic moment, according to Equations (7–24) and (7–36), is

$$m_z = -\frac{M}{2}\frac{v_\phi^2}{B_z} = -\frac{K_\phi}{B_z} \tag{7-37}$$

where $K_\phi$ is the contribution to the kinetic energy from the circular motion. This component of $\mathbf{m}$ is along the field $\mathbf{B}$; note the minus sign (see Figure (7–5). The component of $\mathbf{m}$ along the symmetry axis of $\mathbf{B}$ given by Equation (7–37) is nearly time-independent, as we will prove later.

For a magnetic field that slowly increases with $z$ (i.e., the field lines slowly converge, as in Figure 7–6), the particle moves in a helix with slowly decreasing radius. The particle's motion along the field decelerates, and it is eventually reflected. This magnetic field configuration constitutes a *magnetic mirror*.

To understand the mirror mechanism, we consider a slowly varying, cylindrically symmetric magnetic field $\mathbf{B} = B_\imath\hat{\boldsymbol{\imath}} + B_z\hat{\mathbf{z}}$. Since the magnetic force is perpendicular to both $\mathbf{B}$ and $\mathbf{v}$, we see from Figure 7–6 that there is a force component in the negative $z$ direction. In terms of cylindrical components the $z$ component of the equation of motion, Equation (7–35), is then

$$M \frac{dv_z}{dt} = -qv_\phi B_\imath \tag{7-38}$$

We can relate the component $B_\imath$ to $B_z$ by evaluating $\int \mathbf{B}\cdot d\mathbf{S} = 0$ on the

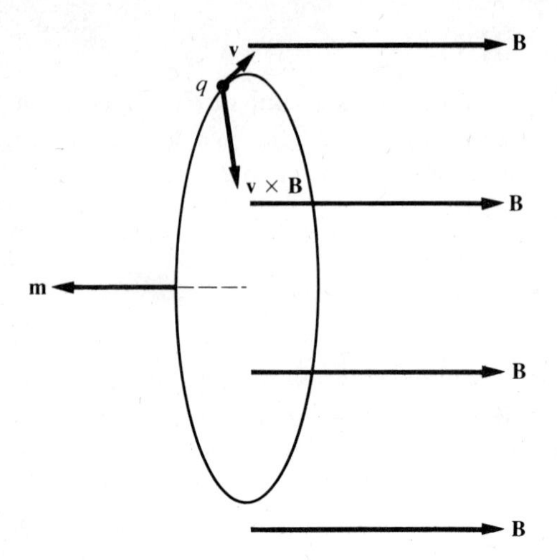


**FIGURE 7-5** Charged particle in circular motion about the direction of a uniform magnetic field **B**. It gives a magnetic moment **m** whose direction is opposite to that of **B**.

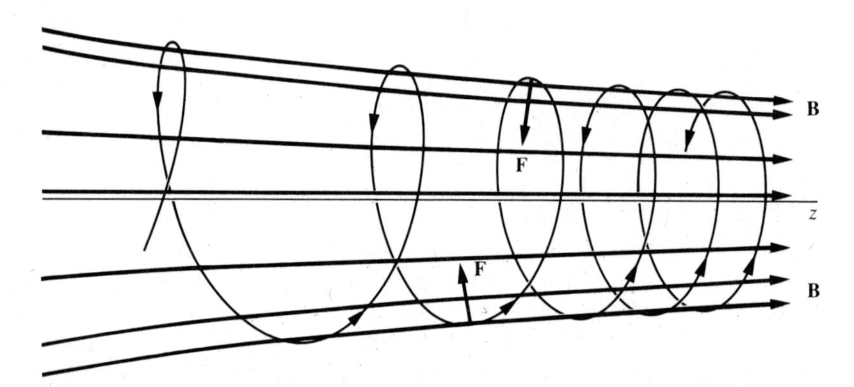


**FIGURE 7-6** Trajectory of a charged particle in a magnetic mirror. The trajectory is a slowly converging helix.

surface of a thin cylindrical prism centered on the axis of a converging magnetic field; see Figure 7–7. Assuming that any variation of $B_z$ with $r$ is small, we find, for small $r$,

$$B_z(z + dz)\pi r^2 - B_z(z)\pi r^2 + 2\pi r\, dz\, B_r = 0 \qquad \textbf{(7-39)}$$

which gives the relation

$$B_r = -\frac{r}{2}\frac{dB_z}{dz} \qquad \textbf{(7-40)}$$

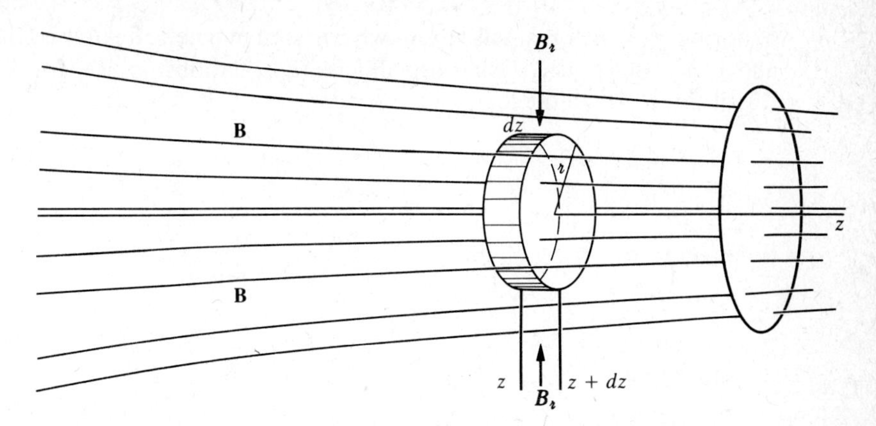


**FIGURE 7–7** Relating the radial component of the magnetic field to the change in the $z$ component of the magnetic field. The relation is obtained by evaluating $\int \mathbf{B} \cdot d\mathbf{S}$ over a thin cylindrical prism.

For a $B_z$ component whose strength increases with increasing $z$, the radial component $B_\imath$ is directed inward. Inserting Equation (7–40) in the equation of motion, we obtain

$$M \frac{dv_z}{dt} = \frac{q}{2}(v_\phi \imath) \frac{dB_z}{dz} \tag{7–41}$$

Eliminating $\imath$ by using $\imath = v_\phi/\omega$ and substituting $\omega$ from Equation (7–36),

$$\imath = -\frac{v_\phi M}{qB_z} \tag{7–42}$$

we obtain

$$F_z = M \frac{dv_z}{dt} = -\frac{M}{2} \frac{v_\phi^2}{B_z} \frac{dB_z}{dz} = -m_z \frac{dB}{dz} \tag{7–43}$$

which agrees with Equation (7–33) if $m_z$ = constant. For $dB_z/dz > 0$ the force $F_z$ is negative, with the exception of particles moving exactly on the axis ($v_\phi = 0$ in that circumstance) for which the reflection force vanishes.

Multiplying through by $dz = v_z\,dt$, we can cast the differential equation in the form

$$v_z\,dv_z = -\frac{v_\phi^2}{2} \frac{dB_z}{B_z} \tag{7–44}$$

Since a magnetic field does no work on a charged particle, the kinetic energy

$$K = \frac{M}{2}(v_z^2 + v_\phi^2) \tag{7–45}$$

(ignoring $v_r^2$, which is small in a slowly varying magnetic field) is conserved, and $v_z\,dv_z = -v_\phi\,dv_\phi$. Using this differential relationship, we can convert Equation (7–44) to the form

$$2\frac{dv_\phi}{v_\phi} = \frac{dB_z}{B_z} \tag{7-46}$$

By integration

$$\ln v_\phi^2 = \ln B_z + \text{constant}$$

or equivalently,

$$\frac{v_\phi^2}{B_z} = \text{constant} \tag{7-47}$$

Hence the $z$ component of the magnetic moment in Equation (7–37) is time-independent.

The magnetic flux through a cross section of the helical orbit is

$$\mathscr{F} = \pi r^2 B_z = \frac{\pi M^2}{q^2}\frac{v_\phi^2}{B_z} \tag{7-48}$$

where we have substituted Equation (7–42). Thus by Equation (7–47) this magnetic flux is also a constant of the motion. The constancy of $\mathscr{F}$ or $m_z$ for motion in a slowly varying magnetic field is called *adiabatic invariance.* The constancy of the flux means that the same lines of **B** link the orbit during the motion; that is, the particle moves on the surface of a flux tube.

Great magnetic storms on the sun associated with sunspot activity, known as solar flares, release large numbers of electrons and protons. As these particles encounter the earth, some become trapped in the earth's magnetic field through the mirror mechanism. Because the **B** lines converge toward the magnetic poles, the charged particles are reflected. These trapped particles form "belts" around the earth, called *Van Allen belts.* Figure 7–8 shows the Van Allen belts along with the observed magnetic field lines of the earth, which are distorted by the solar wind.

The mirror property is used in attempts to confine plasma (ionized matter) into a "magnetic bottle," as shown in Figure 7–9; the bottle has a magnetic mirror at each end. If the plasma is to be at the very high temperatures needed for nuclear fusion reactions, it cannot be allowed to touch material walls. Plasma particles with small $v_\phi$ will quickly leave from the ends. The remaining particles are subject to collective instabilities, which cause the magnetic bottle to leak. Recent improvements in mirror design using double mirrors at each end and pumping in additional particles have led to a revival of interest in such machines. The ends can also be connected together so that the bottle forms a torus. The tokamak fusion device is based on this configuration.

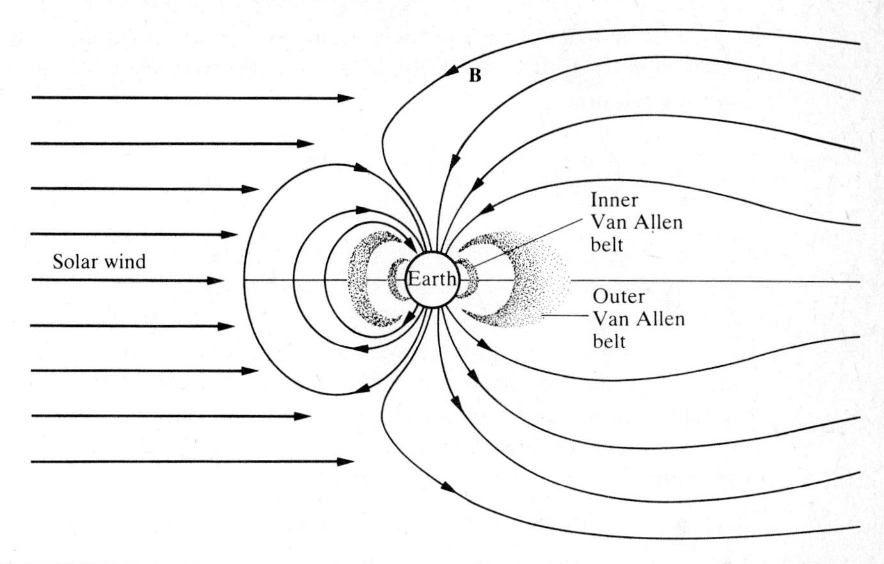


**FIGURE 7-8** Van Allen radiation belts. The inner belt is located approximately 10,000 km from the center of the earth, and the outer belt is approximately 22,500 km from the center of the earth. These distances vary greatly with solar activity.

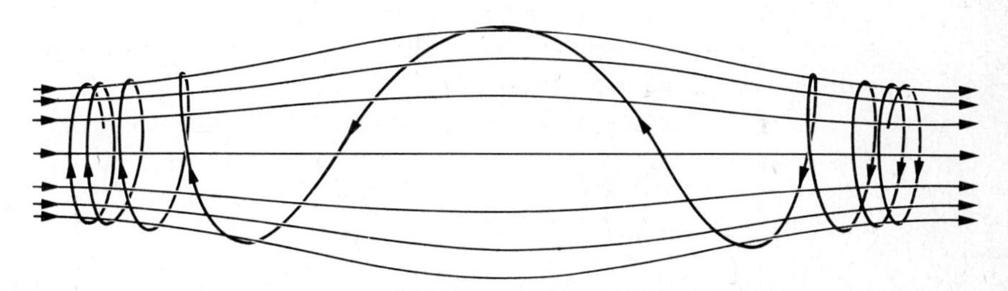

**FIGURE 7-9** Trajectory of a charged particle in a magnetic bottle

## 7-4 INTRINSIC MAGNETIC DIPOLE MOMENTS

For a particle or a system of particles, the magnetic dipole moment is often proportional to the angular momentum. In these cases the ratio of the magnetic dipole moment to the angular momentum is called the *gyromagnetic ratio.* As an example, we compute the gyromagnetic ratio for a system of identical charged particles. For a system of identical particles the angular momentum of the system is

$$\mathbf{L} = \sum_i \mathbf{r}'_i \times M\mathbf{v}_i = M \sum_i \mathbf{r}'_i \times \mathbf{v}_i \tag{7-49}$$

where $M$ is the mass of each particle. Using our previous result for the magnetic moment in Equation (7–20), assuming each particle has charge $q$, we obtain the relation

$$\mathbf{m} = \left(\frac{q}{2M}\right)\mathbf{L} \tag{7-50}$$

The proportionality factor $(q/2M)$ in the relation between **m** and **L** is known as the *classical gyromagnetic ratio*. The ratio of the true magnetic moment to the classical result of Equation (7–50) is called the *g factor*.

In addition to possessing magnetic moments of classical origin owing to the flow of electric charges, the constituent particles of matter (electrons, protons, and neutrons) each possess an intrinsic magnetic moment and an intrinsic angular momentum (called *spin*). For electrons the proportionality of the spin and the magnetic moment is conventionally expressed as

$$\mathbf{m} = g\left(-\frac{e}{2m_e}\right)\mathbf{S} \tag{7-51}$$

where the $g$ factor is the correction to the classical relation in Equation (7–50), **S** is the spin vector, and $e$ is the magnitude of the electron charge. The Dirac relativistic quantum theory of the electron predicts $g = 2$. There are very small corrections to the Dirac value predicted by the theory of quantum electrodynamics, which are in excellent agreement with the experimental value, as discussed in Section 7–4B.

For a proton (or neutron) the magnetic moment is represented by

$$\mathbf{m} = g\left(\frac{e}{2M}\right)\mathbf{S} \tag{7-52}$$

where $M$ is a proton or neutron mass. The $g$ factors are $g \approx 2.7$ for the proton and $g \approx -1.9$ for the neutron. The electron mass is about 2000 times less than the proton (or neutron) mass, and so the electron magnetic moment is much larger than the nuclear moments. The intrinsic magnetic moment of the electron is especially important in the phenomenon of magnetism.

For an electron, proton, or neutron the measurement of the spin component along a given axis (say the $z$ axis) yields two possible values,

$$S_z = \pm\frac{1}{2}\hbar \tag{7-53}$$

where $\hbar = h/2\pi$ and $h$ is Planck's constant ($\hbar = 1.054 \times 10^{-34}$ J·s). The fact that spin has discrete values is a consequence of quantum theory. For an electron the magnetic moment along this axis is thus

$$m_z = \pm\frac{g}{2}\left(\frac{e\hbar}{2m_e}\right) \tag{7-54}$$

The quantity

$$\mu_B \equiv \frac{e\hbar}{2m_e} \tag{7-55}$$

is known as the *electronic* (*Bohr*) *magneton*; it has the numerical value

$$\mu_B = \frac{(1.6 \times 10^{-19})(1.054 \times 10^{-34})}{2(9.11 \times 10^{-31})} = 0.93 \times 10^{-23}\ \text{A·m}^2 \tag{7-56}$$

In a magnetic field the intrinsic spin of a body, in this case an electron, precesses around the direction of the field. To see this result, we recall from Equation (7–25) that the torque acting on its intrinsic magnetic moment of Equation (7–52) is

$$\mathbf{N} = \mathbf{m} \times \mathbf{B} = -\frac{ge}{2m_e}\mathbf{S} \times \mathbf{B} \tag{7-57}$$

Then from the law $\mathbf{N} = \dot{\mathbf{L}}$ of Newtonian mechanics, where $\mathbf{L} = \mathbf{S}$ is the spin angular momentum and $\dot{\mathbf{L}} = d\mathbf{L}/dt$, we have

$$\dot{\mathbf{S}} = \boldsymbol{\omega}_s \times \mathbf{S} \tag{7-58}$$

where

**Angular Velocity of Precession**

$$\boldsymbol{\omega}_s = \frac{ge}{2m_e}\mathbf{B} \tag{7-59}$$

For a constant magnetic field Equation (7–58) describes the precessional motion of the intrinsic electron spin vector $\mathbf{S}$ with angular velocity $\omega_s$.

## 7-4A Magnetic Spin Resonance

One of the most useful tools in the scientific repertoire is the induction of *magnetic spin resonances.* The electron spin resonance and nuclear magnetic spin resonance (NMR) techniques are used in such diverse areas as mineral prospecting, the chemical analyses of complex compounds, and in NMR medical scanning. A typical setup for a spin resonance experiment is shown in Figure 7–10. The intrinsic magnetic moments of the particles precess about the vertical $z$ axis with angular velocity $\omega_s = geB_z/(2m)$. A small oscillating magnetic field is now introduced along the $x$ axis with frequency $\omega$. If the oscillating frequency $\omega$ is equal to the spin precession frequency $\omega_s$, the spin component along $B_z$ will oscillate with large amplitude. A sharp corresponding increase in susceptibility and energy loss in the oscillating circuit is seen. The equation of motion, Equation (7–57), with time-varying $B_x(t)$ is

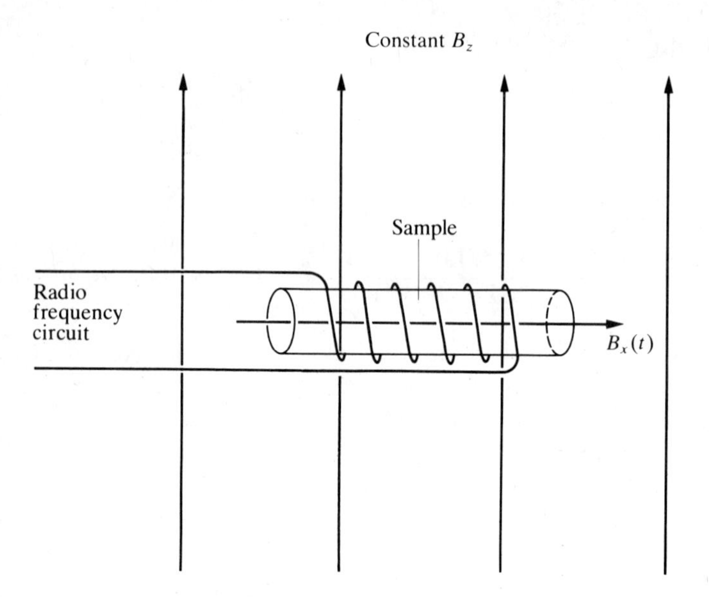


**FIGURE 7-10** Schematic of a magnetic spin resonance apparatus

difficult to solve. To see our way through this problem, we employ a trick first used by I. Rabi in solving the same problem in quantum mechanics.

The oscillating field $B_x(t)$ can be written as the sum of two counter-rotating magnetic fields $\mathbf{b}_+$ and $\mathbf{b}_-$, as shown in Figure 7–11. The magnitude of each $\mathbf{b}_\pm$ is one-half the magnitude of $B_x$. Only the component $\mathbf{b}_+$, which rotates in the direction of the spin precession $\boldsymbol{\omega}_s$, induces a large spin in the $xy$ plane when $\omega \approx \omega_s$, as we will see shortly. The component rotating in the opposite direction has a much smaller effect and can be ignored. In the frame rotating with the vector $\mathbf{b}_+$ the equation of motion considerably simplifies. The $\mathbf{b}_+$ vector rotates with angular velocity $\omega\hat{\mathbf{z}}$, so in the rotating frame Equation (7–58) becomes

$$\dot{\mathbf{S}} = \frac{ge}{2m}\mathbf{B} \times \mathbf{S} = \frac{\delta \mathbf{S}}{\delta t} + \omega\hat{\mathbf{z}} \times \mathbf{S} \qquad \textbf{(7-60)}$$

where $\mathbf{B} = B_z\hat{\mathbf{z}} + b_+\hat{\mathbf{x}}_1$ and $\delta\mathbf{S}/\delta t$ is the time derivative of the spin in the magnetic field rest frame. The unit vector $\hat{\mathbf{x}}_1$ rotates with angular velocity $\omega$ in the $xy$ plane.

Expressing Equation (7–60) in component form, we obtain

$$\frac{\delta S_1}{\delta t} = -(\omega_s - \omega)S_2 \qquad \textbf{(7-61a)}$$

$$\frac{\delta S_2}{\delta t} = (\omega_s - \omega)S_1 - \omega_1 S_z \qquad \textbf{(7-61b)}$$

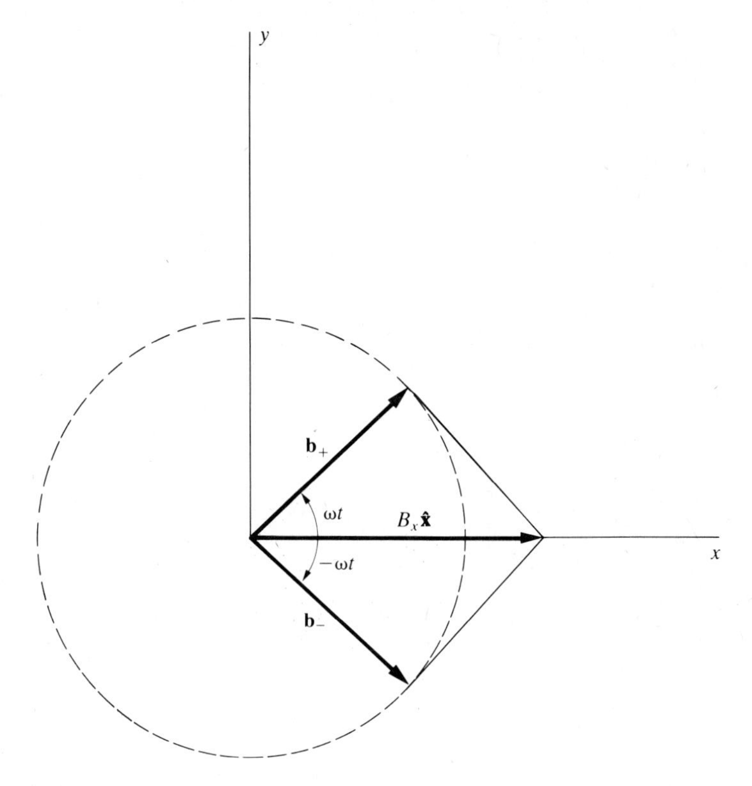


**FIGURE 7-11** Oscillating magnetic field $B_x(t)\hat{\mathbf{x}}$ represented by the superposition of two rotating fields $\mathbf{b}_+$ and $\mathbf{b}_-$

$$\frac{\delta S_z}{\delta t} = \omega_1 S_2 \tag{7-61c}$$

where $\mathbf{S} = S_1\hat{\mathbf{x}}_1 + S_2\hat{\mathbf{x}}_2 + S_z\hat{\mathbf{z}}$ and $\omega_1 \equiv geb_+/(2m)$. These equations are now linear with constant coefficients and easily solved. A spin $\mathbf{S}$ of the form $\mathbf{S}_0 e^{pt}$ is found to be a solution only if $p$ satisfies a cubic equation, with roots $p = 0, i\nu, -i\nu$. Thus the general real solution has the form $S_k = A_k + B_k \sin \nu t + C_k \cos \nu t$. Substituting this form into Equations (7–61), we arrive at the solution

$$S_1 = \omega_1 A + \frac{\omega_s - \omega}{\nu} \beta \cos(\nu t + \Phi) \tag{7-62a}$$

$$S_2 = \beta \sin(\nu t + \Phi) \tag{7-62b}$$

$$S_z = (\omega_s - \omega_1)A - \frac{\omega_1}{\nu} \beta \cos(\nu t + \Phi) \tag{7-62c}$$

where

$$\nu^2 = \omega_1^2 + (\omega_s - \omega)^2 \tag{7-62d}$$

Here we have defined $A_1 = \omega_1 A$, $B_2 = \beta \sin \Phi$, and $C_2 = \beta \cos \Phi$. We can verify that the total spin value is a constant given by

$$S^2 = S_1^2 + S_2^2 + S_z^2 = (\nu A)^2 + \beta^2 \tag{7-63}$$

The nature of the solution of Equations (7–62) is evident if we observe that there is a linear combination of the $S_1$ and $S_z$ components that is time-independent:

$$\omega_1 S_1 + (\omega_s - \omega)S_z = \nu^2 A \tag{7-64}$$

This relation has the form

$$\hat{\mathbf{n}} \cdot \mathbf{S} = S_n \tag{7-65a}$$

where

$$\hat{\mathbf{n}} = \frac{\omega_1}{\nu}\hat{\mathbf{x}}_1 + \frac{\omega_s - \omega}{\nu}\hat{\mathbf{z}} \tag{7-65b}$$

and

$$S_n = \nu A \tag{7-65c}$$

So $\mathbf{S}$ has a fixed projection on the axis $\hat{\mathbf{n}}$. Since it also has a fixed length, $\mathbf{S}$ must be simply precessing about the fixed axis $\hat{\mathbf{n}}$. According to Equations (7–62), this precession has the angular frequency $\nu$. The motion of $\mathbf{S}$ in the laboratory frame is the combination of a precession at rate $\omega$ about the $z$ axis and a precession at rate $\nu$ about the axis $\hat{\mathbf{n}}$. The angular velocity $\boldsymbol{\omega}_{\text{LAB}}$ of $\mathbf{S}$ in the laboratory frame is

$$\boldsymbol{\omega}_{\text{LAB}} = \nu\hat{\mathbf{n}} + \omega\hat{\mathbf{z}} \tag{7-66}$$

When $\omega = \omega_s$, the rotation axis is $\hat{\mathbf{n}} = \hat{\mathbf{x}}_1$, orthogonal to $\hat{\mathbf{z}}$, so the precession about $\hat{\mathbf{n}}$ has a large effect on $\mathbf{S}$. Usually, $b_+ \ll B_z$, and the spin axis remains nearly vertical until $\omega$ is nearly equal to $\omega_s$. Note that the counterrotating field $b_-$ can never have this resonance, as one would see by replacing $\omega \to -\omega$ in our solution.

The physical sample in a resonance experiment consists of a very large number of atomic or nuclear magnetic moments that are randomly oriented in the absence of an external magnetic field. Because of the paramagnetic effect (see Section 7–6B), the static $B_z$ field will induce a small net magnetic moment along the $z$ axis. Since Equation (7–60) is linear in $\mathbf{S}$, it applies to the vector sum of the spins and hence to the macroscopic magnetic moment, which thus has the motion given in Equation (7–62) and is described as follows: As a consequence of the oscillating field $\mathbf{b}_+$, the $z$ component of the magnetization oscillates at a frequency $\nu$ with an amplitude that at resonance, $\omega = \omega_s$, equals its equilibrium value. Owing to interactions with the spins, there is a relaxation to the equilibrium state in which the magnetization is aligned with the static magnetic field. The oscillating circuit must therefore provide energy to keep the magnetization tilted away from the static field. To the extent to which the magnetization is inclined away from the $z$ axis toward the $\mathbf{b}_+$ rotating axis, there will also be an increased susceptibility and hence an increased inductance in the oscillating circuit. Both of these effects are maximum when $\omega = \omega_s$.

For a known $B_z$ resonant frequencies characteristic of a particular substance can be identified. Conversely, if the $g$ factor is known, magnetic resonance provides a simple and very accurate measurement of the magnetic field.

### 7-4B The $g - 2$ Experiment

The equation of motion, Equation (7–35), for a charged particle moving in a magnetic field has the same form as the spin precession equation of motion, Equation (7–58), with the replacement of $\mathbf{S}$ by $\mathbf{v}$ and $\boldsymbol{\omega}_s = ge\mathbf{B}/(2m_e)$ by $\omega_c = -q\mathbf{B}/M$. The cyclotron frequency can be observed by the magnetic resonance technique as well, with resonant frequency $\omega_c$. By simultaneous excitation of these two resonances, we can measure the quantity

$$\frac{\omega_s - \omega_c}{\omega_c} = \frac{1}{2}(g - 2) \tag{7-67}$$

as the observed beat frequency. As we mentioned previously, the Dirac equation of the electron predicts that $g = 2$, and the small departures from this equality provide a sensitive test of the theory of quantum electrodynamics. The most recent measurements of the electron $g$ factor give

$$g = 2.00231930468 \pm 0.00000000004 \tag{7-68}$$

which agrees well with theory.

## 7-5 MACROSCOPIC THEORY OF MAGNETIC MEDIA

In developing the macroscopic theory of magnetic materials, we consider a small volume $V$ that contains a large number of magnetic dipoles. (This development parallels the development of dielectric theory in Section 4–2.) The total dipole moment in a small volume $\Delta V$, divided by $\Delta V$, defines the macroscopic magnetization vector:

**Magnetization**
$$\mathbf{M} = \frac{1}{\Delta V} \sum_i \mathbf{m}_i \tag{7-69}$$

In terms of $\mathbf{M}$ the continuous limit for the magnetic dipole moment of a volume element $dV$ is

$$d\mathbf{m} \equiv \mathbf{M}(\mathbf{r}')\, dV' \tag{7-70}$$

By Equation (7–10) the differential vector potential due to this differential dipole moment at $\mathbf{r}'$ is

$$d\mathbf{A}(\mathbf{r}) = \frac{\mu_0}{4\pi} \frac{d\mathbf{m} \times \mathbf{R}}{R^3} \tag{7-71}$$

where $\mathbf{R} = \mathbf{r} - \mathbf{r}'$. Thus the total vector potential from a finite volume of magnetized material is

$$\mathbf{A}(\mathbf{r}) = \frac{\mu_0}{4\pi} \int \frac{\mathbf{M}(\mathbf{r}') \times \mathbf{R}}{R^3}\, dV' \tag{7-72}$$

This equation can be manipulated into a form that we can interpret in terms of equivalent magnetization currents. First, we substitute the relation $\nabla_{\mathbf{r}'}(1/R) = \mathbf{R}/R^3$ in the integrand:

$$\mathbf{A}(\mathbf{r}) = \frac{\mu_0}{4\pi} \int \mathbf{M}(\mathbf{r}') \times \nabla_{\mathbf{r}'} \left(\frac{1}{R}\right) dV' \tag{7-73}$$

Next, using the identity

$$\nabla_{\mathbf{r}'} \times \left(\frac{\mathbf{M}}{R}\right) = \frac{\nabla_{\mathbf{r}'} \times \mathbf{M}}{R} - \mathbf{M} \times \nabla_{\mathbf{r}'} \left(\frac{1}{R}\right) \tag{7-74}$$

we rewrite the integrand as

$$\mathbf{A}(\mathbf{r}) = -\frac{\mu_0}{4\pi} \int \nabla_{\mathbf{r}'} \times \left(\frac{\mathbf{M}}{R}\right) dV' + \frac{\mu_0}{4\pi} \int \frac{\nabla_{\mathbf{r}'} \times \mathbf{M}}{R}\, dV' \tag{7-75}$$

Finally, corollary I to the divergence theorem, Equation (2–15), applied to the vector $\mathbf{M}/R$,

$$\int \nabla \times \left(\frac{\mathbf{M}}{R}\right) dV' = \oint \hat{\mathbf{n}} \times \left(\frac{\mathbf{M}}{R}\right) dS' \tag{7-76}$$

allows us to express the vector potential as

$$\mathbf{A}(\mathbf{r}) = \frac{\mu_0}{4\pi} \oint \frac{\mathbf{M} \times \hat{\mathbf{n}}}{R}\, dS' + \frac{\mu_0}{4\pi} \int \frac{\nabla \times \mathbf{M}}{R}\, dV' \tag{7-77}$$

This expression is of the form

$$\mathbf{A}(\mathbf{r}) = \frac{\mu_0}{4\pi}\oint \frac{\mathbf{j}_M}{R}\,dS' + \frac{\mu_0}{4\pi}\int \frac{\mathbf{J}_M}{R}\,dV' \qquad \text{(7–78)}$$

where $\mathbf{j}_M$ and $\mathbf{J}_M$ are effective surface and volume magnetization current densities, respectively, defined as follows:

**Magnetization Current Densities**

$$\mathbf{j}_M = \mathbf{M} \times \hat{\mathbf{n}} \qquad \mathbf{J}_M = \nabla \times \mathbf{M} \qquad \text{(7–79)}$$

These current densities give the same macroscopic magnetic field **B** as the average field produced by the microscopic magnetic dipoles. These currents were first proposed by Ampère and are often termed Amperian current densities; a more modern terminology is magnetization current densities.

As a simple example, we consider an infinite cylinder of uniformly magnetized material, as shown in Figure 7–12. Since **M** is uniform, $\nabla \times \mathbf{M} = 0$, and there is no volume current density. The magnetization current density on the surface is $\mathbf{j}_M = \mathbf{M} \times \hat{\mathbf{n}}$, which is azimuthal around the cylinder with magnitude $j_M = M$. This surface current will produce a magnetic field similar to that of an infinite solenoid, namely, uniform inside the cylinder and zero outside. We can apply Ampère's law to determine the **B** field that results from the magnetization surface current. As in Section 5–17, we employ a rectangular circuit that intersects the cylinder, giving a magnetic field inside the cylinder of

$$\mathbf{B} = \mu_0 j_M \hat{\mathbf{z}} = \mu_0 M \hat{\mathbf{z}} \qquad \text{(7–80)}$$

By comparison with the coil-wound solenoid result, Equation (5–147), we can make the correspondence

$$M = nI \qquad \text{(7–81)}$$

Thus the magnetization is equivalent to the current per unit length.

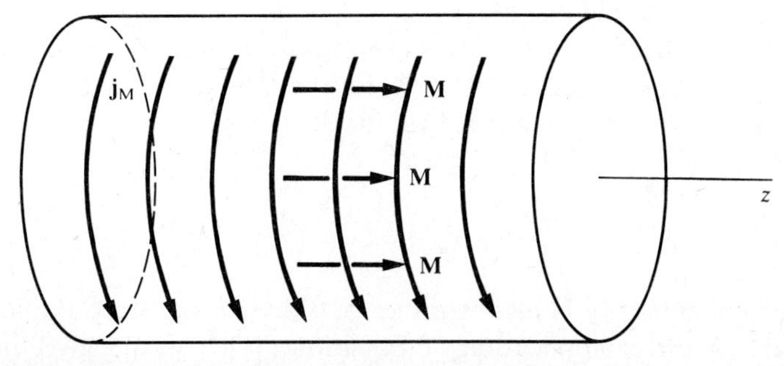


**FIGURE 7–12** Equivalent surface currents for calculating the magnetic field outside an infinite cylinder of uniformly magnetized material

When magnetization currents are present, we must modify Ampère's law to include the magnetization sources. In differential form the modification is

$$\nabla \times \mathbf{B} = \mu_0(\mathbf{J} + \nabla \times \mathbf{M}) \tag{7-82}$$

We can rearrange the equation as

$$\nabla \times \left(\frac{\mathbf{B}}{\mu_0} - \mathbf{M}\right) = \mathbf{J} \tag{7-83}$$

The form of this equation invites the introduction of an auxiliary magnetic vector

$$\mathbf{H} = \frac{\mathbf{B}}{\mu_0} - \mathbf{M} \tag{7-84}$$

or inversely, we have

$$\mathbf{B} = \mu_0(\mathbf{H} + \mathbf{M}) \tag{7-85}$$

The **H** field is sometimes called the *magnetic intensity*. Where there is no magnetic material, $\mathbf{B} = \mu_0\mathbf{H}$.

In terms of **H** the differential form of Ampère's law becomes

**Ampère's Law for H (Differential Form)**

$$\nabla \times \mathbf{H} = \mathbf{J} \tag{7-86}$$

For steady currents the integral form of Equation (7–86) is

**Ampère's Law for H**

$$\oint \mathbf{H} \cdot d\mathbf{r} = I \tag{7-87}$$

In physical situations with azimuthal symmetry, we can use this form of Ampère's law to find **H** directly.

To illustrate these concepts, we again consider a rod with constant magnetization $\mathbf{M} = M\hat{\mathbf{z}}$ along its axis. For an infinitely long rod $\mathbf{B} = \mu_0\mathbf{M}$ from Equation (7–80), and the **H** field inside the rod is

$$\mathbf{H} = \frac{\mu_0\mathbf{M}}{\mu_0} - \mathbf{M} = 0 \tag{7-88}$$

The magnetic intensity **H** also vanishes outside the rod, since **B** and **M** are zero there. A direct application of Equation (7–87) would give the same result and hence an alternate method to obtain Equation (7–80).

For a rod of finite length we can use the correspondence with a finite solenoid to deduce **B**. The magnetic field component $B_z$ on the axis is a maximum at the middle of the solenoid and even at this position does not quite reach the field strength of an infinite solenoid. Thus the **H** field inside the rod on the axis $H_z = (B_z/\mu_0) - M$ will always be less than **H** of an infinite rod, that is,

$$H_z < 0 \tag{7-89}$$

Consequently, the direction of **H** is opposite to that of **B** inside the rod, as illustrated in Figure 7–13.

The fact that **H** and **B** are antiparallel on the axis inside the rod and parallel everywhere outside the rod is necessary for Ampère's law, Equation (7–87), to be satisfied. Since there are no currents,

$$\oint \mathbf{H} \cdot d\mathbf{r} = 0 \tag{7-90}$$

When the line integral is taken along a field line, the contribution inside the rod must cancel the contribution outside.

In a magnetic material with no electric fields and no conduction current ($\mathbf{J} = 0$), Equation (7–86) becomes

$$\nabla \times \mathbf{H} = 0 \tag{7-91}$$

following the methods of electrostatics, where $\nabla \times \mathbf{E} = 0$, a curl-free **H** field can be derived from a magnetic scalar potential:

$$\mathbf{H} = -\nabla\Phi_{\mathrm{M}} \tag{7-92}$$

We can define a magnetic charge density

$$\rho_{\mathrm{M}} = \nabla \cdot \mathbf{H} \tag{7-93}$$

as the source of **H**. From Equation (7–84)

$$\rho_{\mathrm{M}} = \frac{1}{\mu_0} \nabla \cdot \mathbf{B} - \nabla \cdot \mathbf{M} = -\nabla \cdot \mathbf{M} \tag{7-94}$$

By using a correspondence with the treatment of dielectics in Section 4–2, we can deduce that the magnetic surface charge density is

$$\sigma_{\mathrm{M}} = \mathbf{M} \cdot \hat{\mathbf{n}} \tag{7-95}$$

Following this correspondence with polarization charge densities, the magnetic scalar potential is

$$\Phi_{\mathrm{M}} = \frac{1}{4\pi} \int \frac{\sigma_{\mathrm{M}}\, dS'}{R} + \frac{1}{4\pi} \int \frac{\rho_{\mathrm{M}}\, dV'}{R} \tag{7-96}$$

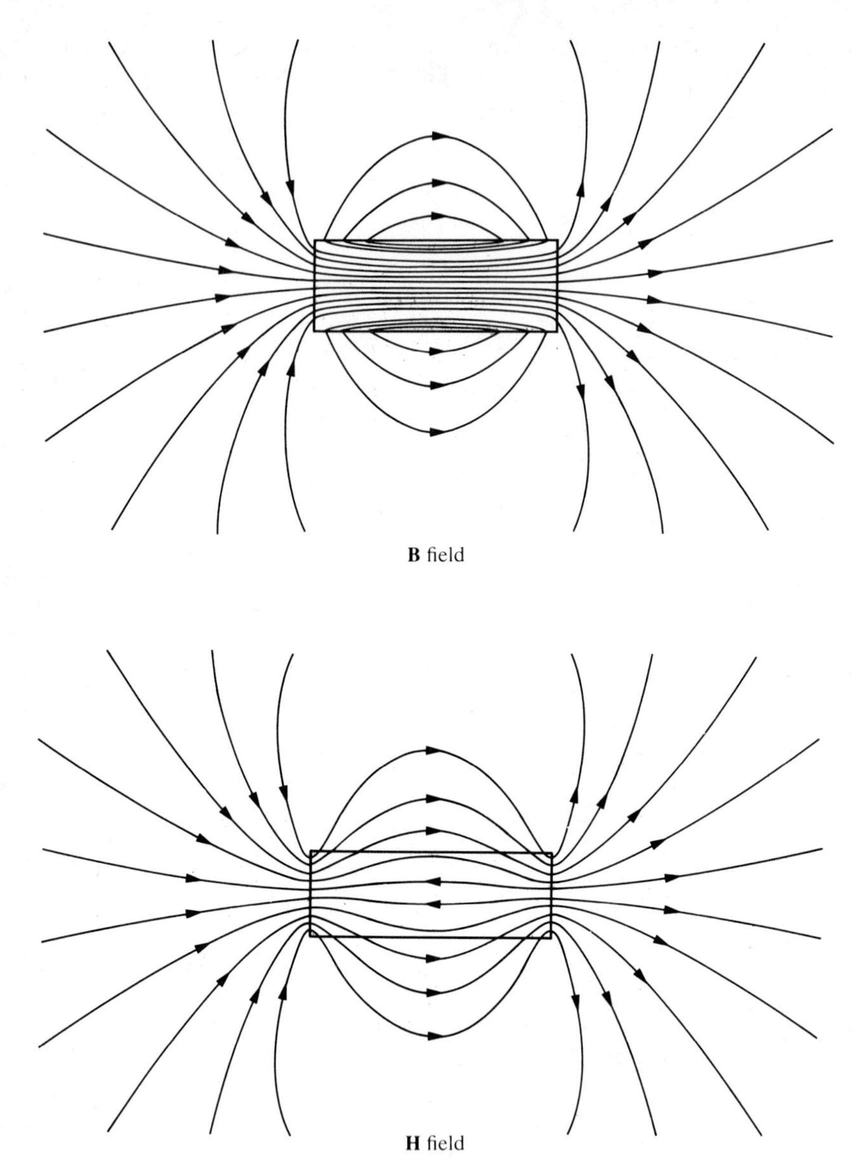


**FIGURE 7-13** Fields **B** and **H** for a rod with constant magnetization along its axis

Whenever there are no conduction currents, the **H** field can be calculated as if there were magnetic charge sources. This calculation in no way conflicts with the absence of magnetic charge sources for **B**.

We now reconsider the uniformly magnetized rod example, this time in terms of the **H** field's magnetic charges. In this case $\rho_M = 0$ since the rod is uniformly magnetized, and $\sigma_M = \pm M$ on the ends. If the rod is infinitely long, $\mathbf{H} = 0$ since the only charges are infinitely far away. For a finite rod $\mathbf{H}$ reverses at the ends, where its sources are located.

If the magnetization is proportional to $\mathbf{H}$ with a scalar proportionality constant

$$\mathbf{M} = \chi_M \mathbf{H} \tag{7-97}$$

the material is called magnetically linear and isotropic. The dimensionless quantity $\chi_M$ is called the *magnetic susceptibility*. This constitutive equation holds for a large class of magnetic materials. For many materials the susceptibility is constant for weak magnetic fields, although ferromagnetic materials, such as iron, are prominent exceptions.

From Equation (7–85) we have

$$\mathbf{B} = \mu_0(1 + \chi_M)\mathbf{H} \tag{7-98}$$

We also introduce the quantities

$$\mu = \mu_0(1 + \chi_M) \tag{7-99}$$

called the *permeability*, and

$$\kappa_M = \frac{\mu}{\mu_0} = 1 + \chi_M \tag{7-100}$$

called the *relative permeability*. In terms of these quantities

$$\mathbf{B} = \mu \mathbf{H} = \mu_0 \kappa_M \mathbf{H} \tag{7-101}$$

Equations (7–86) and (7–101) give

$$\nabla \times \mathbf{B} = \mu \mathbf{J} \tag{7-102}$$

Hence the Biot-Savart law and Ampère's law for $\mathbf{B}$ are modified only by the replacement of $\mu_0$ by $\mu$. A given current will now generate a magnetic field $\mu/\mu_0$ times larger, which can be an increase of a factor of several thousand for a ferromagnetic material. The coefficients of inductance will likewise increase. For an infinite solenoid with a linear magnetic material filling the core, the self-inductance per unit length is, by Equation (6–63),

$$L = \mu \pi a^2 n^2 \tag{7-103}$$

When there is a sharp discontinuity between different materials, an often useful technique is to obtain separate solutions for the magnetic field on each side and match them by using the proper boundary conditions. The equation $\oint \mathbf{B} \cdot d\mathbf{S} = 0$ applied to a prism of infinitesimal height centered in the interface, as shown in Figure 7–14(a), implies that the normal component

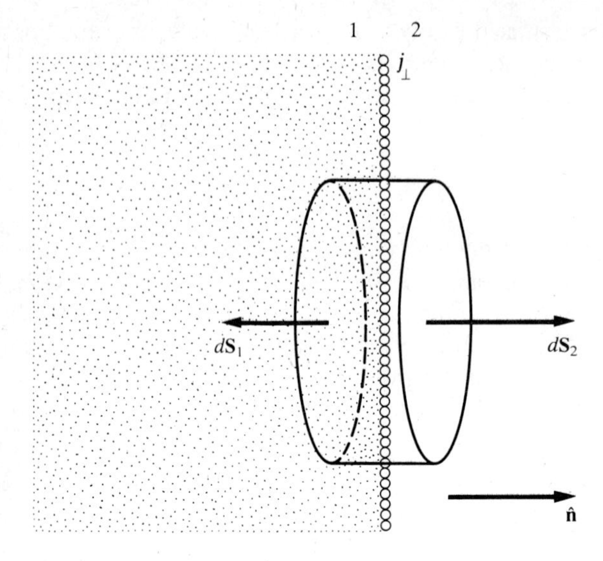


(a) Normal **B** field

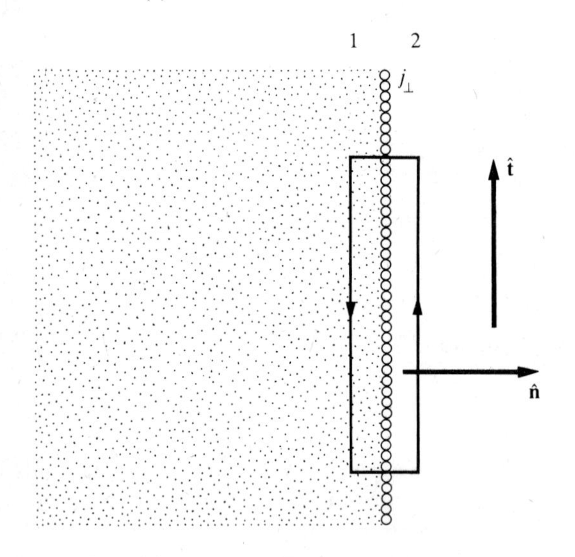


(b) Tangential **H** field

**FIGURE 7-14** Boundary conditions for the magnetic field at an interface. The boundary conditions for the normal magnetic field are determined by applying $\oint \mathbf{B} \cdot d\mathbf{S}$ to the surface in (a), while applying Ampère's law to the circuit in (b) gives the boundary condition on the tangential magnetic induction field.

of **B** is continuous:

| **Boundary Condition on Normal B** | $$B_{1n} = B_{2n} \tag{7-104}$$ |
|---|---|

Applying Ampère's law to a long rectangular circuit of infinitesimal width, as illustrated in Figure 7–14(b), we find

| **Boundary Condition on Tangential H** | $$H_{2t} - H_{1t} = j_\perp \tag{7-105}$$ |
|---|---|

where $j_\perp$ is the surface current density component along the $\hat{\mathbf{n}} \times \hat{\mathbf{t}}$ direction. From Figure 7–14(b) the directions of $H$ and $j_\perp$ correspond to the right-hand rule.

Using the boundary conditions of Equations (7–104) and (7–105) and the magnetic scalar potential defined in Equation (7–92), we can apply the methods for solving electrostatic boundary value problems, presented in Chapter 3, to solve analogous magnetostatic problems.

## 7-6 MICROSCOPIC MAGNETIZATION

In this section we discuss the microscopic causes of magnetization. All matter reacts in some way to a magnetic field. In a schematic experiment we consider a small sample of initially unmagnetized matter suspended in an inhomogeneous magnetic field, as shown in Figure 7–15, with the sample located on the symmetry axis of the field. The field is produced by an electromagnet with asymmetric pole shapes; the magnetic field **B** is stronger near the pointed pole. When the current in the electromagnet is turned on, the sample will either be attracted into or repelled from the strong-field region. The magnetic properties of the material are classified according to this behavior. A diamagnetic material is weakly repulsed, a paramagnetic material is weakly attracted, and a ferromagnetic material is strongly attracted to the high-field region.

We suppose that a uniform magnetization is induced in the sample, with $\mathbf{M} = M_z\hat{\mathbf{z}}$ along the symmetry axis of **B**. The magnetic moment of the sample is then

$$\mathbf{m} = M_z V \hat{\mathbf{z}} \tag{7-106}$$

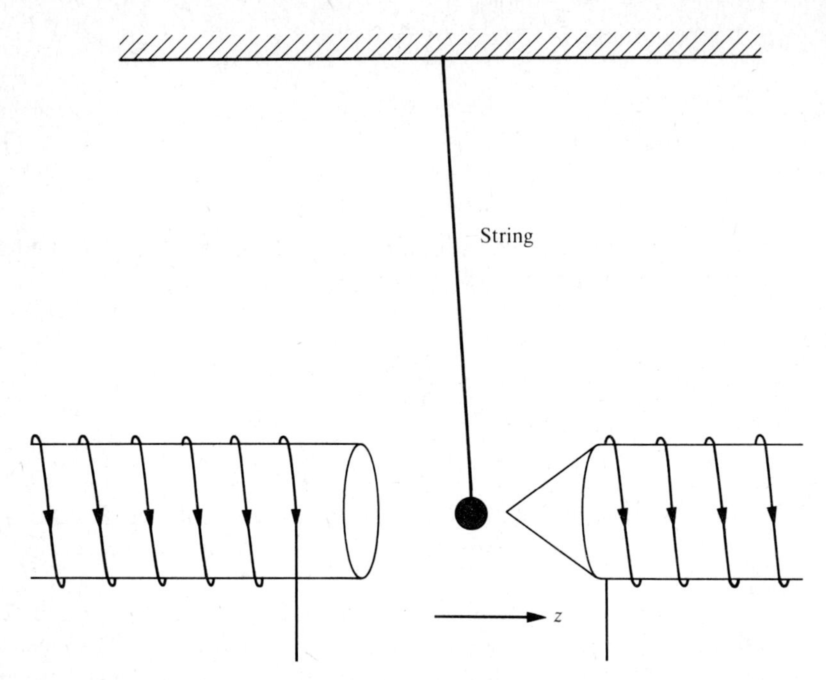


**FIGURE 7-15** Initially unmagnetized sample placed in an inhomogeneous magnetic field. This experiment is used to determine the magnetic properties of the sample.

where $V$ is the volume of the sample. From Equation (7–33) the $z$ component of the force on the sample due to the external field is

$$F_z = M_z V \frac{dB_z}{dz} \tag{7-107}$$

By the constitutive equation, (7–97),

$$M_z = \chi_M H_z = \chi_M \frac{B_z}{\mu} \tag{7-108}$$

we can express the force as

$$F_z = \frac{\chi_M}{\mu} V B_z \frac{dB_z}{dz} \tag{7-109}$$

For $dB_z/dz > 0$, as in Figure 7–15, the force $F_z$ acts to move the sample away from the large-field region if $\chi_M < 0$ and toward the stronger-field region if $\chi_M > 0$. By the classification in terms of the force, diamagnetic materials have negative susceptibilities, while paramagnetic and ferromagnetic materials have positive susceptibilities.

## 7-6A Diamagnetism

*Diamagnetism* was discovered by Faraday in 1836 when he noted that bismuth was repelled from the poles of a permanent magnet. This magnetic effect is quite weak, giving susceptibilities of the order of $-10^{-5}$ for solids.

Diamagnetism is a result of the reaction of atomic currents to an external field. All matter has diamagnetic properties, though these properties may be overshadowed by other kinds of magnetism. Since $\chi_M < 0$ for diamagnetic materials, the induced magnetic dipoles are opposite in direction to the applied magnetic field.

To see the origin of diamagnetism, let us consider a system of identical charged particles in a uniform magnetic field **B**. The equation of motion for the $i$th particle is

$$M \frac{d^2\mathbf{r}_i}{dt^2} = \mathbf{F}_i + q\mathbf{v}_i \times \mathbf{B} \quad \textbf{(7–110)}$$

where $\mathbf{F}_i$ is the force in the absence of the magnetic field. We can transform this equation to a coordinate system that rotates with angular velocity $\boldsymbol{\omega}$. Denoting the time derivative of **r** in this rotating frame by $\delta\mathbf{r}/\delta t$, we write the equation of motion as

$$M \frac{\delta^2\mathbf{r}_i}{\delta t^2} = \mathbf{F}_i + q\mathbf{v}_i \times \mathbf{B} - 2M\boldsymbol{\omega} \times \mathbf{v}_i - M\boldsymbol{\omega} \times (\boldsymbol{\omega} \times \mathbf{r}_i) \quad \textbf{(7–111)}$$

If we make the choice

**Larmor Frequency**

$$\boldsymbol{\omega} = \boldsymbol{\omega}_L = -\frac{q}{2M}\mathbf{B} \quad \textbf{(7–112)}$$

and neglect the term quadratic in $\boldsymbol{\omega}$ (i.e., quadratic in **B**), the $\mathbf{v}_i \times \mathbf{B}$ and $\boldsymbol{\omega} \times \mathbf{v}_i$ terms cancel, and the equation of motion becomes the same as in the original system with no magnetic field.

The result we have obtained is known as Larmor's theorem, which states that the motion with a weak magnetic field is the zero-field solution with an added rotation with angular velocity $\omega_L$ about the axis of the field. Since $\omega_L$ is inversely proportional to the mass, the rotation frequency is much higher for electrons than for nuclei. Subsequently, we take into account only the electronic effect; with $q = -e$ Equation (7–112) becomes

$$\boldsymbol{\omega}_L = \frac{e}{2m_e}\mathbf{B} \quad \textbf{(7–113)}$$

The current density in a spherically symmetric atom due to the Larmor rotation in a uniform magnetic field $\mathbf{B} = B\hat{\mathbf{z}}$ is

$$\mathbf{J} = \rho(r)\boldsymbol{\omega}_{\mathrm{L}} \times \mathbf{r} = \frac{eB}{2m_{\mathrm{e}}}\rho(r)\hat{\mathbf{z}} \times \mathbf{r} \tag{7-114}$$

The corresponding induced atomic dipole moment is, from Equations (7–11) and (7–114),

$$\mathbf{m} = \frac{eB}{4m_{\mathrm{e}}}\int \rho(r')\mathbf{r}' \times (\hat{\mathbf{z}} \times \mathbf{r}')\, dV' \tag{7-115}$$

We write the cross product in the integrand in terms of its Cartesian components:

$$\mathbf{r}' \times (\hat{\mathbf{z}} \times \mathbf{r}') = \hat{\mathbf{z}}r'^2 - \mathbf{r}'(\hat{\mathbf{z}} \cdot \mathbf{r}') = \hat{\mathbf{z}}(x'^2 + y'^2) - \hat{\mathbf{x}}(x'z') - \hat{\mathbf{y}}(y'z')$$

From the spherical symmetry of $\rho(r')$ the $x'z'$ and $y'z'$ terms will integrate to zero, and

$$\int \rho(r')x'^2\, dV' = \int \rho(r')y'^2\, dV' = \int \rho(r')z'^2\, dV'$$

$$= \frac{1}{3}\int \rho(r')r'^2\, dV' \tag{7-116}$$

Thus the magnetic moment is

$$\mathbf{m} = \frac{e\mathbf{B}}{6m_{\mathrm{e}}}\int \rho(r')r'^2\, dV' \tag{7-117}$$

For an atom of $Z$ electrons the integrated electronic charge is

$$\int \rho(r')\, dV' = -Ze \tag{7-118}$$

The mean-square radius of the atom is defined as

$$\langle r^2 \rangle = -\frac{1}{Ze}\int \rho(r')r'^2\, dV' \tag{7-119}$$

In terms of $\langle r^2 \rangle$ we can express $\mathbf{m}$ as

$$\mathbf{m} = -\frac{Ze^2\mathbf{B}}{6m_{\mathrm{e}}}\langle r^2 \rangle \tag{7-120}$$

Since a static dipole moment does not depend on the coordinate origin [see Equation (7–16)], the moment for a sample of $n_{\mathrm{v}}$ atoms per unit volume is $n_{\mathrm{v}}$ times the magnetic moment of a single atom, which gives a magnetization vector

$$\mathbf{M} = n_v\mathbf{m} = -\frac{Ze^2\mathbf{B}}{6m_e}\langle r^2\rangle n_v \tag{7–121}$$

Since $\mathbf{M} = \chi_M\mathbf{H} = \chi_M\mathbf{B}/\mu$, the magnetic susceptibility is

$$\chi_M = -\frac{\mu Z e^2 n_v \langle r^2\rangle}{6m_e} \tag{7–122}$$

We observe that $\chi_M$ is negative, as anticipated. The numerical value of $\chi_M$ is much less than 1, so we can approximate $\mu = \mu_0(1 + \chi_M)$ by $\mu_0$ in this formula.

For the element bismuth, $Z = 83$, $A = 209$, and the mass density is $\rho = 9.75 \times 10^3$ kg/m$^3$. The number of bismuth atoms per cubic meter [see Equation (4–78)] is

$$n_v = \frac{\rho N_A}{A} = \frac{(9.75 \times 10^3 \text{ kg/m}^3)(6.02 \times 10^{26} \text{ atoms/mol})}{209 \text{ kg/mol}}$$
$$= 2.8 \times 10^{28} \text{ atoms/m}^3 \tag{7–123}$$

The diamagnetic susceptibility of bismuth is then

$$\chi_M = -\frac{(4\pi \times 10^{-7})(83)(1.6 \times 10^{-19})^2(2.8 \times 10^{28})\langle r^2\rangle}{6(9.1 \times 10^{-31})}$$
$$= -1.4 \times 10^{16}\langle r^2\rangle \tag{7–124}$$

If we crudely estimate that

$$\langle r^2\rangle \approx (10^{-10} \text{ m})^2$$

we obtain

$$\chi_M \approx -1.4 \times 10^{-4}$$

This value is an overestimate since much of the atomic charge lies within the typical atomic dimension of $10^{-10}$ m. The experimentally measured susceptibility for bismuth is

$$\chi_M^{\text{exp}} = -1.7 \times 10^{-5}$$

which suggests that the root-mean-square radius of a bismuth atom is about $0.3 \times 10^{-10}$ m.

## 7-6B Paramagnetism

*Paramagnetism* is due to the alignment of intrinsic magnetic dipoles in the presence of an external field. Thermal agitation prevents complete alignment and results in the susceptibility being temperature-dependent.

The calculation of the paramagnetic susceptibility is completely analogous to the calculation of the orientational electric polarizability in Section 4–5D. The orientation energy $U = -\mathbf{p}\cdot\mathbf{E}$ is replaced by $U = -\mathbf{m}\cdot\mathbf{B}$.

The average $\cos\theta$, where $\theta$ is the angle between **m** and **B**, can be inferred from Equation (4–107):

$$\langle\cos\theta\rangle = \frac{mB}{3kT} \qquad \text{(7-125)}$$

The magnetization is

$$M = n_v m\langle\cos\theta\rangle = n_v\frac{m^2 B}{3kT} \qquad \text{(7-126)}$$

This expression leads to a susceptibility of

$$\chi_M = \frac{M}{H} = \frac{\mu_0 n_v m^2}{3kT} \qquad \text{(7-127)}$$

where we have approximated $H = B/\mu_0$ in the denominator since $\chi_M$ turns out to be very small. For very large magnetic fields all of the intrinsic dipole moments will align with the field; this phenomenon is known as *saturation.* In quantum theory the dipole energy is not a continuous function of the angle between **m** and **B** but consists of discrete levels depending on discrete spin components. The correct quantum result for small magnetic fields is

$$\chi_M = \frac{\mu_0 n_v m^2}{3kT}\left(\frac{S+1}{S}\right) \qquad \text{(7-128)}$$

where $S$ is the intrinsic spin of the atom or molecule in units of $\hbar$.

In many materials the spins of all of the electrons are paired, so their intrinsic magnetic moments cancel; in those cases only the diamagnetic effect is seen. For molecular gases a rare exception to this rule is molecular oxygen, in which the spins of two electrons add to give

$$m = 2\mu_B \qquad S = 1 \text{ (in units of } \hbar)$$

For $O_2$ gas at 1 atm and at room temperature ($T = 20°C = 293$ K), the number of molecules per cubic meter can be determined from the ideal gas law:

$$n_v = \frac{P}{kT} = \frac{10^5}{(1.4 \times 10^{-23})(293)} = 2.4 \times 10^{25} \text{ molecules/m}^3$$

where $P = 10^5$ N/m$^2$ is the atmospheric pressure and $k$ is Boltzmann's constant. The predicted susceptibility from Equation (7–128) is then

$$\chi_M(O_2) \approx \frac{8}{3}\frac{\mu_0 n_v \mu_B^2}{kT} \approx 1.7 \times 10^{-6} \qquad \text{(7-129)}$$

This value agrees well with the experimental value of

$$\chi_M^{exp}(O_2) \approx 2.1 \times 10^{-6}$$

For comparison we note that the calculated diamagnetic effect in oxygen is smaller by a factor of several hundred. A convincing demonstration of the paramagnetic property of oxygen is provided by pouring liquid oxygen over the poles of the magnet; the liquid is trapped between the poles much like iron filings. A similar demonstration with liquid nitrogen yields no observable effect.

## 7-6C Ferromagnetism

Certain substances, principally the transition elements iron, cobalt, and nickel but also some rare earth elements and some compounds such as chromium oxide ($CrO_2$), exhibit a particularly interesting form of magnetism known as *ferromagnetism*. Much modern technology (such as motors, generators, tape recorders, and transformers) uses the ferromagnetic property of iron.

In ferromagnetic materials the magnetic susceptibility is typically $10^3$ to $10^4$, in contrast to paramagnetic materials, where $\chi_M$ is of order $10^{-5}$. The difference is due to a large spin alignment torque between the valence electrons of different atoms in ferromagnetic materials. This *exchange interaction* is quantum mechanical in origin and involves both the crystal and atomic properties. The alignment torque is far greater than that due to purely magnetic interactions. Ferromagnetism diminishes as the temperature rises and vanishes at and above the *Curie temperature*. For iron this temperature is about 770°C.

The magnetic effect in iron comes mostly from two unpaired electron spins in each atom. With a large enough applied field all of the associated Bohr magnetons add to give a saturation magnetization of

$$\begin{aligned} M_s &= n_v(2\mu_B) = \frac{N_A \rho(2\mu_B)}{A} \\ &= \frac{(6.02 \times 10^{26})(7.9 \times 10^3)}{(55.6)}(2)(0.93 \times 10^{-23}) \\ &= 1.59 \times 10^6 \text{ A/m} \end{aligned}$$

The corresponding magnetic field due to this magnetization is

$$B_s = \mu_0 M_s = 2.0 \text{ T}$$

This value is close to the measured saturation field of 2.15 T, the difference presumably due to small orbital magnetic corrections.

The Einstein–de Haas effect is a clear experimental demonstration that electron spin alignment is responsible for ferromagnetism. An iron rod is suspended by a thin fiber so that it can easily rotate, as illustrated in Figure 7–16. With the rod at rest a solenoid field is used to saturate the rod. When the solenoid current is reversed, the magnetization in the cylinder is reversed, and all of the unpaired electron spins should reverse. The change in angular momentum due to the spin flips appears as a rotation of the whole rod. From Equation (7–51) the $g$ factor for the electrons in the iron rod is then

$$g = \left(\frac{2m_e}{e}\right)\frac{\Delta m}{\Delta L} \tag{7-130}$$

where $\Delta m$ is the change in magnetic moment of the rod and $\Delta L$ is the final angular momentum of the rod. The experimental result for the right-hand side of Equation (7–130) is nearly 2, implying that the magnetization is associated with electron spin.

The magnetization in an unsaturated ferromagnet is not everywhere the same; it is uniform only over regions roughly $10^{-3}$ cm in size, each

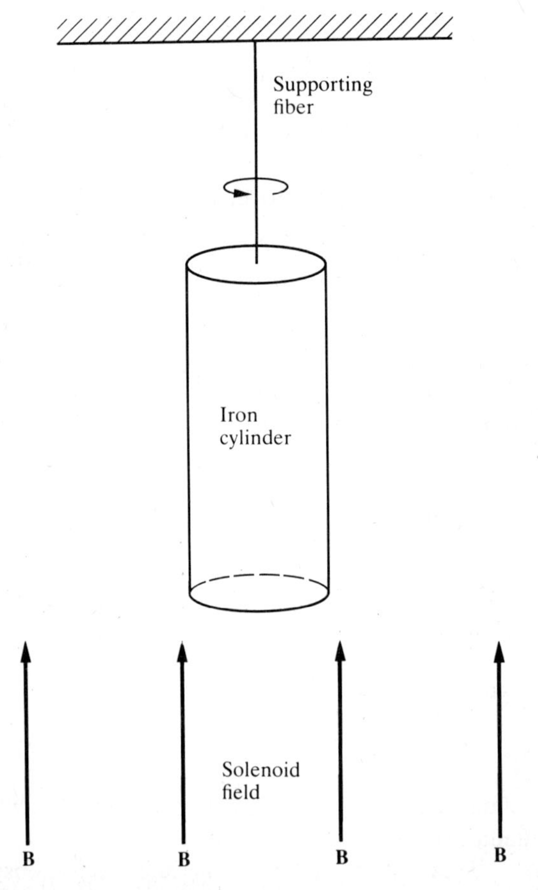


**FIGURE 7–16** Schematic of the Einstein–de Haas effect experiment

containing about $10^{14}$ atoms, called *domains.* All atomic magnetic moments in a domain are aligned in the same direction. This domain structure can be understood by an energy argument. One might naively expect that the lowest energy state (and hence the most stable) would have every magnetic moment aligned in the same direction. Every piece of iron would then always be a saturated permanent magnet. This situation is not, however, the most eco-

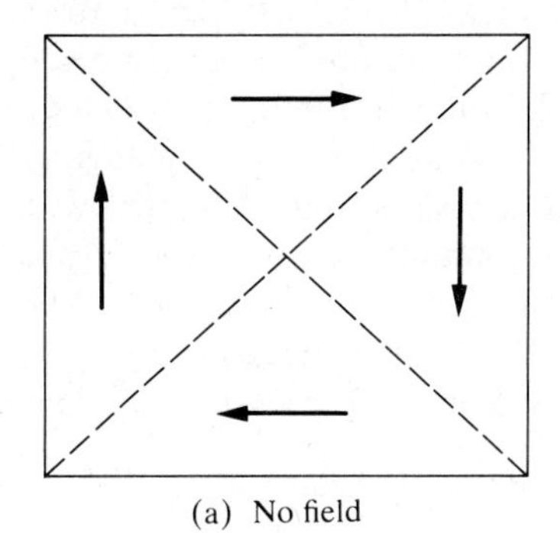

(a) No field

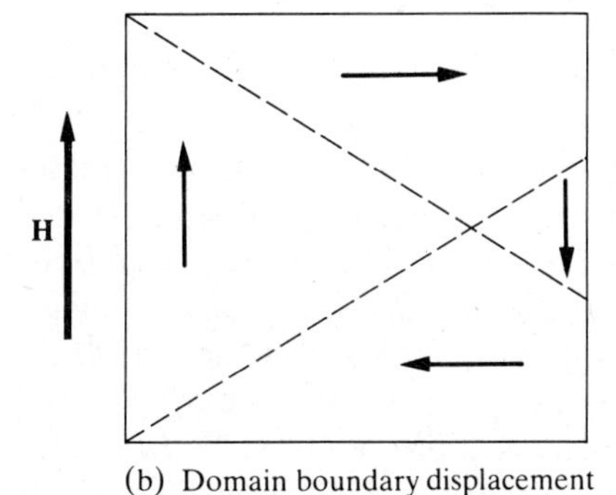


(b) Domain boundary displacement

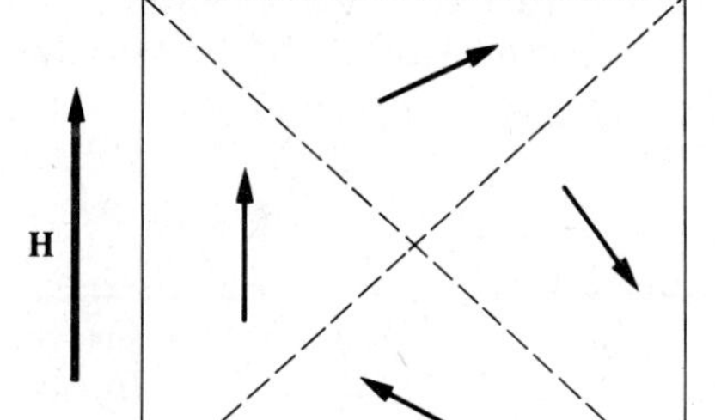


(c) Domain rotation

**FIGURE 7-17** Magnetic domains (a) with no external magnetic field, (b) responding to applied magnetic field by displacing domain boundaries, and (c) responding to applied magnetic field by rotating domains

nomical energetically, since the aligned atoms would produce a large magnetic field energy outside the sample. With the volume divided into domains saturated in different directions, the external field energy is lower. For example, Figure 7–17(a) shows four domains giving zero net magnetization and hence a small external field. Beyond a certain point division into domains does not decrease the energy further since there is an energy per unit area (surface tension) of domain boundaries. This situation arises because in the boundary, or the transition region from one domain to the next (Bloch wall), the spins of neighboring atoms are not exactly aligned, and hence the Bloch wall, considered by itself, is not in the lowest possible energy state. When an external magnetic field is applied, the domains either rotate or deform, and a net magnetization results that is aligned with the external field such that the energy is a minimum; see Figures 7–17(b) and 7–17(c).

A ferromagnetic material can be magnetized by passing a current through a wire wound around the sample. We consider the example of an iron torus with a current $I$ through a winding of $N$ turns. The $H$ field in the iron is determined by Ampère's law:

$$H_\phi = \frac{NI}{\ell} \tag{7–131}$$

where $\ell$ is the circumference at the radius of interest. By changes in $I$, $H_\phi$ can be varied.

As the applied $H$ field increases, the magnetization increases (dashed line in Figure 7–18) until all domains are aligned; that is, saturation is reached. When $H$ is decreased below the saturation field, the magnetization does not retrace its path; and when $H$ reaches zero, there is a *remanent magnetic field.* The reversed $H$ field required to reduce the magnetic field to zero is known as the *coercive field.* The magnetization curve traced out by a cycle in $H$ is called a *hysteresis loop.* The hysteresis loops of $M$ versus $H$ and $B$ versus $H$ are shown in Figure 7–18. We see that the $B$ field depends on the $H$ field but also on the previous history of the sample. A representative sample of magnetic parameters (remanence and coercive force) is given in Table 7–1. Iron of ordinary purity has a coercive field of about 50 A/m and can be magnetized by the earth's field, which has $H \approx 50$ A/m.

**TABLE 7-1** Magnetic parameters

| **Material** | **Remanence, $B$(T)** | **Coercive Field, $H$(A/m)** |
|---|---|---|
| Alnico | | |
| 5 (51% Fe, 24% Co, 14% Ni, 8% Al, 3% Cu) | 1.25 | 44,000 |
| Silicon steel | 1.20 | 4 |
| Iron | 0.4 | 50 |
| Supermalloy | 0.5 | 0.32 |

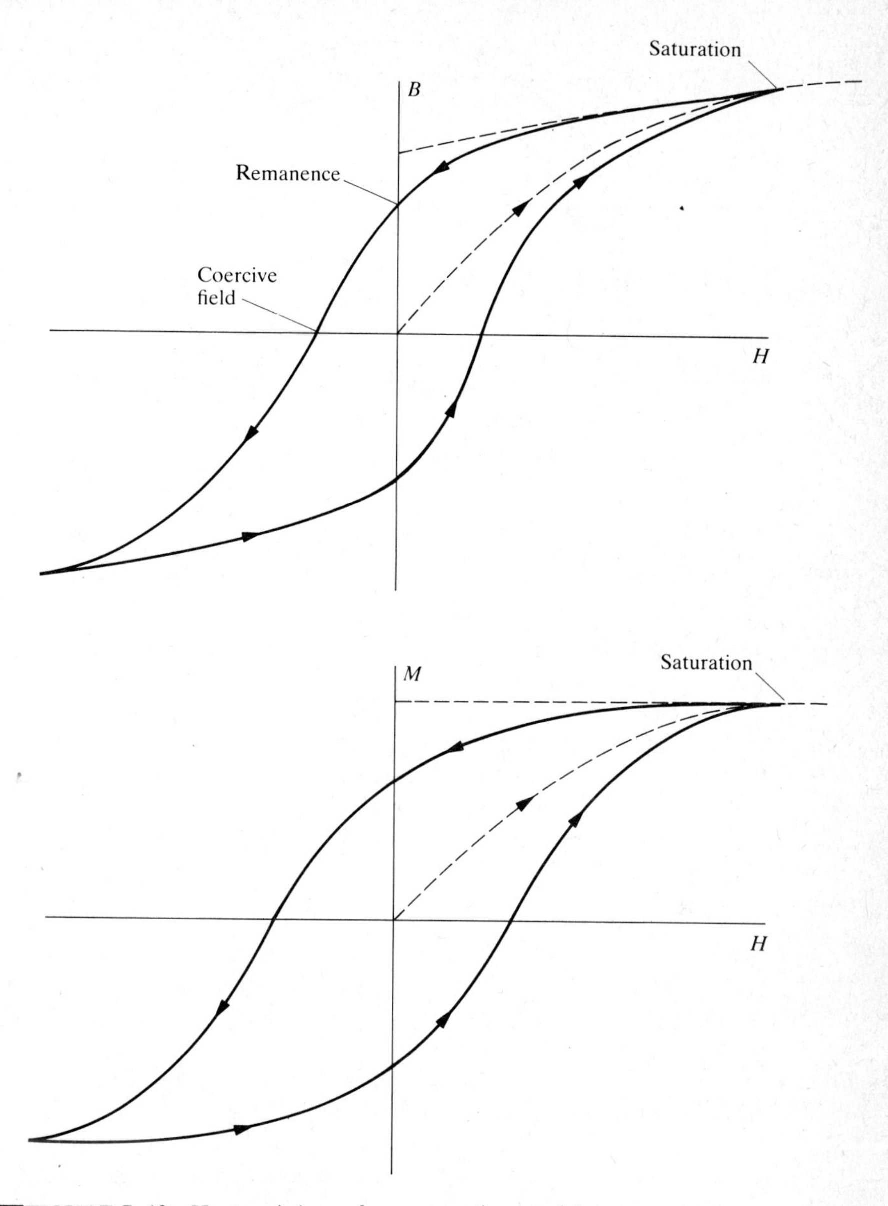


**FIGURE 7-18** Hysteresis loops for a magnetic material

A method by which the hysteresis curve can be mapped out uses a ferromagnetic toroid (Rowland ring), illustrated in Figure 7–19. The $H$ field is the independent variable determined by the current in the primary winding. A secondary winding of $N_s$ turns is connected to a galvanometer measuring the total charge passing through the circuit. By an analysis similar to that of the flip coil result of Equation (6–21), we find that a change $dB$ in the magnetic field in the ring induces a charge $dQ = (S/R)\, dB$ to flow through the

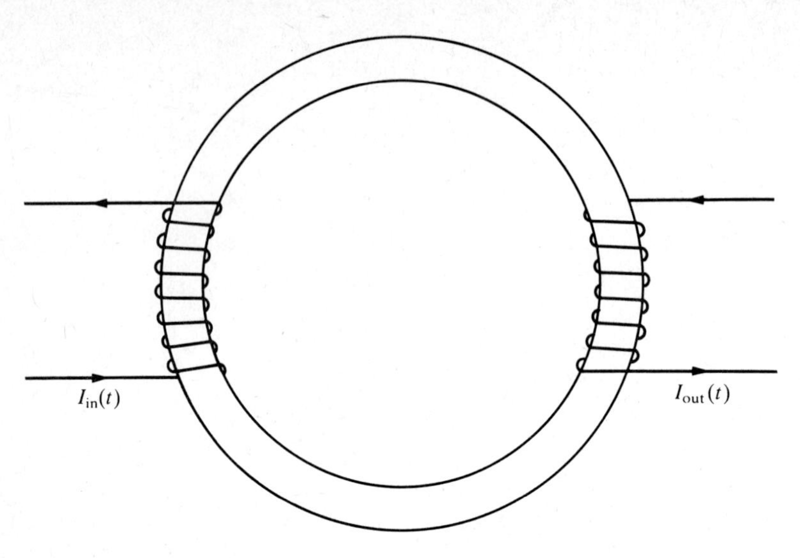


**FIGURE 7-19** Rowland ring used to map out the hysteresis curve

galvanometer, where $S$ is the effective cross-sectional area of the toroid and $R$ is the resistance of the secondary circuit. Starting from an unmagnetized sample, the magnetic field at any time is then

$$B(t) = \frac{R}{S} Q(t) \tag{7-132}$$

Mapping out $H(t)$ with $B(t)$ over a period yields a hysteresis curve. Domain movement is an irreversible process since acoustic waves and heat are generated by the motion, dissipating magnetic energy. This irreversibility of the hysteresis curve is characteristic of ferromagnetism.

If we look in fine detail at a portion of a hysteresis curve, a steplike structure is seen instead of a smooth curve. This fine structure is produced by individual domain wall movements and is known as the *Barkhausen effect.* As a domain moves from one configuration to another, there is a tiny jump in the flux in the Rowland ring.

To relate the hysteresis curve to energy loss, we again consider the Rowland ring. If the current $I$ is increased by $dI$ in an interval $dt$, an emf will be induced equal to

$$\mathscr{E} = \frac{R\,dQ}{dt} = NS\frac{dB}{dt} \tag{7-133}$$

in a direction opposing the current increase. In the time interval $dt$ the work done against the induced emf is

$$dW = \mathcal{E}\, dQ = \mathcal{E} I\, dt = NIS\, dB \tag{7-134}$$

Since the current is related to the $H$ field by Equation (7–131), $H\ell = NI$, we have

$$dW = H\, dB(S\ell) \tag{7-135}$$

The differential work done per unit volume is thus

$$\frac{dW}{V} = H\, dB \tag{7-136}$$

For a reversible curve the net work done in a cycle of $H$ is zero, but for a ferromagnetic material the net work per unit volume is just the area enclosed by the hysteresis curve. This loss is clearly greater for a material with a large coercive field and remanence.

We conclude this section by collecting some of the results of this chapter that bear on the phenomenon of permanent magnetism. Certain ferromagnetic materials can be permanently magnetized; once the internal domains are aligned, they remain aligned under the influence of stray fields such as the fields of other magnets. The best permanent magnets have large remanence and coercive fields. We can approximate a bar magnet as a uniformly saturated magnetic cylinder. The magnetic field due to this cylinder is equivalent to that produced by either an effective solenoidal magnetization current or magnetic $H$ charges on the ends of the cylinder. In both cases cutting the cylinder crosswise results in two magnets of the same strength as before. In the vicinity of the end of a very long bar magnet, we can imagine the near isolation of a single magnetic pole. The force between two such poles is nearly inverse-square.

We usually define the north pole of a magnet as the end from which the **B** field emerges. Two magnetic dipoles repel in all orientations except when the north and south poles are aligned or when they lie side by side antiparallel, as seen in Figure 7–4. A permanent magnet will attract "softer" ferromagnetic materials by aligning their domains parallel to the magnet's field. The magnetic field of the earth is approximately a dipole away from the magnetic core. The magnetic pole in the northern hemisphere of the earth is actually a south pole by the preceding definition; the north pole of a compass needle will point toward it.

Permanent magnets find wide application in circumstances where a static field of about 1 T is required, since it is often inconvenient to produce high fields by currents. An example is the stereo speaker. The two wires entering each speaker carry a current proportional to the amplitude of the desired sound. The current passes through a coil in front of the speaker's permanent magnet. The Lorentz force on this coil moves the speaker cone, which in turn generates an acoustic wave by moving the adjacent air.

## 7-7 MACROSCOPIC FIELDS AND MAGNETIC MATERIALS

In Equations (7–104) and (7–105) we saw that at a sharp discontinuity between different magnetic materials, the normal **B** field, $B_n$, is continuous and the tangential **H** field, $H_t$, jumps by the surface conduction current flowing perpendicular to $H_t$. We now apply these boundary conditions to the boundary of ferromagnetic material for which $j_\perp = 0$ and $\mu/\mu_0$ is quite large. Except for the particular case of an exactly normal **B** field, continuity in $H_t = B_t/\mu$ implies that within the iron the dominant field component is the magnetic field $B_t$ parallel to the surface of the iron. In the limit $\mu/\mu_\circ \to \infty$ a finite $B_t$ in the iron implies that the other $B$ field components inside and outside the iron are negligible. Thus to a good approximation, the magnetic field is confined in the iron.

A very important practical application is the calculation of magnetic fields in a magnetic "circuit" consisting of iron paths, air gaps, and current windings. Such considerations apply to the design of transformers, motors, generators, and electromagnets. A typical magnetic field calculation involves a rectangular iron structure with $N$ turns of wire carrying current $I$ on one side and a gap in the other, as illustrated in Figure 7–20. The iron path length is $\ell$, and the gap width is $\delta$. Ampère's law, Equation (7–87), implies

$$H_i\ell + H_a\delta = NI \qquad \textbf{(7-137)}$$

where $H_i$ and $H_a$ are the $H$ field magnitudes in the iron and the air, respectively. At the gap the **B** field remains constant if fringing can be neglected,

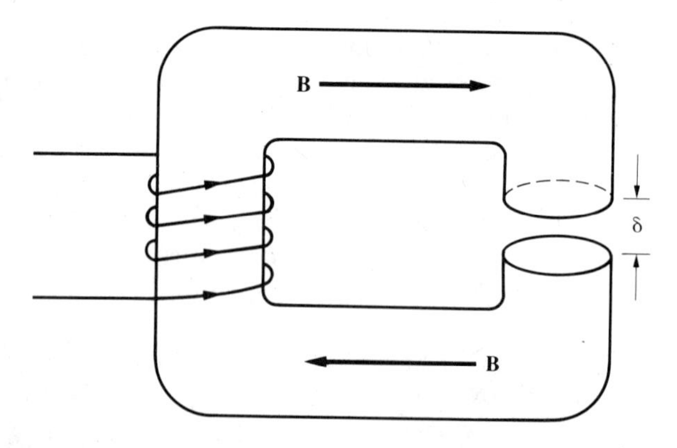


**FIGURE 7-20** Rectangular iron structure with a gap. The magnetic field in the gap of this circuit is calculated as an application of the boundary conditions derived in this section.

so $B = \mu_0 H_a = \mu H_i$. Solving for the magnetic field, we obtain

$$B = \frac{NI}{(\ell/\mu) + (\delta/\mu_0)} \tag{7-138}$$

For small air gaps $[\delta \leqslant (\mu_0/\mu)\ell]$ the magnetic field in such magnetic circuits is of the order $\mu/\mu_0$ times larger than the field in an air-core solenoid of the same length and number of windings. This field enhancement in the air gap is roughly a factor of several thousand.

A systematic approach to magnetic circuit analysis is based on an analogy to electric current flow. As summarized in Table 7–2, the quantities **B**, **H**, and $\mu$ obey equations analogous to the electrical quantities **J**, **E**, and $g$. The magnetic analog of Ohm's law is

$$\mathscr{E}_m = NI = \mathscr{F}\mathscr{R}_m \tag{7-139}$$

where $\mathscr{E}_m$ is the magnetomotive force, $\mathscr{F}$ is the magnetic **B** flux, and $\mathscr{R}_m$ is the reluctance, given by

$$\mathscr{R}_m = \sum_k \frac{\ell_k}{\mu_k S_k} \tag{7-140}$$

Here the subscript $k$ denotes an element of the circuit of length $\ell_k$, permeability $\mu_k$, and cross-sectional area $S_k$. The sum is over all elements of the circuit considered.

Applying Equation (7–139) to the magnetic circuit of Figure 7–20 (with the iron segment of length $\ell$, area $S$, and permeability $\mu$ and the air gap

**TABLE 7–2** Electrical and magnetic circuit relations

| Electrical | Magnetic |
|---|---|
| $\nabla \cdot \mathbf{J} = 0$ | $\nabla \cdot \mathbf{B} = 0$ |
| $\mathbf{J} = g\mathbf{E}$ | $\mathbf{B} = \mu\mathbf{H}$ |
| $\mathscr{E} = \oint \mathbf{E} \cdot d\mathbf{r}$ | $\mathscr{E}_m = \oint \mathbf{H} \cdot d\mathbf{r} = NI$ |
| $\mathscr{E} = \oint \frac{\mathbf{J} \cdot d\mathbf{r}\, S}{gS} = IR$ | $\mathscr{E}_m = \oint \frac{\mathbf{B} \cdot d\mathbf{r}\, S}{\mu S} = \mathscr{F}\mathscr{R}_m$ |
| $R = \sum_k \frac{\ell_k}{g_k S_k}$ | $\mathscr{R}_m = \sum_k \frac{\ell_k}{\mu_k S_k}$ |
| $S$ = cross-sectional area | $S$ = cross-sectional area |
| $\mathscr{E}$ = electromotive force | $\mathscr{E}_m$ = magnetomotive force |
| $R$ = resistance | $\mathscr{R}_m$ = reluctance |
| $I$ = electric current | $\mathscr{F}$ = magnetic flux |
| $\mathscr{E} = RI$ | $\mathscr{E}_m = NI = \mathscr{R}_m\mathscr{F}$ |

of length $\delta$), we obtain

$$BS\left(\frac{\ell}{\mu S} + \frac{\delta}{\mu_0 S}\right) = NI \tag{7-141}$$

This result is identical to Equation (7–138). The magnetic circuit analogy can be applied only when $\mu_i \gg \mu_0$ since $\mu_0 \neq 0$ outside a magnetic material, unlike the electrical case where $g = 0$ outside the conductor.

In the usual case the magnetic permeability $\mu$ changes with $H$ so that we do not know what $\mu$ to use in Equation (7–141) until $H$ is known. To handle this situation, we use $H_i = B_i/\mu$ to rewrite Equation (7–141) as

$$H_i = \frac{NI/\ell}{1 + (\mu/\mu_0)(\delta/\ell)} \tag{7-142}$$

and then solve by successive approximation, as follows: If the gap $\delta$ is small, the zeroth approximation for $H_i$ is

$$H_i^{(0)} = \frac{NI}{\ell}$$

Using the hysteresis curve, we calculate the corresponding $\mu_i^{(0)}$ from $H_i^{(0)}/B$. We iterate again, using $H_i^{(1)}$ calculated from Equation (7–142) and using the hysteresis curve to determine $\mu_i^{(1)}$. This cycle can be continued until the desired accuracy is achieved. Usually one or two iterations suffice.

Magnetic circuits like that in Figure 7–20 are known as *electromagnets* (as opposed to permanent magnets). The utility of an electromagnet is that its strength can be easily varied. For example, a large electromagnet mounted on a boom can lift heavy iron objects and deposit them in a new location. Clearly, a good electromagnet should have a low coercive force so that a relatively small current can magnetize and demagnetize it.

The force exerted on moving charged particles by the magnetic field of an electromagnet can be used to deflect them and thus to bend and focus beams of charged particles. For fast particles, such as those produced by particle accelerators, electromagnets are used instead of electrostatic devices because for practically achievable fields the magnetic (Lorentz) force on a fast charged particle can be much larger than the electric force. Magnets for deflection have two pole faces, as in Figure 7–20, and have an approximately constant magnetic field between the faces. These magnets are used to bend and disperse charged particle beams in particle spectrometers and also to bend the beam of a synchrotron (circular particle accelerator) into a closed orbit (electric fields along the beam direction are used to accelerate the particles). Focusing magnets have four faces, with the poles arranged in alternating sequence north-south-north-south, analogous to Figure 3–10. At the center of the gap the magnetic field is zero, and a charged particle moving through the magnet at the center is undeflected. The $q\mathbf{v} \times \mathbf{B}$ force on a charged particle on an off-center trajectory is directed toward the center in certain regions and in other regions is directed away from the center. So that a net focusing effect is achieved, a similar magnet is located behind the first with reversed magnetic polarities. The sequence of focusing-defocusing (or vice versa) leads to net focusing, as with optical lenses.

In a general multipole expansion in two dimensions for a potential (either $\Phi$ or $\mathbf{A}$), terms proportional to both $\imath^n$ and $\imath^{-n}$ occur. If the sources of the potential are near $\imath = 0$, the potential at large $\imath$ is given by a sum of $\imath^{-n}$ terms. If the sources are at large $\imath$, the potential at small $\imath$ is given by a sum of $\imath^n$ terms. In particular, for sources of vector potential at large $\imath$ the dipole ($n = 1$) term has a linear $\imath$ dependence, and the magnetic field is constant. For this reason bending magnets are called dipole magnets. Similarly, focusing magnets are called quadrupole magnets, since their fields correspond to the quadrupole ($n = 2$) term in the multipole expansion. The magnetic field of a magnetic quadrupole lens is similar to the electric field of an electric quadrupole lens, given in Equation (3–108).

## SUMMARY

### Important Concepts and Equations

$$\mathbf{m} = \frac{1}{2}\int (\mathbf{r}' \times \mathbf{J})\, dV'$$

*The magnetic dipole moment* **m** *for a current distribution* **J**.

$$\mathbf{A} = \frac{\mu_0}{4\pi}\left(\frac{\mathbf{m} \times \mathbf{r}}{r^3}\right)$$

*The vector potential* **A** *for a magnetic dipole moment* **m**.

$$\mathbf{B} = \frac{\mu_0}{4\pi}\left[\frac{3\mathbf{r}(\mathbf{r} \cdot \mathbf{m}) - r^2\mathbf{m}}{r^5}\right]$$

*The magnetic field* **B** *at* $r \neq 0$ *for a magnetic dipole moment* **m**.

$$\mathbf{m} = IS\hat{\mathbf{n}}$$

*The magnetic dipole moment* **m** *for a planar circuit with current* $I$ *and area* $S$; $\hat{\mathbf{n}}$ *is the unit normal vector determined by the right-hand rule.*

$$\mathbf{N} = \mathbf{m} \times \mathbf{B}$$

*The torque* **N** *on a current distribution is the cross product of the magnetic dipole moment* **m** *and the magnetic field* **B**.

$$U = -\mathbf{m} \cdot \mathbf{B}$$

*The potential energy* $U$ *for a magnetic dipole* **m** *in a magnetic field* **B**.

$$U = -\frac{\mu_0}{4\pi}\left[\frac{3(\mathbf{m}_1 \cdot \mathbf{r})(\mathbf{m}_2 \cdot \mathbf{r}) - r^2\mathbf{m}_1 \cdot \mathbf{m}_2}{r^5}\right]$$

*The energy of interaction* $U$ *between two magnetic dipoles with* $\mathbf{m}_1$ *at* $\mathbf{r} = 0$ *and* $\mathbf{m}_2$ *at* $\mathbf{r}$.

$$\omega_c = -\frac{q}{M}\mathbf{B}$$

*The angular velocity of a particle of mass M and charge q in a magnetic field* **B***;* $\omega_c$ *is known as the cyclotron frequency.*

Adiabatic invariants

*Adiabatic invariants are constants of motion for a periodic system that remain constant when the system is slowly changed so that it is no longer quite periodic. The magnetic flux through a cross section of the helical orbit and the magnetic moment of a particle in a slowly varying magnetic field are adiabatic invariants.*

$$\mathbf{m} = \frac{q}{2M}\mathbf{L}$$

*The orbital magnetic moment* **m** *of a particle of charge q and mass M is proportional to the orbital angular momentum* **L**. *The proportionality factor* $q/(2M)$ *is the classical gyromagnetic ratio.*

$$\mathbf{m} = g\left(-\frac{e}{2m_e}\right)\mathbf{S}$$

*The intrinsic electron magnetic moment* **m** *is due to the intrinsic angular momentum (spin)* **S** *of the electron. The g factor for the electron is approximately 2.*

$$m_z = \pm\frac{1}{2}\mu_B \qquad \mu_B \equiv \frac{e\hbar}{2m_e}$$

*Because the spin along a given axis can have only values of* $\pm\hbar/2$, *the magnetic moment of an electron along that axis also has only two values. The Bohr magneton* $\mu_B$ *is* $0.93 \times 10^{-23}$ A·m$^2$.

$$\omega_s = \frac{ge}{2m_e}\mathbf{B}$$

*The angular velocity* $\boldsymbol{\omega}_s$ *of precession of the intrinsic electron spin vector* **S** *in a constant magnetic field* **B**. *The electron g factor is approximately 2.*

Magnetic spin resonance

*In a constant magnetic field a particle spin vector precesses with the angular velocity* $\boldsymbol{\omega}_s$. *If a small oscillating magnetic field is introduced perpendicular to the static field with frequency* $\omega_s$, *the spins will flip, causing a change in susceptibility. This process is known as magnetic spin resonance.*

$g - 2$ experiment

*The electron g factor can be determined by simultaneous excitation of precession* ($\omega_s$) *and cyclotron* ($\omega_c$) *resonances. By measurement of the beat frequency*

$$\frac{\omega_s - \omega_c}{\omega_c} = \frac{1}{2}(g - 2)$$

*the g factor can be determined.*

$$\mathbf{M} = \frac{1}{\Delta V} \sum_i m_i$$

*The magnetization* **M** *is the total dipole moment in a small volume* $\Delta V$ *divided by* $\Delta V$.

$$\mathbf{j}_M = \mathbf{M} \times \hat{\mathbf{n}} \qquad \mathbf{J}_M = \nabla \times \mathbf{M}$$

*The magnetization current densities,* $\mathbf{j}_M$ *and* $\mathbf{J}_M$, *are the effective surface and volume magnetization current densities, respectively, which give the same macroscopic magnetic field as the average field produced by the microscopic dipoles.*

$$\mathbf{H} = \frac{\mathbf{B}}{\mu_0} - \mathbf{M}$$

*The magnetic intensity* **H** *is the difference between the magnetic field* **B** *divided by* $\mu_0$ *and the magnetization* **M**.

$\nabla \times \mathbf{H} = \mathbf{J}$ (Ampère's law for magnetic material, differential form)

*The curl of the magnetic intensity* $\nabla \times \mathbf{H}$ *is equal to the current density* **J**.

$\oint \mathbf{H} \cdot d\mathbf{r} = I$ (Ampère's law for magnetic material, integral form)

*The line integral of magnetic intensity around a closed circuit is equal to the current I passing through a surface bounded by the closed circuit.*

$$\mathbf{H} = -\nabla\Phi_M$$

*The magnetic intensity* **H** *in a magnetic material with no electric fields and no conduction current* ($\mathbf{J} = 0$) *can be derived from a magnetic scalar potential* $\Phi_M$. *The sources of* **H** *are the effective magnetic volume charge density* $\rho_M = \nabla \cdot \mathbf{M}$ *and the effective magnetic surface charge density* $\sigma_M = \mathbf{M} \cdot \hat{\mathbf{n}}$.

$$\mathbf{M} = \chi_M \mathbf{H}$$

*A linear isotropic material has a magnetization* **M** *given by the product of the dimensionless constant* $\chi_M$, *called the magnetic susceptibility, and the magnetic intensity* **H**.

$$\mu = \mu_0(1 + \chi_M) \qquad \kappa_M = \frac{\mu}{\mu_0} = 1 + \chi_M$$

*The basic quantities used to describe magnetic phenomena are the magnetic permeability* $\mu$, *the magnetic susceptibility* $\chi_M$, *and the relative permeability* $\kappa_M$.

$\mathbf{B} = \mu\mathbf{H} = \mu_0\kappa_M\mathbf{H}$

*The magnetic field* **B** *and the magnetic intensity* **H** *are proportional in a linear isotropic magnetic medium.*

$\nabla \times \mathbf{B} = \mu\mathbf{J}$

*Within a linear isotropic material the source equation for* **B** *must be modified by replacing* $\mu_0$ *with* $\mu$.

$B_{1n} = B_{2n} \qquad H_{2t} - H_{1t} = j_\perp$

*Boundary conditions on the normal* **B** *field and the tangential* **H** *field at a boundary interface.*

Diamagnetism

*Materials that have magnetic dipoles induced opposite to the applied magnetic field are diamagnetic. All matter has diamagnetic properties, with* $\chi_M$ *on the order of* $-10^{-5}$ *for solids.*

Paramagnetism

*Paramagnetism is due to the alignment of intrinsic magnetic dipoles in the presence of an external field. Thermal agitation prevents complete alignment and results in the susceptibility being temperature-dependent;* $\chi_M$ *is typically of the order* $10^{-5}$ *for solids.*

Ferromagnetism

*Ferromagnetic properties result from a large spin alignment torque between the valence electrons of different atoms in ferromagnetic materials. The forces that align these spins are known as exchange forces and are quantum mechanical in nature. Here* $\chi_M$ *is typically on the order* $10^3$ *to* $10^4$.

Remanence field

*The magnetic field that remains after a saturated ferromagnetic sample has been removed from an applied magnetic field.*

Coercive field

*The H field that must be applied to a ferromagnet to demagnetize its remanence field.*

Hysteresis loop

*The magnetization curve traced out by a cycle in H is called a hysteresis loop. The magnetic field in a ferromagnet depends not only on the H field but also on the previous history of the sample.*

# PROBLEMS

## Section 7-1: Magnetic Moments and the Earth's Magnetic Field

**7-1** Calculate the magnetic moments of the following charge distributions that rotate about the $z$ axis with angular velocity $\omega$; in each case the charge density is uniform, the total charge is $Q$, and the mass is $M$. Compute the angular momentum $L$ in each case, and find the ratio of magnetic moment $m$ to $L$.
(i) A spherical shell of radius $a$.
(ii) A solid sphere of radius $a$.
(iii) A thin disk of radius $a$ in the $xy$ plane.
(iv) A cylindrical shell of radius $a$ and length $\ell$.
(v) A solid circular cylinder of radius $a$ and length $\ell$.

**7-2** Show that the magnetic dipole field must have the delta function contribution at $\mathbf{r} = 0$ given by Equation (7–14). (*Hint*: Recall that $\mathbf{B} = \nabla \times \mathbf{A}$, and apply the divergence theorem corollary $\int \mathbf{B}\, dV = \int d\mathbf{S} \times \mathbf{A}$ to a spherical region.)

**7-3** We wish to cancel the earth's magnetism at a point well out in space by a single-turn coil at the earth's magnetic equator, where the earth's radius is $6.37 \times 10^6$ m.
(i) What current will be required in the coil if the magnetic dipole moment of the earth is $8.2 \times 10^{22}$ A·m$^2$?
(ii) The diameter of the wire is 0.2 m. What will be the magnitude of the magnetic field adjacent to the wire?

**7-4** Assume that the earth's magnetic field is caused by a small current loop located at the center of the earth. Given that the field near the pole is 0.6 G and that the radius of the earth is $R = 6.37 \times 10^6$ m, use the Biot-Savart law to calculate the strength of the magnetic moment of the small current loop.

**7-5** A magnetic dipole of moment $\mathbf{m} = m\hat{\mathbf{z}}$ produces a magnetic field

$$\mathbf{B} = \frac{\mu_0}{4\pi}\left[-\frac{\mathbf{m}}{r^3} + \frac{3(\mathbf{m}\cdot\mathbf{r})\mathbf{r}}{r^5}\right]$$

A small wire loop of resistance $R$ and radius $a$ is on the $z$ axis at a distance $d$ from the dipole. The plane of the loop is perpendicular to $\mathbf{m}$. The current that produces $\mathbf{m}$ is now quickly turned off (i.e., $\mathbf{m} \to 0$). Find the direction and the magnitude of the impulse acting on the wire loop. Assume $a \ll d$.

**7-6** An electron is revolving in a circular orbit of radius $r$ with angular velocity $\omega$.
(i) What is the magnetic moment due to its motion?
(ii) A magnetic field $B$, parallel to the axis of revolution of the electron, is turned on. Derive an expression for $\Delta\omega$, the change in angular velocity of the electron. (*Hint*: Use Faraday's law of induction and Newtonian mechanics.)
(iii) Does $\Delta\omega$ increase, decrease, or leave unchanged the magnetic moment?

**7-7** A dipole whose magnetic moment $\mathbf{m}_1$ lies in the $xy$ plane makes a fixed angle $\phi_1$ with respect to the $x$ axis. A coplanar dipole $\mathbf{m}_2$ at fixed distance $d$ on the $x$ axis from the origin is free to rotate in the $xy$ plane. For what angle $\phi_2$ of $\mathbf{m}_2$ with the $x$ axis will the system be in equilibrium? Is the equilibrium stable?

## Section 7-2: Torque on a Magnetic Dipole

**7-8** A circular loop of radius $R$ lies in the $xy$ plane with its center at the origin. The loop carries a current $I$ in the positive sense for a normal to the circuit along the positive $z$ axis. A point dipole $\mathbf{m} = m\hat{\mathbf{y}}$ lies on the positive $x$ axis at a distance $D \gg R$ from the origin. Find the force and the torque on $\mathbf{m}$ due to the current loop.

## Section 7-3: Magnetic Reflection

**7-9** Charged particles with zero azimuthal velocity cannot be confined in a magnetic bottle since they have no magnetic force acting on them. These particles are not the only source of loss in a magnetic bottle. Show that charged particles with transverse velocities given by

$$v_{z_0} > v_{\phi_0}\left(\frac{B_M}{B_0} - 1\right)^{1/2}$$

will also escape, where $v_{\phi_0}$, $v_{z_0}$, and $B_0$ are the velocities and the magnetic field at the midpoint of the bottle and $B_M$ is the magnetic field at the end of the bottle. (*Hint*: Use the conservation of energy and the fact that $\mathbf{m}$ is time-independent.)

**7-10** An infinite current sheet has a surface current density $\mathbf{j} = j\hat{\mathbf{y}}$ in the plane $z = 0$. See Figure 7-21. A point particle of mass $m$ and charge $q$ has initial velocity $\mathbf{v}_0 = -v_0\hat{\mathbf{z}}$ at a height $z = +h$ (assume $v_0 \ll c$).

(i) Find the magnetic field everywhere (ignoring the self-field of the particle).

(ii) For given values of $j$, $q$, $m$, and $h$, what is the largest initial speed $v_0$ such that the particle remains on the same side of the current sheet? Describe the orbit.

(iii) Suppose $v_0$ exceeds the value in part (ii), and the particle can freely pass through the current sheet. Describe the resulting orbit.

(iv) If the motion in part (iii) continues, the orbit will repeat as the particle drifts along the $y$ axis. Find the average *drift* velocity.

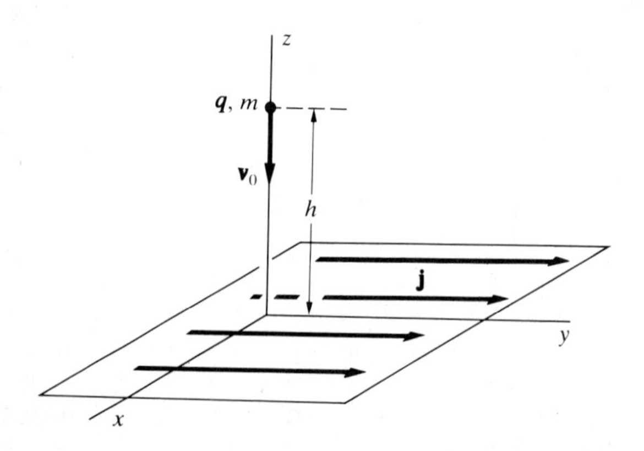


**FIGURE 7-21** Charged particle moving toward a current sheet

## Section 7-4: Magnetic Spin Resonance

**7-11** For actual samples the position of the resonance peaks for both electron spin resonance and nuclear magnetic spin resonances depends on the local field seen by the spins. However, the position of the resonances can be estimated roughly by calculating the resonance frequencies for free electrons and protons. Assuming $g = 2.0$ for the electron and $g = 5.58$ for the proton, find these frequencies for an applied magnetic field of 2.0 T.

## Section 7-5: Magnetization, Equivalent Charge, and Current Distributions

**7-12** A uniformly magnetized sphere has a total magnetic moment of $\frac{4}{3}\pi a^3 M$, where $a$ is the radius and $M$ is the magnetization. Calculate the equivalent surface currents that can replace this sphere as far as external effects are concerned. Show that this current distribution has the same total magnetic moment.

**7-13** An infinitely long cylinder has azimuthal magnetization $\mathbf{M} = M_0(\imath/a)\hat{\boldsymbol{\phi}}$, where $M_0$ is a constant and $a$ is the radius. Find **B** and **H** in the cylinder and outside.

**7-14** A toroid having an iron core of square cross section and permeability $\mu$ is wound with $N$ closely spaced turns of wire carrying a current $I$. Find the magnitude of the magnetization $M$ everywhere inside the iron.

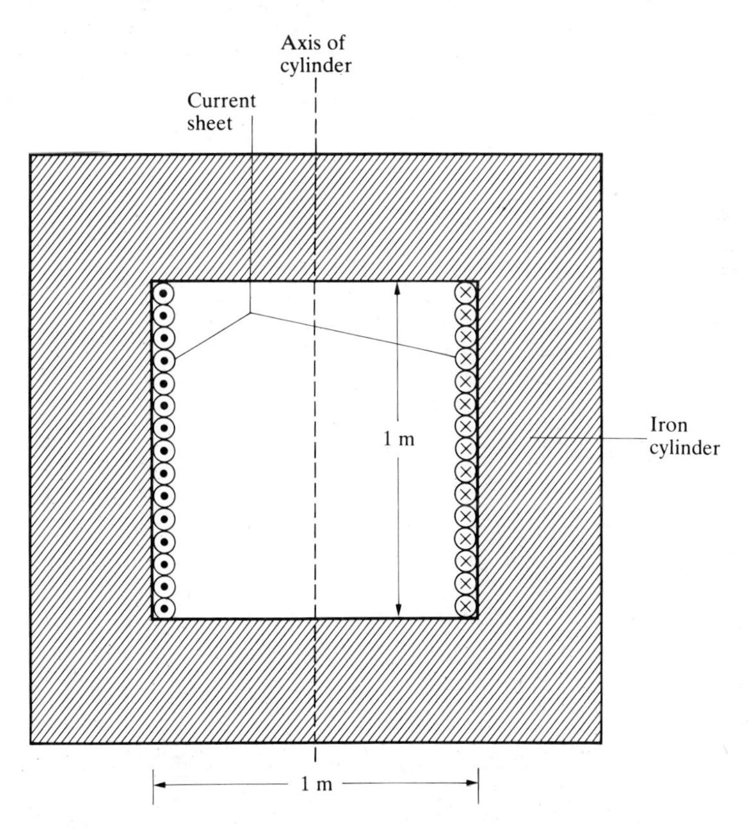


**FIGURE 7-22** Electromagnet with gap

**7-15** An electromagnet consists of a hollow cylinder of iron with closed ends; it is excited by a thin current sheet flowing azimuthally. Figure 7–22 shows a lengthwise section. Assume the permeability of the iron is infinite. The field strength in the interior is 1.0 T.

(i) What total current (ampere-turns) is required?
(ii) What is the total stored energy of the magnet?
(iii) What is the force of attraction between the ends of the cylinder?

**7-16** A cylinder of length $L$ and radius $R$ is uniformly magnetized along its axis. Find the form of the scalar potential on the axis. Show that the magnetic field at the center of a long cylinder is

$$B = \mu_0 M\left(1 - \frac{2R^2}{L^2}\right)$$

**7-17** Let region 1 be free space, where $z < 0$, and let region 2 be dielectric with $\varepsilon_2 = 2\varepsilon_0$, where $z > 0$. At $z = 0$, $\sigma_f = 0.2\ \text{C/m}^2$. For $\mathbf{D}_1 = 3x\hat{\mathbf{x}} + 4y^2\hat{\mathbf{y}} + 3\hat{\mathbf{z}}$ and $\mathbf{H}_2 = 2\hat{\mathbf{x}} + 5y^3\hat{\mathbf{y}} + 5\hat{\mathbf{z}}$, find $\mathbf{E}_2$, $\mathbf{D}_2$, $\mathbf{B}_2$, $\mathbf{E}_1$, $\mathbf{B}_1$, and $\mathbf{H}_1$ at the boundary.

**7-18** A surface current density $\mathbf{j} = j_0\hat{\mathbf{x}}$ amperes per meter exists in the $z = 0$ plane. Region 1 with $\mu_1$ is located in the space $z < 0$, and region 2 with $\mu_2$ is located in the space $z > 0$. The **H** field in region 1 is $H_1 = H_{1x}\hat{\mathbf{x}} + H_{1y}\hat{\mathbf{y}} + H_{1z}\hat{\mathbf{z}}$. Solve for $\mathbf{H}_2$ at the boundary $z = 0$.

**7-19** Find the magnetic field due to a uniformly magnetized permanent magnet (intensity of magnetization **M**) of spherical form (radius $R$) surrounded by a medium of magnetic permeability $\mu$.

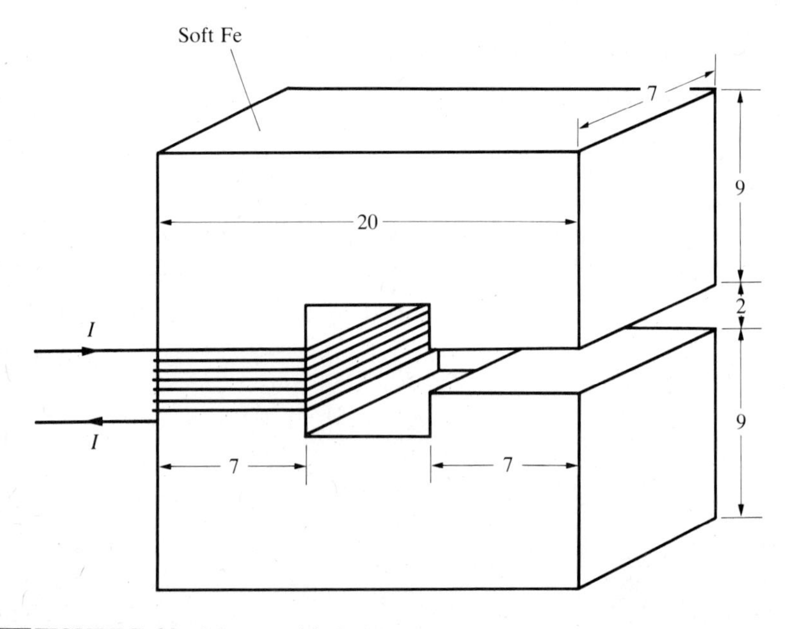


**FIGURE 7-23** Magnet with C shape

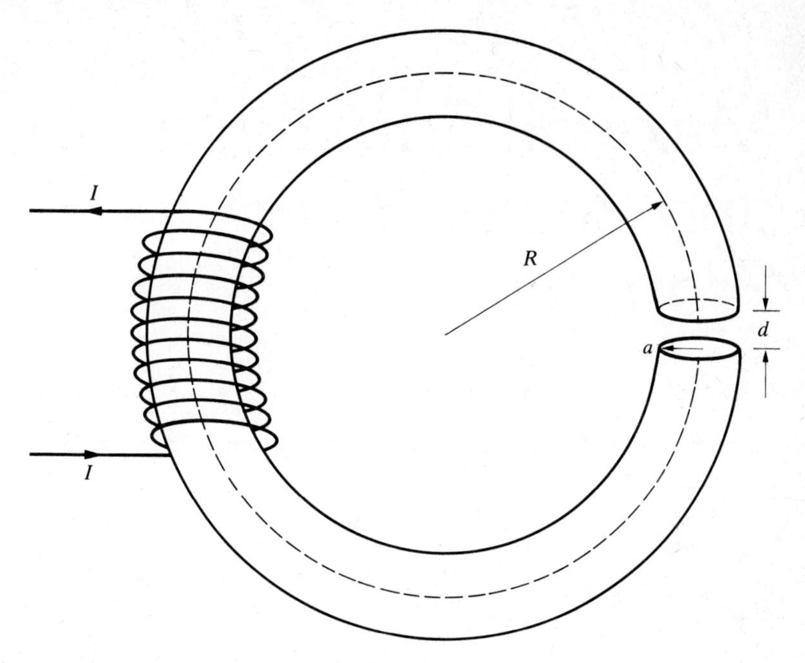


**FIGURE 7-24** Iron toroid with gap

## Section 7-6: Microscopic Magnetization

**7-20** Calculate the diamagnetic susceptibility for nitrogen at 1 atm and 20°C. Compare this value with the experimental value of $-5.0 \times 10^{-9}$.

**7-21** A long solenoid has 1000 turns/m, current $I = 10$ A, and a magnetic field inside of $6 \times 10^{-3}$ T. Is the solenoid filled with diamagnetic or paramagnetic material? Explain.

## Section 7-7: Magnetic Circuits

**7-22** A C magnet is shown in Figure 7–23. All dimensions are in centimeters. The relative permeability of the soft Fe yoke is 3000. A current $I = 1$ A is to produce a field of about 100 G in the gap. How many turns of wire are required?

**7-23** A closely wound toroid of iron has $N$ turns and carries a current $I$. The cross-sectional radius of the toroid is $a$, and the mean radius is $R$, as shown in Figure 7–24. An air gap of mean length $d$ has been cut from the toroid. The current $I$ is sufficiently large so that the magnetization $M$ takes its saturation value $M_s$. Find $H$ and $B$ in the iron and in the gap. When the current is shut off, the magnetization falls to its remanence value $M_r$. Again, find $H$ and $B$ in the iron and in the gap.

# Maxwell Equations and Plane Wave Propagation

CHAPTER CONTENTS

A physical theory is expected to satisfy three criteria. First, the physical quantities must obey a set of local laws, which are usually formulated as differential equations. Second, the state of some region at a certain time and the effects that propagate into the region at later times must be specified by a set of boundary conditions. Finally, any law of the theory should be invariant under Lorentz space-time transformations; that is, its mathematical form must be the same for any observer in any inertial frame, as will be discussed in Chapter 11. The fundamental set of differential equations for electric and magnetic fields, formulated by James Clerk Maxwell in 1865 as a generalization of several empirical physical laws, is a complete theory of electromagnetism. Over a century after its formulation no deviations from its predictions have been discovered, and it is reasonable to believe that Maxwell's theory is exact.

Shortly after formulating the fundamental equations of electromagnetism, Maxwell realized that the theory allowed the possibility of wave solutions for **E** and **B**. The predicted velocity of the electromagnetic waves was the same as the observed velocity of light, within measurement errors. Maxwell suggested that light waves are electromagnetic waves of a particular frequency range. This revolutionary idea put optics on a firm theoretical basis, as a part of electromagnetism. Maxwell's theory of waves was further confirmed in a series of early experiments in 1887 by Heinrich Hertz, who produced waves by electrical means, verified their wave nature by interference, and by comparing their frequency and wavelength, deduced that their speed was the same as that of light waves.

We will motivate Maxwell equations from a semihistorical perspective by using Coulomb's law, the apparent absence of magnetic charges in nature, Faraday's law of induction, and the Biot-Savart law.

We deduce the form of the Maxwell equations first in empty space and then in matter, and we study their wave solutions. Then we derive energy and momentum conservation laws from the Maxwell theory. At the end of the chapter we briefly introduce some quantum effects in electromagnetism.

## 8-1 MAXWELL EQUATIONS IN EMPTY SPACE

In preceding chapters we deduced from empirical laws the following differential relations for electric and magnetic fields in empty space. From Coulomb's law of electrostatics we found

$$\nabla \cdot \mathbf{E}(\mathbf{r}) = \frac{\rho(\mathbf{r})}{\varepsilon_0} \qquad \textbf{(8-1)}$$

and

$$\nabla \times \mathbf{E}(\mathbf{r}) = 0 \qquad \textbf{(8-2)}$$

From the Biot-Savart law of magnetostatics we found

$$\nabla \cdot \mathbf{B}(\mathbf{r}) = 0 \tag{8-3}$$

and

$$\nabla \times \mathbf{B}(\mathbf{r}) = \mu_0 \mathbf{J}(\mathbf{r}) \tag{8-4}$$

From Faraday's law of induction we found

$$\nabla \times \mathbf{E}(\mathbf{r}, t) \equiv -\frac{\partial \mathbf{B}}{\partial t}(\mathbf{r}, t) \tag{8-5}$$

Starting with this set of differential equations, we try to find a consistent generalization for the general time-dependent case that will be valid for all electromagnetic phenomena; this generalization was Maxwell's accomplishment.

Gauss's law in differential form, Equation (8–1), relates the electric field to the electric charge density. We postulate that this source equation has the time-dependent generalization

**Maxwell I**

$$\nabla \cdot \mathbf{E}(\mathbf{r}, t) = \frac{\rho(\mathbf{r}, t)}{\varepsilon_0} \tag{8-6}$$

According to Equation (8–3), the magnetic field has no magnetic charge source. Magnetic fields are found to originate entirely from currents, which are the source term in Equation (8–4). Extensive experimental searches have been made for magnetic charges (called magnetic monopoles), but none have been found. We postulate therefore that Equation (8–3) is generally valid:

**Maxwell II**

$$\nabla \cdot \mathbf{B}(\mathbf{r}, t) = 0 \tag{8-7}$$

The Biot-Savart law for the magnetic field produced by a slowly varying current satisfies Equation (8–7), as is evident from the form of Equation (5–60):

$$\nabla \cdot \mathbf{B} = \nabla \cdot \nabla \times \mathbf{A} = \nabla \times \nabla \cdot A = 0$$

The integrated form of Equation (8–7) is

$$\oint \mathbf{B} \cdot d\mathbf{S} = 0 \tag{8-8}$$

Since there are no sources of magnetic flux, magnetic field lines are continuous, unlike electric field lines, which can begin on positive charges and end on negative charges.

We can show from Equation (8–8) that the magnetic flux is the same through all surfaces that have the same bounding circuit. Consider two surfaces $S_1$ and $S_2$ with normals $\hat{\mathbf{n}}_1$ and $\hat{\mathbf{n}}_2$ defined in the sense of the right-hand rule with respect to the direction about the circuit $C$, as illustrated in Figure 8–1. From Equation (8–8) the net outward flux through the closed surface $S = S_1 + S_2$ is

$$\oint_{S=S_1+S_2} \mathbf{B} \cdot \hat{\mathbf{n}}\, dS = 0 \tag{8–9}$$

Since the outward normal is $\hat{\mathbf{n}}_1$ on $S_1$ and $-\hat{\mathbf{n}}_2$ on $S_2$, Equation (8–9) implies

$$\int_{S_1} \mathbf{B} \cdot \hat{\mathbf{n}}_1\, dS = \int_{S_2} \mathbf{B} \cdot \hat{\mathbf{n}}_2\, dS \tag{8–10}$$

Thus one may speak of the magnetic flux through the circuit and calculate it as the flux through any conveniently chosen surface bounded by the circuit. The flux through a circuit is also referred to as the flux linking the circuit.

The curl of **E** in Equation (8–5) reduces to Equation (8–2) in the static case; we assume that it is correctly specified in all cases by

**Maxwell III**

$$\nabla \times \mathbf{E}(\mathbf{r}, t) = -\frac{\partial \mathbf{B}}{\partial t}(\mathbf{r}, t) \tag{8–11}$$

The divergence of the left-hand side of this equation is identically zero; consistent with this result, the divergence of the right-hand side also vanishes by Equation (8–7).

We finally turn to Equation (8–4) for the curl of **B**. This equation was obtained from the Biot-Savart law based on the assumption of steady currents:

$$\nabla \cdot \mathbf{J}(\mathbf{r}) = 0 \tag{8–12}$$

Taking the divergence of Equation (8–4) reproduces this restriction. But the

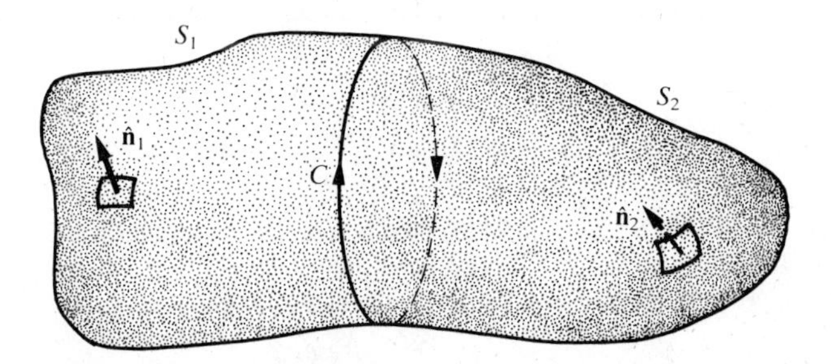


**FIGURE 8–1** Two surfaces with the same bounding circuit. The direction of the surface normals, determined by the right-hand rule, is outward for $S_1$ and inward for $S_2$.

continuity equation, Equation (5–14), requires that

$$\nabla \cdot \mathbf{J}(\mathbf{r}, t) = \frac{\partial \rho(\mathbf{r}, t)}{\partial t} \tag{8-13}$$

so some modification of Equation (8–4) is necessary when $\rho$ is time-dependent. Using Equation (8–6), we can write the continuity equation in the form

$$\nabla \cdot \left[ \mathbf{J}(\mathbf{r}, t) + \varepsilon_0 \frac{\partial \mathbf{E}}{\partial t}(\mathbf{r}, t) \right] = 0 \tag{8-14}$$

This equation suggests that the appropriate modification of Equation (8–4), compatible with the continuity equation, is

**Maxwell IV**

$$\nabla \times \mathbf{B}(\mathbf{r}, t) = \mu_0 \left[ \mathbf{J}(\mathbf{r}, t) + \varepsilon_0 \frac{\partial \mathbf{E}}{\partial t}(\mathbf{r}, t) \right] \tag{8-15}$$

With this modification the divergence of both sides of the equation vanishes.

In the preceding equations $\mathbf{J}$ is the conduction current density and

$$\varepsilon_0 \frac{\partial \mathbf{E}}{\partial t} \equiv \mathbf{J}_\mathrm{D} \tag{8-16}$$

is called the displacement current density. The divergence of the right-hand side of Equation (8–15) vanishes,

$$\nabla \cdot (\mathbf{J} + \mathbf{J}_\mathrm{D}) = 0 \tag{8-17}$$

and so the sum of the charge current and displacement current,

$$I + I_\mathrm{D} = \int (\mathbf{J} + \mathbf{J}_\mathrm{D}) \cdot d\mathbf{S} \tag{8-18}$$

**TABLE 8-1** Maxwell's equations in empty space

| Differential Forms | Integral Forms | |
|---|---|---|
| I. $\nabla \cdot \mathbf{E} = \dfrac{\rho}{\varepsilon_0}$ | I. $\oint_S \mathbf{E} \cdot d\mathbf{S} = \dfrac{Q}{\varepsilon_0}$ | (Gauss's law) |
| II. $\nabla \cdot \mathbf{B} = 0$ | II. $\oint_S \mathbf{B} \cdot d\mathbf{S} = 0$ | |
| III. $\nabla \times \mathbf{E} = -\dfrac{\partial \mathbf{B}}{\partial t}$ | III. $\oint_C \mathbf{E} \cdot d\mathbf{r} = -\dfrac{d}{dt}\int_S \mathbf{B} \cdot d\mathbf{S}$ | (Faraday's law) |
| IV. $\nabla \times \mathbf{B} = \mu_0\left(\mathbf{J} + \varepsilon_0 \dfrac{\partial \mathbf{E}}{\partial t}\right)$ | IV. $\oint_C \mathbf{B} \cdot d\mathbf{r} = \mu_0 \int_S \left(\mathbf{J} + \varepsilon_0 \dfrac{\partial \mathbf{E}}{\partial t}\right) \cdot d\mathbf{S}$ | |

is the same through any open surface bounding a given circuit. The proof of this statement parallels the demonstration of the equality of magnetic flux through different open surfaces bounded by the same circuit.

In this section we have succeeded in writing consistent expressions for the divergence and the curl of both **E** and **B**. The four equations that relate the fields in empty space to the electric charge and current sources are summarized in Table 8–1. They are given in both differential and integral forms. These equations were developed from a set of experimental laws of restricted domain of applicability, but this generalization provides a complete explanation of all electromagnetic phenomena observed thus far. The formulation of these equations by Maxwell represents a triumph of phenomenology without equal in the history of physics.

## 8-2 DISPLACEMENT CURRENT IN A CIRCUIT CONTAINING A CAPACITOR

To better understand the role of the displacement current, we consider a parallel-plate capacitor that is charged by a constant current $I$, as shown in Figure 8–2. If the plates are large in comparison with their separation,

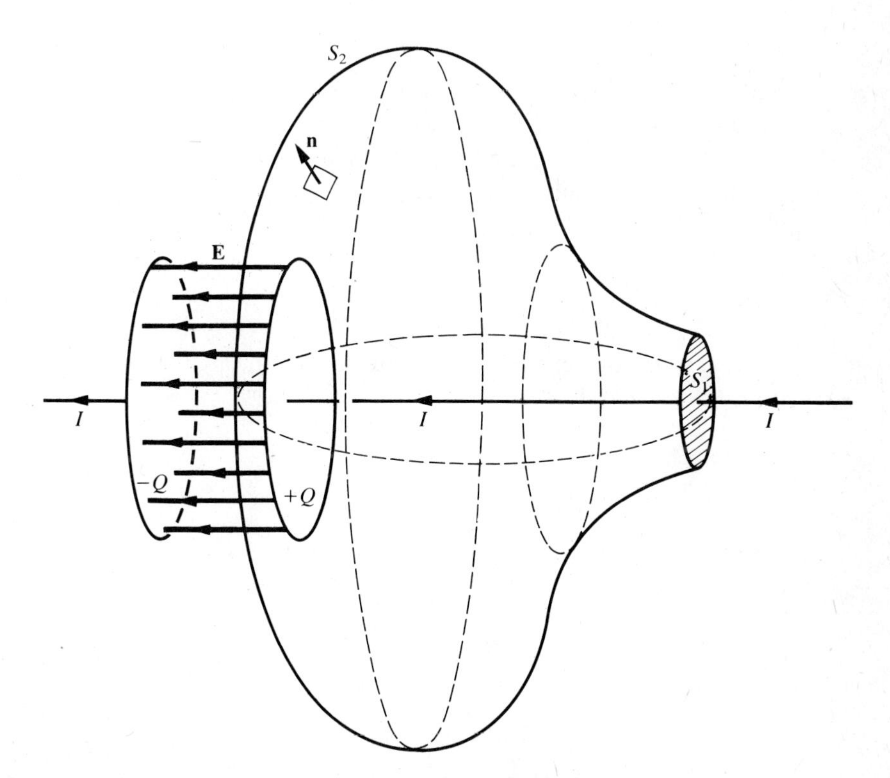


**FIGURE 8–2** Parallel-plate capacitor charged by a constant current $I$

there will be an **E** field only between the plates, and it will be uniform over most of the area of the plates to a good approximation; we assume this condition in the following development.

From Equation (8–15) we obtain the integral relation

$$\oint_C \mathbf{B} \cdot d\mathbf{r} = \mu_0 \int_S \left( \mathbf{J} + \varepsilon_0 \frac{\partial \mathbf{E}}{\partial t} \right) \cdot d\mathbf{S} \tag{8–19}$$

which is the generalized form of Ampère's circuital law. We apply this relation to a circuit $C$ around the wire leading to the capacitor, as illustrated in Figure 8–2. For the capping surface $S_1$ the right-hand side of Equation (8–19) gives $\mu_0 I$; the displacement current term,

$$I_D = \varepsilon_0 \int \frac{\partial \mathbf{E}}{\partial t} \cdot d\mathbf{S}_1 = 0$$

vanishes since **E** is constant in a wire carrying a constant current. For the same circuit $C$ but with the capping surface $S_2$ of Figure 8–2, the conduction current term $I = \int \mathbf{J} \cdot d\mathbf{S}_2$ vanishes, and the right-hand side of Equation (8–19) gives

$$I_D = \varepsilon_0 \int \frac{\partial \mathbf{E}}{\partial t} \cdot d\mathbf{S}_2 \tag{8–20}$$

In the region between the plates **E** increases with time as the plates accumulate charge.

The electric field between the plates can be found by Gauss's law. When we apply Gauss's law to a pillbox volume surrounding the positive plate, we find

$$\mathbf{E} = \frac{Q}{\varepsilon_0 S} \hat{\mathbf{n}} \tag{8–21}$$

where $Q$ is the total charge on the plate and $S$ is the area of the plate. Thus the time derivative of **E** is

$$\frac{\partial \mathbf{E}}{\partial t} = \frac{1}{\varepsilon_0 S} \frac{dQ}{dt} \hat{\mathbf{n}} = \frac{1}{\varepsilon_0 S} I \hat{\mathbf{n}} \tag{8–22}$$

and the total displacement current is

$$I_D = \varepsilon_0 \int \frac{\partial \mathbf{E}}{\partial t} \cdot d\mathbf{S}_2 = I \tag{8–23}$$

Hence for the circuit $C$ we obtain

$$\oint_C \mathbf{B} \cdot d\mathbf{r} = \mu_0 I$$

for either surface $S_1$ or surface $S_2$ when both conduction and displacement currents are taken into account in Ampère's law. This capacitor example shows concretely the necessity for the displacement current term in the fourth Maxwell equation.

## 8-3 PLANE WAVE SOLUTIONS IN EMPTY SPACE

There are solutions of Maxwell's equations that were not suspected before Maxwell's synthesis. We consider wave solutions for **E** and **B** in a region of space with no charges ($\rho = 0$) or currents ($\mathbf{J} = 0$). For simplicity we restrict our attention initially to fields varying in space only with the $z$ coordinate. Maxwell's equations with no source charges or currents are

$$\nabla \cdot \mathbf{E} = 0 \qquad \nabla \times \mathbf{E} = -\frac{\partial \mathbf{B}}{\partial t}$$
$$\nabla \cdot \mathbf{B} = 0 \qquad \nabla \times \mathbf{B} = \mu_0 \varepsilon_0 \frac{\partial \mathbf{E}}{\partial t} \tag{8–24}$$

We seek solutions of the type

$$\mathbf{E} = E(z, t)\hat{\mathbf{e}}_1 \qquad \mathbf{B} = B(z, t)\hat{\mathbf{e}}_2 \tag{8–25}$$

where $\hat{\mathbf{e}}_1$ and $\hat{\mathbf{e}}_2$ are constant unit vectors.

Substituting the trial solutions into Equations (8–24), we find

$$\frac{\partial E}{\partial z}\hat{\mathbf{z}} \cdot \hat{\mathbf{e}}_1 = 0 \qquad \text{(I)}$$
$$\frac{\partial B}{\partial z}\hat{\mathbf{z}} \cdot \hat{\mathbf{e}}_2 = 0 \qquad \text{(II)}$$
$$\frac{\partial E}{\partial z}\hat{\mathbf{z}} \times \hat{\mathbf{e}}_1 = -\frac{\partial B}{\partial t}\hat{\mathbf{e}}_2 \qquad \text{(III)}$$
$$\frac{\partial B}{\partial z}\hat{\mathbf{z}} \times \hat{\mathbf{e}}_2 = \mu_0 \varepsilon_0 \frac{\partial E}{\partial t}\hat{\mathbf{e}}_1 \qquad \text{(IV)} \tag{8–26}$$

From I and II, $\hat{\mathbf{e}}_1$ and $\hat{\mathbf{e}}_2$ must be in the $xy$ plane. Then from III or IV, $\hat{\mathbf{e}}_1$ and $\hat{\mathbf{e}}_2$ must be orthogonal. We choose our coordinate system such that

$$\hat{\mathbf{e}}_1 = \hat{\mathbf{x}} \qquad \hat{\mathbf{e}}_2 = \hat{\mathbf{y}} \tag{8–27}$$

Then III and IV become

$$\frac{\partial E}{\partial z} = -\frac{\partial B}{\partial t} \qquad \frac{\partial B}{\partial z} = -\mu_0 \varepsilon_0 \frac{\partial E}{\partial t} \tag{8–28}$$

These equations imply that changes in the electric field in the neighborhood of a given position are the cause of changes in the magnetic field at this point. Conversely, the magnetic field in the neighborhood of a given position causes the changes of the electric field at that position.

By further differentiation the pair of first-order equations can be converted to a single second-order differential equation for $E$ or $B$, as follows:

**Wave Equations for Fields in Empty Space**

$$\frac{\partial^2 E}{\partial z^2} - \frac{1}{c^2}\frac{\partial^2 E}{\partial t^2} = 0$$
$$\frac{\partial^2 B}{\partial z^2} - \frac{1}{c^2}\frac{\partial^2 B}{\partial t^2} = 0 \qquad \textbf{(8–29)}$$

Here we have defined the constant $c$ as

**Speed of Light in Empty Space**

$$c = \frac{1}{\sqrt{\mu_0 \varepsilon_0}} \qquad \textbf{(8–30)}$$

These equations are known as wave equations since their solutions are waves.

The solutions to the one-dimensional wave equation can be found by integration. We choose a new set of independent variables $\alpha$ and $\beta$ defined by

$$\alpha \equiv z - ct \qquad \beta \equiv z + ct \qquad \textbf{(8–31)}$$

By the chain rule we have

$$\frac{\partial}{\partial z} = \frac{\partial \alpha}{\partial z}\frac{\partial}{\partial \alpha} + \frac{\partial \beta}{\partial z}\frac{\partial}{\partial \beta} = \frac{\partial}{\partial \alpha} + \frac{\partial}{\partial \beta}$$
$$\frac{\partial}{\partial t} = \frac{\partial \alpha}{\partial t}\frac{\partial}{\partial \alpha} + \frac{\partial \beta}{\partial t}\frac{\partial}{\partial \beta} = -c\left(\frac{\partial}{\partial \alpha} - \frac{\partial}{\partial \beta}\right) \qquad \textbf{(8–32)}$$

The wave equation, (8–29), becomes

$$\left[\left(\frac{\partial}{\partial \alpha} + \frac{\partial}{\partial \beta}\right)^2 - \left(\frac{\partial}{\partial \alpha} - \frac{\partial}{\partial \beta}\right)^2\right] E(\alpha, \beta) = 0 \qquad \textbf{(8–33)}$$

or

$$\frac{\partial^2 E}{\partial \alpha\, \partial \beta} = 0 \qquad \textbf{(8–34)}$$

First, we integrate over $\beta$, to obtain

$$\frac{\partial E}{\partial \alpha} = h(\alpha) \qquad \textbf{(8–35)}$$

where $h(\alpha)$ is an arbitrary function of $\alpha$. Note that a "partial integration" of a partial differential equation yields an arbitrary function instead of an integration constant. Next, we integrate over $\alpha$, to get

$$E(\alpha, \beta) = \int^{\alpha} h(\alpha')\, d\alpha' + f_{-}(\beta) \tag{8–36}$$

where $f_{-}(\beta)$ is an arbitrary function from this partial integration. We write the solution in Equation (8–37) in the form

$$E = f_{+}(\alpha) + f_{-}(\beta) \tag{8–37}$$

where

$$f_{+}(\alpha) = \int^{\alpha} h(\alpha')\, d\alpha' \tag{8–38}$$

In terms of the original independent variables $z$ and $t$, the solution for $E(z, t)$ is

$$E(z, t) = f_{+}(z - ct) + f_{-}(z + ct) \tag{8–39}$$

This equation is the general solution to the wave equation, (8–30). This solution must also satisfy Equations (8–28), which give

$$\begin{aligned} \frac{\partial B}{\partial t} &= -[f'_{+}(z - ct) + f'_{-}(z + ct)] \\ \frac{\partial B}{\partial z} &= \frac{1}{c}[f'_{+}(z - ct) - f'_{-}(z + ct)] \end{aligned} \tag{8–40}$$

where primes denote differentiation with respect to the argument of the function. Integrating these equations, we find

$$B(z, t) = \frac{1}{c}[f_{+}(z - ct) - f_{-}(z + ct)] + \text{constant} \tag{8–41}$$

Assuming that static fields are absent, we can drop the integration constant.

If we put these results in Equations (8–25) and (8–27), **E** and **B** fields of the wave solution of Maxwell's equations are

$$\begin{aligned} \mathbf{E}(z, t) &= [f_{+}(z - ct) + f_{-}(z + ct)]\hat{\mathbf{x}} \\ \mathbf{B}(z, t) &= \frac{1}{c}[f_{+}(z - ct) - f_{-}(z + ct)]\hat{\mathbf{y}} \end{aligned} \tag{8–42}$$

The $f_{+}$ term represents a wave moving to the right, to larger values of $z$. To keep up with the wave, we must keep constant the argument of $f_{+}$, the phase $\alpha = z - ct$. At fixed $\alpha$ (i.e., constant phase), as time increases, the coordinate $z$ must also increase, as shown in Figure 8–3(a). So

$$d\alpha = 0 = dz - c\,dt$$

or

**Phase Velocity in Empty Space**

$$v_\mathrm{p} \equiv \frac{dz}{dt} = c \tag{8-43}$$

This velocity, called the phase velocity, is the speed of propagation of the wave in empty space. By a similar argument the $f_-$ term represents waves propagating toward negative $z$ with the phase velocity, as shown in Figure 8–3(b). Because the surface of constant phase $\alpha$ or $\beta$ at fixed time $t$ is a plane ($z$ = constant), these wave solutions are called *plane waves.*

Figure 8–4 shows **E** and **B** as a function of $z$ at a particular time for the $f_+$ solution. The directions of the **E** and **B** fields for such a traveling wave are perpendicular to the direction of propagation and to each other; the

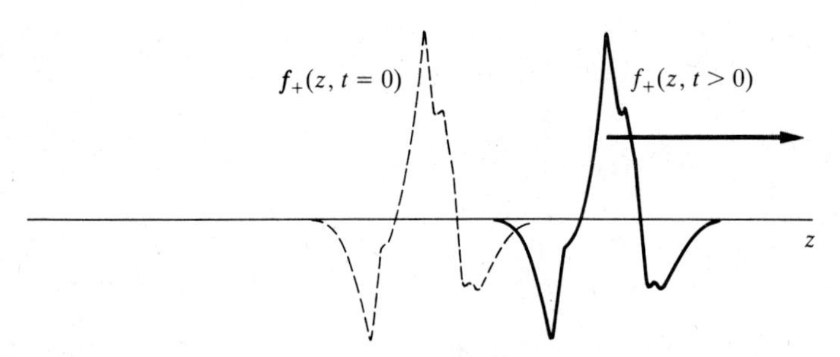


(a) Positive $z$ direction

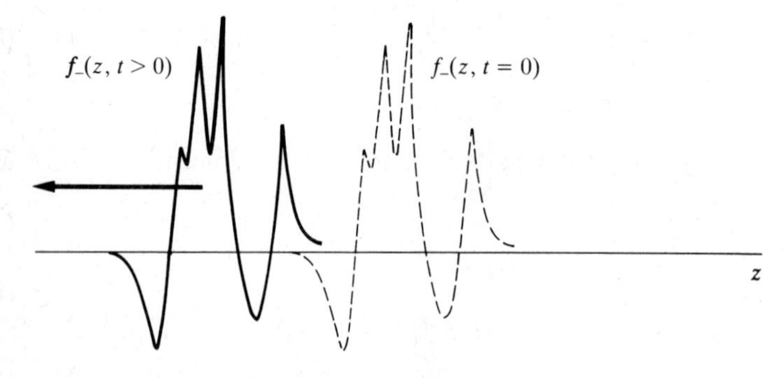


(b) Negative $z$ direction

**FIGURE 8–3** Wave solutions to Maxwell's equations, propagating in the (a) positive $z$ direction and (b) negative $z$ direction

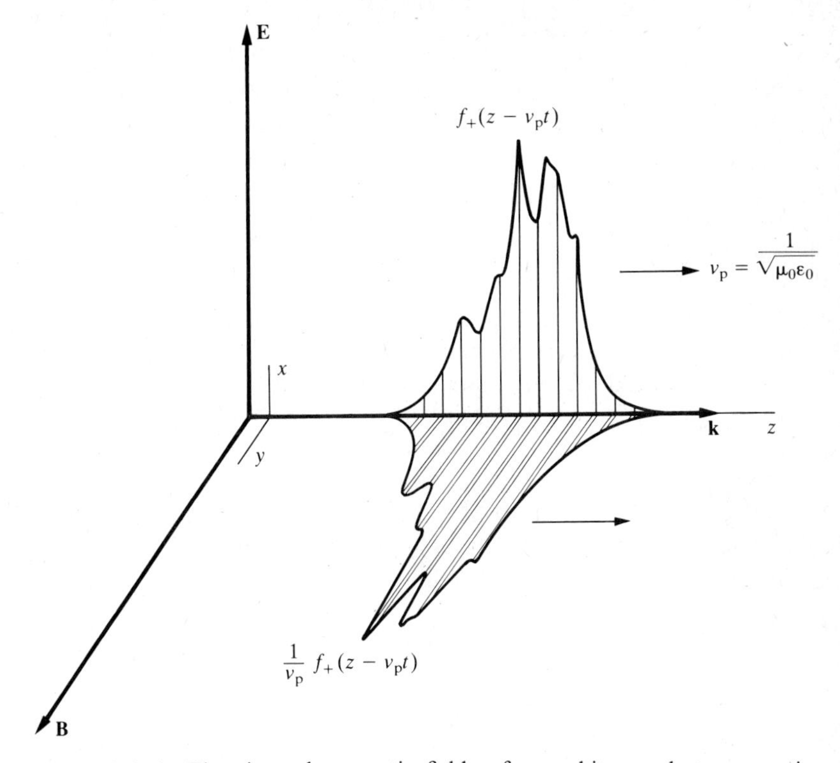


**FIGURE 8–4** Electric and magnetic fields of an arbitrary electromagnetic wave

cross product $\mathbf{E} \times \mathbf{B}$ points in the direction of propagation. These electromagnetic waves are called *transverse waves* because **E** and **B** are perpendicular to the direction of propagation. Since the electric vector points in the $\hat{\mathbf{x}}$ direction, this wave is called a *linearly polarized wave* in the $\hat{\mathbf{x}}$ direction. *Polarization* is usually measured through the response of charges to the electric field of the radiation, so it is the electric field that is used to define the polarization direction.

The speed of propagation of electromagnetic waves in a vacuum, Equation (8–43), is independent of frequency. Maxwell calculated $c$ from $\varepsilon_0$ and $\mu_0$ by using Equation (8–43) and noticed that it was about the same as the measured speed of light; he thus inferred that light was an electromagnetic wave. With the value of $\varepsilon_0$ from capacitance measurements and the defined value of $\mu_0$, the calculated value of the speed of electromagnetic waves is

$$c = \frac{1}{\sqrt{\mu_0 \varepsilon_0}} = (2.9971 \pm 0.0010) \times 10^8 \text{ m/s} \qquad \textbf{(8–44)}$$

[See E. B. Rosa and N. E. Dorsey, *Bureau of Standards Bull.*, 3: 439 (1907).] Recent direct measurements of the wavelength and the frequency of laser light yielded the value

$$c = 299{,}792{,}458.1 \text{ m/s} \tag{8-45}$$

[See, e.g., K. M. Baird, D. S. Smith, and B. G. Whitford, *Optics Comm.*, 31:367 (1979).] The predicted value of $c$ in Equation (8–44) from quasi-static measurements is in excellent accord with Equation (8–45). The speed of light is now measured more accurately than the meter can be, as it used to be defined. The meter is now *defined* by the distance in empty space traversed by light in $(299{,}792{,}458)^{-1}$ s. The speed of light has now become a fundamental unit, and the meter a derived one.

Since a consequence of Maxwell's equations in empty space is that the velocity of electromagnetic radiation is independent of frequency or wavelength, radio waves, radar, X rays, or any other electromagnetic wave should travel in empty space with the same velocity as visible light. An extremely sensitive test of this conclusion is provided by the periodic pulses of radiation coming from the pulsar in the Crab nebula. The fact that X-ray and visible-light pulses arrive at the earth in synchronism after traveling for 2000 years places an extremely strong upper limit on any dependence of the propagation speed on wavelength.

Simple harmonic waves constitute an important class of wave solutions. A *simple harmonic wave* traveling in the positive $z$ direction has the form

$$f_+(z - ct) = \mathscr{E}_+ \cos(kz - ckt + \delta_+) \tag{8-46}$$

Here $\mathscr{E}_+$, $\delta_+$, and $k$ are arbitrary constants. The constant $k$ is called the *wave number* and has units of radians per meter. The angular frequency is

**Angular Frequency**

$$\omega = ck \tag{8-47}$$

which has units of radians per second. The wavelength $\lambda$ of a simple harmonic wave is the spatial period:

**Wavelength**

$$\lambda = \frac{2\pi}{k} \tag{8-48}$$

The period $\tau$ is the temporal period:

**Period**

$$\tau = \frac{2\pi}{\omega} \tag{8-49}$$

The wavelength and period are illustrated in Figure 8–5.

The frequency of a periodic wave is

**Frequency**

$$\nu = \frac{1}{\tau} = \frac{\omega}{2\pi} \tag{8-50}$$

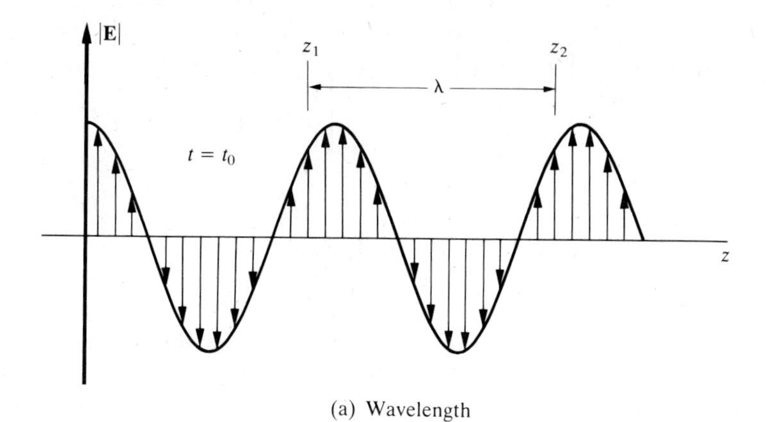


(a) Wavelength

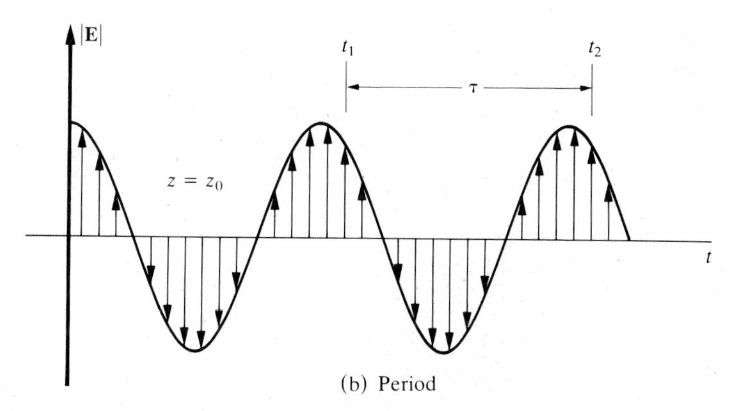


(b) Period

**FIGURE 8–5** Characteristics of simple harmonic waves. (a) The wavelength $\lambda$ is the distance in space in which a wave repeats itself at a given time. (b) The period $\tau$ is the time it takes for a wave to repeat itself at a given position.

with units of hertz = 1 cycle per second. The observed electromagnetic spectrum is illustrated in Figure 8–6; wavelengths of various parts of the spectrum are summarized in Table 8–2. The simple harmonic solutions are called *monochromatic waves*, because visible light of a single color has a definite wavelength.

For a monochromatic wave the general one-dimensional solution for plane polarized waves can be written as

$$\mathbf{E}(\mathbf{r}, t) = \hat{\mathbf{x}}[\mathscr{E}_+ \cos(kz - \omega t + \delta_+) + \mathscr{E}_- \cos(-kz - \omega t + \delta_-)] \tag{8-51}$$

Since $\cos\theta$ is the real part of the complex exponential $e^{i\theta}$,

$$\cos\theta = \mathrm{Re}(e^{i\theta})$$

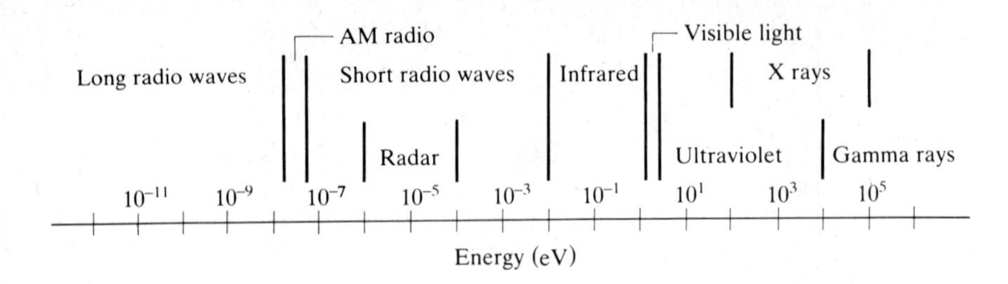


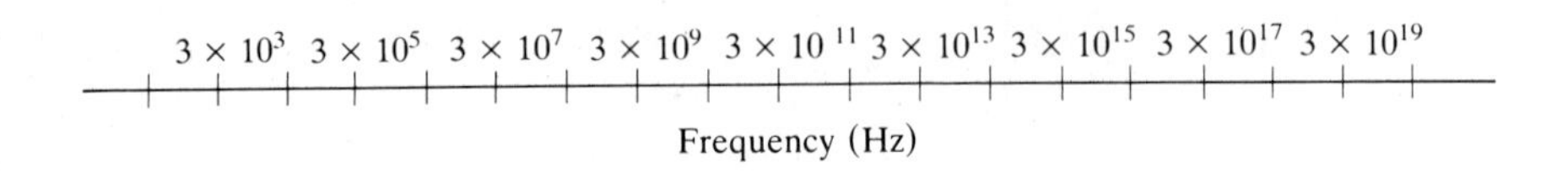


$10^5$ $10^3$ $10^1$ $10^{-1}$ $10^{-3}$ $10^{-5}$ $10^{-7}$ $10^{-9}$ $10^{-11}$

Wavelength (m)

**FIGURE 8-6** Electromagnetic spectrum

we can write Equation (8–51) in complex notation as

$$\mathbf{E}(\mathbf{r}, t) = \text{Re}[\hat{\mathbf{x}}\mathscr{E}_+ e^{i(kz - \omega t + \delta_+)} + \hat{\mathbf{x}}\mathscr{E}_- e^{i(-kz - \omega t + \delta_-)}] \qquad \textbf{(8–52)}$$

If we now define

$$\boldsymbol{\mathscr{E}}_\pm \equiv \hat{\mathbf{x}}\mathscr{E}_\pm e^{i\delta_\pm} \qquad \textbf{(8–53)}$$

we can express Equation (8–52) as

$$\mathbf{E} = \text{Re}[\boldsymbol{\mathscr{E}}_+ e^{i(kz - \omega t)} + \boldsymbol{\mathscr{E}}_- e^{i(-kz - \omega t)}] \qquad \textbf{(8–54)}$$

When the complex representation is used, the physical electromagnetic field vector is always understood to be obtained by taking the real part of the complex form.

Thus far we have considered a linearly polarized plane wave propagating along the $z$ axis in empty space. The most general simple harmonic plane wave has the complex form

$$\mathbf{E}(\mathbf{r}, t) = \boldsymbol{\mathscr{E}} e^{i(\mathbf{k}\cdot\mathbf{r} - \omega t)} \qquad \mathbf{B}(\mathbf{r}, t) = \boldsymbol{\mathscr{B}} e^{i(\mathbf{k}\cdot\mathbf{r} - \omega t)} \qquad \textbf{(8–55)}$$

It represents a wave propagating in the direction

$$\mathbf{k} = k_x\hat{\mathbf{x}} + k_y\hat{\mathbf{y}} + k_z\hat{\mathbf{z}} \qquad \textbf{(8–56)}$$

**TABLE 8-2** Electromagnetic spectrum

| Type of Radiation | $\lambda$ (m) | $\nu$ (Hz) | Energy (eV) |
|---|---|---|---|
| Gamma rays | $0\text{–}10^{-10}$ | $3 \times 10^{18}\text{–}\infty$ | $1.2 \times 10^{4}\text{–}\infty$ |
| $^{60}$Cobalt | $\approx 10^{-12}$ | $2.9 \times 10^{20}$ | $\approx 1.2 \times 10^{6}$ |
| X rays | $10^{-8}\text{–}10^{-11}$ | $3 \times 10^{16}\text{–}3 \times 10^{19}$ | $120\text{–}1.2 \times 10^{5}$ |
| Dental X rays | $\approx 4.9 \times 10^{-11}$ | $\approx 6.1 \times 10^{18}$ | $\approx 2.5 \times 10^{4}$ |
| Ultraviolet | $3.9 \times 10^{-7}\text{–}4 \times 10^{-9}$ | $7.7 \times 10^{14}\text{–}7.5 \times 10^{16}$ | 3.2–310 |
| Visible light | $7.6 \times 10^{-7}\text{–}3.9 \times 10^{-7}$ | $3.9 \times 10^{14}\text{–}7.7 \times 10^{14}$ | 1.6–3.2 |
| Blue (mercury vapor lamp) | $4.36 \times 10^{-7}$ | $6.88 \times 10^{14}$ | 2.84 |
| Yellow (sodium vapor lamp) | $5.89 \times 10^{-7}$ | $5.09 \times 10^{14}$ | 2.11 |
| Red (helium-neon laser) | $6.33 \times 10^{-7}$ | $4.74 \times 10^{14}$ | 1.96 |
| Infrared | $10^{-4}\text{–}7.6 \times 10^{-7}$ | $3 \times 10^{12}\text{–}3.9 \times 10^{14}$ | 0.012–1.6 |
| Blackbody radiation | | | |
| From sun's surface | $\approx 10^{-6}$ | $\approx 3 \times 10^{14}$ | $\approx 1.2$ |
| At room temperature | $2 \times 10^{-5}$ | $\approx 1.5 \times 10^{13}$ | $\approx 0.06$ |
| Microwaves | $0.3\text{–}3 \times 10^{-4}$ | $10^{9}\text{–}10^{12}$ | $4.1 \times 10^{-6}\text{–}4.1 \times 10^{-3}$ |
| 3°K background radiation | $\approx 2 \times 10^{-3}$ | $\approx 1.5 \times 10^{11}$ | $6.2 \times 10^{-4}$ |
| Radar | 0.1 | $3 \times 10^{9}$ | $1.2 \times 10^{-4}$ |
| Microwave oven | 0.12 | $2.45 \times 10^{9}$ | $1.0 \times 10^{-5}$ |
| Interstellar hydrogen line | 0.21 | $1.43 \times 10^{9}$ | $5.9 \times 10^{-6}$ |
| Radio waves | $3 \times 10^{4}\text{–}10^{-3}$ | $3 \times 10^{11}\text{–}10^{4}$ | $4.1 \times 10^{-11}\text{–}1.2 \times 10^{-3}$ |
| UHF TV | 0.75–0.37 | $4 \times 10^{8}\text{–}8.1 \times 10^{8}$ | $1.6 \times 10^{-6}\text{–}3.3 \times 10^{-6}$ |
| VHF TV | 5.5–1.5 | $5.4 \times 10^{7}\text{–}2 \times 10^{8}$ | $2.2 \times 10^{-7}\text{–}8.3 \times 10^{-7}$ |
| FM radio | 100–10 | $3 \times 10^{6}\text{–}3 \times 10^{7}$ | $1.2 \times 10^{-7}\text{–}1.2 \times 10^{-8}$ |
| AM radio | 600–200 | $5 \times 10^{5}\text{–}1.5 \times 10^{6}$ | $2.1 \times 10^{-9}\text{–}6.2 \times 10^{-9}$ |
| ELF (submarine communication) | $\approx 4 \times 10^{6}$ | $\approx 75$ | $\approx 3.1 \times 10^{-13}$ |

Here $\mathscr{E}$ and $\mathscr{B}$ are constant vectors that have complex components. Substituting this solution into the Maxwell equations, we find that the following conditions must be satisfied:

$$\begin{aligned} &\mathbf{k}\cdot\mathscr{E} = 0 && \text{(I)} \\ &\mathbf{k}\cdot\mathscr{B} = 0 && \text{(II)} \\ &\mathbf{k}\times\mathscr{E} = \omega\mathscr{B} && \text{(III)} \\ &\mathbf{k}\times\mathscr{B} = -\frac{\omega}{c^2}\mathscr{E} && \text{(IV)} \end{aligned} \tag{8-57}$$

We can see that $\mathbf{k}$, $\mathscr{E}$, and $\mathscr{B}$ must form an orthogonal set. Taking the cross product of $\mathbf{k}$ with III and using IV to eliminate $\mathbf{k}\times\mathscr{B}$, we obtain

$$\mathbf{k}\times(\mathbf{k}\times\mathscr{E}) = -\frac{\omega^2}{c^2}\mathscr{E} \tag{8-58}$$

Expanding the triple cross product and using I, we find the condition

$$\omega = ck \tag{8-59}$$

Thus the general monochromatic plane wave solution is

$$\begin{aligned} \mathbf{E}(\mathbf{r}, t) &= \mathscr{E}e^{i(\mathbf{k}\cdot\mathbf{r}-\omega t)} \\ \mathbf{B}(\mathbf{r}, t) &= \frac{1}{\omega}\mathbf{k}\times\mathbf{E}(\mathbf{r}, t) \end{aligned} \tag{8-60}$$

where the relation for $\mathbf{B}$ follows from III. This solution is subject to the condition $\mathbf{k}\cdot\mathscr{E} = 0$.

The direction of propagation of a plane wave is defined to be the direction along which the phase of the plane wave has its maximum rate of change. Since $d\alpha = \mathbf{k}\cdot d\mathbf{r} - \omega\, dt$, the direction of propagation of the plane wave in Equation (8–60) is along $\mathbf{k}$. The vector $\mathbf{k}$ is called the *propagation vector* or *wave number vector.*

A superposition of waves of the same amplitude and frequency but traveling in opposite directions is called a *standing wave.* Standing waves arise when waves are reflected by material boundaries, as in wave guides or cavities.

## 8-4 LINEAR AND CIRCULAR POLARIZED PLANE WAVES

The plane wave described by Equation (8–60) is said to be linearly polarized if the real and imaginary parts of $\mathscr{E}$ are in the same direction, as in Equation (8–53). By superposition of two such electric fields that are linearly polarized along orthogonal transverse directions $\boldsymbol{\mathscr{E}}_1$ and $\boldsymbol{\mathscr{E}}_2$, we can construct a more general plane wave solution:

$$\mathbf{E} = (\boldsymbol{\mathscr{E}}_1 + \boldsymbol{\mathscr{E}}_2)e^{i(\mathbf{k}\cdot\mathbf{r} - \omega t)}$$
$$\mathbf{B} = \frac{1}{\omega}\mathbf{k} \times (\boldsymbol{\mathscr{E}}_1 + \boldsymbol{\mathscr{E}}_2)e^{i(\mathbf{k}\cdot\mathbf{r} - \omega t)} \tag{8-61}$$

If $\boldsymbol{\mathscr{E}}_1$ and $\boldsymbol{\mathscr{E}}_2$ have the same complex phase, then Equation (8–61) is again a linearly polarized wave, with polarization at an angle $\theta = \tan^{-1}(\mathscr{E}_2/\mathscr{E}_1)$ from $\boldsymbol{\mathscr{E}}_1$ and with magnitude $\mathscr{E} = (\mathscr{E}_1^2 + \mathscr{E}_2^2)^{1/2}$, as illustrated in Figure 8–7. We note that the magnitude $\mathscr{E}$ of a complex vector $\boldsymbol{\mathscr{E}}$ is given by

$$\mathscr{E} = (\boldsymbol{\mathscr{E}} \cdot \boldsymbol{\mathscr{E}}^*)^{1/2} \tag{8-62}$$

where $\boldsymbol{\mathscr{E}}$ is the complex conjugate of $\boldsymbol{\mathscr{E}}$. When $\boldsymbol{\mathscr{E}}_1$ and $\boldsymbol{\mathscr{E}}_2$ are along orthogonal directions, with equal magnitudes $\mathscr{E}_1 = \mathscr{E}_2$, and have phases that differ by 90°, *circular polarization* results.

To illustrate, let us take $\mathbf{k} = k\hat{\mathbf{z}}$ and

$$\boldsymbol{\mathscr{E}}_1 = \mathscr{E}\hat{\mathbf{x}} \qquad \boldsymbol{\mathscr{E}}_2 = e^{i\pi/2}\mathscr{E}\hat{\mathbf{y}} = i\mathscr{E}\hat{\mathbf{y}} \tag{8-63}$$

with $\mathscr{E}$ real, for which

$$\mathbf{E} = (\hat{\mathbf{x}} - i\hat{\mathbf{y}})\mathscr{E}e^{i(kz - \omega t)} \tag{8-64}$$

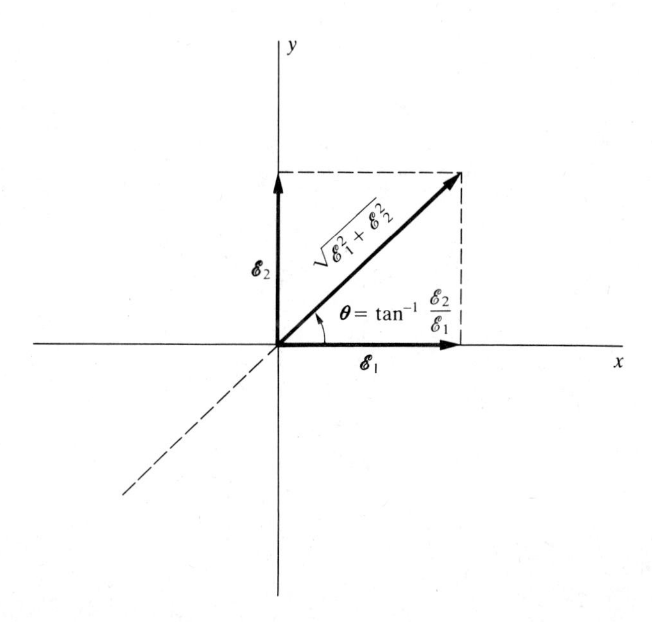


**FIGURE 8–7** Superposition of two linearly polarized waves with the same complex phase

The physical electric field is given by the real part of Equation (8–64). Its components are

$$E_x = \mathscr{E}\cos(kz - \omega t)$$
$$E_y = -\mathscr{E}\sin(kz - \omega t) \tag{8-65}$$

The magnitude of the electric vector is constant,

$$E = \sqrt{E_x^2 + E_y^2} = \mathscr{E} \tag{8-66}$$

but the components oscillate with $z$ and $t$. At a given position the electric vector sweeps out a circle in the $xy$ plane with angular frequency $\omega$, as indicated in Figure 8–8(a). Facing the wave, the electric field rotates counterclockwise. In optics this wave is called a *left circularly polarized wave.* If we had chosen $\boldsymbol{\mathscr{E}}_2 = -i\mathscr{E}\hat{\mathbf{y}}$ instead in Equation (8–63), the rotation would

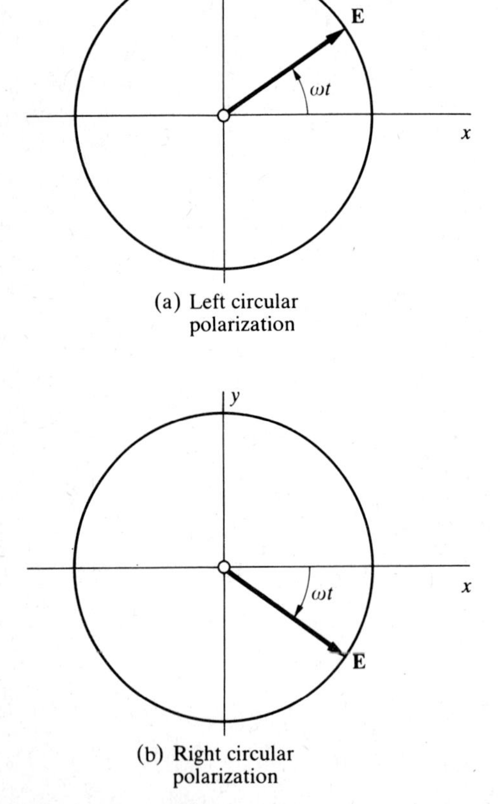


**FIGURE 8–8** Schematic drawing of (a) left circular polarization and (b) right circular polarization

have been clockwise as viewed looking into the oncoming wave; this choice corresponds to *right circular polarization*, as illustrated in Figure 8–8(b).

In summary, the polarized electric fields are the real parts of the following expressions:

**Plane Polarized Along $\hat{\mathbf{x}}$**

$$\mathbf{E} = \mathscr{E}\hat{\mathbf{x}}^{i(kz-\omega t)} \tag{8-67a}$$

**Left Circular Polarized**

$$\mathbf{E} = \mathscr{E}(\hat{\mathbf{x}} + i\hat{\mathbf{y}})e^{i(kz-\omega t)} \tag{8-67b}$$

**Right Circular Polarized**

$$\mathbf{E} = \mathscr{E}(\hat{\mathbf{x}} - i\hat{\mathbf{y}})e^{i(kz-\omega t)} \tag{8-67c}$$

These expressions are for waves propagating toward increasing $z$. In more general circumstances $\mathscr{E}_1$ and $\mathscr{E}_2$ have different phases or magnitudes. The electric vector then sweeps out an ellipse, and the wave is called *elliptically polarized.*

# 8-5 MAXWELL EQUATIONS IN MATTER

For the description of electromagnetism in the presence of matter, the Maxwell equations can be modified to include contributions from charges and currents of the matter. As in Chapter 4, the total charge is separated into the polarization charge density $\rho_P$ and the free charge density $\rho_f$. The polarization charge density is

$$\rho_P = -\nabla \cdot \mathbf{P} \tag{8-68}$$

where the polarization vector $\mathbf{P}$ is the electric dipole moment per unit volume in the dielectric, as discussed in Section 4–2. In a nonstatic electric field the polarization charge density will change with time, and there will be an associated polarization current density $\mathbf{J}_P$. The polarization charge and current densities must satisfy an equation of continuity:

$$\nabla \cdot \mathbf{J}_P = -\frac{\partial \rho_P}{\partial t} = \nabla \cdot \frac{\partial \mathbf{P}}{\partial t} \tag{8-69}$$

Thus we identify the polarization current density:

**Polarization Current Density**

$$\mathbf{J}_P = \frac{\partial \mathbf{P}}{\partial t} \tag{8-70}$$

Another contribution to the total current density is the magnetization current density $\mathbf{J}_\mathrm{M}$, found in Section 7–5 to be

$$\mathbf{J}_\mathrm{M} = \nabla \times \mathbf{M} \tag{8-71}$$

where the magnetism vector $\mathbf{M}$ is the magnetic dipole moment per unit volume. Since $\nabla \cdot \mathbf{J}_\mathrm{M} = 0$ and $\rho_\mathrm{M} = 0$ (i.e., no magnetic charge), the continuity equation is satisfied for these quantities.

A final contribution to the total current density is the current density $\mathbf{J}_\mathrm{f}$ due to the motion of free charges. Including the above matter charge densities and currents in the Maxwell equations of Table 8–1, we arrive at

$$\nabla \cdot \mathbf{E} = \frac{\rho_\mathrm{f} - \nabla \cdot \mathbf{P}}{\varepsilon_0} \qquad \nabla \times \mathbf{E} = -\frac{\partial \mathbf{B}}{\partial t} \qquad \nabla \cdot \mathbf{B} = 0$$
$$\nabla \times \mathbf{B} = \mu_0 \left( \varepsilon_0 \frac{\partial \mathbf{E}}{\partial t} + \frac{\partial \mathbf{P}}{\partial t} + \nabla \times \mathbf{M} + \mathbf{J}_\mathrm{f} \right) \tag{8-72}$$

We can introduce auxiliary fields

$$\mathbf{D} = \varepsilon_0 \mathbf{E} + \mathbf{P} \qquad \mathbf{H} = \frac{1}{\mu_0} \mathbf{B} - \mathbf{M} \tag{8-73}$$

as in Chapters 4 and 7 and recast the first and fourth Maxwell equations as

$$\nabla \cdot \mathbf{D} = \rho_\mathrm{f} \qquad \nabla \times \mathbf{H} = \mathbf{J}_\mathrm{f} + \frac{\partial \mathbf{D}}{\partial t} \tag{8-74}$$

In an isotropic dielectric $\mathbf{P}$ is parallel to $\mathbf{E}$, and in an isotropic magnetic material $\mathbf{M}$ is parallel to $\mathbf{B}$. In such cases we can write the material equations

$$\mathbf{D} = \varepsilon \mathbf{E} \qquad \mathbf{H} = \frac{1}{\mu} \mathbf{B} \tag{8-75}$$

A summary of the Maxwell equations in matter is given in Table 8–3.

When two regions possessing different electromagnetic properties are adjacent, it is often useful to solve for the fields in each and then relate the solutions by the proper *boundary conditions*. These boundary relations have been previously discussed for static electric fields (Section 4–3) and magnetic fields (Section 7–5). As we now show, these boundary conditions are valid for the general case also.

Since the integral form of Maxwell I is the same with time-varying fields, the normal component of $\mathbf{D}$ changes by the free surface charge density, as in Equation (4–51). When we apply Maxwell III (integral form) to the circuit of Figure 4–12(b), we now must include a term proportional to the time rate of change of the magnetic flux. Since the circuit of Figure 4–12(b)

**TABLE 8-3** Maxwell's equations in matter

**Differential Forms**

**I.** $\nabla \cdot \mathbf{D} = \rho_f \qquad \mathbf{D} = \varepsilon_0 \mathbf{E} + \mathbf{P}$

**II.** $\nabla \cdot \mathbf{B} = 0$

**III.** $\nabla \times \mathbf{E} = -\dfrac{\partial \mathbf{B}}{\partial t}$

**IV.** $\nabla \times \mathbf{H} = \mathbf{J}_f + \dfrac{\partial \mathbf{D}}{\partial t} \qquad \mathbf{H} = \dfrac{1}{\mu_0}\mathbf{B} - \mathbf{M}$

**Integral Forms**

**I.** $\oint_S \mathbf{D} \cdot d\mathbf{S} = Q_f$ (Gauss's law for dielectrics)

**II.** $\oint \mathbf{B} \cdot d\mathbf{S} = 0$

**III.** $\oint_C \mathbf{E} \cdot d\mathbf{r} = -\dfrac{d}{dt}\int_S \mathbf{B} \cdot d\mathbf{S}$ (Faraday's law)

**IV.** $\oint_C \mathbf{H} \cdot d\mathbf{r} = \int_S \mathbf{J}_f \cdot d\mathbf{S} + \dfrac{d}{dt}\int_S \mathbf{D} \cdot d\mathbf{S}$ (Ampère's law for magnetic material)

**TABLE 8-4** Boundary conditions

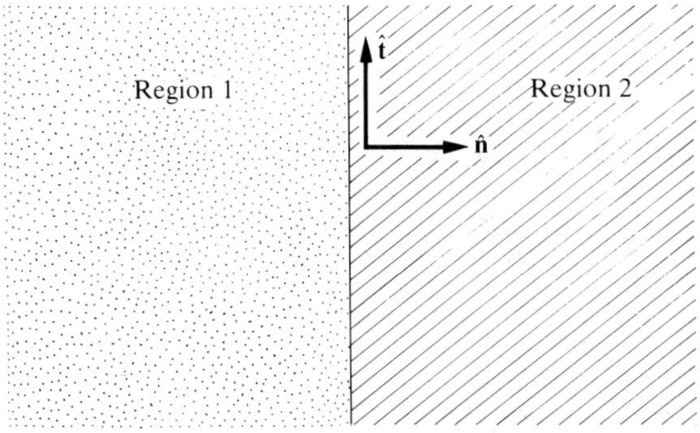

Region 1

Region 2

$\hat{\mathbf{t}}$

$\hat{\mathbf{n}}$

$(\mathbf{D}_2 - \mathbf{D}_1) \cdot \hat{\mathbf{n}} = \sigma_f$

$(\mathbf{B}_2 - \mathbf{B}_1) \cdot \hat{\mathbf{n}} = 0$

$(\mathbf{E}_2 - \mathbf{E}_1) \cdot \hat{\mathbf{t}} = 0$

$(\mathbf{H}_2 - \mathbf{H}_1) \cdot \hat{\mathbf{t}} = j_\perp$ where $j_\perp = \mathbf{j} \cdot (\mathbf{n} \times \mathbf{t})$

is supposed, in the limit, to be thin, the magnetic flux term will vanish, and again we conclude that the tangential **E** field is continuous.

Maxwell II has the same form as in the static case, so as in Equation (7–104), the normal component of **B** is continuous. Application of the integral form of Maxwell IV to the circuit of Figure 7–14(b) again yields the change in the tangential **H** field to be the free surface current, since the displacement flux term is negligible. The boundary conditions are summarized in Table 8–4.

## 8-6 POYNTING'S THEOREM (ENERGY CONSERVATION)

Conservation laws are valuable tools in the analysis of physical systems. Their virtue is that general conclusions can often be drawn without the necessity of a detailed solution. Maxwell's equations lead to conservation of energy, momentum, and angular momentum. In this section we concentrate on the important energy conservation result known as Poynting's theorem (after John Henry Poynting, 1852–1914).

We begin by taking the dot product of Maxwell IV with **E**,

$$\mathbf{E}\cdot\nabla\times\mathbf{H} = \mathbf{E}\cdot\frac{\partial\mathbf{D}}{\partial t} + \mathbf{E}\cdot\mathbf{J} \tag{8-76}$$

and then dot Maxwell III with **H**,

$$\mathbf{H}\cdot\nabla\times\mathbf{E} = -\mathbf{H}\cdot\frac{\partial\mathbf{B}}{\partial t} \tag{8-77}$$

Subtracting Equation (8–77) from (8–76) and using the identity for the divergence of a cross product, we obtain

$$\mathbf{E}\cdot\mathbf{J} + \left(\mathbf{E}\cdot\frac{\partial\mathbf{D}}{\partial t} + \mathbf{H}\cdot\frac{\partial\mathbf{B}}{\partial t}\right) = -\nabla\cdot(\mathbf{E}\times\mathbf{H}) \tag{8-78}$$

Integrating over a fixed finite volume $V$ yields

**Poynting's Theorem**

$$\int_V \mathbf{E}\cdot\mathbf{J}\,dV + \int_V\left(\mathbf{E}\cdot\frac{\partial\mathbf{D}}{\partial t} + \mathbf{H}\cdot\frac{\partial\mathbf{B}}{\partial t}\right)dV = -\int_S \mathbf{E}\times\mathbf{H}\cdot d\mathbf{S} \tag{8-79}$$

where we have used the divergence theorem to evaluate the volume integral of $\nabla\cdot(\mathbf{E}\times\mathbf{H})$. As previously shown in Section 5–11, the volume integral of $\mathbf{E}\cdot\mathbf{J}$ is the power given to electric charges by the electric field:

$$\int_V \mathbf{E}\cdot\mathbf{J}\,dV = \frac{dK}{dt} \tag{8-80}$$

where $K$ is the kinetic energy of the charges within $V$.

The right-hand side of Equation (8–79) represents a net energy flux carried across the boundary surface by the electromagnetic fields. The following vector, which governs the rate and direction of the flow of field energy, is known as Poynting's vector.

**Poynting's Vector**

$$\mathbf{Y} \equiv \mathbf{E} \times \mathbf{H} \tag{8-81}$$

The units of $\mathbf{Y}$ are watts per meter squared. The second term in Equation (8–79), which also has the units of power, represents the time rate of change of the energy stored in the electromagnetic fields within $V$. The corresponding time rate of change of the energy density is

$$\frac{\partial u}{\partial t} = \mathbf{E} \cdot \frac{\partial \mathbf{D}}{\partial t} + \mathbf{H} \cdot \frac{\partial \mathbf{B}}{\partial t} \tag{8-82}$$

For an electromagnetically linear, isotropic material (in which $\varepsilon$ and $\mu$ are constants independent of the fields), we have

$$\frac{\partial u}{\partial t} = \frac{\partial}{\partial t}\left(\frac{1}{2}\varepsilon E^2 + \frac{1}{2}\frac{B^2}{\mu}\right) \tag{8-83a}$$

and hence

$$u = \frac{1}{2}\varepsilon E^2 + \frac{1}{2}\frac{B^2}{\mu} \tag{8-83b}$$

which is the sum of the electric and magnetic energy densities. Thus we see that Poynting's theorem in Equation (8–79) is an equation expressing energy conservation. The power delivered to the electric charges plus the time rate of change of the field energy is balanced by the electromagnetic power entering the volume.

## 8-6A Energy Flow in Conduction

Previously we have discussed the important notion of current conduction in a metallic wire in terms of the conduction electrons. We will demonstrate that the energy that ends up as joule heating in fact is carried by the electromagnetic field outside the wire. This field energy flows into the wire laterally and turns into heat energy. For simplicity we consider a DC current $I$ flowing along a long straight wire of radius $a$, as shown in Figure 8–9. For steady conduction the fields are time-independent, so Equation (8–79) reads

$$\int_V \mathbf{E} \cdot \mathbf{J}\, dV = -\oint_S \mathbf{E} \times \mathbf{H} \cdot d\mathbf{S} \tag{8-84}$$

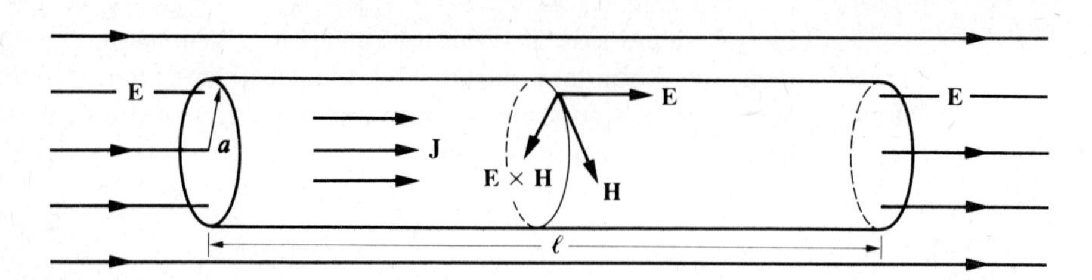


**FIGURE 8-9** Poynting vector $\mathbf{E} \times \mathbf{H}$ at the surface of a current-carrying wire

The left side measures the transformation of field energy into kinetic energy by collisions of conduction electrons with ions, while the right side represents the flow of field energy into the wire segment.

We consider a length $\ell$ of the wire. If the potential change along this length, taken as the $z$ axis, is $V_0$, the average electric field in the wire is

$$\mathbf{E} = \frac{V_0}{\ell}\hat{\mathbf{z}} \tag{8-85}$$

Evaluating the left side of Equation (8–84), we obtain

$$\int_V \mathbf{E} \cdot \mathbf{J}\, dV = \frac{V_0}{\ell}\left(\frac{I}{\pi a^2}\right)\pi a^2 \ell = V_0 I \tag{8-86}$$

which is the usual result for joule heating in the wire. To evaluate the right side of Equation (8–84), we need the magnetic field on the surface of the wire; Ampère's law gives it to be

$$\mathbf{H} = \frac{I}{2\pi a}\hat{\boldsymbol{\phi}} \tag{8-87}$$

The resulting Poynting vector is

$$\mathbf{Y} = \mathbf{E} \times \mathbf{H} = -\frac{V_0 I}{2\pi a \ell}\hat{\mathbf{r}} \tag{8-88}$$

which is the energy flux passing radially into the wire. In integrating over the surface of the wire segment, we observe that no field energy enters the ends; it enters only through the curved cylindrical surface. The rate of transport of energy through the lateral surface is then

$$-\oint \mathbf{E} \times \mathbf{H} \cdot d\mathbf{S} = \frac{V_0 I}{2\pi a \ell}(2\pi a \ell) = V_0 I \tag{8-89}$$

Hence Poynting's theorem is satisfied; the electromagnetic energy flow into the wire from its sides is converted into kinetic (heat) energy within the wire.

### 8-6B Energy Carried by a Plane Wave

For radiation applications studying the energy flow in a plane wave moving in empty space is useful. By Equation (8–60), $\mathbf{H} = \hat{\mathbf{k}} \times \mathbf{E}/(\mu_0 c)$, and the Poynting vector is

$$\mathbf{Y} = \mathbf{E} \times \mathbf{H} = \frac{1}{\mu_0 c}\mathbf{E} \times (\hat{\mathbf{k}} \times \mathbf{E}) = \frac{E^2}{\mu_0 c}\hat{\mathbf{k}} \tag{8-90}$$

whose magnitude is the power per unit area. This power is transferred in the direction of propagation $\hat{\mathbf{k}}$. Integrating over a unit area with normal along $\hat{\mathbf{k}}$ yields

$$\frac{\text{power}}{\text{area}} = \frac{[E(t)]^2}{\mu_0 c} \tag{8-91}$$

For $\mathbf{E} = \mathscr{E} \cos(\mathbf{k} \cdot \mathbf{r} - \omega t)$ the time-averaged power is then

$$\frac{\text{time-averaged power}}{\text{area}} = \frac{\mathscr{E}^2}{2\mu_0 c} \tag{8-92}$$

This time-averaged power is independent of the location in space, so if we consider a fixed volume, as much energy leaves as enters, on the average. Referring to the Poynting theorem, Equation (8–79), we see that the time-averaged energy density must be constant in space.

For a plane wave in a linear medium the magnetic energy density equals the electric energy density:

$$u_B = \frac{B^2}{2\mu_0} = \frac{(\hat{\mathbf{k}} \times \mathbf{E})^2}{2\mu_0 c^2} = \frac{\varepsilon_0 E^2}{2} = u_E \tag{8-93}$$

The time average is clearly position-independent.

## 8-7 MAXWELL STRESS TENSOR (MOMENTUM CONSERVATION)

In the preceding section we used Maxwell's equations to derive an energy conservation law, which involved the energy density and the energy flux (the Poynting vector $\mathbf{Y} = \mathbf{E} \times \mathbf{H}$) of the electromagnetic field. As one might expect, momentum is also carried by the field, and Maxwell's equations can be used to establish an amended form of Newton's law, where the total momentum is the sum of the momenta of particles and the momentum of the field. Energy is a scalar quantity, and its flux is a vector (the Poynting vector). Momentum is a conserved vector quantity, and its flux is a tensor with two indices (the stress tensor).

The Lorentz force per unit volume **f** is

$$\mathbf{f} = \rho\mathbf{E} + \mathbf{J} \times \mathbf{B} \tag{8-94}$$

We begin by eliminating the sources $\rho$ and **J**, using Maxwell I and Maxwell IV in empty space, giving

$$\mathbf{f} = \varepsilon_0 \mathbf{E}\,\nabla\cdot\mathbf{E} + \frac{1}{\mu_0}(\nabla \times \mathbf{B}) \times \mathbf{B} - \varepsilon_0 \frac{\partial \mathbf{E}}{\partial t} \times \mathbf{B} \tag{8-95}$$

Using the relation

$$\frac{\partial}{\partial t}(\mathbf{E} \times \mathbf{B}) = \frac{\partial \mathbf{E}}{\partial t} \times \mathbf{B} + \mathbf{E} \times \frac{\partial \mathbf{B}}{\partial t} = \frac{\partial \mathbf{E}}{\partial t} \times \mathbf{B} - \mathbf{E} \times (\nabla \times \mathbf{E}) \tag{8-96}$$

we can rewrite Equation (8–95) as

$$\mathbf{f} = \varepsilon_0[\mathbf{E}\,\nabla\cdot\mathbf{E} + (\nabla \times \mathbf{E}) \times \mathbf{E}] + \frac{1}{\mu_0}[\mathbf{B}\,\nabla\cdot\mathbf{B} + (\nabla \times \mathbf{B}) \times \mathbf{B}]$$
$$- \varepsilon_0 \frac{\partial}{\partial t}(\mathbf{E} \times \mathbf{B}) \tag{8-97}$$

where $\mathbf{B}\,\nabla\cdot\mathbf{B}/\mu_0 = 0$ is added for reasons of symmetry.

The identity

$$[(\nabla \times \mathbf{E}) \times \mathbf{E}]_i = E_j\left(\frac{\partial E_i}{\partial x_j} - \frac{\partial E_j}{\partial x_i}\right) \tag{8-98}$$

follows directly from expansion in Cartesian coordinates. Here and in the rest of this section a sum is understood over the repeated index $j$. With this identity and a similar one involving **B**, Equation (8–97) becomes

$$f_i = \varepsilon_0\left(E_i\frac{\partial E_j}{\partial x_j} + E_j\frac{\partial E_i}{\partial x_j} - E_j\frac{\partial E_j}{\partial x_i}\right) + \frac{1}{\mu_0}\left(B_j\frac{\partial B_i}{\partial x_j} + B_j\frac{\partial B_i}{\partial x_j}\right.$$
$$\left. - B_j\frac{\partial B_j}{\partial x_i}\right) - \varepsilon_0\frac{\partial}{\partial t}(\mathbf{E} \times \mathbf{B})_i \tag{8-99a}$$

This relation can be reexpressed as

$$f_i = \varepsilon_0\frac{\partial}{\partial x_j}\left(E_iE_j - \frac{1}{2}\delta_{ij}E^2\right) + \frac{1}{\mu_0}\frac{\partial}{\partial x_j}\left(B_iB_j - \frac{1}{2}\delta_{ij}B^2\right)$$
$$- \varepsilon_0\frac{\partial}{\partial t}(\mathbf{E} \times \mathbf{B})_i \tag{8-99b}$$

where $\delta_{ij}$ is the Kronecker delta ($\delta_{ij} = 1$ for $i = j$ and $\delta_{ij} = 0$ for $i \neq j$).

The Maxwell stress tensor is defined as follows:

**Maxwell Stress Tensor**

$$T_{ij} \equiv \varepsilon_0 E_i E_j + \frac{1}{\mu_0} B_i B_j - \frac{1}{2}\delta_{ij}\left(\varepsilon_0 E^2 + \frac{1}{\mu_0} B^2\right) \tag{8-100}$$

The field momentum volume density is defined as follows:

**Field Momentum Density**

$$\mathbf{p}_{\text{field}} = \varepsilon_0 \mathbf{E} \times \mathbf{B} \tag{8-101}$$

In terms of these quantities Equation (8–99b) is

$$f_i = \frac{\partial T_{ij}}{\partial x_j} - \frac{\partial (\rho_{\text{field}})_i}{\partial t} \tag{8-102}$$

When Equation (8–102) is integrated over a finite volume, the momentum conservation law is obtained. With only electromagnetic forces acting, the momentum change of the charges is related by Newton's second law to the volume force by

$$\frac{d(\mathbf{P}_{\text{charges}})_i}{dt} = \mathbf{F}_i = \int \mathbf{f}_i \, dV \tag{8-103}$$

From the divergence theorem the volume integral of the gradient of the Maxwell stress tensor in Equation (8–102) can be converted to a surface integral:

$$\int \frac{\partial T_{ij}}{\partial x_j} dV = \int \nabla \cdot \mathbf{T}_i \, dV = \oint \mathbf{T}_i \cdot d\mathbf{S} = \oint T_{ij} \, dS_j \tag{8-104}$$

Here we applied the usual divergence theorem by thinking of $T_{ij}$ as three separate vectors $\mathbf{T}_i$ having components $(\mathbf{T}_i)_j = T_{ij}$. The volume integral of Equation (8–102) now becomes

**Momentum Conservation**

$$\frac{d}{dt}(\mathbf{P}_{\text{charges}} + \mathbf{P}_{\text{field}})_i = \oint T_{ij} \, dS_j \tag{8-105}$$

The capitalized momenta are the volume-integrated momentum densities. We see that the total rate of change of momentum has been expressed as a stress force acting over the bounding surface.

## 8-7A Electrostatic Stress Tensor

From Equation (8–100) the electric part of the stress tensor is

$$T_{ij} = \varepsilon_0 \left( E_i E_j - \frac{1}{2} \delta_{ij} E^2 \right) \tag{8-106}$$

We can gain a simple geometric interpretation by choosing the $z$ axis ($x_3 = z$) along the local normal to the surface and choosing the $x$ axis ($x_1 = x$) so that the electric field at the surface lies in the $xz$ plane and makes an angle $\theta$ with the surface normal, as shown in Figure 8–10.

The electric field components are then

$$E_x = E \sin\theta \qquad E_y = 0 \qquad E_z = E \cos\theta \tag{8-107}$$

Since the normal to the surface is along the $z$ axis, $dS_1 = dS_2 = 0$. The contributing stress tensor components to Equation (8–105) are then $T_{xz}$, $T_{yz}$, and $T_{zz}$, but $T_{yz} = 0$ from Equations (8–106) and (8–107). Hence the stress force per unit area is

$$\mathbf{T} = T_{xz}\hat{\mathbf{x}} + T_{zz}\hat{\mathbf{z}} \tag{8-108}$$

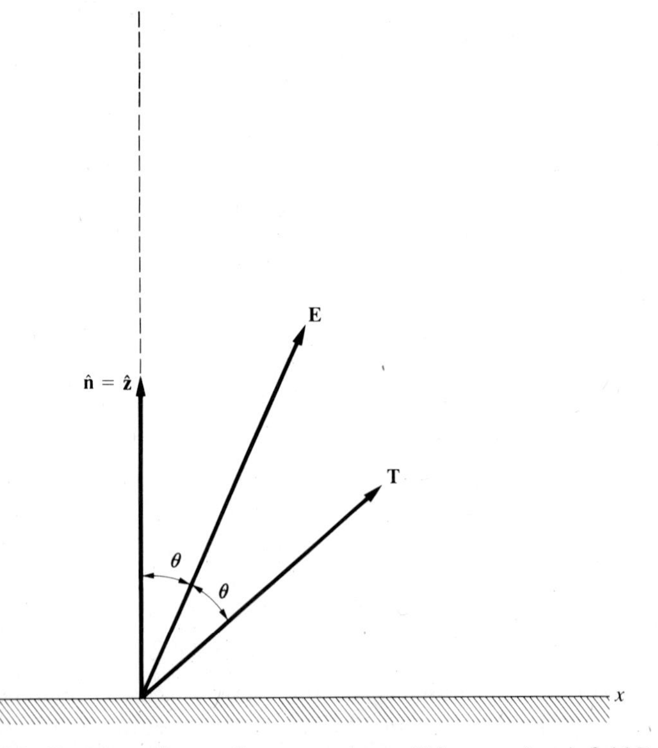


**FIGURE 8-10** Maxwell stress force per unit area **T** from an electric field **E** acting on a surface

where from Equations (8–106) and (8–107)

$$T_{xz} = \frac{1}{2}\varepsilon_0 E^2 \sin 2\theta$$
$$T_{zz} = \varepsilon_0 E^2\left(\cos^2\theta - \frac{1}{2}\right) = \frac{1}{2}\varepsilon_0 E^2 \cos 2\theta \tag{8–109}$$

The resulting stress force is

$$\mathbf{T} = \frac{1}{2}\varepsilon_0 E^2(\hat{\mathbf{x}} \sin 2\theta + \hat{\mathbf{z}} \cos 2\theta) \tag{8–110}$$

Note that the electric field vector bisects the angle that **T** makes with the surface normal, as shown in Figure 8–10.

The stress can be described as the combination of tension of the lines of force and repulsion between them. This conclusion is shown clearly by special cases. If **E** is parallel (or antiparallel) to $\hat{\mathbf{n}}$, the stress force is in the $\hat{\mathbf{n}}$ direction, corresponding to tension of the lines of force. If **E** is perpendicular to $\hat{\mathbf{n}}$, the stress is opposite to $\hat{\mathbf{n}}$, corresponding to repulsion between the lines of force. In Figure 8–11 the field lines of two equal and of two equal but opposite charges are shown. Taking the bounding surface to be the median plane, we see how the force across the plane according to the stress tensor gives attraction and repulsion, respectively. This intuitive interpretation was first given by Faraday in terms of elastic properties of the lines of force.

The magnetic term of the stress tensor has exactly the same form as the electric stress tensor, with the magnetic field replacing the electric field and $\mu_0$ replacing $1/\varepsilon_0$. Thus the same geometric interpretation applies to it.

## 8-7B Momentum Carried by a Plane Wave

Comparing the field momentum density of Equation (8–101) with the Poynting vector **Y** of Equation (8–81), we find

$$\mathbf{p}_{\text{field}} = \frac{1}{c^2}\mathbf{Y} \tag{8–111}$$

Since the wave propagates in empty space at velocity $c$, the momentum carried through a unit area each second is

$$\text{momentum flux} = c\mathbf{p}_{\text{field}} = \frac{1}{c}\mathbf{Y} \tag{8–112}$$

The ratio of the Poynting vector to the momentum flux tells us the ratio of energy to momentum carried by a plane wave. From Equation (8–112) we see that

$$\text{energy flux} = c(\text{momentum flux}) \tag{8–113}$$

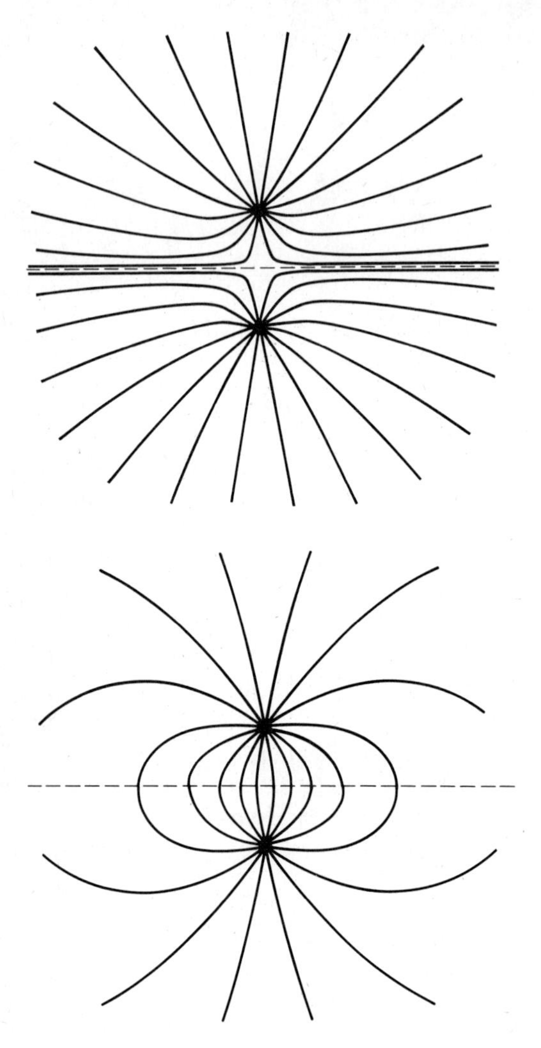

**FIGURE 8-11** Electric field lines from systems of two equal charges of the opposite and the same sign

This result is interesting for several reasons. In quantum physics electromagnetic radiation is emitted and absorbed in quantized bits called *photons*. These photons carry energy, momentum, and angular momentum. On the average, they must behave as predicted by Maxwell's equations, so we can conclude that the ratio of energy to momentum of a photon must also satisfy Equation (8–113). Another aspect of Equation (8–113) will become evident in Chapter 11 when we discuss the relativistic relationship between energy and momentum; the relation in Equation (8–113) implies that light quanta must be massless.

If an object absorbs the momentum flux of an electromagnetic wave, it will experience a pressure (i.e., force per unit area), by Newton's law. This radiation pressure is, on the average, by Equation (8–113),

$$\text{radiation pressure} = \frac{1}{c}(\text{energy flux}) \tag{8-114}$$

Ordinarily, this radiation pressure is quite small. For example, the energy flux of sunlight is roughly 1 $kW/m^2$, so the force on a sunlit square meter is $3 \times 10^{-6}$ N, which is the weight of a 0.3-mg object at the earth's surface. Radiation pressure can be important for small objects because they have a large area-to-volume ratio.

# 8-8 SOME QUANTUM EFFECTS IN ELECTROMAGNETISM

The Maxwell equations as presented in this chapter provide a complete understanding of electricity and magnetism at the classical level. The theory also applies at the atomic and subatomic levels, provided that the electromagnetic potentials are expressed as quantum operators that can create or destroy photons, the quanta of the electromagnetic field. In the following sections we briefly discuss some interesting quantum effects in electromagnetism.

## 8-8A Electric Charge Quantization and Conservation

Some profound aspects of electromagnetism are outside the scope of Maxwell's theory. Electric charge is found only in integral multiples of the electron's charge, except for quarks, the constituents of protons and neutrons, which have charges $\frac{2}{3}$ and $-\frac{1}{3}$ of $e$. Another fact unexplained by the Maxwell theory is that the proton and the electron have exactly opposite charges, which is the reason that atoms are electrically neutral. Progress in understanding electric charge quantization has been made with grand unified theories (GUTs), which unite electromagnetic, weak, and strong interactions.

Charge conservation is observed to hold in all interactions, down to the finest subatomic scale. Even in a very violent collision between elementary particles (e.g., a proton hitting another proton), the final net charge is the same as the net charge of the two initial particles; no net charge is created or destroyed in the collision.

## 8-8B Superconductors

The phenomenon of superconductivity is characterized by zero electrical resistance. Because of the infinite conductivity, the magnetic flux linking a superconductor cannot change (since a time-varying magnetic flux would induce an electric field, which would cause an infinite current to flow); hence an applied magnetic field will not penetrate the superconductor. A sufficiently strong magnetic field will, however, destroy the superconducting state.

The magnetic flux inside the bulk of a superconductor is, in fact, found to be zero. This phenomenon is called the *Meissner effect*. A superconductor at sufficiently high temperature is normal (non-superconducting) and is penetrated by an applied magnetic field; but if the temperature is then lowered until the sample becomes superconducting, the field will be expelled. Surface currents arise in the precise amount and configuration to cancel the previous internal magnetic field.

A theoretical description of the superconducting state was first provided by J. Bardeen, L. N. Cooper, and J. R. Schrieffer [*Phys. Rev.*, 108: 1175 (1957)]. In this theory electrons near the Fermi surface are correlated in pairs (called Cooper pairs) through interactions with the crystal lattice. These correlation forces are quite weak, and moderately high temperatures (20 K or more) or magnetic fields will disrupt the superconducting state.

The energy of a superconducting loop has, in addition to the normal $\frac{1}{2}LI^2$ term, a term that is a periodic function of the magnetic flux linking the loop. The period is the so-called fluxoid unit $\mathscr{F}_0$,

**Superconductor Fluxoid Unit**

$$\mathscr{F}_0 = \frac{h}{2e} \tag{8-115}$$

where $h$ is Planck's constant. If the loop is not strongly perturbed by outside effects, its energy has minima at integral multiples of $\mathscr{F}_0$, and the linking flux will take on a value $n\mathscr{F}_0$. The charge of $2e$ in the denominator of Equation (8–115) is that of one Cooper pair of electrons. Numerically, the unit of the superconducting fluxoid is (in webers, Wb)

$$\mathscr{F}_{\text{min}} = \frac{6.626 \times 10^{-34}}{2(1.6 \times 10^{-19})} = 2.07 \times 10^{-15} \text{ Wb} \tag{8-116}$$

If the magnetic field is confined to a hole of diameter 0.1 mm, this value corresponds to about 1% of the earth's field, a readily detectable value.

## 8-8C Magnetic Monopoles

There is no theoretical reason why isolated magnetic poles, "magnetic charges," could not exist. A *monopole* of magnetic charge $g$ is the source of a radial magnetic field

$$\mathbf{B} = \frac{\mu_0 g}{4\pi r^2}\,\hat{\mathbf{r}} \tag{8-117}$$

in analogy to Coulomb's law for electric charges. A consistent quantum theory of electric and magnetic charges was first given by P. A. M. Dirac in 1931. The fundamental result is that the product of any monopole charge $g$

with any electric charge $q$ can take on only values that are integral multiples of $h/\mu_0$; that is,

**Dirac Quantization Condition**

$$gq = n\frac{h}{\mu_0} \tag{8–118}$$

where $n$ is an integer. As a consequence, if a magnetic monopole exists, then all electric charges must be quantized, of the form

$$q = n\left(\frac{h}{\mu_0 g}\right) \tag{8–119}$$

Since all electric charges of free particles are observed to be integral multiples of the electron charge magnitude $e$, we can identify $h/(\mu_0 g)$ with $e$. Thus Equation (8–118) says that the magnetic charge of any monopole must be an integral multiple of the "unit"

$$g = \frac{h}{\mu_0 e} = 3.29 \times 10^{-9}\ \text{A·m} \tag{8–120}$$

The grand unified theories of electromagnetic, weak, and strong interactions predict that monopoles exist with mass

$$m_{\text{monopole}} \approx 10^{16} m_{\text{proton}} \approx 10^{-8}\ \text{g}$$

These monopoles might have been produced in the very early stages of the creation of the universe. The observation of such magnetic monopoles would support the unification of forces. These theories also predict that the proton is unstable, with a lifetime of about $10^{32}$ years, which is just at the level of present experimental limits on proton decay. If monopoles exist, we must add to Maxwell's equations magnetic source terms. This modification is straightforward, and the results are given in Table 8–5. The magnetic charge

**TABLE 8–5** Maxwell's equations in empty space with both electrical and magnetic sources

| Differential Forms | Integral Forms |
|---|---|
| I. $\nabla \cdot \mathbf{E} = \dfrac{\rho_e}{\varepsilon_0}$ | I. $\oint_S \mathbf{E} \cdot d\mathbf{S} = \dfrac{Q_e}{\varepsilon_0}$ |
| II. $\nabla \cdot \mathbf{B} = \mu_0 \rho_m$ | II. $\oint_S \mathbf{B} \cdot d\mathbf{S} = \mu_0 Q_m$ |
| III. $\nabla \times \mathbf{E} = -\dfrac{\partial \mathbf{B}}{\partial t} - \mu_0 \mathbf{J}_m$ | III. $\oint_C \mathbf{E} \cdot d\mathbf{r} = -\dfrac{d}{dt}\int_S \mathbf{B} \cdot d\mathbf{S} - \mu_0 \int_S \mathbf{J}_m \cdot d\mathbf{S}$ |
| IV. $\nabla \times \mathbf{B} = \mu_0\left(\mathbf{J}_e + \varepsilon_0 \dfrac{\partial \mathbf{E}}{\partial t}\right)$ | IV. $\oint_C \mathbf{B} \cdot d\mathbf{r} = \mu_0 \int_S \mathbf{J}_e \cdot d\mathbf{S} + \mu_0 \varepsilon_0 \dfrac{d}{dt}\int_S \mathbf{E} \cdot d\mathbf{S}$ |

density $\rho_m$ and the magnetic current density $\mathbf{J}_m$ satisfy a continuity equation as their electric counterparts do.

## 8-8D Monopole Passing Through a Superconducting Ring

The magnetic flux linking a superconducting loop cannot be changed by the effects of electric currents. However, as we will show in this section, a magnetic monopole passing through the loop will change the number of superconducting fluxoids by two. The change of current induced in the loop corresponding to this change of flux provides a unique experimental signature to search for magnetic monopoles.

Suppose that a monopole passes through a superconducting loop $C$ bounded by a surface $S$ in the $z = 0$ plane, as in Figure 8–12. From Table 8–5 the integral form of Maxwell III with magnetic sources is

$$\oint_C \mathbf{E} \cdot d\mathbf{r} = -\frac{d\mathcal{F}}{dt} - \mu_0 \int_S \mathbf{J}_m \cdot d\mathbf{S} \tag{8-121}$$

and the monopole current is

$$\mathbf{J}_m = g\mathbf{v}\delta^3(\mathbf{r} - \mathbf{v}t) \tag{8-122}$$

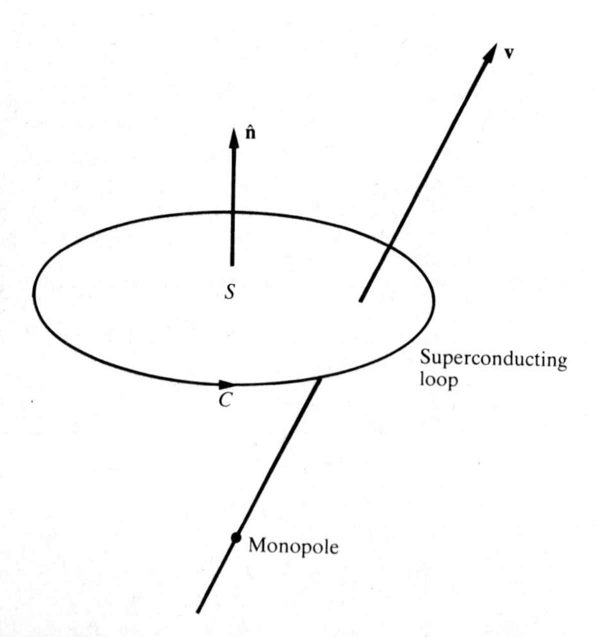


**FIGURE 8-12** Magnetic monopole passing through a superconducting circuit $C$ by a surface $S$

In evaluation of the surface integral in Equation (8–121) over the plane of the loop, the $x$ and $y$ factors of the delta function integrate to unity, and one finds

$$\int \mathbf{J}_{\mathrm{m}} \cdot d\mathbf{S} = g v_z \delta(-v_z t) = g\delta(t) \qquad \textbf{(8–123)}$$

The $\delta(t)$ of this result reflects the fact that the monopole intersects the plane of the loop at time $t = 0$, and so $\mathbf{J}_{\mathrm{m}}$ on that plane is nonvanishing only at $t = 0$. The left-hand side of Equation (8–121) vanishes because within the superconductor $\mathbf{E} = 0$. The result is

$$\frac{d\mathscr{F}}{dt} = -\mu_0 g\delta(t) \qquad \textbf{(8–124)}$$

Integrating from $t = -\infty$, we obtain

$$\mathscr{F}(t) = \mathscr{F}(-\infty) - \mu_0 g\theta(t) \qquad \textbf{(8–125)}$$

Thus in passing through the superconducting loop, the monopole causes a total flux change $\Delta\mathscr{F} = \mathscr{F}(+\infty) - \mathscr{F}(-\infty)$ of

$$\Delta\mathscr{F} = -\mu_0 g = -\frac{h}{e} \qquad \textbf{(8–126)}$$

where we have used Equation (8–120) for $g$. Hence any trajectory of a magnetic charge $g$ that passes through the ring will result in a two-fluxoid change (one superconductor fluxoid $= h/2e$), while one that misses the ring will produce no flux change. This flux change is accompanied by a change in the persistent current flowing on the superconductor, which can be measured.

We emphasize that at finite times, when the monopole is near the loop, the relation between the current $I$ in the loop and the flux $\mathscr{F}$ through the loop is not that of an isolated loop. The current $I$ is directly observable, but the flux $\mathscr{F}$ is not, and so the abrupt jump of $\mathscr{F}$ at $t = 0$ implied by Equation (8–124) is not observable. In fact, an arbitrary surface capping the loop can be used in the derivation, and so the time at which the monopole crosses the surface, the time of the abrupt jump of $\mathscr{F}$, can be arbitrary. To state this condition more generally, in the absence of monopoles the relation $\nabla \cdot \mathbf{B} = 0$ implies that the flux through any surface capping a loop is the same (the magnetic lines are endless), and so without ambiguity one can speak simply of the flux through the loop [see Equation (8–10)]. In the presence of monopoles this statement is no longer true.

We now consider explicitly the case of a monopole moving along the symmetry axis of the ring; see Figure 8–13. From Equation (8–117) the magnetic flux of the monopole through the ring is

$$\mathscr{F}_g = \frac{\mu_0 g}{4\pi} \int_S \frac{\hat{\mathbf{R}} \cdot \hat{\mathbf{n}}\, dS}{R^2} \qquad \textbf{(8–127)}$$

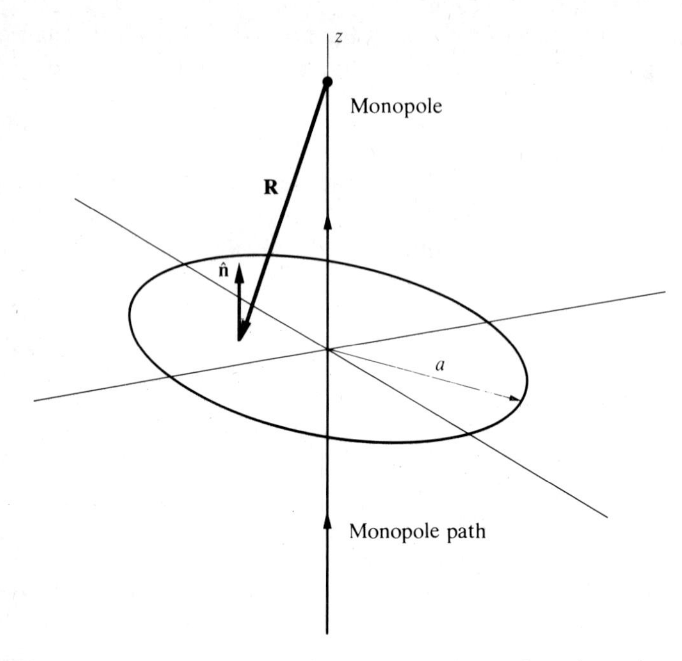


**FIGURE 8-13** Geometry for a monopole moving along the symmetry axis of a superconducting ring

where **R** is the vector from the monopole to a point on the disk enclosed by the ring and the integration is over the area enclosed by the ring. By a calculation parallel to that in Section 1–8, we obtain

$$\mathscr{F}_g = \frac{\mu_0 g}{2} \frac{z}{(z^2 + \imath'^2)^{1/2}}\bigg|_{\imath'=0}^{\imath'=a} \tag{8-128a}$$

which gives

$$\mathscr{F}_g = \frac{\mu_0 g}{2}\left[1 - 2\theta(z) + \frac{z}{(z^2 + a^2)^{1/2}}\right] \tag{8-128b}$$

Here $\theta(z)$ is the theta function.

The total flux through the surface is

$$\mathscr{F} = \mathscr{F}_g + \mathscr{F}_{\mathrm{I}} \tag{8-129}$$

where $\mathscr{F}_{\mathrm{I}}$ is the induced flux,

$$\mathscr{F}_{\mathrm{I}} = -LI \tag{8-130}$$

Here $I$ is the induced supercurrent, and $L$ is the self-inductance of the ring.

If we set $\mathscr{F} = 0$ at $z = -\infty$, Equation (8–124) gives

$$\mathscr{F} = -\mu_0 g \theta(z) \tag{8–131}$$

since the monopole intersects $z = 0$ at time $t = 0$.

Combining Equations (8–128) through (8–131), we find the induced supercurrent

$$I = \frac{\mu_0 g}{2L}\left[1 + \frac{z}{(z^2 + a^2)^{1/2}}\right] \tag{8–132}$$

Figure 8–14 plots the induced supercurrent $I$ in the ring as a function of the monopole position. The induced current rises from zero when $z = -\infty$ to the value

$$I = \frac{\mu_0 g}{L} \tag{8–133}$$

when $z = +\infty$. The result, Equation (8–132), for $I(z)$ is independent of the choice of surface bounded by the ring.

## 8-8E Josephson Junction

As a final example of quantum effects, we consider the *Josephson junction.* In 1962 B. D. Josephson pointed out that a superconducting circuit cut by a thin insulator, as schematically illustrated in Figure 8–15, has remarkable quantum properties. For an applied voltage across the insulator of

$$V(t) = V_0 + V_1 \sin \omega t \tag{8–134}$$

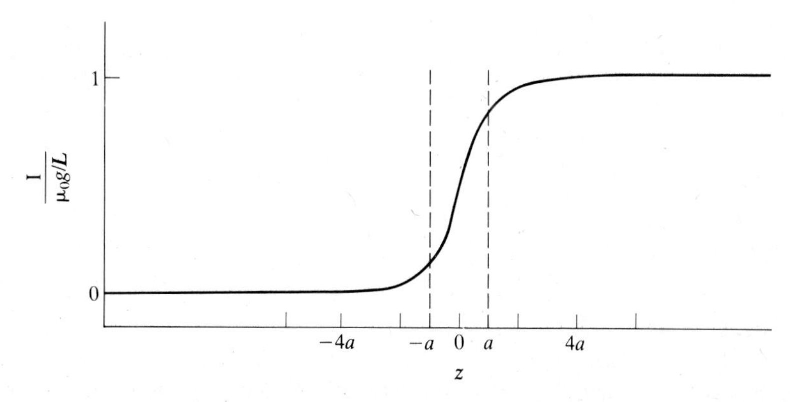


**FIGURE 8–14** Induced supercurrent in a superconducting ring due to a monopole as a function of the monopole position on the axis of the ring

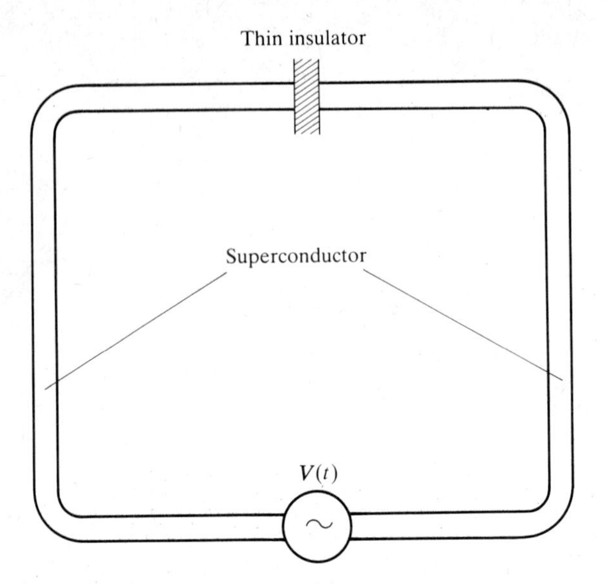


**FIGURE 8-15** Josephson junction circuit

a current will cross the insulator if and only if the angular frequency $\omega$ of the oscillating voltage is related to the constant voltage $V_0$ by

$$\omega = \frac{e}{\pi h} V_0 \tag{8-135}$$

At a potential of 1 mV the angular frequency $\omega$ is $7.7 \times 10^{10}$ rad/s, which is in the microwave range. Since frequencies can be very accurately measured, the Josephson junction now constitutes the most accurate method of measuring voltage, and the volt is now standardized by the Josephson effect.

The most accurate method of measuring small magnetic fields is also based on the Josephson effect. If two Josephson junctions are put in parallel, as in Figure 8–16, the current depends on the magnetic flux $\mathscr{F}$ from an external field that threads the loop of the circuit containing the junctions. The current is given by

$$I = I_0 \cos\left(\frac{2\pi e}{h} \mathscr{F}\right) \tag{8-136}$$

By observation of changes in the current, changes in magnetic flux on the order of $h/e$ can be observed.

A commercially available apparatus for magnetic field measurement based on this principle is the SQUID (acronym for superconducting quantum interference device). SQUIDs are used in medical physics to measure magnetic fields as small as $10^{-11}$ T in organs of the human body. These fields are

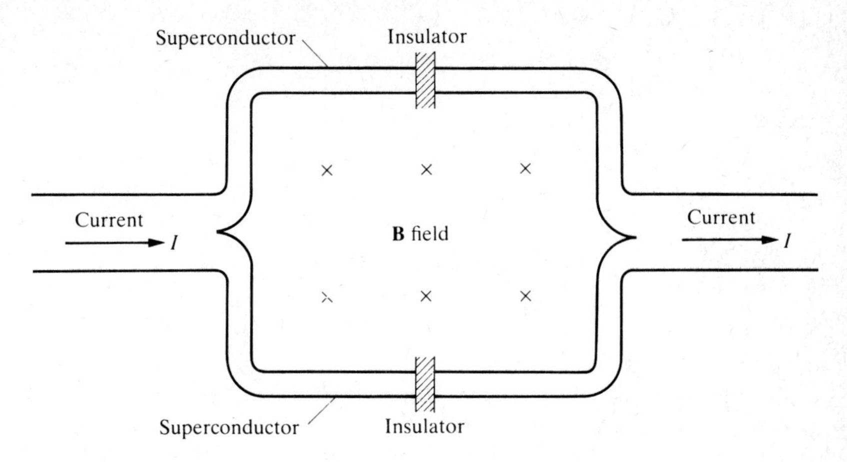


**FIGURE 8-16** Two Josephson junctions in parallel

produced by nerve and muscle electric currents and by particles of magnetized iron. The magnetic study of heart action, the magnetocardiogram (the analog of the electrocardiogram), has the advantage that the magnetic effects come directly from the area of interest and not from changes in skin properties.

## SUMMARY

### Important Concepts and Equations

Maxwell's equations in empty space

*The fundamental set of differential equations describing electric and magnetic fields in empty space. These equations, along with their integral forms, are listed in Table* 8–1.

Results from Maxwell's equations: magnetic flux through a circuit

*The magnetic flux is the same through all surfaces that have the same bounding circuit.*

Results from Maxwell's equations: displacement current

*The sum of the displacement current* $\int (\varepsilon_0\, \partial \mathbf{E}/\partial t) \cdot dS$ *and the conduction current* $\int \mathbf{J} \cdot d\mathbf{S}$ *is the same through any open surface bounding a given circuit.*

Results from Maxwell's equations: wave solutions

*Maxwell's equations allow wave solutions for electric and magnetic fields. These electromagnetic waves, which include the visible spectrum, all propagate in empty space at the speed of light,* $c \approx 3 \times 10^8$ m/s.

$\omega = ck \qquad \tau = \dfrac{2\pi}{\omega}$ (simple harmonic waves)

$\lambda = \dfrac{2\pi}{k} \qquad \nu = \dfrac{1}{\tau}$

*The basic quantities used to describe harmonic electromagnetic waves—the angular frequency $\omega$, the speed of propagation $c$ (in empty space), the wave number $k$, the wavelength $\lambda$, the period $\tau$, and the frequency $\nu$—in terms of one another.*

Standing waves

*A superposition of waves of the same amplitude and frequency but traveling in opposite directions.*

$\mathbf{E} = \mathscr{E}\hat{\mathbf{x}}^{i(kz-\omega t)}$ (linear polarization)

*A plane wave that has the electric field in a fixed direction $\hat{\boldsymbol{\mathscr{E}}}$, transverse to the propagation vector* **k**, *is linearly polarized. The sum of two linearly polarized waves with the same complex phase is a linearly polarized wave.*

$\mathbf{E} = \mathscr{E}(\hat{\mathbf{x}} + i\hat{\mathbf{y}})e^{i(kz-\omega t)}$ (left circular polarization)

$\mathbf{E} = \mathscr{E}(\hat{\mathbf{x}} - i\hat{\mathbf{y}})e^{i(kz-\omega t)}$ (right circular polarization)

*The superposition of two linearly polarized plane waves with equal magnitudes but having phases that differ by 90° results in circular polarization. At a given position the electric vector sweeps out a circle with angular frequency $\omega$. Left circularly polarized waves have electric fields rotating counterclockwise when facing the wave, while right circularly polarized waves have electric fields rotating clockwise when facing the wave.*

Elliptical polarization

*The superposition of two linearly polarized plane waves with different magnitudes or phases results in an elliptically polarized wave. The electric field vector at a given position sweeps out an ellipse.*

Maxwell's equations in matter

*Maxwell's equations must be modified when one is calculating fields in matter. The fields from the polarization charge and current densities must be included. With the introduction of the auxiliary fields* $\mathbf{D} = \varepsilon_0\mathbf{E} + \mathbf{P}$ and $\mathbf{H} = (1/\mu_0)\mathbf{B} - \mathbf{M}$, the *equations in matter can be written in the differential and integral forms of Table* 8–3.

$\mathbf{Y} = \mathbf{E} \times \mathbf{H}$ (Poynting vector)

*This vector is the rate and direction of energy flow per unit area;* **Y** *is the electromagnetic energy flux density.*

$$\frac{d}{dt}(K + U) = -\oint_S \mathbf{Y} \cdot d\mathbf{S} \qquad \text{(Poynting's theorem)}$$

$$\frac{dK}{dt} = \int_V \mathbf{E} \cdot \mathbf{J}\, dV \qquad \frac{dU}{dt} = \frac{d}{dt}\int_V u\, dV$$

*The time rate of change of the kinetic energy K of charges and the electromagnetic field energy U within a volume V equals the rate at which energy flows across the bounding surface of V. For a linear isotropic material the energy density u is* $u = \frac{1}{2}\varepsilon E^2 + (1/2\mu)B^2$.

$$T_{ij} = \varepsilon_0 E_i E_j + \frac{1}{\mu_0} B_i B_j - \frac{1}{2}\delta_{ij}\left(\varepsilon_0 E^2 + \frac{1}{\mu_0} B^2\right) \qquad \text{(Maxwell stress tensor)}$$

*A tensor quantity involving electromagnetic fields in terms of which the momentum conservation law is expressed;* $T_{ij}$ *is the flux density of electromagnetic momentum.*

$$\mathbf{p}_{\text{field}} = \varepsilon_0 \mathbf{E} \times \mathbf{B} \qquad \text{(field momentum density)}$$

*Momentum per unit volume carried by electromagnetic fields.*

$$\frac{d(\mathbf{P})_i}{dt} = \oint T_{ij}\, dS_j \qquad \text{(momentum conservation)}$$

$$\mathbf{P} = \mathbf{P}_{\text{charges}} + \mathbf{P}_{\text{field}} \qquad \mathbf{P} = \int \mathbf{p}\, dV$$

*The time rate of change of the total momentum of the charges and fields is given by a surface integral of the Maxwell stress tensor* $T_{ij}$.

Meissner effect

*There is no magnetic field inside a superconductor.*

$$\mathscr{F}_0 = \frac{h}{2e}$$

*The magnetic flux linking a superconducting loop takes on integer multiples of the basic unit, the fluxoid* $\mathscr{F}_0$.

$$gq = n\frac{h}{\mu_0} \qquad \text{(Dirac quantization condition)}$$

*The product of any magnetic monopole charge g and the electric charge e can only assume integral multiples of* $h/\mu_0$.

Magnetic monopoles

*Magnetic monopoles, if they exist, must be rare. The Dirac quantization condition gives a value of* $g = h/\mu_0 e = 3.29 \times 10^{-9}$ *A·m for the magnetic monopole charge. For regions containing monopoles Maxwell's equation must be modified, as in Table 8–5.*

# PROBLEMS

## Section 8-1: Maxwell's Equations in Empty Space

**8-1** Starting with Maxwell's equations for a vacuum with charges and currents, derive the equation of continuity for electric charge density and current density.

**8-2** Consider a static magnetic field given in rectangular coordinates by

$$\mathbf{B} = \frac{B_0(x\hat{\mathbf{x}} - y\hat{\mathbf{y}})}{a}$$

(i) Show that this field obeys Maxwell's equations in free space.
(ii) Sketch the field lines, and indicate where filamentary currents would be placed to produce such a field.
(iii) If the magnetic field $B_0(t)$ is slowly varying in time, what electric field would a stationary observer at location $(x, y)$ measure? Displacement current can be neglected.

**8-3** Consider the following electric and magnetic fields in empty space in the region $0 \leqslant z \leqslant a$:

$$E_x = -B_0\omega\left(\frac{a}{\pi}\right)\sin\left(\frac{\pi z}{a}\right)\sin(ky - \omega t)$$

$$B_z = B_0 k\left(\frac{a}{\pi}\right)\sin\left(\frac{\pi z}{a}\right)\sin(ky - \omega t)$$

$$B_y = B_0 \cos\left(\frac{\pi z}{a}\right)\cos(ky - \omega t)$$

All other field components are zero; $B_0$, $a$, $k$, and $\omega$ are constants.
(i) Find the condition for which these fields satisfy Maxwell's equations.
(ii) Assuming that the fields are zero for $z \leqslant 0$ and that there is a perfectly conducting plate in the $z = 0$ plane, find the free surface charge density and the free surface current density on the plate as functions of $x$, $y$, and $t$.

## Section 8-2: Displacement Current

**8-4** A parallel-plate capacitor consists of two circular disks of radius $a$ and separation $d \ll a$. The capacitor is connected to an AC voltage, and the charge on the positive plate is $q = q_0 \cos \omega t$. The lines of magnetic field induced by the displacement current are circles centered on the axis of symmetry. Find the magnetic field at any position between the plates.

**8-5** A spherical capacitor has an emf $\mathscr{E} = V_0 \cos \omega t$ applied to plates consisting of concentric spherical shells of radii $a$ and $b$. Show that the displacement current between the plates is exactly equal to the conduction current entering and leaving the capacitor.

**8-6** A parallel-plate capacitor consists of two circular metal disks of radius $a$ and separation $d$. The plates are connected to a battery that provides a constant current $I$. As the battery moves charge from one disk to the other, the surface charge density $\sigma(t)$ increases steadily with time. Find the magnetic field $\mathbf{B}$ between the plates for both $\imath < a$ and $\imath > a$.

Assume that $\sigma(t)$ is uniform over the inner faces of the plates, and neglect fringing of the **E** lines.

## Section 8-3: Plane Waves in Empty Space

**8-7** Making use of the change of variables of Equations (8–31), directly integrate the first-order differential equations in Equations (8–28), and obtain the solutions of Equations (8–42).

**8-8** The Crab nebula is 2000 light years distant. Pulses of electromagnetic radiation arrive each $\frac{1}{30}$ s. Given that the visible pulses ($\lambda = 500$ nm) and X-ray pulses ($\lambda = 1$ nm) are synchronized within $\frac{1}{100}$ s, place an upper limit on the wavelength dependence $dc/d\lambda$ of the propagation speed in this wavelength region.

**8-9** The electric field of an electromagnetic wave in vacuum is given by

$$E_x = 0 \qquad E_y = 30\cos(2\pi \times 10^8 t - \tfrac{2}{3}\pi x) \qquad E_z = 0$$

where $E$ is in volts per meter, $t$ is in seconds, and $x$ is in meters.
(i) Determine the frequency $\nu$.
(ii) Determine the wavelength $\lambda$.
(iii) Determine the direction of propagation of the wave.
(iv) Determine the direction of the magnetic field.

**8-10** The electric field of a plane electromagnetic wave in empty space is given by

$$\mathbf{E} = 1000\hat{\mathbf{x}} \exp\left\{i\left[\frac{(2\hat{\mathbf{y}} - \hat{\mathbf{z}})\cdot\mathbf{r}}{100} - \omega t\right]\right\}$$

in volts per meter. Calculate the associated magnetic field. Find the wavelength and the frequency of the wave.

**8-11** A plane electromagnetic wave traveling in a vacuum is given by $\mathbf{E} = \hat{\mathbf{y}}E_0 e^{i(kz-\omega t)}$, where $E_0$ is real. A circular loop of radius $a$, $N$ turns, and resistance $R$ is located with its center at the origin. The loop is oriented so that a diameter lies along the $z$ axis and the plane of the loop makes an angle $\theta$ with the $y$ axis. Find the emf induced in the loop as a function of time. Assume that $a \ll \lambda$.

## Section 8-4: Polarized Plane Waves

**8-12** Two plane waves with opposite circular polarizations and the same $\omega$, $k$, and amplitude $\mathscr{E}$ are superposed. Show that the resulting wave is linearly polarized and has amplitude $2\mathscr{E}$.

**8-13** Show that when two linearly polarized plane waves that have polarizations $\boldsymbol{\mathscr{E}}_1$ and $\boldsymbol{\mathscr{E}}_2$ along orthogonal directions with phases that differ by 90° are superposed, the electric field vector sweeps out an elliptical path.

**8-14** Starting with Maxwell's equations in empty space, show that one can obtain transverse electromagnetic waves with **E** and **B** parallel by proceeding as follows:
(i) Show that if **E** and **B** are derived from a vector potential **A** as $\mathbf{B} = \nabla \times \mathbf{A}$ and $\mathbf{E} = -\partial\mathbf{A}/\partial t$, then Maxwell's equations are satisfied in empty space if $\nabla\cdot\mathbf{A} = 0$ and **A** satisfies the wave equation. For harmonic time dependence, the space-dependent part of the vector potential satisfies $(\nabla^2 + k^2)\mathbf{A}_\omega = 0$.

(ii) Show that the vector

$$\mathbf{F}_\omega = \mathbf{A}_\omega + \frac{\nabla \times \mathbf{A}_\omega}{k}$$

is also a solution to the wave equation in part (i) and therefore is an allowed vector potential.

(iii) Prove that $\nabla \times \mathbf{F}_\omega = k\mathbf{F}_\omega$, thus establishing that the magnetic field $\nabla \times \mathbf{F}_\omega$ and the electric field $-\partial \mathbf{F}/\partial t$ are parallel to each other and perpendicular to the direction of propagation $\mathbf{k}$.

**8-15** A vector potential for a plane wave polarized in the $x$ direction is given by $\mathbf{A} = A\hat{\mathbf{x}} \sin kz \cos \omega t$.

(i) Use this vector potential to construct a new potential $\mathbf{F}$, defined in Problem 8-14, describing a transverse wave with parallel $\mathbf{E}$ and $\mathbf{B}$ fields.

(ii) From this solution calculate $\mathbf{E}$ and $\mathbf{B}$ to show explicitly that they are parallel.

(iii) Finally, show that this solution corresponds to two circularly polarized waves propagating in opposite directions.

## Section 8-6: Plane Wave Propagation in Isotropic Matter

**8-16** An electromagnetic wave with electric field given by

$$E_y = E_0 e^{i(kz - \omega t)} \qquad E_x = E_z = 0$$

propagates in a uniform medium consisting of $n_V$ free electrons per unit volume. All other charges in the medium are fixed and do not affect the wave. Write Maxwell's equations for the fields in the medium, and show that they can be satisfied by $\omega^2 > n_V e^2 (m_e \varepsilon_0)$. Find the magnetic field and the wavelength of the electromagnetic wave for a given (allowable) $\omega$. Neglect the magnetic force on the electrons.

## Section 8-7: Maxwell Stress Tensor

**8-17** For two equal charges of the same sign and for two equal charges of opposite sign, integrate the Maxwell stress force per unit area over the median plane, and show that the Coulomb force law is obtained.

**8-18** For what orientation of the electric field lines to a surface does the stress force become purely a sheer force (i.e., the force across the surface is parallel to the surface)?

## Section 8-8: Magnetic Monopoles

**8-19** Obtain the vector potential for a magnetic monopole from $\mathbf{B} = \nabla \times \mathbf{A}$ by assuming that the vector potential has only a $\phi$ component of the form

$$A_\phi \approx \frac{1}{r} f(\theta)$$

Verify your results by recalculating the $\mathbf{B}$ field from the potential you obtain.

# Optics and Waveguides

## CHAPTER CONTENTS

The description of optical phenomena is an important application of the theory of electromagnetism, which we take up in this chapter. The propagation of electromagnetic waves is affected by the presence of matter. The propagation of waves through a medium with handedness is considered, for which optical rotation of the plane of the polarization occurs. The boundary conditions on the waves at the interface between two media (e.g., air to glass) determine reflected and refracted waves in terms of the incident wave. The predictions are given in terms of macroscopic quantities such as the complex index of refraction, which represents average effects of the microscopic interactions of the waves with molecules. The standard formulas of optics, such as Snell's law, Fresnel equations, and Brewster's angle, are then straightforwardly derived. As an interesting example, the theory of the rainbow is presented. In the final section wave propagation in metallic guides is considered.

## 9-1 PLANE WAVE PROPAGATION IN ISOTROPIC MATTER

Consider a homogeneous isotropic medium of dielectric constant $\varepsilon$, permeability $\mu$, and conductivity $g$. With the material equations $\mathbf{J} = g\mathbf{E}$, $\mathbf{D} = \varepsilon\mathbf{E}$, and $\mathbf{B} = \mu\mathbf{H}$, Maxwell's equations take the form

$$\begin{aligned}
\nabla\cdot\mathbf{E} &= \frac{\rho_f}{\varepsilon} && \text{(I}'\text{)}\\
\nabla\cdot\mathbf{H} &= 0 && \text{(II}'\text{)}\\
\nabla\times\mathbf{E} &= -\mu\frac{\partial\mathbf{H}}{\partial t} && \text{(III}'\text{)}\\
\nabla\times\mathbf{H} &= g\mathbf{E} + \varepsilon\frac{\partial\mathbf{E}}{\partial t} && \text{(IV}'\text{)}
\end{aligned} \tag{9-1}$$

If we take the divergence of IV′, the time derivative of I′, and eliminate $\nabla\cdot\partial\mathbf{E}/\partial t$ between the resulting two equations, we obtain

$$\frac{\partial\rho_f}{\partial t} + \frac{g}{\varepsilon}\rho_f = 0 \tag{9-2}$$

This equation has the solution

$$\rho_f(\mathbf{r}, t) = \rho_f(\mathbf{r}, 0)e^{-t/\tau} \tag{9-3}$$

where $\tau = \varepsilon/g$ (for copper $\tau \approx 2 \times 10^{-19}$ s). Even for poor conductors $\tau$ is fairly small, so $\rho_f$ is effectively always zero.

We might comment that the situation is not quite this straightforward [see H. Ohanian, *Am. J. Phys.*, 51: 1020 (1983)], since in finite conductors charge oscillations can also occur; these waves, however, damp out rapidly. In addition, $g$ and $\varepsilon$ generally are functions of frequency, and more correctly, we should consider each frequency component separately. For our subsequent considerations of plane waves in matter, we take $\rho_f = 0$ in I′, which is usually appropriate for either a conductor or an uncharged dielectric.

We now try plane wave solutions to Equations (9–1):

$$\mathbf{E}(\mathbf{r}, t) = \boldsymbol{\mathscr{E}} e^{i(\mathbf{k}\cdot\mathbf{r} - \omega t)} \qquad \mathbf{B}(\mathbf{r}, t) = \boldsymbol{\mathscr{B}} e^{i(\mathbf{k}\cdot\mathbf{r} - \omega t)} \tag{9–4}$$

From I′ with $\rho_f = 0$, we find

$$\mathbf{k} \cdot \boldsymbol{\mathscr{E}} = 0 \tag{9–5}$$

With III′ we can solve for $\boldsymbol{\mathscr{B}}$:

$$\omega \boldsymbol{\mathscr{B}} = \mathbf{k} \times \boldsymbol{\mathscr{E}} \tag{9–6}$$

Hence both $\boldsymbol{\mathscr{E}}$ and $\boldsymbol{\mathscr{B}}$ are transverse to the propagation direction $\mathbf{k}$. The condition II′ is fulfilled by this solution for $\boldsymbol{\mathscr{B}}$. From IV′ we obtain

$$\boldsymbol{\mathscr{E}} \left( \varepsilon\mu + \frac{i\mu g}{\omega} \right) = -\frac{1}{\omega} \mathbf{k} \times \boldsymbol{\mathscr{B}}$$

Since by Equations (9–5) and (9–6), $\mathbf{k} \times (\mathbf{k} \times \boldsymbol{\mathscr{B}}) = -k^2 \boldsymbol{\mathscr{B}}$, we obtain the relation

$$\frac{k^2}{\omega^2} = \varepsilon\mu + \frac{i\mu g}{\omega} \tag{9–7}$$

We define an index of refraction $n$ as

$$n \equiv \frac{kc}{\omega} \tag{9–8}$$

By Equation (9–7) we find

**Complex Index of Refraction**

$$n = \left[ \frac{\varepsilon\mu + (i\mu g/\omega)}{\varepsilon_0 \mu_0} \right]^{1/2} \tag{9–9}$$

This index of refraction is a complex number. Its complex nature is due to the explicit $i$ in Equation (9–9) and due to thc generally complex material constants $\varepsilon$, $\mu$, and $g$ (which are frequency-dependent), as we observed in Chapter 4.

In a dielectric the conductivity is small and $\mu \approx \mu_0$; hence the index of refraction from Equation (9–9) is as follows:

**Dielectric Index of Refraction**

$$n \approx \sqrt{\frac{\varepsilon}{\varepsilon_0}} = \sqrt{\kappa} \tag{9–10}$$

If $\omega$ is far above or below the natural atomic or molecular angular frequencies, the dielectric constant $\kappa$ is real (see Section 4–5) and $n$ is real.

For wave propagation in a good conductor the conductivity term in Equation (9–9) dominates, and we have

**Index of Refraction of a Good Conductor**

$$n \approx e^{i\pi/4}\sqrt{\frac{g\mu}{\omega\mu_0\varepsilon_0}} = (1 + i)\sqrt{\frac{g\mu}{2\omega\mu_0\varepsilon_0}} \tag{9–11}$$

At low frequencies $\omega$, the conductivity $g$ is real and $n$ has equal real and imaginary parts.

We write $n$ in terms of its real and imaginary parts:

$$n = \mathrm{Re}(n) + i\,\mathrm{Im}(n) \tag{9–12}$$

Then from Equation (9–8) the plane wave solution for $\mathbf{E}$ in Equation (9–4) becomes

$$\mathbf{E}(\mathbf{r}, t) = \boldsymbol{\mathscr{E}} \exp\left[-(\mathrm{Im}\,n)\frac{\omega}{c}\hat{\mathbf{k}}\cdot\mathbf{r}\right]\exp\left[i\omega\left(\frac{\mathrm{Re}\,n}{c}\hat{\mathbf{k}}\cdot\mathbf{r} - t\right)\right] \tag{9–13}$$

With the positive $z$ axis along the direction of propagation $\hat{\mathbf{k}}$, the condition for constant phase of the wave is as follows:

**Phase Velocity in Matter**

$$v_p \equiv \frac{dz}{dt} = \frac{c}{\mathrm{Re}\,n} \tag{9–14}$$

which is the phase velocity in matter. Since $\mathrm{Im}\,n > 0$ for all normal substances, the wave is attenuated by a factor

$$\frac{|\mathbf{E}(z, t)|}{|\mathbf{E}(0, t)|} = e^{-\mathrm{Im}\,n(\omega/c)z} \tag{9–15}$$

so that the wave damps by a factor $1/e$ in a distance

$$\delta = \frac{c}{(\mathrm{Im}\,n)\omega} \tag{9–16}$$

called the *skin depth*. For a good conductor we use Equation (9–11) to obtain

$$\delta \approx \sqrt{\frac{2}{g\mu\omega}} \tag{9-17}$$

The skin depth for microwaves of frequency $\nu = 3 \times 10^9$ Hz propagating into silver, which has a conductivity $g \approx 3 \times 10^7$ $(\Omega\cdot\text{m})^{-1}$ at this frequency, is $\delta \approx 5 \times 10^{-6}$ cm. A thin silver plate of at least this thickness is deposited on the inside surfaces of waveguides for microwave transmission to minimize the ohmic heating losses. Insofar as the electrical characteristics of the waveguide are concerned, the coated waveguide functions as well as solid silver since the electromagnetic fields do not penetrate much beyond a depth $\delta$.

The conductivity of seawater is $g \approx 4.3\,(\Omega\cdot\text{m})^{-1}$. At an extremely low frequency (ELF) of 75 Hz the corresponding skin depth is $\delta \approx 28$ m. Submarines are able to receive messages at ELF frequencies in the 72–80 Hz range at depths of 100 m and even deeper with especially sensitive receiving circuits. For comparison we note that at AM radio frequencies ($10^6$ Hz) the skin depth is only about 2.5 cm in seawater.

## 9-2 APPLICATION: OPTICAL ACTIVITY

The study of the interaction of light with matter has yielded vast amounts of information about atomic and molecular structure. An interesting example is the phenomenon of optical activity discovered in the early part of the nineteenth century. In an optically active substance the electric field of a beam of linearly polarized light undergoes a rotation around the ray, proportional to the distance traveled in the substance along the ray. Optical activity can be observed with the following simple experiment. Common colorless corn syrup is placed in a beaker. A Polaroid sheet is placed on one side of the beaker and illuminated by a flashlight. A second Polaroid sheet is placed on the opposite side of the beaker. As the analyzer sheet is rotated with respect to the polarizer sheet, changing colors are observed. This change is caused by the optical rotation, which is inversely proportional to the wavelength of the incident light.

A number of materials exhibit optical activity. Solids such as quartz crystals and liquids such as sugar solutions and turpentine are examples. Remarkably, many organic substances rotate polarized light, which means that these substances have an intrinsic handedness.

Although a complete understanding of optical activity requires a quantum explanation, there is a simple classical description of the phenomenon. Typically, the molecular structure of an optically active substance has a handedness. Suppose the electric field $\mathbf{E}_i$ of an incident wave is parallel to the axis of a helical molecule, as shown in Figure 9–1. Electrons will oscillate along the direction of $\mathbf{E}_i$; the charge separations resulting from this

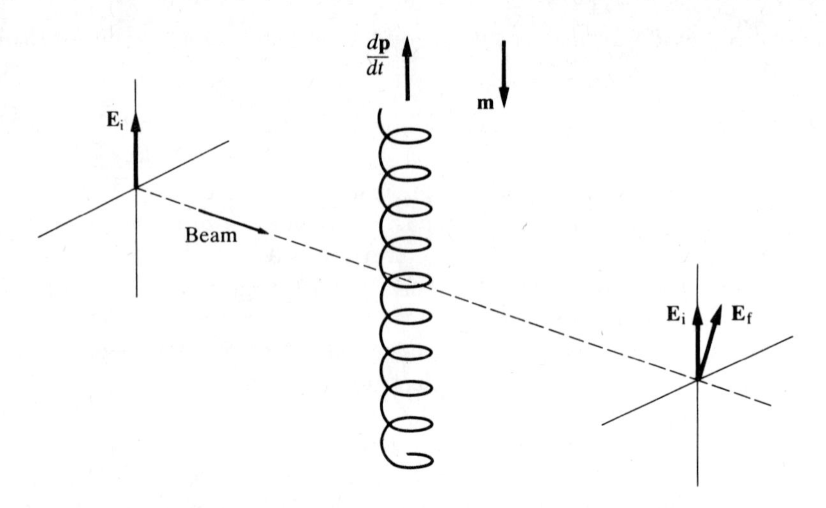


**FIGURE 9-1** Optical rotation of a plane polarized wave passing through a medium containing helical molecules

field give rise to a time-varying electric dipole moment $\mathbf{p}(t)$ parallel to the axis of the molecule helix. To move parallel to the axis of the helix, the electrons must circulate around the helix; thus there is a magnetic moment $\mathbf{m}(t)$ proportional to $d\mathbf{p}(t)/dt$. In terms of macroscopic quantities the polarization current $\partial\mathbf{P}/\partial t$ causes a magnetization $\mathbf{M}$. Thus we can characterize an optically active medium by the material equation

$$\mathbf{M} = \eta \frac{\partial \mathbf{P}}{\partial t} \tag{9-18}$$

where $\eta$ is a proportionality constant with units of length. The sign of $\eta$ depends on the handedness of the material. From the material equation of a dielectric,

$$\mathbf{P} = \varepsilon_0(\kappa - 1)\mathbf{E} \tag{9-19}$$

the magnetization is given by

$$\mathbf{M} = \varepsilon_0(\kappa - 1)\eta \frac{\partial \mathbf{E}}{\partial t} \tag{9-20}$$

Thus the $\mathbf{D}$ and $\mathbf{H}$ fields for a homogeneous, isotropic, optically active medium are

$$\mathbf{D} = \varepsilon_0 \kappa \mathbf{E} \qquad \mathbf{H} = \frac{1}{\mu_0}\mathbf{B} - \varepsilon_0(\kappa - 1)\eta \frac{\partial \mathbf{E}}{\partial t} \tag{9-21}$$

With these relations Maxwell's equations for optically active materials can be expressed as

$$\nabla \cdot \mathbf{E} = 0 \tag{9-22a}$$

$$\nabla \cdot \mathbf{B} = 0 \tag{9-22b}$$

$$\nabla \times \mathbf{E} = -\frac{\partial \mathbf{B}}{\partial t} \tag{9-22c}$$

$$\nabla \times \left(\mathbf{B} - \frac{\kappa - 1}{c^2}\,\eta\,\frac{\partial \mathbf{E}}{\partial t}\right) = \frac{\kappa}{c^2}\frac{\partial \mathbf{E}}{\partial t} \tag{9-22d}$$

We look for simple harmonic plane wave solutions of the form [see Equation (9–13)]

$$\mathbf{E}(z, t) = \boldsymbol{\mathcal{E}} e^{i\omega[(n/c)z - t]} \tag{9-23}$$

with the wave propagating in the direction of the positive $z$ axis. From Eq. (9–22c) the **B** field is

$$\mathbf{B}(z, t) = \frac{n}{c}\hat{\mathbf{z}} \times \boldsymbol{\mathcal{E}} e^{i\omega[(nz/c) - t]} \tag{9-24}$$

Equations (9–22a) and (9–22b) require that $\boldsymbol{\mathcal{E}}$ be perpendicular to $\hat{\mathbf{z}}$. Finally, Equation (9–22d) yields the relationship

$$\boldsymbol{\mathcal{E}} = iF\hat{\mathbf{z}} \times \boldsymbol{\mathcal{E}} \tag{9-25}$$

where the quantity $F$ is defined by

$$F = \eta(\kappa - 1)\frac{n\omega}{c}\frac{1}{n^2 - \kappa} \tag{9-26}$$

With $\boldsymbol{\mathcal{E}} = \mathcal{E}_x\hat{\mathbf{x}} + \mathcal{E}_y\hat{\mathbf{y}}$, the condition in Equation (9–25) can be written as

$$\mathcal{E}_x = -iF\mathcal{E}_y \qquad \mathcal{E}_y = iF\mathcal{E}_x \tag{9-27}$$

This condition implies that $F = \pm 1$, for which

$$\boldsymbol{\mathcal{E}} = \mathcal{E}_x(\hat{\mathbf{x}} \pm i\hat{\mathbf{y}}) \tag{9-28}$$

Thus only circularly polarized plane waves propagate without change of polarization through an optically active substance.

Imposing the condition $F = \pm 1$ on Equation (9–26), we obtain

$$n^2 = \kappa \pm \eta\left(\frac{n}{c}\right)\omega(\kappa - 1) \tag{9-29}$$

For small $\eta$ we can approximate $n$ as

$$n \approx \sqrt{\kappa} \pm \frac{\eta n \omega(\kappa - 1)}{2c\sqrt{\kappa}} \approx \sqrt{\kappa} \pm \frac{\eta \omega(\kappa - 1)}{2c} \qquad \textbf{(9–30)}$$

Referring to Equation (9–28), we note that $F = +1$ ($F = -1$) corresponds to left (right) circularly polarized waves. Hence we have

$$n_L - n_R \approx \frac{\eta(\kappa - 1)\omega}{c} \qquad \textbf{(9–31)}$$

The difference in the left and right indices of refraction is proportional to the angular frequency $\omega$ (i.e., inversely proportional to the wavelength); the sign of the difference is determined by the sign of $\eta$.

We are now able to give a quantitative classical description of optical rotation. The appropriate physical states are left- and right-handed circularly polarized waves. For propagation in the positive $z$ direction the left and right circularly polarized components of the electric field can be written as $\mathbf{E} = \mathbf{E}_L + \mathbf{E}_R$, where we have

**Left and Right Circular Polarized Waves**

$$\mathbf{E}_L = \frac{1}{2}\mathscr{E}(\hat{\mathbf{x}} + i\hat{\mathbf{y}})e^{i[(n_L/c)z - t]\omega}$$

$$\mathbf{E}_R = \frac{1}{2}\mathscr{E}(\hat{\mathbf{x}} - i\hat{\mathbf{y}})e^{i[(n_R/c)z - t]\omega} \qquad \textbf{(9–32)}$$

and $n_L$ and $n_R$ are the corresponding indices of refraction. The amplitudes $\mathscr{E}$ have been chosen to be equal so that the electric field at $z = 0$ is plane-polarized in the $\hat{\mathbf{x}}$ direction:

$$\mathbf{E}(z = 0, t) = \mathscr{E}\hat{\mathbf{X}}e^{-i\omega t}$$

This expression represents a plane polarized wave entering the medium at $z = 0$. The left and right circularly polarized waves propagate at different speeds in the substance. At a distance $z$ into the medium the total electric field is

$$\mathbf{E}(z, t) = \frac{1}{2}\mathscr{E}[\hat{\mathbf{x}}(e^{in_L(\omega/c)z} + e^{in_R(\omega/c)z}) + i\hat{\mathbf{y}}(e^{in_L(\omega/c)z} - e^{in_R(\omega/c)z})]e^{-i\omega t} \qquad \textbf{(9–33)}$$

To show that the wave is plane-polarized at $z$ in a new direction, we first express $n_L$ and $n_R$ in terms of the sum and difference $n_L \pm n_R$ as

$$n_L = \frac{1}{2}(n_L + n_R) + \frac{1}{2}(n_L - n_R)$$

$$n_R = \frac{1}{2}(n_L + n_R) - \frac{1}{2}(n_L - n_R) \qquad \textbf{(9–34)}$$

With this replacement we are able to recast Equation (9–33) in the form

$$\mathbf{E}(z, t) = \boldsymbol{\mathscr{E}} e^{(i/2)(n_L + n_R)(\omega/c)z} e^{-i\omega t} \tag{9-35}$$

with

$$\boldsymbol{\mathscr{E}} = (\hat{\mathbf{x}} \cos \beta - \hat{\mathbf{y}} \sin \beta)\mathscr{E} \qquad \beta = \frac{1}{2}(n_L - n_R)\left(\frac{\omega}{c}\right)z \tag{9-36}$$

Anywhere along the $z$ axis the resultant **E** vector is linearly polarized, its orientation depending on $z$. The rotation per unit length of propagation,

$$\frac{\beta}{z} = \frac{1}{2}(n_L - n_R)\frac{\omega}{c} \tag{9-37}$$

is called the *specific rotatory power*. It is positive—that is, the plane of polarization rotates clockwise—if $n_L > n_R$. Some representative values of $\beta/z$ are given in Table 9–1. Since $\omega/c = 2\pi/\lambda$, where $\lambda$ is the incident wavelength, Equation (9–37) can be alternatively expressed as

$$\frac{\beta}{z} = \frac{\pi}{\lambda}(n_L - n_R) \tag{9-38}$$

As a numerical example, we return to the case of corn syrup and solve for $n_L - n_R$ from the value of $\beta/z$ in Table 9–1:

$$\begin{aligned} n_L - n_R &\approx (12°/\text{cm})\left(\frac{1 \text{ rad}}{57.3°}\right)\left(\frac{100 \text{ cm}}{\text{m}}\right)(5 \times 10^{-7} \text{ m}) \\ &\approx 3.3 \times 10^{-6} \end{aligned} \tag{9-39}$$

A similar calculation for a cholesteric liquid crystal gives

$$n_L - n_R \approx 0.1 \tag{9-40}$$

Note that although Maxwell's equations in empty space have no a priori preferred handedness, the equations and solutions to physical prob-

**TABLE 9–1** Values of specific rotatory power for selected substances

| Substance | Specific Rotatory Power (°/cm) in Green Light |
|---|---|
| Quartz crystal | 220 |
| Corn syrup | 12 |
| Turpentine | −3.7 |
| Cholesteric liquid crystal | 400,000 |

lems may have a left-right asymmetry. In optical activity the handedness is determined by the handedness of the microscopic structure.

## 9-3 REFLECTION AND REFRACTION

At the interface between two homogeneous optical materials, a plane wave splits into two components, a transmitted (or refracted) wave in the second medium and a reflected wave propagating back into the first medium. This situation is illustrated in Figure 9–2.

The electric and magnetic fields for each of these waves will be assumed to have the form of plane waves in their respective media. We then match these waves by using the boundary conditions of Table 8–4. From the mere existence of these boundary conditions on the interface, the laws of

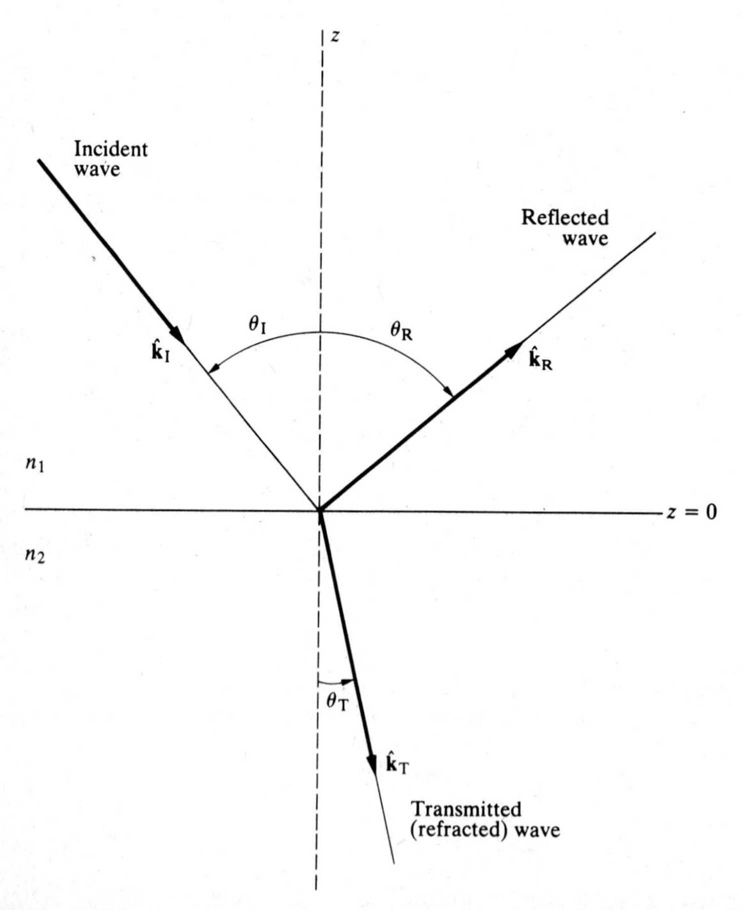


**FIGURE 9-2** Components of plane wave. An incident plane wave in a medium of index $n_1$ results in a reflected wave in $n_1$ and a transmitted wave in the medium of index $n_2$.

reflection and refraction are established independent of any dynamic details.

We assume the following plane wave forms for the incident, reflected and transmitted waves, respectively:

$$\begin{aligned}\mathbf{E}_{\mathrm{I}} &= \boldsymbol{\mathscr{E}}_{\mathrm{I}} e^{i(\mathbf{k}_{\mathrm{I}}\cdot\mathbf{r}-\omega_{\mathrm{I}}t)}\\ \mathbf{E}_{\mathrm{R}} &= \boldsymbol{\mathscr{E}}_{\mathrm{R}} e^{i(\mathbf{k}_{\mathrm{R}}\cdot\mathbf{r}-\omega_{\mathrm{R}}t)}\\ \mathbf{E}_{\mathrm{T}} &= \boldsymbol{\mathscr{E}}_{\mathrm{T}} e^{i(\mathbf{k}_{\mathrm{T}}\cdot\mathbf{r}-\omega_{\mathrm{T}}t)}\end{aligned} \tag{9–41}$$

where in each case, by Equations (9–6) and (9–8), the magnetic field is

$$\mathbf{B} = \frac{n}{c}\hat{\mathbf{k}} \times \mathbf{E} \tag{9–42}$$

The three electric and three magnetic field vectors must satisfy the boundary conditions of Table 8–4 at *every point* on the plane interface *at all times*. If the boundary conditions hold at one point and at one time, they will not hold at all times unless the phases of the waves in Equation (9–41) are equal. The equality of phases requires that the three angular frequencies must be the same:

$$\omega_{\mathrm{I}} = \omega_{\mathrm{R}} = \omega_{\mathrm{T}} \equiv \omega \tag{9–43}$$

Equality of the phases for all points on the interface implies that

$$\mathbf{k}_{\mathrm{I}}\cdot\mathbf{r} = \mathbf{k}_{\mathrm{R}}\cdot\mathbf{r} = \mathbf{k}_{\mathrm{T}}\cdot\mathbf{r} \tag{9–44}$$

for all $\mathbf{r}$ on the interface. If we take the plane media interface to lie in the $z = 0$ plane ($\mathbf{r} = x\hat{\mathbf{x}} + y\hat{\mathbf{y}}$), the equalities of Equations (9–44) become

$$k_{\mathrm{I}} \sin\theta_{\mathrm{I}} = k_{\mathrm{R}} \sin\theta_{\mathrm{R}} \tag{9–45a}$$

$$k_{\mathrm{I}} \sin\theta_{\mathrm{I}} = k_{\mathrm{T}} \sin\theta_{\mathrm{T}} \tag{9–45b}$$

Since by Equation (9–8), $k = (\omega/c)n$, the wave numbers are related by Equation (9–43) as

$$k_{\mathrm{I}} = k_{\mathrm{R}} \equiv k_1 \tag{9–46a}$$

$$\frac{k_1}{n_1} = \frac{k_{\mathrm{T}}}{n_2} \tag{9–46b}$$

Combining Equations (9–45a) and (9–46a), we see that the reflected angle must equal the incident angle:

**Law of Reflection**

$$\theta_{\mathrm{R}} = \theta_{\mathrm{I}} \tag{9–47}$$

When Equation (9–45b) is substituted into Equation (9–46b), we obtain

**Snell's Law of Refraction**

$$n_1 \sin \theta_\mathrm{I} = n_2 \sin \theta_\mathrm{T} \tag{9–48}$$

which is Snell's law of refraction. The reflection and refraction laws are valid for many wave motions, such as sound waves. How the incident intensity splits up between reflected and transmitted waves will depend on the media and the incident wave direction, which we discuss next.

## 9-3A Fresnel Equations

By the necessary equality of the phases of the fields of Equations (9–41), the boundary conditions of Table 8–4 are reduced to relations among the constant vectors $\boldsymbol{\mathcal{E}}_\mathrm{I}$, $\boldsymbol{\mathcal{E}}_\mathrm{R}$, and $\boldsymbol{\mathcal{E}}_\mathrm{T}$. We now will determine the necessary relations between these constants. That there must be a solution is not immediately obvious, because there are four boundary conditions and we can have the freedom of fixing only two ratios. However, it turns out that for this application not all the boundary conditions are independent and only two of them provide new constraints. We will also see that these field vectors are relatively real. For simplicity we consider only nonmagnetic materials so that $\mu_1 = \mu_2 = \mu_0$, which is usually the case for most materials that light can pass through.

The boundary conditions of Table 8–4 applied to the situation are

$$(\varepsilon_1 \boldsymbol{\mathcal{E}}_\mathrm{I} + \varepsilon_1 \boldsymbol{\mathcal{E}}_\mathrm{R}) \cdot \hat{\mathbf{n}} = \varepsilon_2 \boldsymbol{\mathcal{E}}_\mathrm{T} \cdot \hat{\mathbf{n}} \tag{9–49a}$$

$$(\boldsymbol{\mathcal{B}}_\mathrm{I} + \boldsymbol{\mathcal{B}}_\mathrm{R}) \cdot \hat{\mathbf{n}} = \boldsymbol{\mathcal{B}}_\mathrm{T} \cdot \hat{\mathbf{n}} \tag{9–49b}$$

$$(\boldsymbol{\mathcal{E}}_\mathrm{I} + \boldsymbol{\mathcal{E}}_\mathrm{R}) \cdot \hat{\mathbf{t}} = \boldsymbol{\mathcal{E}}_\mathrm{T} \cdot \hat{\mathbf{t}} \tag{9–49c}$$

$$(\boldsymbol{\mathcal{B}}_\mathrm{I} + \boldsymbol{\mathcal{B}}_\mathrm{R}) \cdot \hat{\mathbf{t}} = \boldsymbol{\mathcal{B}}_\mathrm{T} \cdot \hat{\mathbf{t}} \tag{9–49d}$$

where $\hat{\mathbf{n}}$ is the normal unit vector to the boundary, $\hat{\mathbf{t}}$ is a tangential unit vector, and

$$\boldsymbol{\mathcal{B}} = \frac{n}{c} \hat{\mathbf{k}} \times \boldsymbol{\mathcal{E}} \tag{9–49e}$$

From Equation (9–44) we observe that $\mathbf{k}_\mathrm{I}$, $\mathbf{k}_\mathrm{R}$, and $\mathbf{k}_\mathrm{T}$ have the same projection onto the $z = 0$ plane. Hence the incident, reflected, and transmitted waves all lie in a plane, perpendicular to the media interface, called the *scattering plane*. Since the waves are all transverse, we can conveniently consider two cases: (1) the incident electric vector is perpendicular to the scattering plane, and (2) the incident electric vector is in the scattering plane. The general situation is a superposition of these two cases.

As we mentioned in Chapter 8, the polarization direction can be identified with the electric field direction. We will on occasion use the terms *electric field direction* and *polarization* interchangeably. It is the electric field of a wave that more strongly acts on electric charges, since the magnetic force is proportional to the charge velocity, which is usually much smaller than $c$.

**Electric Vector Perpendicular to the Scattering Plane.** If the incident electric field is normal to the scattering plane, then the electric fields of the reflected and transmitted waves must have their electric fields similarly oriented. If they were not, the electric fields in the scattering plane could not simultaneously satisfy continuity of tangential **E** and normal **D**. For electric fields normal to the scattering plane the electric and magnetic field amplitudes are shown in Figure 9–3. The boundary conditions of Equations

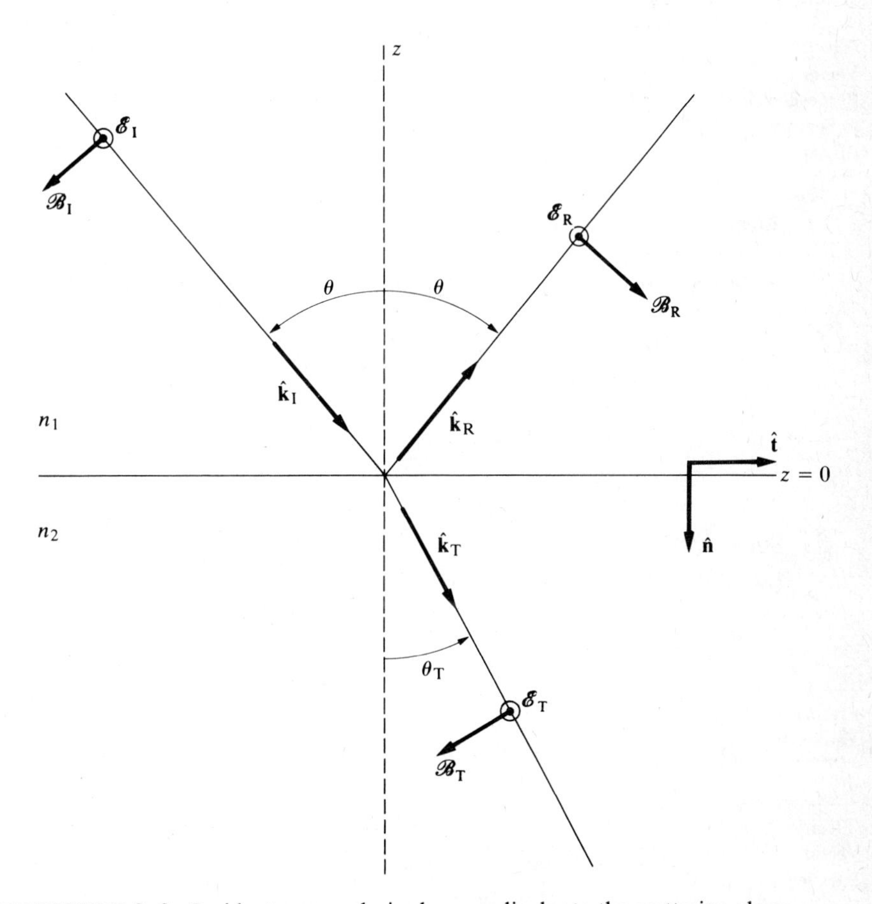


**FIGURE 9–3** Incident wave polarized perpendicular to the scattering plane

(9–49) applied to this case are, respectively,

$$E_z = 0 \tag{9-50a}$$

$$n_1 \mathscr{E}_\mathrm{I} \sin\theta + n_1 \mathscr{E}_\mathrm{R} \sin\theta = n_2 \mathscr{E}_\mathrm{T} \sin\theta_\mathrm{T} \tag{9-50b}$$

$$\mathscr{E}_\mathrm{I} + \mathscr{E}_\mathrm{R} = \mathscr{E}_\mathrm{T} \tag{9-50c}$$

$$-n_1 \mathscr{E}_\mathrm{I} \cos\theta + n_1 \mathscr{E}_\mathrm{R} \cos\theta = -n_2 \mathscr{E}_\mathrm{T} \cos\theta_\mathrm{T} \tag{9-50d}$$

Only two of these equations have new content, since Snell's law, Equation (9–48), combined with Equation (9–50b) just reproduces Equation (9–50c). Solving for the ratios $\mathscr{E}_\mathrm{R}/\mathscr{E}_\mathrm{I}$ and $\mathscr{E}_\mathrm{T}/\mathscr{E}_\mathrm{I}$ from Equations (9–50c) and (9–50d) yields the following Fresnel equations:

**Fresnel Equations**

$$\left(\frac{\mathscr{E}_\mathrm{T}}{\mathscr{E}_\mathrm{I}}\right)_\perp = \frac{2n_1 \cos\theta}{n_1 \cos\theta + n_2 \cos\theta_\mathrm{T}} \tag{9-51a}$$

**(Electric Vector Perpendicular to Scattering Plane)**

$$\left(\frac{\mathscr{E}_\mathrm{R}}{\mathscr{E}_\mathrm{I}}\right)_\perp = \frac{n_1 \cos\theta - n_2 \cos\theta_\mathrm{T}}{n_1 \cos\theta + n_2 \cos\theta_\mathrm{T}} \tag{9-51b}$$

**Electric Vector in the Scattering Plane.** When the **E** vectors lie within (parallel to) the scattering plane, as shown in Figure 9–4, the boundary conditions of Equations (9–49) become, respectively,

$$\varepsilon_1 \mathscr{E}_\mathrm{I} \sin\theta - \varepsilon_1 \mathscr{E}_\mathrm{R} \sin\theta = \varepsilon_2 \mathscr{E}_\mathrm{T} \sin\theta_\mathrm{T} \tag{9-52a}$$

$$B_z = 0 \tag{9-52b}$$

$$-\mathscr{E}_\mathrm{I} \cos\theta - \mathscr{E}_\mathrm{R} \cos\theta = -\mathscr{E}_\mathrm{T} \cos\theta_\mathrm{T} \tag{9-52c}$$

$$\mathscr{B}_\mathrm{I} - \mathscr{B}_\mathrm{R} = \mathscr{B}_\mathrm{T} \tag{9-52d}$$

Since $\varepsilon_2/\varepsilon_1 = (n_2/n_1)^2$ and $\mathscr{B} = (n/c)\mathscr{E}$, we note that Snell's law shows that Equations (9–52a) and (9–52d) are equivalent. So, again, there are two independent boundary conditions to determine the two ratios, which yield the following Fresnel equations:

**Fresnel Equations**

$$\left(\frac{\mathscr{E}_\mathrm{T}}{\mathscr{E}_\mathrm{I}}\right)_{/\!/} = \frac{2n_1 \cos\theta}{n_1 \cos\theta_\mathrm{T} + n_2 \cos\theta} \tag{9-53a}$$

**(Electric Vector Parallel to Scattering Plane)**

$$\left(\frac{\mathscr{E}_\mathrm{R}}{\mathscr{E}_\mathrm{I}}\right)_{/\!/} = \frac{n_1 \cos\theta_\mathrm{T} - n_2 \cos\theta}{n_1 \cos\theta_\mathrm{T} + n_2 \cos\theta} \tag{9-53b}$$

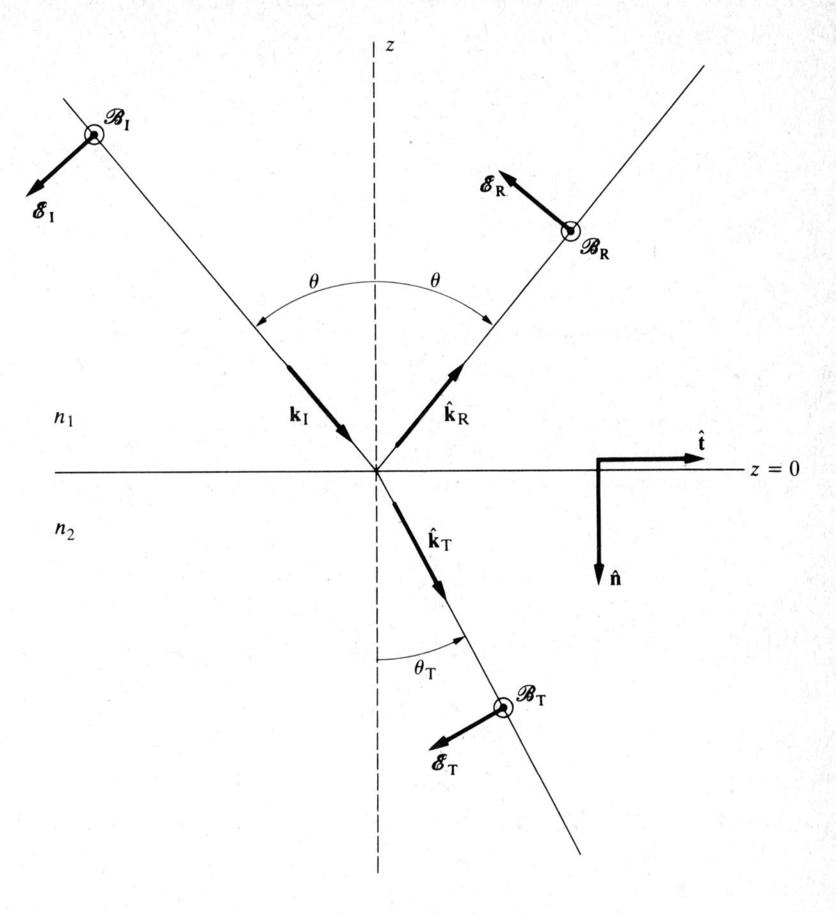


**FIGURE 9-4** Incident wave polarized in the scattering plane

If the incident radiation strikes the boundary normally ($\theta = \theta_T = 0$), the two Fresnel cases reduce to the same result:

**Normal Incidence**

$$\frac{\mathscr{E}_T}{\mathscr{E}_I} = \frac{2n_1}{n_1 + n_2}$$
$$\frac{\mathscr{E}_R}{\mathscr{E}_I} = \frac{n_1 - n_2}{n_1 + n_2} \tag{9-54}$$

The reflected wave has opposite sign to the incident wave when $n_2$ is larger than $n_1$. This well-known phase reversal produces the central dark spot in the Newton's rings demonstration.

Another interesting special case is that of *grazing incidence*. As $\theta \to \pi/2$, in either Fresnel case, we see by Equation (9–51b) or Equation

(9–53b) that

$$\mathscr{E}_R \rightarrow -\mathscr{E}_I \qquad \mathscr{E}_T = 0 \tag{9-55}$$

Hence for grazing incidence all of the light is reflected. This effect is easily observed by using a piece of glass. For normal incidence the reflection is weak, but the glass acts like a perfect mirror for grazing incidence.

## 9-3B Reflectivity and Transmissivity: Polarization and Total Reflection

We now examine how the energy of the incident wave is divided between the two secondary waves. From Poynting's vector, Equation (8–81), the *energy flux*, or intensity (in watts per meter squared), crossing the media boundary is

$$Y_z = Y\cos\theta = \frac{n}{2\mu_0 c}\mathscr{E}^2\cos\theta \tag{9-56}$$

The ratio of the reflected to incident energy flux is called the *reflectivity*:

$$R = \frac{Y_{zR}}{Y_{zI}} = \left(\frac{\mathscr{E}_R}{\mathscr{E}_I}\right)^2 \tag{9-57}$$

The *transmissivity* is similarly defined as the ratio of transmitted to incident energy flux:

$$T = \frac{Y_{zT}}{Y_{zI}} = \frac{n_2\cos\theta_T}{n_1\cos\theta}\left(\frac{\mathscr{E}_T}{\mathscr{E}_I}\right)^2 \tag{9-58}$$

The reflectivity and transmissivity depend on the initial electric field direction or polarization. For the two cases of Equations (9–51) and (9–53) we obtain from Equations (9–57) and (9–58) the results

$$R_\perp = \left(\frac{n_1\cos\theta - n_2\cos\theta_T}{n_1\cos\theta + n_2\cos\theta_T}\right)^2 \tag{9-59a}$$

$$T_\perp = \frac{4n_1n_2\cos\theta\cos\theta_T}{(n_1\cos\theta + n_2\cos\theta_T)^2} \tag{9-59b}$$

$$R_{/\!/} = \left(\frac{n_1\cos\theta_T - n_2\cos\theta}{n_1\cos\theta_T + n_2\cos\theta}\right)^2 \tag{9-60a}$$

$$T_{/\!/} = \frac{4n_1n_2\cos\theta\cos\theta_T}{(n_1\cos\theta_T + n_2\cos\theta)^2} \tag{9-60b}$$

We see immediately from these equations that for either polarization energy is conserved, since

$$R + T = 1 \tag{9-61}$$

**Reflection from a More Dense Medium.** In Figure 9 – 5 we plot the reflectivities $R_\perp$ and $R_{//}$ as a function of the incident angle $\theta$ for reflection from the interface between air, $n_1 = 1$, and glass, with $n_2 = 1.5$. For angles between 30° and 80° relatively little light is reflected when the electric field lies in the scattering plane. At Brewster's angle, defined as

**Brewster's Angle**

$$\tan \theta_B = \frac{n_2}{n_1} \tag{9-62}$$

$R_{//}$ exactly vanishes, as found by using Equations (9–60a) and (9–48).

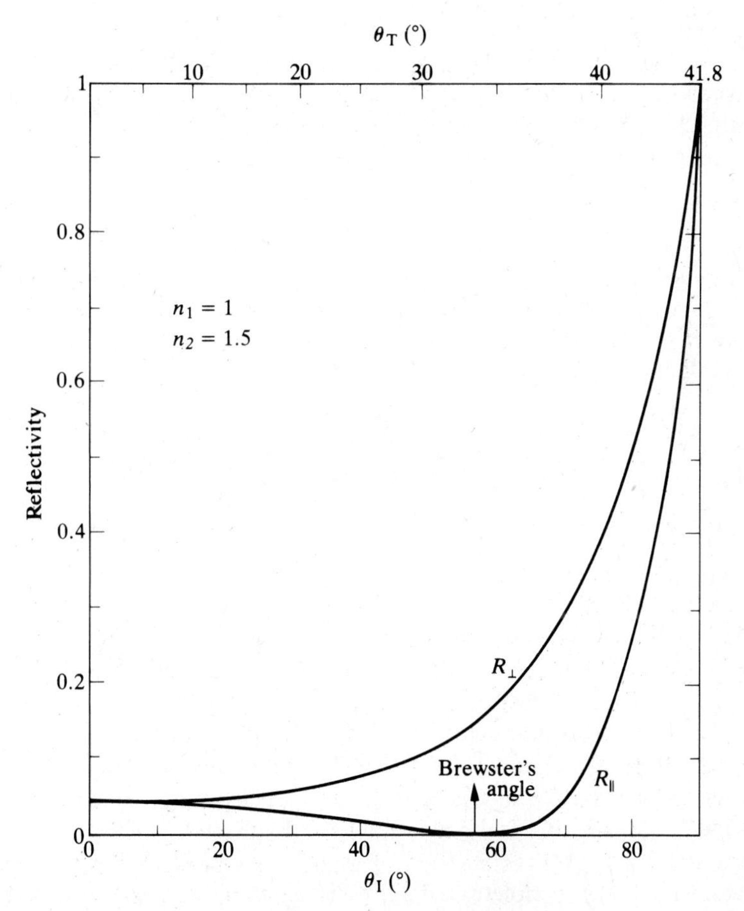


**FIGURE 9–5** Reflectivities for polarization perpendicular to the plane, $R_\perp$, and parallel to the scattering plane, $R_{//}$, for reflection from a medium of larger index

A common method for plane polarizing light is to pass it through a stack of thin glass plates. Since $R_{//}/R_{\perp}$ is less than unity, the transmitted light will emphasize the parallel component. Unpolarized light sent through a stack of five plates at the Brewster's angle will result in a 5-to-1 ratio of parallel-to-perpendicular polarization.

Perhaps the easiest and most common method of plane-polarizing a light beam is to pass it through a material having greatly different absorptions in two directions. Sheets of organic polymers with long-chain molecules brought into almost complete alignment by stretching or other treatment are known as Polaroid sheets. The ratio of transmissivities of a Polaroid can be as high as 100:1 with about 80% transmission in one direction and less than 1% at right angles. This difference is related to the large polarizability along the polymer chain direction. As we will establish in the next chapter, radiation from an oscillating electric dipole is proportional to its polarizability, and the result is large absorption of oscillating electric fields in this direction. It is only the light polarized perpendicular to the polymer chains that passes through.

When sunlight reflects from a surface, such as a body of water, the reflected light is primarily polarized normal to the scattering plane (i.e., parallel to the water). If one wears polaroid glasses oriented to eliminate this component, the reflection glare is largely eliminated.

**Reflection from a Less Dense Medium.** In Figure 9–6 we plot the reflectivity predicted by Equations (9–59a) and (9–60a) for $n_1 = 1.5$ and $n_2 = 1.0$. As a function of $\theta$, both $R_{//}$ and $R_{\perp}$ become unity above $\theta_T = 90°$. This phenomenon is *total reflection.* We also see from Figure 9–6 that there is a Brewster's angle where light is reflected only if its electric field vector is perpendicular to the scattering plane.

At normal incidence $\theta = \theta_T = 0$, and the reflectivity is

$$R_{\perp} = R_{//} = \left(\frac{n-1}{n+1}\right)^2 = 0.04$$

for $n = 1.5$ from either the front or the back surface of a window. Hence with reflection from the front and back sides of a glass plate, about 8% of the incident energy is reflected normally. For precision optical lenses with many lens elements, this reflection loss is serious, and furthermore, it degrades the image by spurious reflections. A thin coating of $\frac{1}{4}$-wavelength thickness of a material having an index equal to the geometric mean of the two media indices will entirely eliminate reflections of light of this wavelength. The reflectivity from the coating surface and the lens surface is the same if the coating has index $\sqrt{n}$, where $n$ is the glass index. The quarter-wavelength thickness of the coating then gives destructive interference between the two reflected waves in the air. A complete calculation is left as an exercise.

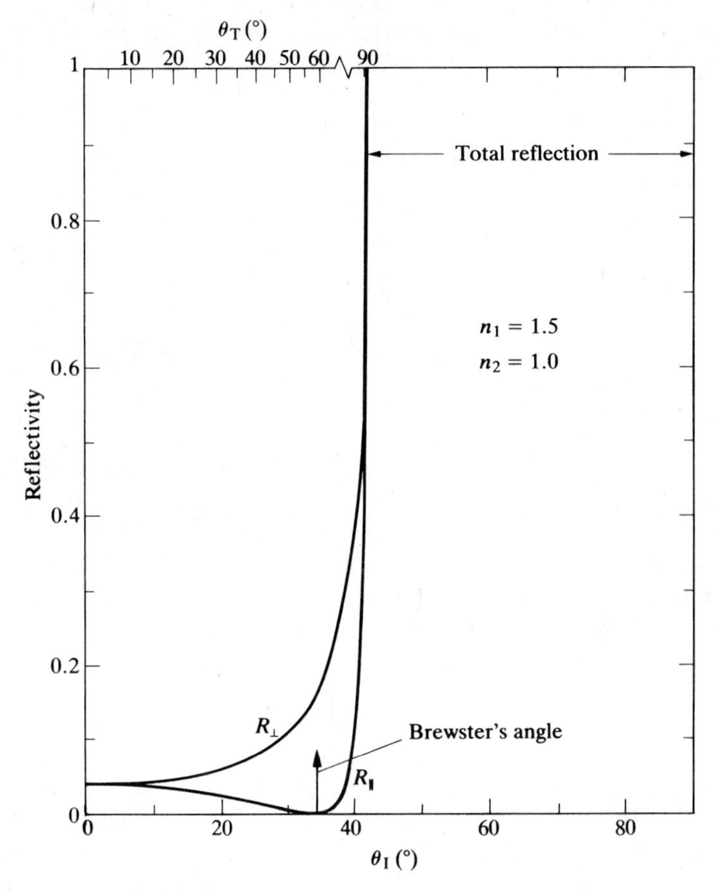


**FIGURE 9–6** Reflectivities $R_\perp$ and $R_{//}$ for reflection from a medium of smaller refractive index

**Total Reflection.** As the angle of incidence increases, the transmitted-ray angle also increases, equaling 90° when $\sin\theta = 1/n$. With further increase in $\theta$ the transmitted ray disappears, and the reflected ray carries all of the incident intensity. This situation is total reflection. Actually, the transmitted wave does not really disappear completely, for it is required to satisfy the boundary conditions; it damps away exponentially from the media plane and carries energy only tangential to this plane. It is called an *evanescent wave.*

From Snell's law, $\sin\theta_T = n\sin\theta$, we find

$$\cos\theta_T = \sqrt{1-\sin^2\theta_T} = \sqrt{1-n^2\sin^2\theta} = \pm i\sqrt{n^2\sin^2\theta - 1} \tag{9-63}$$

Although $\theta_T$ is no longer a real angle, the boundary conditions of Equation (9–44) involve the wave propagation vector

$$\mathbf{k}_T = k_T(\hat{\mathbf{x}} \sin \theta_T - \hat{\mathbf{z}} \cos \theta_T) \tag{9-64}$$

where $(x, z)$ is taken to be the scattering plane. By Equations (9–41) and (9–46) the transmitted electric field is

$$\begin{aligned}\mathbf{E}_T &= \mathscr{E}_T \exp\left[i\left(\frac{\omega}{c}\right)(nx \sin \theta \pm iz\sqrt{n^2 \sin^2 \theta - 1} - ct)\right] \\ &= \mathscr{E}_T \exp\left[i\left(\frac{\omega}{c}\right)(nx \sin \theta - ct)\right] \exp\left[\pm\left(\frac{\omega}{c}\right)\sqrt{n^2 \sin^2 \theta - 1}\, z\right]\end{aligned} \tag{9-65}$$

We must choose the positive sign in this exponential; otherwise, an infinite amount of energy would reside at $z = -\infty$. The solution thus represents a wave propagating along the $x$ axis with a phase velocity characteristic of the adjoining medium. This wave is confined to a depth $\lambda/(2\pi\sqrt{n^2 \sin^2 \theta - 1})$ in the $z$ direction measured from the media boundary, where $\lambda$ is the free-space wavelength. The damping of the wave is a direct consequence of imaginary $\cos \theta_T$. The existence of this surface wave can be verified by putting a second glass plate within a distance less than $\lambda$ behind the glass-air interface. Radiation then leaks into the plate from the incident region.

In the totally reflecting case, when $n \sin \theta \geqslant 1$, Equations (9–51) and (9–53) imply that the ratio of reflected to incident electric fields for either polarization case is of the form $\mathscr{E}_R/\mathscr{E}_I = (\alpha - i\beta)/(\alpha + i\beta)$, which is a complex number of modulus unity. The reflectivities are thus unity, as expected.

In our discussion of refraction and reflection we have assumed a sinusoidal time dependence of the waves, that is, a steady-state situation where the incident wave has always been there. The incident radiation interacts with the atoms of the media, which in turn radiate and change the incident wave. Until the atoms are set in motion, the leading edge of the pulse moves unchanged in direction and speed. Thus a small amount of light will pass through as if there were no material present. After a sufficient time has passed, radiation from the atoms cancels, by destructive interference, the incident wave within the new medium and builds up the reflected and refracted rays. This process is worked out in detail in the next chapter.

## 9-3C Crystal Optics

Up to this point we have confined ourselves to propagation of waves in isotropic media, whose optical properties are the same in all directions. Liquids as well as amorphous solids like plastic or glass are usually treated as isotropic since their molecules are situated and oriented randomly. However, in many optical crystals the physical properties are different in different directions. This optical anisotropy is called *birefringence* or *double refraction.*

Most crystals are not magnetic, so we take $\mu = \mu_0$. Even for weak electric fields the induced polarization will not generally be parallel to the imposed electric field. This situation is described by defining, as mentioned in Chapter 4, a permittivity tensor $\varepsilon_{ij}$ such that

$$D_i = \sum_{j=1}^{3} \varepsilon_{ij} E_j \tag{9-66}$$

Only if $\varepsilon_{ij} = \varepsilon\delta_{ij}$ is $\mathbf{D}$ always proportional to $\mathbf{E}$. The $\varepsilon_{ij}$ tensor is real and symmetric, as can be proved by using an energy argument. Owing to the symmetry of $\varepsilon_{ij}$, there exists a rotated coordinate system in which the off-diagonal elements vanish, and we can write, in this coordinate system,

$$\mathbf{D} = \varepsilon_1 E_1 \hat{\mathbf{x}}_1 + \varepsilon_2 E_2 \hat{\mathbf{x}}_2 + \varepsilon_3 E_3 \hat{\mathbf{x}}_3 \tag{9-67}$$

where $\hat{\mathbf{x}}_1$, $\hat{\mathbf{x}}_2$, and $\hat{\mathbf{x}}_3$ are the unit vectors along the new coordinate axes. These axes for which $\varepsilon_{ij}$ is diagonal are known as the *principal axes*. Cubic crystals are optically isotropic, so $\varepsilon_1 = \varepsilon_2 = \varepsilon_3$. In crystals of trigonal, tetragonal, and hexagonal systems, $\varepsilon_1 = \varepsilon_2$ and the crystal is optically isotropic in one plane. The normal to this plane is called the *crystal axis*. Such crystals are called uniaxial. Finally, for orthorhombic, monoclinic, and triclinic systems, $\varepsilon_1 \neq \varepsilon_2 \neq \varepsilon_3$; such crystals are called biaxial.

We next look for plane wave solutions in a crystal medium. As usual, we try

$$\begin{aligned} \mathbf{E} &= \mathscr{E} e^{i(\mathbf{k}\cdot\mathbf{r} - \omega t)} \\ \mathbf{D} &= \mathscr{D} e^{i(\mathbf{k}\cdot\mathbf{r} - \omega t)} \\ \mathbf{B} &= \mathscr{B} e^{i(\mathbf{k}\cdot\mathbf{r} - \omega t)} \end{aligned} \tag{9-68}$$

Substitution into Maxwell's equations in matter, Table 8–3, gives

$$\begin{aligned} \mathbf{k}\cdot\mathscr{D} &= 0 \qquad \mathbf{k}\cdot\mathscr{B} = 0 \\ \mathbf{k}\times\mathscr{E} &= \omega\mathscr{B} \qquad \mathbf{k}\times\mathscr{B} = -\mu_0\omega\mathscr{D} \end{aligned} \tag{9-69}$$

From Equations (9–69) we observe that $\mathbf{k}$, $\mathscr{D}$, and $\mathscr{B}$ form an orthogonal triad and that $\mathscr{E}$ lies in the plane of $\mathbf{k}$ and $\mathscr{D}$. The situation is illustrated in Figure 9–7.

Elimination of $\mathscr{B}$ in Equations (9–69) leads to

$$\mu_0 c^2 \mathscr{D} = n^2[\mathscr{E} - (\hat{\mathbf{k}}\cdot\mathscr{E})\hat{\mathbf{k}}] \tag{9-70}$$

where, as usual, $n = ck/\omega$. In the principal coordinate system of Equation (9–67) we have then

$$c^2\mu_0\varepsilon_i\mathscr{E}_i = n^2[\mathscr{E}_i - \hat{\mathbf{k}}_i\hat{\mathbf{k}}_j\mathscr{E}_j] \tag{9-71}$$

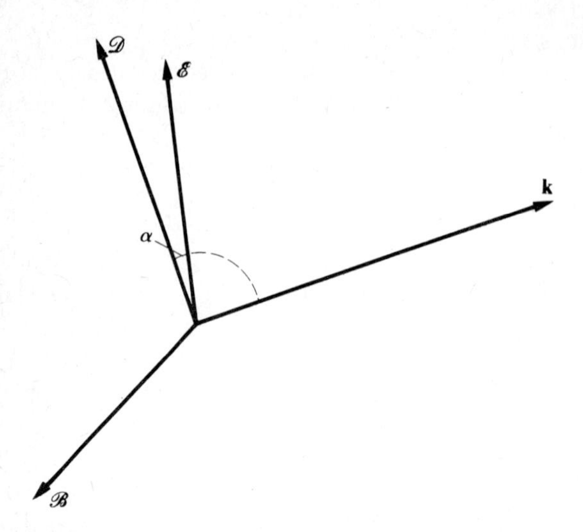


**FIGURE 9-7** Plane wave propagating in a crystalline material. The electric $\mathscr{E}$, magnetic $\mathscr{B}$, and displacement $\mathscr{D}$ field amplitudes are shown in relation to the propagation direction **k**.

This system of homogeneous equations for $\mathscr{E}_i$ has nontrivial solutions ($\mathscr{E}_i$ not all vanishing) only if $n^2$ satisfies the condition

$$\sum_{i=1}^{3} \frac{n_i^2 \hat{k}_i^2}{n^2 - n_i^2} = 0 \tag{9-72}$$

where

$$n_i \equiv \frac{c}{\sqrt{\mu_0 \varepsilon_i}} \tag{9-73}$$

Equation (9–72) has two positive solutions for $n^2$, and the corresponding ratios of electric fields are real. Hence *there can be two plane polarized waves traveling with different velocities in any given direction.* This property is the fundamental result of crystal optics. The algebraic details are left as an exercise.

To illustrate these ideas, we describe a simple experiment. A parallel beam of monochromatic light is passed through a Polaroid polarizer and then through a second Polaroid whose axis is at a right angle to the first. The light between the Polaroid sheets is plane-polarized, and essentially none of the light passes through the second sheet. Now a thin, plane parallel plate cut from a crystalline substance is placed between the two Polaroid sheets, and in general, light now passes through the system. Rotation of the second sheet does not eliminate the transmitted light. In contrast, if the crystal plate is rotated, there are four positions, each separated by 90°, for which the light through the second Polaroid vanishes. From this result we may surmise that the light emerging from the plate is not plane-polarized except when the incident plane polarized light lies along two orthogonal axes. Moreover, for every direction of propagation there are just two waves that preserve their polarization state while traveling through the medium. In general, the wave emerging from the plate is elliptically polarized; by an appropriate choice of the incident polarization direction and thickness of the plate, circular polarization can result. Such a plate that converts plane polarization to circular polarization is known as a *quarter-wave plate.*

For waves incident on a crystal plate at an angle, the two transmitted waves will refract at different angles. Upon emerging from the plate, the waves are parallel to the incident direction but laterally displaced. The resulting image will be doubled—hence the name *double refraction.* Since the bow angles are independent of the drop radius, the rainbow is a distinct phenomenon.

## 9-4 APPLICATION: THE RAINBOW

The natural phenomenon of the rainbow has fascinated people throughout history; it has appeared in literature since at least biblical times and has intrigued scientists since Aristotle. The explanation is a nice exercise in the application of the material developed in the previous section. [See also R. W. Robinett, *Phys. Teach.* (*USA*), 21: 388 (1983).]

Under suitable conditions, when there are water droplets in the air and the sun's angle is sufficiently low, a rainbow may be observed. As shown in Figure 9–8, one may sometimes see a larger secondary bow in addition to the primary bow.

The deflection of a light ray necessary for the formation of the primary bow is given by two refractions and an internal reflection in the water droplets, as illustrated in Figure 9–9. The angle from the incident sunlight to the primary bow $\theta_P$ is, from the geometry of Figure 9–9,

$$\tfrac{1}{2}\theta_P = 2\theta_T - \theta_I \qquad \textbf{(9–74)}$$

while Snell's law gives $\sin\theta_I = n \sin\theta_T$, where $n$ is the index of refraction of water. Thus $\theta_P$ can be considered to be a function of $\theta_I$ alone.

As the angle $\theta_I$ varies from 0° to 180°, $\theta_P$ has a maximum, as shown in Figure 9–10. A large range of incident rays emerge near this maximum bow angle. At this value of $\theta_P$ the deflected light is thus most intense, producing the primary bow. The rainbow angle is determined by the maximum of Equation (9–74),

$$0 = \frac{d\theta_P}{d\theta_I} = 4\frac{d\theta_T}{d\theta_I} - 2 \qquad \textbf{(9–75)}$$

Differentiation of Snell's law gives

$$\cos\theta_I = n\cos\theta_T \frac{d\theta_T}{d\theta_I} \qquad \textbf{(9–76)}$$

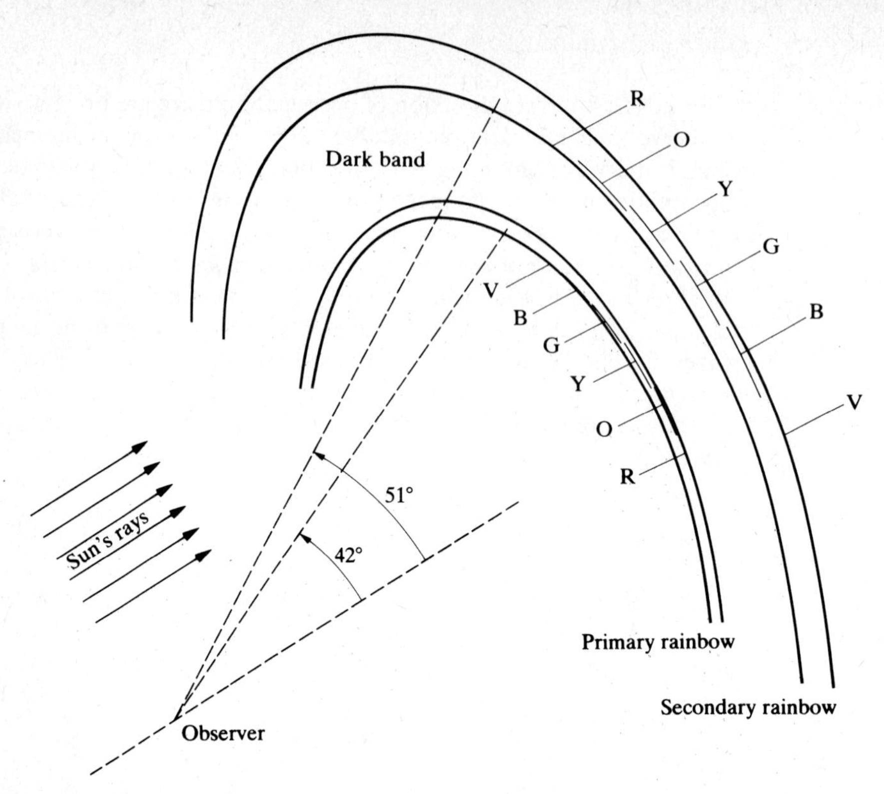


FIGURE 9-8 Primary and secondary rainbows, shown in relation to the sun and the observer

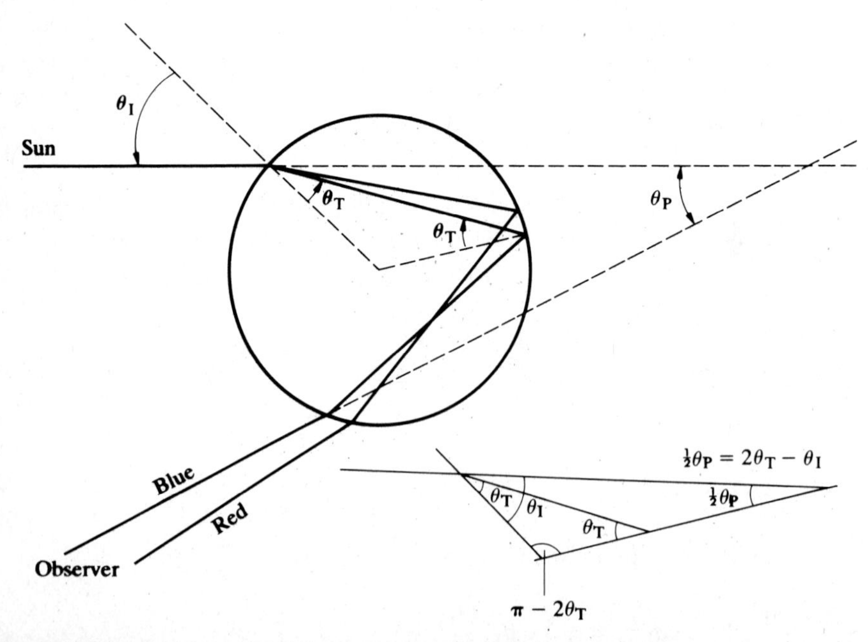


FIGURE 9-9 Primary-rainbow geometry for light in a water drop

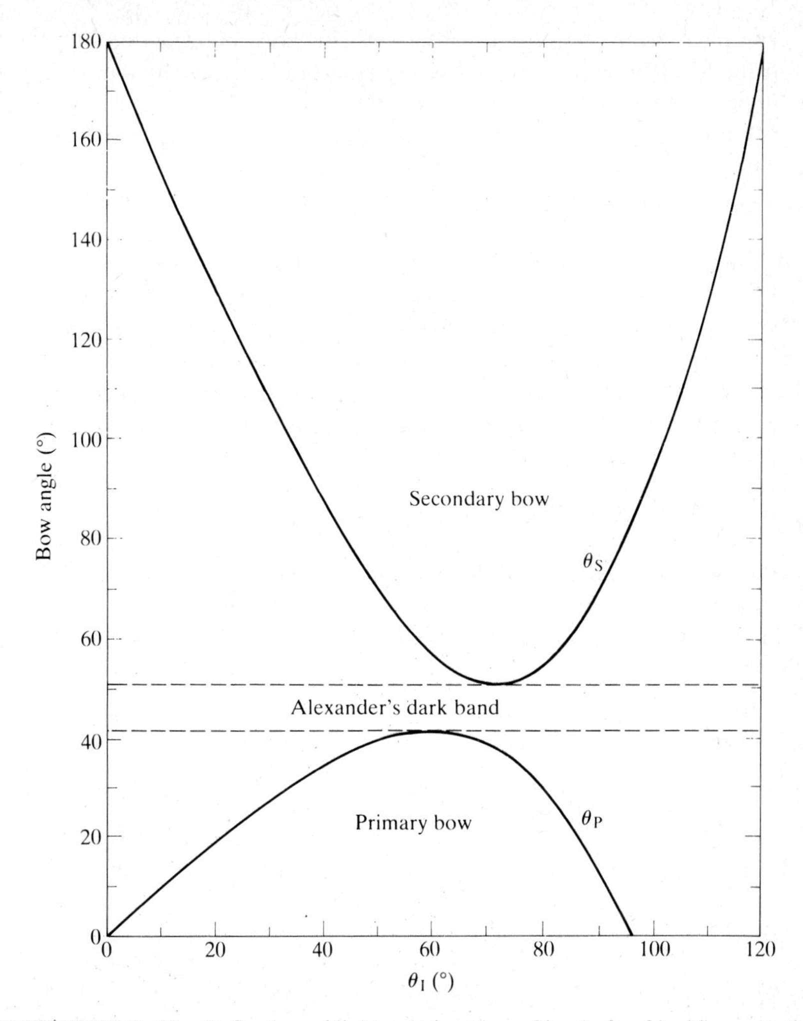


**FIGURE 9–10** Deflection of light as a function of angle $\theta_I$ of incidence on the drop. The extrema give the bow angles for the primary and secondary rainbows.

Solving these equations for $\theta_I$ and $\theta_T$, we obtain

$$\sin\theta_I = \sqrt{\frac{4-n^2}{3}} \qquad \sin\theta_T = \frac{1}{n}\sqrt{\frac{4-n^2}{3}} \tag{9-77}$$

as the maximum deflection. Using the value $n = \frac{4}{3}$ for the refractive index of water leads to the value

$$\theta_P^{max} = 42° \tag{9-78}$$

If the sun's elevation is greater than 42°, the primary bow will be below the horizon and will not be seen from the ground. From an airplane complete circle bows can be observed. The index of refraction of water varies from 1.344 (violet) to 1.331 (red), giving the bow an angular width of 1.8°, approximately a thumb's width at arm's length. The blue end of the spectrum is refracted most, and so it is lower in the sky. All the light rays following the path of the primary bow of Figure 9–9 have a bow angle $\theta_P$ less than 42°, so the sky inside the bow will be brighter than the sky outside the bow.

The internal reflection angle $\theta_T = 40.2°$ for $n = \frac{4}{3}$ is remarkably close to Brewster's angle $\theta_B = 36.9°$ as computed from Equation (9–62), and hence the reflected light is almost completely polarized perpendicular to the scattering plane. Thus the bow light is polarized along the arc of the bow. One can verify this prediction by using Polaroid sunglasses.

The secondary bow is formed by light that makes two internal reflections in the drop, as in Figure 9–11. For this second bow the angle $\theta_S$ is given by

$$\theta_S = \pi + 2(\theta_I - 3\theta_T) \qquad \textbf{(9–79)}$$

As a function of $\theta_I$, the secondary angle $\theta_S$ has an extremum, but this time a minimum, as shown in Figure 9–10. At the minimum angle the scattering will be strongest; the secondary bow angle turns out to be an angle

$$\theta_S^{min} = 51.1° \qquad \textbf{(9–80)}$$

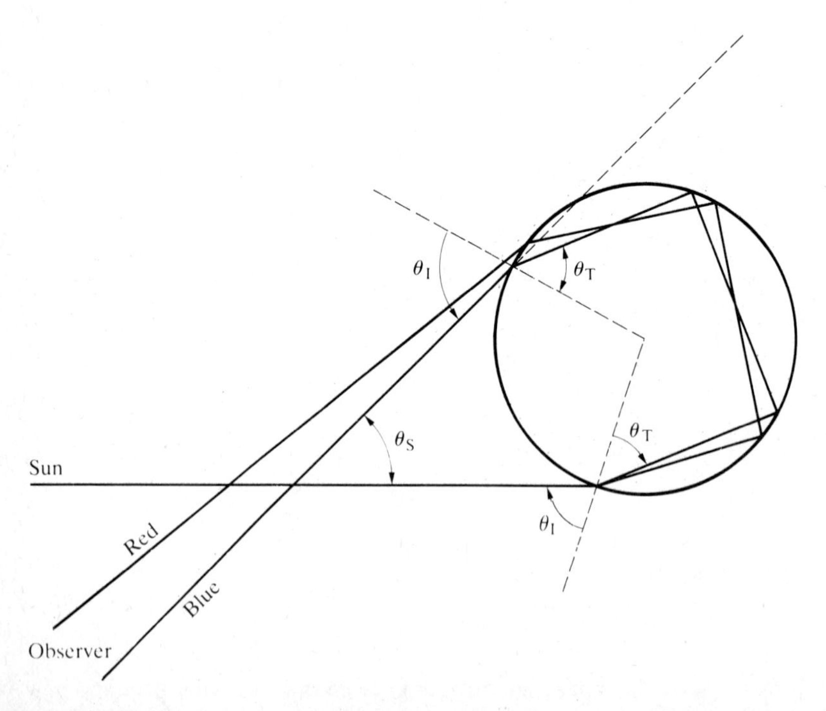


**FIGURE 9–11** Secondary-rainbow geometry for light in a water drop

Because there are two internal reflections, the secondary bow is fainter than the primary and hence is often not seen. As Figure 9–11 illustrates, the bow colors are reversed from the primary case. Since the primary-bow angle is the maximum single-reflection angle and the secondary-bow angle is the minimum double-reflection angle, the space between 42° and 51° is relatively dark. This dark region is known as Alexander's dark band.

## 9-5 WAVE PULSES AND GROUP VELOCITY

In this and the preceding chapter we mainly considered monochromatic waves, that is, waves having a simple harmonic time variation (i.e., a definite frequency). The wave equation requires the wave to also have a simple harmonic spatial variation. These waves have no beginning or end, and so they may not accurately represent an actual wave that has a finite extent. The corresponding phase velocity $v_p$ is not usually directly observable.

In practice, a pulse, or *wave group*, is formed; its size and its progress through space can be measured. A pulse is a combination of waves having different wavelengths and frequencies. A spectrum of monochromatic components add constructively or destructively in space to form the pulse. In materials, usually the different components propagate at different velocities. We will see that the pulse may approximately retain its form, but its *group velocity* is generally slower than the phase velocity.

A general superposition of monochromatic waves traveling along the $z$ direction can be expressed as

$$\begin{aligned} \mathbf{E}(z, t) &= \int \mathscr{E}(k) e^{i(kz - \omega t)}\, dk \\ \mathbf{B}(z, t) &= \int \mathscr{B}(k) e^{i(kz - \omega t)}\, dk \end{aligned} \tag{9–81}$$

where $\mathscr{E}(k)$ and $\mathscr{B}(k)$ are in general complex functions of $k$ and the integration extends over all wave numbers (i.e., $-\infty < k < +\infty$). Each frequency may correspond to a different phase velocity described by the dispersion relation $\omega = \omega(k)$. This function satisfies $\omega(-k) = \omega(k)$ since the phase velocity $v_p = \omega(k)/k$ for a wave moving toward positive $z$ has opposite sign to the $v_p$ for a wave moving toward negative $z$. In order to satisfy Maxwell's equations, the amplitudes for each wave number must satisfy

$$\hat{\mathbf{z}} \cdot \mathscr{E}(k) = \hat{\mathbf{z}} \cdot \mathscr{B}(k) = 0 \qquad \mathscr{B}(k) = \frac{n}{c} \hat{\mathbf{z}} \times \mathscr{E}(k) \tag{9–82}$$

The general wave packet in Equation (9–81) contains the monochromatic solution of wave number $k_0$ as a special case in which $\mathscr{E}(k)$ is given by the delta function,

$$\boldsymbol{\mathscr{E}}(k) = \boldsymbol{\mathscr{E}}_0\delta(k - k_0) \tag{9-83}$$

When Equation (9–83) is inserted into Equation (9–81) and the $dk$ integration carried out, Equation (9–81) reduces to

$$\mathbf{E}(z, t) = \boldsymbol{\mathscr{E}}_0 e^{i(k_0 z - \omega_0 t)} \tag{9-84}$$

where $\omega_0 = (c/n)k_0$. For a superposition of wave numbers distributed about some central value $k_0$, a wave pulse of finite extent can result.

Assume for simplicity that $\boldsymbol{\mathscr{E}}(k)$ is plane-polarized along the same direction for all wave numbers and that $\mathscr{E}(k)$ is nonzero only in an interval $\Delta k$ near $k = k_0$, as in Figure 9–12(a). First, let us investigate the shape of the wave pulse at $t = 0$:

$$E(z, 0) = \int \mathscr{E}(k)e^{ikz}\, dk \tag{9-85}$$

Making the separation

$$k = k_0 + (k - k_0) \qquad e^{ikz} = e^{ik_0 z}e^{i(k - k_0)z} \tag{9-86}$$

we find that

$$E(z, 0) = e^{ik_0 z}\int_{-\infty}^{+\infty} \mathscr{E}(k)e^{i(k - k_0)z}\, dk \tag{9-87}$$

The integral in Equation (9–87) represents a superposition of the amplitudes $\mathscr{E}(k)$ weighted by the complex exponential factor. At $z = 0$ we assume that the $\mathscr{E}(k)$ add to give a large value; such is the case for the $\mathscr{E}(k)$ in Figure 9–12(a). For nonzero $z$ the phase factor introduces destructive interferences, which become more complete as $z$ becomes large. Constructive interference occurs only if the argument of the exponential is less than unity. Since $|k - k_0| \leqslant \Delta k$, then the size $\Delta z$ of the pulse is related to $\Delta k$ by

$$(\Delta z)(\Delta k) \approx 1 \tag{9-88}$$

For a Gaussian distribution in wave number the modulation shape in coordinate space is also Gaussian, as illustrated in Figure 9–12(b).

Only when the spatial length of the pulse is of the order of the carrier wavelength does the pulse become significantly nonmonochromatic. For light, state-of-the-art, minimum pulse lengths now approach $2 \times 10^{-14}$ s in duration or $6 \times 10^{-4}$ cm in length, still much larger than light wavelengths, and hence are still essentially monochromatic.

We have examined the shape of a wave pulse at the fixed time $t = 0$; now let us see how the pulse propagates in time. For waves in empty space

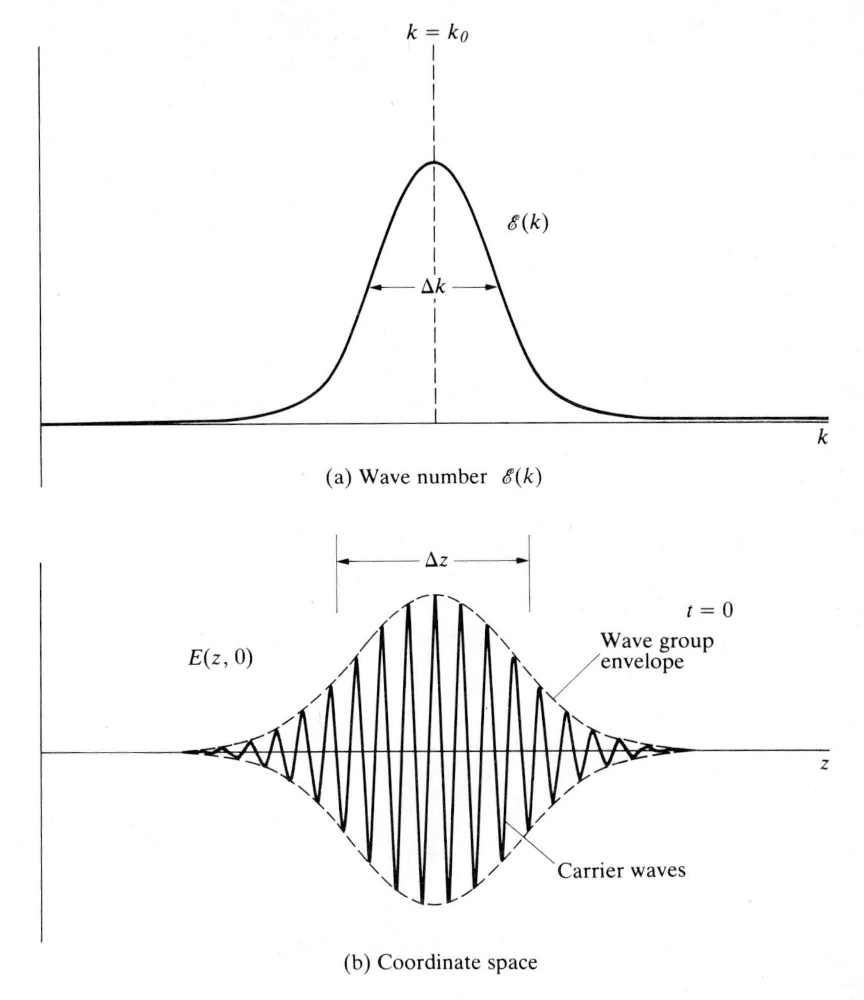


**FIGURE 9–12** Wave group, for (a) a given distribution of waves in wave number $\mathscr{E}(k)$ and (b) the corresponding distribution in coordinate space

$\omega = c|k|$, and the electric field, from Equation (9–81), is

$$E(z, t) = \int \mathscr{E}(k) e^{ik(z-ct)}\, dk \tag{9-89}$$

where $\mathscr{E}(k < 0) = 0$ for a wave moving in the $+z$ direction. From the observation that $E(z, t)$ is only a function of $z - ct$,

$$E(z, t) = E(z - ct) \tag{9-90}$$

we may conclude that the pulse propagates without change of shape in empty space with velocity $c$. Alternatively, if $\mathscr{E}(k > 0) = 0$, the pulse would propagate toward negative $z$.

When a wave propagates in a nonabsorptive material of index $n$, we have seen in the first section of this chapter that

$$\frac{\omega}{k} = v_{\text{p}} = \frac{c}{n} \tag{9–91}$$

where the phase velocity $v_{\text{p}}$ (and hence also $n$) can depend on the wave number,

$$v_{\text{p}} = v_{\text{p}}(k) \tag{9–92}$$

as discussed in Chapter 4. By Equations (9–91) and (9–92) the frequency is a function of $k$, so a superposition of harmonic electric field waves is

$$E(z, t) = \int \mathscr{E}(k)e^{i[kz - \omega(k)t]}\, dk \tag{9–93}$$

Again assuming that $\mathscr{E}(k)$ is localized about $k = k_0 > 0$, as in Figure 9–12(a), $\omega(k)$ can be approximated in this vicinity by

$$\omega(k) \approx \omega_0 + (k - k_0)v_{\text{g}} \tag{9–94}$$

where the *group velocity* is

$$v_{\text{g}} = \left.\frac{d\omega}{dk}\right|_{k_0} \tag{9–95}$$

The phase of Equation (9–93) can then be written as

$$\begin{aligned} kz - \omega(k)t &\approx [k_0 + (k - k_0)]z - [\omega_0 + v_{\text{g}}(k - k_0)]t \\ &= k_0 z - \omega_0 t + (k - k_0)(z - v_{\text{g}}t) \end{aligned} \tag{9–96}$$

The electric field of Equation (9–93) becomes

$$E(z, t) = e^{i(k_0 z - \omega_0 t)}G(z - v_{\text{g}}t) \tag{9–97a}$$

with

$$G(z - v_{\text{g}}t) = \int \mathscr{E}(k)e^{i(k - k_0)(z - v_{\text{g}}t)}\, dk \tag{9–97b}$$

The wave pulse has been expressed as a *carrier plane wave* modulated by a factor $G$ that is large only near $z \approx v_{\text{g}}t$ and is a function only of $z - v_{\text{g}}t$. In this approximation the wave pulse propagates without distortion with the group velocity $v_{\text{g}}$. If higher terms in the expansion in Equation (9–94) are important, the group will distort as it propagates. In a vacuum, $\omega = ck$ and $v_{\text{g}} = d\omega/dk = c$, so the group moves at the speed of light, as we saw previously in Equation (9–90). For ordinary transparent substances in the optical range of frequencies, $v_g < v_p$ [i.e., $\omega(k)$ is an increasing function of $k$]; this situation is *normal dispersion*. For most transparent substances the index of refraction increases through the visible region so that blue light refracts more strongly than red.

If we could observe the details of the pulse, and not just the envelope, we would see (in a normally dispersive material) the carrier wave constantly moving through the pulse and passing out the front.

## 9-6 WAVEGUIDES

A plane wave propagating in an unbounded medium can have any frequency, and its phase velocity is always $v_p = \omega/k = c/n$. When boundaries are present, the variety of possible waves is often severely restricted. An important practical case is the propagation of waves within a tube. We will consider here conducting tubes that have wide application in microwave electronics. A closely related situation is the theory of optical fibers, which are important for telecommunication circuits.

A typical *waveguide* is shown in Figure 9–13. It consists of a long cylinder of arbitrary but constant cross section. We will assume that the

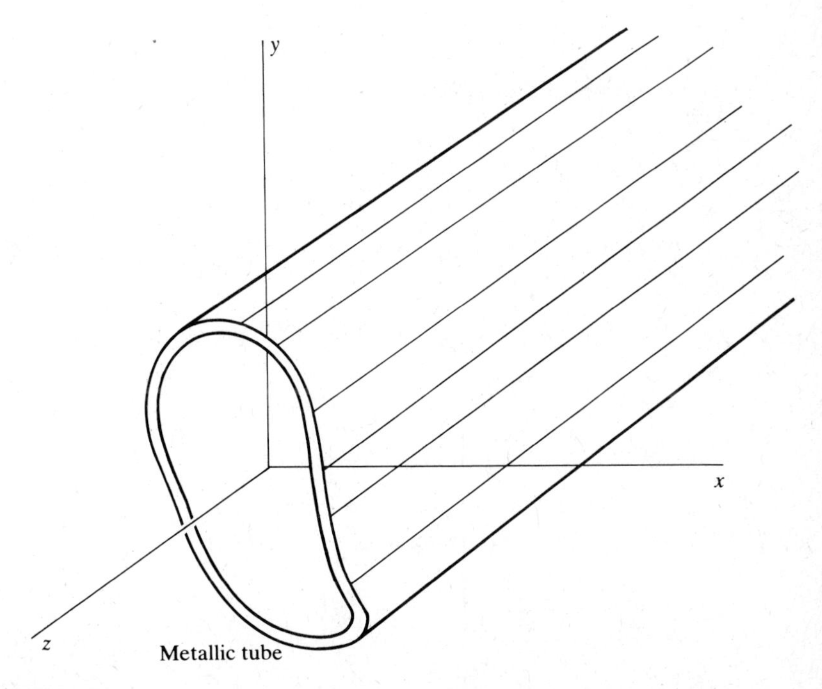


**FIGURE 9–13** Cylindrical waveguide consisting of a hollow metallic tube with axis parallel to the $z$ axis

walls are perfectly conducting so that there are no electromagnetic fields within the metal, although surface charges and currents are generally present on the walls. The tangential and normal boundary conditions to be satisfied at the conducting surfaces are

$$E_t = 0 \tag{9-98a}$$

$$B_n = 0 \tag{9-98b}$$

$$E_n = \frac{\sigma}{\varepsilon_0} \tag{9-98c}$$

$$B_t = \mu_0 j \tag{9-98d}$$

These boundary conditions are of two types. The first two are usually used to determine the shape of the fields. If the strength of the field is known, the second two can be used to calculate the induced charges and currents on the conduction boundaries.

We look for solutions to Maxwell's equations that represent waves propagating down the axis of the waveguide, here taken as the $z$ axis. We assume solutions of the form

$$\mathbf{E} = \boldsymbol{\mathscr{E}}(x, y)e^{i(kz - \omega t)} \tag{9-99a}$$

$$\mathbf{B} = \boldsymbol{\mathscr{B}}(x, y)e^{i(kz - \omega t)} \tag{9-99b}$$

As a first step we use these solutions to express Maxwell's equations in Cartesian component form within the waveguide:

$$\nabla \cdot \mathbf{E} = 0 \qquad \frac{\partial \mathscr{E}_x}{\partial x} + \frac{\partial \mathscr{E}_y}{\partial y} + ik\mathscr{E}_z = 0 \tag{9-100a}$$

$$\nabla \cdot \mathbf{B} = 0 \qquad \frac{\partial \mathscr{B}_x}{\partial x} + \frac{\partial \mathscr{B}_y}{\partial y} + ik\mathscr{B}_z = 0 \tag{9-100b}$$

$$\nabla \times \mathbf{E} = -\frac{\partial \mathbf{B}}{\partial t} \qquad \begin{cases} \dfrac{\partial \mathscr{E}_z}{\partial y} - ik\mathscr{E}_y = i\omega\mathscr{B}_x & \text{(9-100c)} \\[2ex] ik\mathscr{E}_x - \dfrac{\partial \mathscr{E}_z}{\partial x} = i\omega\mathscr{B}_y & \text{(9-100d)} \\[2ex] \dfrac{\partial \mathscr{E}_y}{\partial x} - \dfrac{\partial \mathscr{E}_x}{\partial y} = i\omega\mathscr{B}_z & \text{(9-100e)} \end{cases}$$

$$\nabla \times \mathbf{B} = \frac{1}{c^2}\frac{\partial \mathbf{E}}{\partial t} \qquad \begin{cases} \dfrac{\partial \mathscr{B}_z}{\partial y} - ik\mathscr{B}_y = -\dfrac{i\omega}{c^2}\mathscr{E}_x & \text{(9-100f)} \\[2ex] ik\mathscr{B}_x - \dfrac{\partial \mathscr{B}_z}{\partial x} = -\dfrac{i\omega}{c^2}\mathscr{E}_y & \text{(9-100g)} \\[2ex] \dfrac{\partial \mathscr{B}_y}{\partial x} - \dfrac{\partial \mathscr{B}_x}{\partial y} = -\dfrac{i\omega}{c^2}\mathscr{E}_z & \text{(9-100h)} \end{cases}$$

Inspection of these equations reveals that the transverse field components can all be expressed in terms of the longitudinal (or $z$ component) fields $\mathscr{E}_z$ and $\mathscr{B}_z$. We use Equations (9–100d) and (9–100f) to solve for $\mathscr{E}_x$ and $\mathscr{B}_y$:

$$\mathscr{E}_x = \frac{i}{\gamma^2}\left(k\frac{\partial \mathscr{E}_z}{\partial x} + \omega\frac{\partial \mathscr{B}_z}{\partial y}\right) \tag{9–101a}$$

$$\mathscr{B}_y = \frac{i}{\gamma^2}\left(k\frac{\partial \mathscr{B}_z}{\partial y} + \frac{\omega}{c^2}\frac{\partial \mathscr{E}_z}{\partial x}\right) \tag{9–101b}$$

Then we solve for $\mathscr{E}_y$ and $\mathscr{B}_x$ by using Equations (9–100c) and (9–100g):

$$\mathscr{E}_y = \frac{i}{\gamma^2}\left(k\frac{\partial \mathscr{E}_z}{\partial y} - \omega\frac{\partial \mathscr{B}_z}{\partial x}\right) \tag{9–101c}$$

$$\mathscr{B}_x = \frac{i}{\gamma^2}\left(k\frac{\partial \mathscr{B}_z}{\partial x} - \frac{\omega}{c^2}\frac{\partial \mathscr{E}_z}{\partial y}\right) \tag{9–101d}$$

Here we have introduced the quantity

$$\gamma^2 = \frac{\omega^2}{c^2} - k^2 \tag{9–102}$$

When a waveguide solution is found to satisfy all of the boundary conditions, $\gamma$ will be limited to a definite sequence of values. As a result of Equations (9–101), we can concentrate on the two components $\mathscr{E}_z$ and $\mathscr{B}_z$.

If we substitute the expressions in Equations (9–101) into the remaining Maxwell equations, (9–100), we find that in each case $\mathscr{E}_z$ or $\mathscr{B}_z$ satisfies the equation

$$\left(\frac{\partial^2}{\partial x^2} + \frac{\partial^2}{\partial y^2} + \gamma^2\right)\psi = 0 \tag{9–103}$$

where $\psi = \mathscr{E}_z$ or $\mathscr{B}_z$. Depending on the cross-sectional shape of the waveguide, we might use various coordinate systems for the transverse variables, so in general, $\psi$ satisfies

$$(\nabla_{\mathrm{T}}^2 + \gamma^2)\psi = 0 \tag{9–104}$$

Here the operator $\nabla_{\mathrm{T}}^2$ is the Laplacian for the two transverse dimensions.

Waveguides very commonly have a rectangular cross section, as shown in Figure 9–14. The solution to Equation (9–103) by the separation-of-variables techniquc with $\psi(x, y) = X(x)Y(y)$, as in Section 3–4, is

$$\psi(x, y) = (C_1 \sin \alpha x + C_2 \cos \alpha x)(C_3 \sin \beta y + C_4 \cos \beta y) \tag{9–105}$$

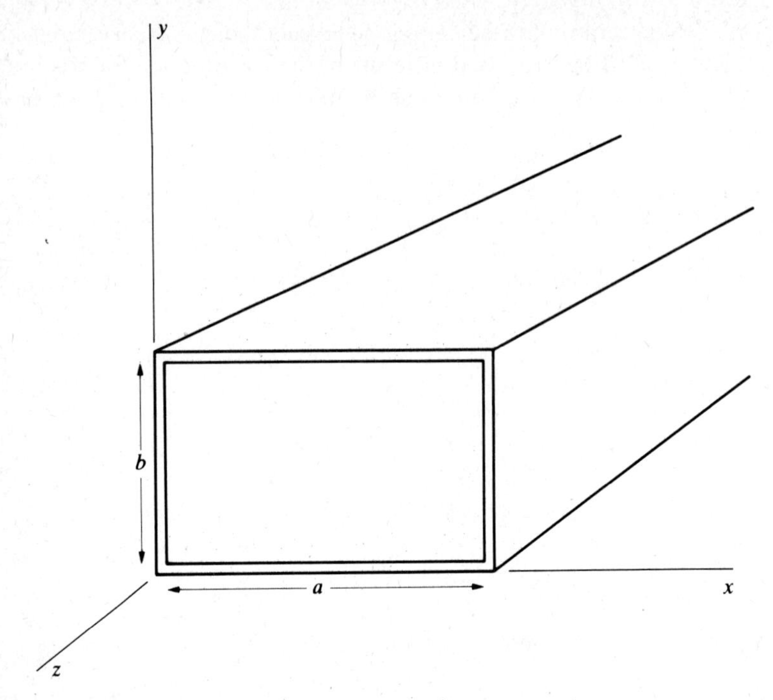


**FIGURE 9-14** Waveguide with rectangular cross section

where

$$\alpha^2 + \beta^2 = \gamma^2 = \frac{\omega^2}{c^2} - k^2 \tag{9-106}$$

So that the boundary conditions are satisfied, the proper choice of the constants $C_i$ for each field leads to

$$\mathscr{E}_z = \mathscr{E} \sin \alpha x \sin \beta y \tag{9-107a}$$

$$\mathscr{B}_z = \mathscr{B} \cos \alpha x \cos \beta y \tag{9-107b}$$

as we will verify. First, $\mathscr{E}_z$ must vanish on all of the side surfaces by Equation (9–98a). This condition requires that

$$\alpha = \frac{m\pi}{a} \qquad \beta = \frac{n\pi}{b}, \qquad m \text{ and } n \text{ positive integers} \tag{9-108}$$

Hence by Equation (9–106)

$$\gamma_{mn}^2 = \pi^2\left[\left(\frac{m}{a}\right)^2 + \left(\frac{n}{b}\right)^2\right] = \left(\frac{\omega}{c}\right)^2 - k^2 \tag{9-109}$$

Wave modes are conventionally denoted by their $(m, n)$ values.

We use Equations (9–107) in Equations (9–101) to compute the transverse fields to verify that they satisfy the boundary conditions:

$$\mathscr{E}_x = \frac{i}{\gamma^2}(k\mathscr{E}\alpha - \omega\mathscr{B}\beta)\cos\alpha x \sin\beta y \tag{9–110a}$$

$$\mathscr{E}_y = \frac{i}{\gamma^2}(k\mathscr{E}\beta + \omega\mathscr{B}\alpha)\sin\alpha x \cos\beta y \tag{9–110b}$$

$$\mathscr{B}_x = \frac{i}{\gamma^2}\left(-k\mathscr{B}\alpha - \frac{\omega}{c^2}\mathscr{E}\beta\right)\sin\alpha x \cos\beta y \tag{9–110c}$$

$$\mathscr{B}_y = \frac{i}{\gamma^2}\left(-k\mathscr{B}\beta + \frac{\omega}{c^2}\mathscr{E}\alpha\right)\cos\alpha x \sin\beta y \tag{9–110d}$$

With the choices for $\alpha$ and $\beta$ of Equation (9–108) we observe that the boundary conditions, Equations (9–98a) and (9–98b), are indeed satisfied on all the waveguide surfaces.

Because the electric and magnetic fields appear linearly in Equations (9–101) or (9–110), the general solution is a superposition of the three special cases where either $\mathscr{E}_z$, $\mathscr{B}_z$, or both varnish.

## 9-6A Transverse Magnetic (TM) Modes

In this special case $\mathscr{B}_z = 0$, and the magnetic field is transverse:

$$\mathscr{E}_z^{\text{TM}} = \mathscr{E}\sin\alpha x \sin\beta y \qquad \mathscr{B}_z^{\text{TM}} = 0 \tag{9–111a}$$

The transverse fields are then

$$\mathscr{E}_x^{\text{TM}} = \frac{i}{\gamma^2}k\alpha\mathscr{E}\cos\alpha x \sin\beta y \qquad \mathscr{E}_y^{\text{TM}} = \frac{i}{\gamma^2}k\beta\mathscr{E}\sin\alpha x \cos\beta y \tag{9–111b}$$

$$\mathscr{B}_x^{\text{TM}} = -\frac{i\omega\beta}{\gamma^2 c^2}\mathscr{E}\sin\alpha x \cos\beta y \qquad \mathscr{B}_y^{\text{TM}} = \frac{i\omega\alpha}{\gamma^2 c^2}\mathscr{E}\cos\alpha x \sin\beta y \tag{9–111c}$$

For the TM mode neither $\alpha$ nor $\beta$ can vanish to avoid having $\mathscr{E}_z = 0$, so the lowest value of $\gamma_{mn}$ is

$$\gamma_{11} = \pi\left(\frac{1}{a^2} + \frac{1}{b^2}\right)^{1/2} \tag{9–112}$$

## 9-6B Transverse Electric (TE) Modes

When $\mathscr{E}_z = 0$, the electric field is transverse:

$$\mathscr{E}_z^{\text{TE}} = 0 \qquad \mathscr{B}_z^{\text{TE}} = \mathscr{B}\cos\alpha x \cos\beta y \tag{9–113a}$$

The corresponding transverse field components are

$$\mathscr{E}_x^{\text{TE}} = -\frac{i\omega\beta}{\gamma^2}\mathscr{B}\cos\alpha x\sin\beta y \qquad \mathscr{E}_y^{\text{TE}} = \frac{i\omega\alpha}{\gamma^2}\mathscr{B}\sin\alpha x\cos\beta y \tag{9-113b}$$

$$\mathscr{B}_x^{\text{TE}} = -\frac{ik\alpha}{\gamma^2}\mathscr{B}\sin\alpha x\cos\beta y \qquad \mathscr{B}_y^{\text{TE}} = -\frac{ik\beta}{\gamma^2}\mathscr{B}\cos\alpha x\sin\beta y \tag{9-113c}$$

If $\alpha = \beta = 0$, we do not have an interesting solution. To see this result, consider the integrated Maxwell equation,

$$\oint_C \mathbf{E}\cdot d\mathbf{r} = i\omega\int_S \mathbf{B}\cdot d\mathbf{S}$$

applied to a cross-sectional plane. If the circuit $C$ lies in the waveguide wall, $\mathbf{E} = 0$, and we have

$$\int_S \mathscr{B}_z\,dx\,dy = 0 \tag{9-114}$$

For $\alpha = \beta = 0$, $\mathscr{B}_z^{\text{TE}} =$ constant by Equation (9–113a), and hence

$$\mathscr{B}_z = 0 \tag{9-115}$$

Since both $\mathscr{E}_z$ and $\mathscr{B}_z$ are zero, we have a TEM mode (see the next subsection), which exists only if the waveguide consists of concentric conductors.

Unlike the TM modes, TE modes can have either $\alpha$ *or* $\beta$ vanishing. If $a > b$, the lowest allowed $\gamma$ value is

$$\gamma_{10}^{\text{TE}} = \frac{\pi}{a} \tag{9-116}$$

In this important case the transverse fields in the waveguide are

**$TE_{10}$ Field Configuration**

$$\mathscr{E}_x^{\text{TE}} = \mathscr{B}_y^{\text{TE}} = 0 \qquad \mathscr{E}_y^{\text{TE}} = \frac{i\omega a}{\pi}\mathscr{B}\sin\frac{\pi x}{a}$$

$$\mathscr{B}_x^{\text{TE}} = -\frac{ika}{\pi}\mathscr{B}\sin\frac{\pi x}{a} \tag{9-117}$$

These fields are illustrated schematically in Figure 9–15.

The reason for the importance of the $TE_{10}$ mode is evident from Equation (9–109). For a given $\gamma$

$$k^2 = \left(\frac{\omega}{c}\right)^2 - \gamma_{mn}^2 \tag{9-118}$$

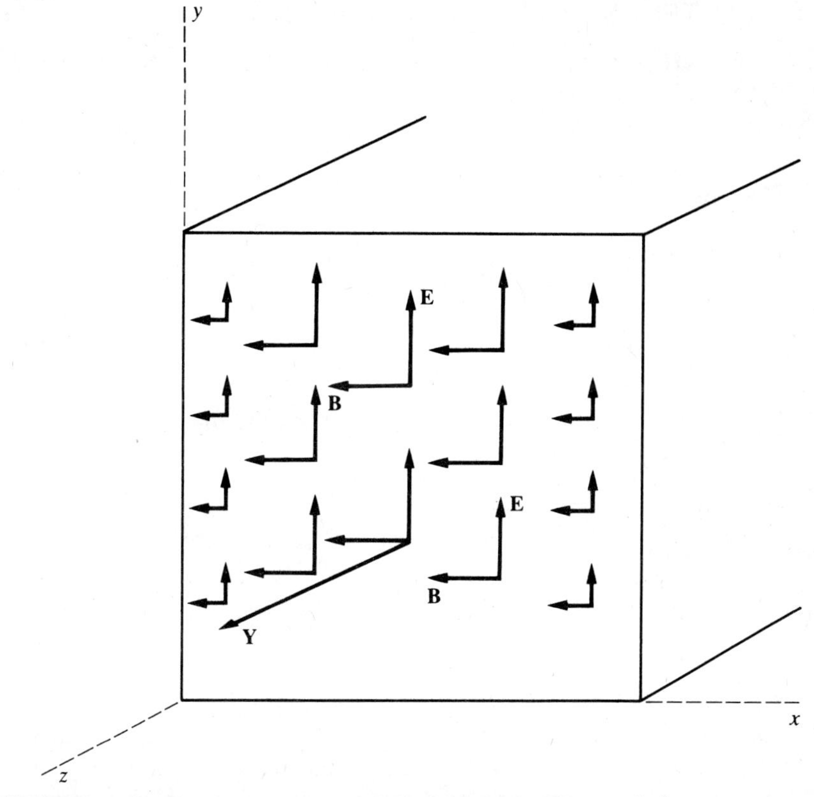


**FIGURE 9-15** Electric and magnetic fields in the $TE_{10}$ mode for a square waveguide

so that $k^2$ becomes negative below a certain frequency $\omega_c$ and waves do not propagate down the guide. This critical frequency is

$$\omega_c = \gamma_{mn} c \tag{9-119}$$

For a given $\omega$ only modes for which $\omega > \omega_c$ can propagate; if $\omega < \gamma_{11} c$ (but larger than $\gamma_{10} c$), only pure $TE_{10}$ modes can propagate.

An interesting aspect of waveguide propagation is that $\omega$ and $k$ are not related by a simple ratio as in a plane wave. From Equation (9–118) the phase velocity is given by

$$v_p = \frac{\omega}{k} = c\sqrt{1 + \left(\frac{\gamma_{mn}}{k}\right)^2} > c \tag{9-120}$$

a number greater than the speed of light. There is no conflict with relativity, though, since information can only be transferred by groups of waves that travel at

$$v_g = \frac{d\omega}{dk} = \frac{c}{\sqrt{1 + (\gamma_{mn}/k)^2}} < c \tag{9-121}$$

## 9-6C TEM Modes: Transmission Lines

Under certain conditions waves that are both transverse electric and magnetic can be produced in waveguides, but only if they consist of two conductors. By Equations (9–101) we see that if $\mathscr{E}_z = \mathscr{B}_z = 0$, the only hope of having nonzero transverse fields is if $\gamma = 0$ or $\omega = ck$. If we refer to the original equations, (9–100), TEM fields must satisfy

$$\frac{\partial \mathscr{E}_x}{\partial x} = -\frac{\partial \mathscr{E}_y}{\partial y} \tag{9-122a}$$

$$\frac{\partial \mathscr{E}_y}{\partial x} = \frac{\partial \mathscr{E}_x}{\partial y} \tag{9-122b}$$

and similar equations for $\mathscr{B}_x$ and $\mathscr{B}_y$. We also find that

$$\omega = ck \tag{9-123a}$$

$$\mathscr{B}_x = \frac{-\mathscr{E}_y}{c} \tag{9-123b}$$

$$\mathscr{B}_y = \frac{\mathscr{E}_x}{c} \tag{9-123c}$$

Equation (9–123a) shows that the TEM wave propagates with $v_p = \omega/k = c$, the speed of light, just as if it were a plane wave in an unbounded region. Equations (9–123b) and (9–123c) show that the electric and magnetic fields are also the same as a freely propagating wave. To establish the existence of TEM waves, we only have to verify that they satisfy the required boundary conditions.

The electric field of Equations (9–122) can be derived from a scalar potential,

$$\boldsymbol{\mathscr{E}} = -\nabla\Phi \tag{9-124}$$

which, when substituted into Equations (9–122), shows that $\Phi$ satisfies Laplace's equation,

$$\nabla_T^2 \phi = 0 \tag{9-125}$$

Since the electric field must be normal to the conductor, the conductor is an equipotential, and so any electrostatic transverse electric field $\boldsymbol{\mathscr{E}}$ can be the electric field of a TEM wave.

We saw in Section 2–8 that a static electric field within a hollow conductor must vanish, so a waveguide of the form of Figure 9–13 or 9–14 cannot have TEM waves. A coaxial cable with two conductors can have TEM waves as well as TM and TE waves. By use of frequencies below the lowest TE mode, a coaxial cable can be made to carry only TEM waves. Waveguides with two conductors are generally known as *transmission lines.*

# SUMMARY

## Important Concepts and Equations

$$n = \frac{kc}{\omega} = \left[\frac{\varepsilon\mu + (i\mu g/\omega)}{\varepsilon_0\mu_0}\right]^{1/2}$$

*Complex index of refraction n for wave propagation with frequency ω.*

$$n \approx \sqrt{\frac{\varepsilon}{\varepsilon_0}} = \sqrt{\kappa}$$

*Dielectric index of refraction.*

$$n \approx (1 + i)\sqrt{\frac{g\mu}{2\omega\mu_0\varepsilon_0}}$$

*Index of refraction of a good conductor.*

$$\mathbf{E}(z, t) = \mathscr{E} \exp\left[-(\text{Im } n)\frac{\omega}{c} z\right] \exp\left[i\omega\left(\frac{\text{Re } n}{c} z - t\right)\right]$$

*Plane wave solution for the electric field propagating in the z direction in a material with index of refraction n.*

$$v_p = \frac{c}{\text{Re } n}$$

*Phase velocity in matter.*

$$\delta = \frac{c}{(\text{Im } n)\omega} \approx \sqrt{\frac{2}{g\mu\omega}}$$

*Skin depth δ of an electromagnetic field penetration into a good conductor.*

$$\mathbf{B} = \frac{n}{c}\hat{\mathbf{k}} \times \mathbf{E}$$

*Relation of* **E** *and* **B** *for plane wave propagation in the direction* $\hat{\mathbf{k}}$ *in an isotropic medium with complex index of refraction n.*

$$\mathbf{E}_L = \tfrac{1}{2}\mathscr{E}(\hat{\mathbf{x}} + i\hat{\mathbf{y}})e^{i[(n_L/c)z - t]\omega}$$

*Left circularly polarized wave.*

$$\mathbf{E}_R = \tfrac{1}{2}\mathscr{E}(\hat{\mathbf{x}} - i\hat{\mathbf{y}})e^{i[n_L/c)z - t]\omega}$$

*Right circularly polarized wave.*

$$\theta_R = \theta_I \qquad \text{(law of reflection)}$$

*The angles $\theta_I$ and $\theta_R$ that the incident and reflective wave directions make with the normal to the interface between two media are equal.*

$n_1 \sin \theta_I = n_2 \sin \theta_T$ (Snell's law of refraction)

*Relation between the angles $\theta_I$ and $\theta_T$ that the incident and transmitted wave directions make with the normal to the interface, from a medium with refractive index $n_1$ to one with index $n_2$.*

$$\left(\frac{\mathscr{E}_T}{\mathscr{E}_I}\right)_\perp = \frac{2n_1 \cos \theta}{n_1 \cos \theta + n_2 \cos \theta_T} \qquad \left(\frac{\mathscr{E}_R}{\mathscr{E}_I}\right)_\perp = \frac{n_1 \cos \theta - n_2 \cos \theta_T}{n_1 \cos \theta + n_2 \cos \theta_T}$$

*Fresnel equations relating the electric field amplitudes of incident, transmitted, and reflected waves, for an electric vector perpendicular to the scattering plane, from a medium with refractive index $n_1$ to one of index $n_2$.*

$$\left(\frac{\mathscr{E}_T}{\mathscr{E}_I}\right)_{/\!/} = \frac{2n_1 \cos \theta}{n_1 \cos \theta_T + n_2 \cos \theta} \qquad \left(\frac{\mathscr{E}_R}{\mathscr{E}_I}\right)_{/\!/} = \frac{n_1 \cos \theta_T - n_2 \cos \theta}{n_1 \cos \theta_T + n_2 \cos \theta}$$

*Fresnel equations for an electric vector parallel to the scattering plane.*

$$\frac{\mathscr{E}_T}{\mathscr{E}_I} = \frac{2n_1}{n_1 + n_2} \qquad \frac{\mathscr{E}_R}{\mathscr{E}_I} = \frac{n_1 - n_2}{n_1 + n_2}$$

*Fresnel equations for normal incidence.*

$$\mathscr{E}_R \to -\mathscr{E}_I \qquad \mathscr{E}_T = 0$$

*Fresnel equations for grazing incidence.*

$$Y_z = \frac{n}{2\mu_0 c} \mathscr{E}^2 \cos \theta$$

*Energy flux (in watts per meter squared) crossing a media boundary.*

$$R = \left(\frac{\mathscr{E}_R}{\mathscr{E}_I}\right)^2$$

*Reflectivity: the ratio of reflected to incident energy flux.*

$$T = \frac{n_2 \cos \theta_T}{n_1 \cos \theta}\left(\frac{\mathscr{E}_T}{\mathscr{E}_I}\right)^2 \qquad (R + T = 1)$$

*Transmissivity: the ratio of transmitted to incident energy flux.*

$$\tan \theta_B = \frac{n_2}{n_1} \qquad \text{(Brewster's angle)}$$

*Angle of the incident wave direction with the normal to the interface at which there is no reflection of the electric field component that lies in the scattering plane.*

$$E(z, t) = e^{i(k_0 z - \omega_0 t)} \int \mathscr{E}(k) e^{i(k - k_0)(z - v_g t)}\, dk$$

*Electric field of a wave group with carrier wave number $k_0$ and frequency $\omega_0$.*

$$v_g = \frac{d\omega}{dk} \quad \text{at} \quad k = k_0$$

*The group velocity at which the envelope of a superposition of waves propagates in a dispersive medium; $k_0$ is the dominant wave number of the wave group.*

$$(\Delta z)(\Delta k) \approx 1$$

*Relation of the size of a wave pulse to the spread $\Delta k$ in wave number.*

$$\mathscr{E}_z = \mathscr{E} \sin \alpha x \sin \beta y \qquad \mathscr{B}_z = \mathscr{B} \cos \alpha x \cos \beta y$$

$$\mathscr{E}_x = \frac{i}{\gamma^2}(k\mathscr{E}\alpha - \omega\mathscr{B}\beta) \cos \alpha x \sin \beta y \qquad \mathscr{E}_y = \frac{i}{\gamma^2}(k\mathscr{E}\beta + \omega\mathscr{B}\alpha) \sin \alpha x \cos \beta y$$

$$\mathscr{B}_x = \frac{i}{\gamma^2}\left(-k\mathscr{B}\alpha - \frac{\omega}{c^2}\mathscr{E}\beta\right) \sin \alpha x \cos \beta y \quad \mathscr{B}_y = \frac{i}{\gamma^2}\left(-k\mathscr{B}\beta + \frac{\omega}{c^2}\mathscr{E}\alpha\right) \cos \alpha x \sin \beta y$$

*Rectangular waveguide solutions for electric and magnetic fields, with z as the propagation direction and*

$$\alpha^2 + \beta^2 = \gamma^2 = \frac{\omega^2}{c^2} - k^2$$

$$\alpha = \frac{m\pi}{a} \qquad \beta = \frac{n\pi}{b}$$

*where a and b are the transverse x and y dimensions and m and n are integers.*

$\mathscr{B}_z = 0$ (TM modes)

*Transverse magnetic field.*

$\mathscr{E}_z = 0$ (TE modes)

*Transverse electric field.*

$\mathscr{B}_z = \mathscr{E}_z = 0$ (TEM modes)

*Transverse electric and magnetic modes (transmission lines).*

# PROBLEMS

## Section 9-1: Plane Waves in Isotropic Matter

**9-1** Using Maxwell's equations for a conducting medium, Equations (9–1), demonstrate that the electric field satisfies the *telegrapher's equation*,

$$\nabla^2\mathbf{E} - \mu\varepsilon\frac{\partial^2\mathbf{E}}{\partial t^2} - \mu g\frac{\partial\mathbf{E}}{\partial t} = 0$$

and that the magnetic field satisfies a similar equation. You may assume that the medium locally remains neutral. Show that plane wave solutions such as Equation (9–4) satisfy the complex-index-of-refraction relation of Equation (9–9).

**9-2** In a metal there are plane wave solutions to Maxwell's equations with the form

$$E_x = E_0 e^{i(\omega t - kz)}$$

where $k$ is a complex number. For low frequencies

$$k = (1 + i)\sqrt{\frac{g\omega}{2\varepsilon_0 c^2}}$$

(i) Write an expression for the magnetic field **B** associated with such a wave.
(ii) What is the angle between **E** and **B**?

**9-3** An electromagnetic wave propagates through a gas of $N$ free electrons per unit volume. Neglecting damping, show that the index of refraction is given by

$$n^2 = 1 - \frac{\omega_p^2}{\omega^2}$$

where $\omega_p$ is the *plasma frequency*, $\omega_p = \sqrt{Ne^2/(\varepsilon_0 m_e)}$. In the ionosphere a typical maximum density of free electrons is $N = 10^5/\text{cm}^3$. Show that the ionosphere is transparent to TV signals ($\omega \approx 10^9$ Hz) but that AM waves ($\omega \approx 10^5$ Hz) are completely reflected.

## Section 9-2: Optical Activity

**9-4** Transverse electromagnetic waves propagate along the $z$ axis through an electron gas of $N$ electrons per unit volume. A uniform magnetic field **B** is parallel to the propagation direction. Neglecting damping, show that only circularly polarized waves, $\hat{\mathbf{x}} \pm i\hat{\mathbf{y}}$, propagate without change and that the index of refraction is

$$n_\mp^2 = 1 - \frac{\omega_p^2}{\omega(\omega \mp \omega_B)}$$

where $\omega_p$ is the plasma frequency, $\omega_p = \sqrt{Ne^2/(\varepsilon_0 m)}$, and $\omega_B = eB/m$ is the cyclotron frequency. Start with the equation of motion,

$$m\dot{\mathbf{v}} + e\mathbf{v} \times \mathbf{B} = -e(\hat{\mathbf{x}} \pm i\hat{\mathbf{y}})\mathscr{E}e^{-i\omega t}$$

and show that the solution is

$$\mathbf{r} = \frac{e}{m\omega(\omega \mp \omega_B)}(\hat{\mathbf{x}} \pm i\hat{\mathbf{y}})\mathscr{E}e^{-i\omega t}$$

The upper sign indicates left-hand circular polarization. For $\omega_p = \omega_B$, plot $n_\pm^2$ as a function of $\omega/\omega_p$ and note the frequency regions in which circularly polarized waves will be reflected or transmitted in such a plasma.

## Section 9-3: Reflection and Refraction

**9-5** Derive the dielectric Fresnel equations directly for the special case of normal incidence.

**9-6** You dive down to a depth $h$ in a large lake. If you now look straight up, what do you see? Assume the index of water is $\frac{4}{3}$.

**9-7** Compute the ratio of reflected intensity to incident intensity for a plane wave incident normally on a conductor of complex conductivity $g_c = g_c(0)/(1 - i\omega\tau)$ and $\varepsilon = \varepsilon_0$. Make a plot of the reflectivity as a function of $\omega$ for copper, using the values $g_c(0) = 6 \times 10^7$, $\tau = 2.5 \times 10^{-4}$ s, $N = 8.5 \times 10^{28}/\text{m}^3$.

**9-8** A thin dielectric film of thickness $\delta$ and index of refraction $n_2$ lies between media of indices of refraction $n_1$ and $n_3$, as shown in Figure 9–16. A light wave of frequency $\omega$ is incident normally from the left. Show that if there is no reflected wave, then $\delta = (N/4)\lambda$ (where $N$ is an odd integer and $\lambda$ is the wavelength in $n_2$) and that $n_2$ must be the geometric mean of $n_1$ and $n_3$. This phenomenon is of great utility in reducing light losses in optical equipment with many glass surfaces.

**9-9** Compute the generalized Brewster's angle for which the reflected intensity vanishes for the electric vector in the scattering plane. The incident and reflected waves are in a medium of index $n_1$, and the transmitted wave is in a medium of index $n_2$. Show that the incident plus transmitted angles add to 90°.

**9-10** Unpolarized light is incident on a dielectric interface at the Brewster's angle. Show that the ratio of transmissivities is

$$\frac{T_{/\!/}}{T_\perp} = \left(\frac{n_1^2 + n_2^2}{2n_1 n_2}\right)^2$$

and that this ratio is greater than unity for $n_2 \neq n_1$. For a stack of five glass plates with $n = 1.5$, show that $T_{/\!/}$ is five times $T_\perp$.

**9-11** For a simple crystal

$$D_i = \varepsilon_{ij} E_j$$

where $\varepsilon_{ij}$ are constants and a sum over $j$ is understood. Poynting's theorem still implies that the time derivative of the electric energy density is $\mathbf{E} \cdot \partial\, \mathbf{D}/\partial t$. If there is a crystal energy density, it must be of the form $\frac{1}{2}\varepsilon_{ij} E_i E_j$. Using these expressions, show that the $\varepsilon_{ij}$ tensor is symmetric.

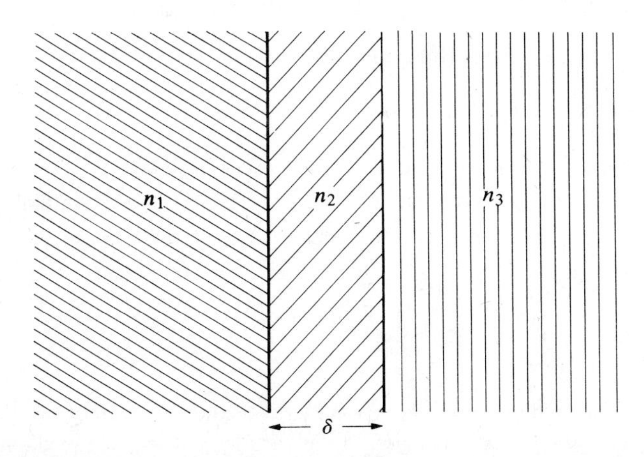


**FIGURE 9-16** Antireflection coating

**9-12** From Equation (9–71) we have $n_i^2\mathscr{E}_i = n^2(\mathscr{E}_i - \hat{\mathbf{k}}_i\hat{\mathbf{k}}_j\mathscr{E}_j)$, where $n_i = c/\sqrt{\mu_0\varepsilon_i}$. By multiplying by $\hat{\mathbf{k}}_i$ and summing, show that

$$\sum_{i=1}^{3} \frac{n^2\hat{\mathbf{k}}_i^2}{n^2 - n_i^2} = 1$$

Subtract $1 = \sum_{i=1}^{3} \hat{\mathbf{k}}_i^2$ to obtain

$$\sum_{i=1}^{3} \frac{n_i^2\hat{\mathbf{k}}_i^2}{n^2 - n_i^2} = 0$$

Solve for $n^2$, and show that the two solutions for $n^2$ are always real and positive.

**9-13** Light is incident normally on a crystalline-plane parallel plate of thickness $d$. Waves polarized along the two axes in the plate travel at velocities $c/n_1$ and $c/n_2$. Find the conditions under which incident plane polarized light emerges as circularly polarized light.

## Section 9-4: Rainbow

**9-14** Show that the fraction of light forming a primary bow is $(1 - R)^2R$, where $R$ is the reflectivity. Evaluate this expression for each polarization at the primary-bow angle. What is the absolute maximum this fraction can be? Repeat these calculations for the secondary bow, where the corresponding fraction is $(1-R)^2R^2$.

**9-15** Show that an explicit expression for the primary-rainbow angle $\theta_{\mathrm{P}}$ in terms of the index of refraction of the water drop is

$$\sin\frac{\theta_{\mathrm{P}}}{2} = \frac{1}{n^2}\left(\frac{4 - n^2}{3}\right)^{3/2}$$

**9-16** Derive the double-internal-scattering-angle formula of Equation (9–79). Find the minimum of $\theta_{\mathrm{S}}$, and compute the corresponding incident and transmitted angles.

## Section 9-5: Wave Pulses and Group Velocity

**9-17** A one-dimensional wave packet propagating along the $z$ axis has

$$\mathscr{E}(k) = \mathscr{E}_0 \exp[-4(k - k_0)^2/(\Delta k)^2],$$

where $\Delta k$ is the full width in wave number. Show that the envelope of the spatial wave group is also of Gaussian shape and that the product of $\Delta z$ and $\Delta k$ is a constant.

**9-18** In wave mechanics a one-dimensional wave packet for a relativistic particle in empty space has the form

$$\psi(x, t) = \frac{1}{\sqrt{2\pi}}\int_{-\infty}^{+\infty} dk\, a(k)e^{i[kx - \omega(k)t]}$$

where the momentum and the energy are given by $p = \hbar k$ and $E = \hbar\omega = \sqrt{c^2p^2 + (mc^2)^2}$. Calculate the phase and group velocities. Is the vacuum a dispersive medium for a material particle?

**9-19** The index of refraction of a material at X-ray frequencies $\omega$ is given by (see Problem 9–3) $n^2 = 1 - (\omega_p^2/\omega^2)$. Find the phase and group velocities. One of these velocities is greater than $c$. Which one is it?

**9-20** Show that for a left-handed, circular polarized wave propagating in an electron gas parallel to a magnetic field (see Problem 9–4), the index of refraction at low frequencies is $n_-^2 \approx \omega_p^2/(\omega\omega_B)$, and the group and phase velocities are

$$v_g = 2v_p = \frac{2c}{\omega_p}\sqrt{\omega_B\omega}$$

A thunderstorm in one hemisphere of our planet will generate radiation pulses that can travel along the dipole magnetic field lines and be detected in the other hemisphere. Because of the above dispersion formula, the high frequencies arrive before the lower ones. These waves have frequencies below $10^5$ Hz and, when picked up by a radio receiver, are called *whistlers* because of their characteristic falling tone. For a magnetic field of about 0.5 G, propagation distances of about $10^4$ km, and on average of $2 \times 10^4$ electrons/cm$^3$, show that the time scale of the frequency falloff is several seconds.

## Section 9-6: Waveguides

**9-21** A hollow waveguide with perfectly conducting walls has a square cross section $a$ on a side, as shown in Figure 9–17(a).

(i) What is the lowest frequency that can be transmitted by the guide?
(ii) Is this mode degenerate? Specify the designation(s).
(iii) What are the components of the electric field for this mode?
(iv) If the guide is bisected diagonally, as shown in Figure 9–17(b), by the conducting plane $x = y$, what is the lowest frequency that can be transmitted? Explain.

**9-22** For an oscillating tangential magnetic field of frequency $\omega$ outside a semi-infinite conductor of conductivity $g$, use the telegrapher's equation (see Problem 9–1) to find the magnetic field as a function of depth $z$. Try a solution of the form

$$B(z, t) = B(0)e^{-z/\delta}e^{ikz}e^{-i\omega t}$$

For good conductors, show that $k\delta \approx 1$ and the skin depth $\delta$ is given by $\delta = \sqrt{2/(\mu g\omega)}$.

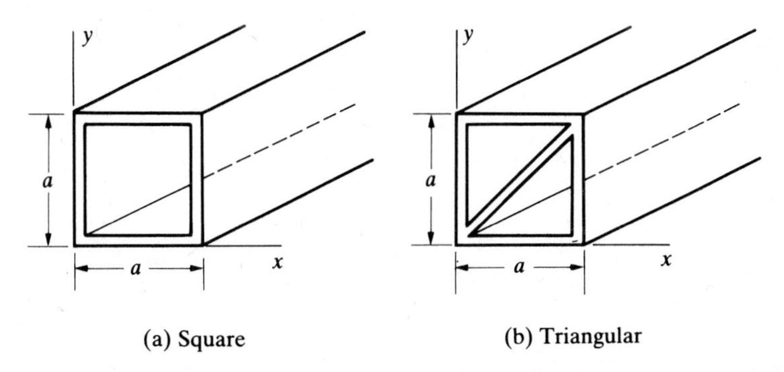


**FIGURE 9-17** Waveguides of (a) square and (b) triangular cross sections

**9-23** In the text we obtained the fields inside a waveguide under the assumption of perfectly conducting walls. The boundary conditions then imply time-varying surface charges and currents on the walls. For actual waveguides the walls have a finite conductivity, and the fields will penetrate about a skin depth $\delta$ into the wall material. The associated currents will cause ohmic heating, and some of the energy carried by the waveguide will be lost. Using the skin depth result of Problem 9–22, compute the current density as a function of depth, and compare the integrated value with that expected by an effective surface current. If $B(0)$ is the surface tangential field over a surface element $dS$, show that the power dissipated per unit area is

$$\frac{dP}{dS} = \frac{B(0)^2}{4}\sqrt{\frac{2\mu\omega}{g}}$$

The power density delivered to the conductor is $J^2/g$, where $J(z)$ is the physical (real) current density. For the $TE_{10}$ mode of a square waveguide, evaluate the fraction of power lost per unit length of the waveguide.

**9-24** Extremely low frequency (ELF) radiation is produced by lightning strokes. Both the ionosphere and the earth reflect these waves, since the wave frequency is much less than the plasma frequency (see Problem 9–3). The earth and the ionosphere form a concentric spherical cavity. Roughly estimate the lowest few standing wave frequencies by approximating the spherical surfaces by cylindrical ones. These standing waves are known as *Schumann resonances.*

# 10

# Radiation

CHAPTER CONTENTS

Electromagnetic wave solutions to the Maxwell equations were shown to exist in Chapter 8, but the source of the waves was not addressed there. Our purpose now is to discuss how electromagnetic waves are generated by the acceleration of charges. When a charge accelerates, kinks appear in the electric field lines from the charge. These kinks propagate radially outward from the charge with the speed of light. Far from the charge the kink is observed as a transverse wave pulse. In a limited region of space the pulse approximates a plane wave. This physical picture of the origin of waves is developed first, and then a mathematical treatment of wave generation is undertaken, using vector and scalar potentials. The wave equations for the potentials are derived from Maxwell's equations, and radiation solutions are obtained. Simple types of radiation sources of practical importance, such as electric and magnetic dipole radiation, are considered at length. Pulsars are discussed as an example of magnetic dipole radiation. Other interesting aspects of radiation theory are taken up, such as the forces required to maintain radiation, the radiation from a moving point charge, and the microscopic explanation of reflection and refraction of waves. We conclude with a statement of the wave equations for **E** and **B** fields for given charge and current sources. These equations will be used in the next chapter to establish the relativistic source transformations.

## 10-1 WAVE GENERATION

An electric charge at rest has radially directed electric field lines. A uniformly moving charge carries its electric field lines along with it, and for slow, steady motion ($v \ll c$) these radial lines are spherically symmetric. In other words, the field lines from a slowly moving charge do not depend on its velocity and are directed away from the particle's position. If the charge is accelerated from rest to a velocity $\Delta u$ in a time $\Delta t$, a kink will be formed in the field lines, which propagates outward at the speed of light, as will be discussed shortly. [For the original discussion, see J. J. Thomson, *Electricity and Matter* (New Haven, Conn.: Yale University Press, 1904).] This kink becomes a wave pulse at large distances from the charge.

In Figure 10–1 a charge $q$ is located at the origin until $t = 0$, at which time it has a constant acceleration $a$ along the $z$ axis for a time interval $\Delta t$. In this interval the charge attains a speed $\Delta u = a\,\Delta t$ while moving a distance $z_0 = \frac{1}{2}a(\Delta t)^2$. At times $t > \Delta t$ the charge moves uniformly at speed $\Delta u$. Consider a time $t \gg \Delta t$ when the charge is at position $z_1$. Beyond a radius $ct$ the electric field vector points back to the origin since electromagnetic interactions can only propagate with the speed of light. For radii less than $c(t - \Delta t)$ the electric field points to the present position of the uniformly moving charge. Because **E** field lines can only terminate on charges, and there is no charge in this system except $q$, the field lines must connect somehow in the shell of thickness $c\,\Delta t$.

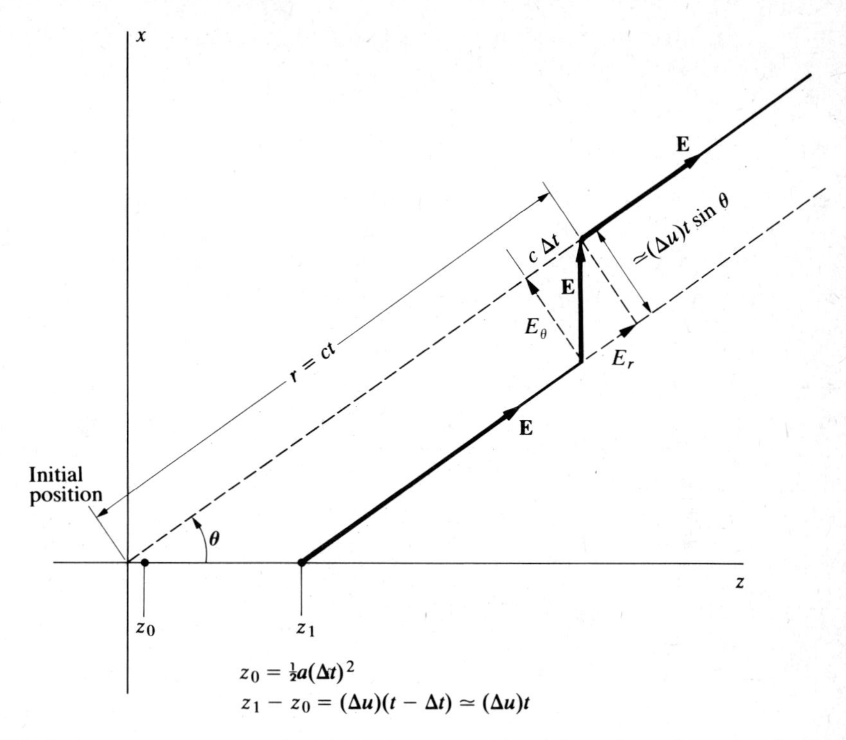


**FIGURE 10–1** Electric field kink due to accelerated motion of an electric charge

From the geometry of Figure 10–1 we see that neglecting the distance $z_0$ it took to accelerate the charge compared with $z_1$, the ratio of $E_\theta$ to $E_r$ in the kink region is given by

$$\frac{E_\theta}{E_r} \approx \frac{(\Delta u)t \sin\theta}{c(\Delta t)} = a\left(\frac{r}{c^2}\right)\sin\theta \tag{10-1}$$

where we have used $a = \Delta u/\Delta t$ and $r = ct$. As time goes on and the kink moves outward, the ratio $E_\theta/E_r$ grows larger, and the electric field in the kink becomes nearly transverse to the radial direction. The radial component is always given by Coulomb's law, which requires the radial flux to be constant; hence

$$E_r = \frac{q}{4\pi\varepsilon_0 r^2} \qquad E_\theta = \frac{aq}{4\pi\varepsilon_0 c^2 r}\sin\theta \tag{10-2}$$

In the forward ($\theta = 0$) or backward ($\theta = \pi$) directions the field lines remain straight, so we expect no radiation of the charge along the line of acceleration.

As we have previously seen in Equation (8–60), a wave propagating along direction $\hat{\mathbf{r}}$ must have an associated magnetic field $\mathbf{B} = (1/c)\hat{\mathbf{r}} \times \mathbf{E}$. For the kink region we thus find

$$B_\phi = \frac{aq}{4\pi\varepsilon_0 c^3} \frac{\sin\theta}{r} \tag{10-3}$$

where $\phi$ is the azimuthal angle defined in the usual way for spherical coordinates. The power carried outward through an infinitesimal spherical area element by the kink is given by the product of the Poynting vector $\mathbf{Y}$ (see Section 8–6) and the area element at radius $r = ct$,

$$dP = \mathbf{Y} \cdot \hat{\mathbf{r}} r^2 \, d\Omega \tag{10-4}$$

The radial intensity of the radiation is

$$\begin{aligned} \mathbf{Y} \cdot \hat{\mathbf{r}} &= \frac{1}{\mu_0} \mathbf{E} \times \mathbf{B} \cdot \hat{\mathbf{r}} = \frac{1}{\mu_0} \hat{\mathbf{r}} \times \mathbf{E} \cdot \mathbf{B} \\ &= \varepsilon_0 c E_\theta^2 = \frac{a^2 q^2 \sin^2\theta}{(4\pi)^2 \varepsilon_0 c^3 r^2} \end{aligned} \tag{10-5}$$

which has a $1/r^2$ dependence. Integrating Equation (10–5) over the sphere and multiplying the result by $\Delta t$ gives the total energy radiated, $\Delta U$, to be

$$\Delta U = P\,\Delta t = \left[ \frac{a^2 q^2}{(4\pi)^2 \varepsilon_0 c^3} 2\pi \int_0^\pi \sin^3\theta \, d\theta \right] \Delta t$$

or

**Larmor Formula**

$$\Delta U = \frac{q^2}{4\pi\varepsilon_0} \left( \frac{2a^2}{3c^3} \right) \Delta t \tag{10-6}$$

The total energy radiated in the pulse depends on the product of the square of the acceleration and its duration.

As a numerical example, we imagine a charge of 1 C accelerating at 100 times the gravitational acceleration at the earth's surface for 100 s. By Equation (10–6) the energy radiated is less than $10^{-7}$ J. Radiators for radio broadcasting, for instance, must thus involve either much larger charges (or currents) or much larger accelerations.

The preceding arguments can be generalized to relativistic velocities and arbitrary accelerations. [See Jack H. Tessman and Joseph T. Finnell, Jr., *Am. J. Phys.*, 35: 523 (1967).] Figure 10–2 shows the resulting radiation pattern of $\mathbf{Y}$ at both a low velocity ($u \ll c$) and a high velocity for the same acceleration. At high velocity the radiated power is larger, and it is collimated toward the direction of motion.

In an antenna for transmitting radio or TV waves the radiating charges move as an oscillating current within a conductor driven by an

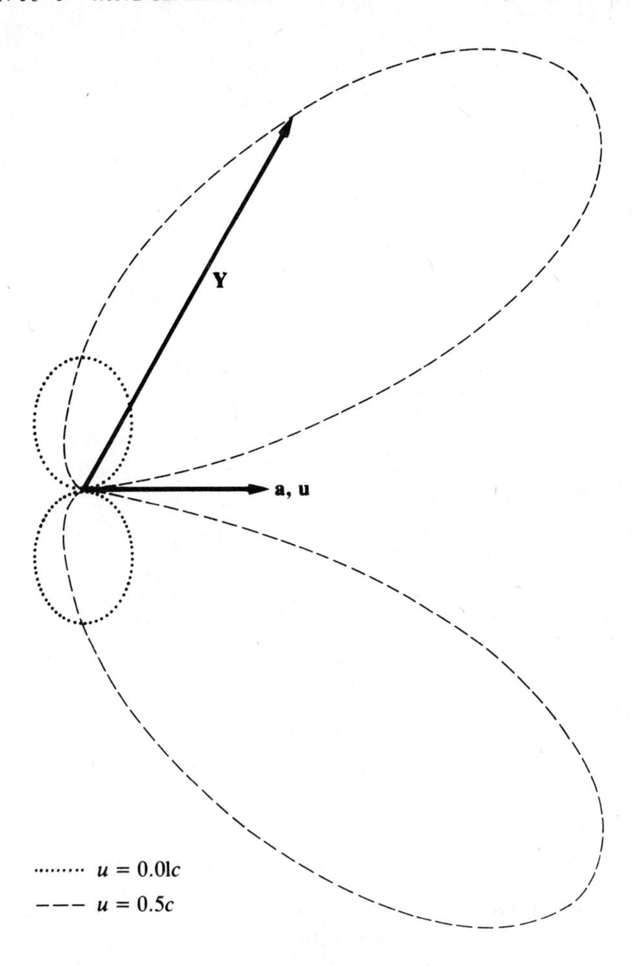


**FIGURE 10–2** Power distribution for a given acceleration **a** parallel to the charge velocity **u** when $u = 0.01c$ and $u = 0.5c$

external voltage source, which supplies the power to the radiation field. In a synchrotron radiation source charges move in circular or spiral orbits in a magnetic field. The centripetal acceleration of the charges causes the radiation.

An interesting application is the excitation of an atom by an incident plane wave. The oscillating electric field of the plane wave causes the electrons and the nucleus of the atom to oscillate at the same frequency as the incident wave and gives rise to an electric dipole moment, which we denote by $qz(t)$. Except for very intense incident waves, magnetic forces are negligible, since the oscillation velocities are small. The accelerations of the electrons are large compared with that of the nucleus, because of their mass ratio. A secondary radiation wavelet due to the acceleration of the electron is emitted from the atom, as illustrated in Figure 10–3.

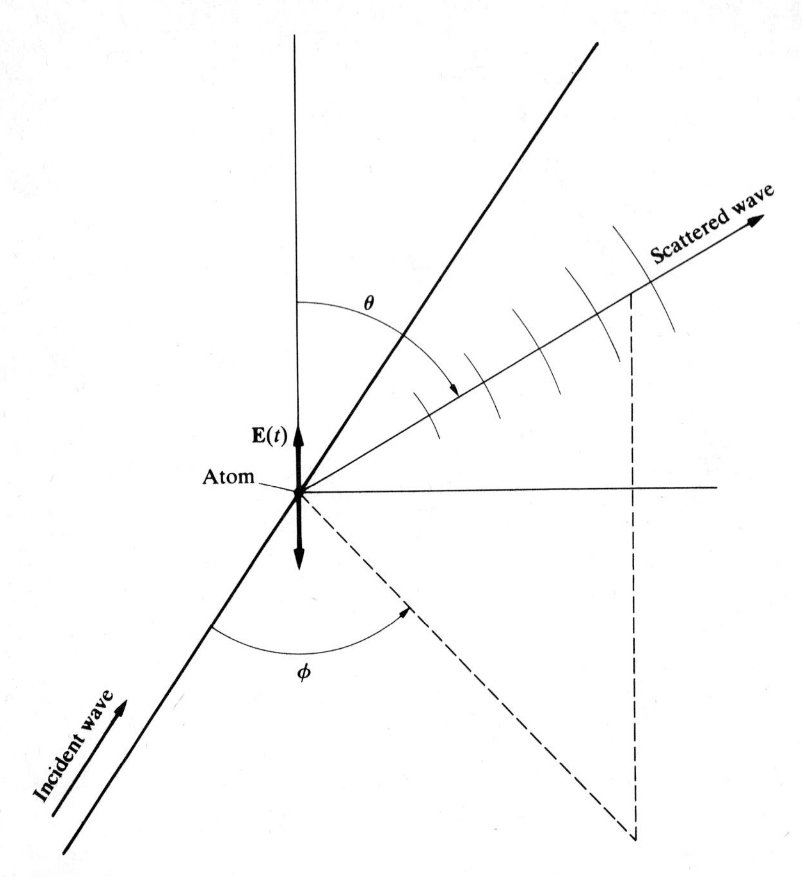


**FIGURE 10–3** Electromagnetic wave incident on an atom. The atomic electrons are accelerated, and the atom emits scattered radiation.

If the wave frequency $\omega$ of the radiation is much lower than the natural atomic frequency $\omega_0$, the inertial term $m\ddot{z}$ in the equation of motion can be neglected. Then the atom oscillates quasi-statically with the restoring force $m_e\omega_0^2 z(t)$ always balancing the applied force $qE_z(t)$ at the atom:

$$qE_z(t) = q\mathscr{E} \cos \omega t = m_e\omega_0^2 z(t) \tag{10–7}$$

Here $\mathscr{E}$ is the electric field strength of the incident wave at the atom. The electron's acceleration is then

$$a = \ddot{z} = -\frac{q\omega^2}{m_e\omega_0^2} \cos \omega t \tag{10–8}$$

with charge $q = -e$. The instantaneous power radiated by the atom found from Equation (10–6) is

$$P = \frac{dU}{dt} = \frac{1}{4\pi\varepsilon_0}\frac{2e^4\mathscr{E}^2}{3c^3m_e^2}\left(\frac{\omega}{\omega_0}\right)^4 \cos^2 \omega t$$

The average power is obtained by taking the time average of the instantaneous power over one period,

$$\begin{aligned}\langle P\rangle &= \frac{1}{4\pi\varepsilon_0}\frac{2e^4\mathscr{E}^2}{3c^2m_e^2}\left(\frac{\omega}{\omega_0}\right)^4 \frac{\omega}{2\pi}\int_0^{2\pi/\omega} \cos^2 \omega t\, dt \\ &= \frac{1}{4\pi\varepsilon_0}\frac{e^4\mathscr{E}^2}{3c^2m_e^2}\left(\frac{\omega}{\omega_0}\right)^4 \end{aligned} \qquad \textbf{(10–9)}$$

The $\omega^4$ dependence (or equivalently $1/\lambda^4$) of the radiated power spectrum of Equation (10–9) is the famous law derived by Lord Rayleigh. It explains, among other things, the blue sky, as follows: The sun's rays fall on atmospheric atoms, which in turn radiate light. The atoms in the earth's atmosphere are randomly distributed, so their radiation adds *incoherently*, and the power from $N$ molecules is just $N$ times Equation (10–9). Since $\omega_{\text{blue}} \approx 1.8\omega_{\text{red}}$, the radiated power of blue light is about ten times that of red, so the sky appears blue. By similar reasoning the sun at sunset or sunrise (or the moon at moonset or moonrise) is red. At these times the particularly long path through the atmosphere results in much of the blue light being scattered out of the beam of light from the sun to us, leaving an excess of red. Sunsets are distinctly more red when viewed from an earth satellite orbit since there is then twice the usual atmospheric path. The reddening of the sun at sunset is enhanced by fine particles from volcanic eruptions or forest fires, since dust in the atmosphere increases the scattering of the sunlight.

The kink picture gives a heuristic explanation of radiation from an accelerating charge. For further understanding of radiation a more mathematical approach must be pursued. For this purpose electromagnetic potentials provide the simplest framework. In the next section we derive wave equations for the potentials from the Maxwell theory, and then in the subsequent section we discuss radiation solutions to these wave equations.

## 10-2 WAVE EQUATIONS FOR THE VECTOR AND SCALAR POTENTIALS

For static charges the electric field satisfies $\nabla \times \mathbf{E} = 0$ and can be represented by a scalar potential $\Phi$ via $\mathbf{E} = -\nabla\Phi$. For steady currents the magnetic field can be expressed in terms of a vector potential $\mathbf{A}$ via

$$\mathbf{B} = \nabla \times \mathbf{A} \qquad \textbf{(10–10)}$$

For nonsteady fields we can still represent the magnetic field by Equation (10–10) since $\nabla \cdot \mathbf{B} = 0$ is valid generally, as long as there are no magnetic monopoles. The vector potential $\mathbf{A}$ is then a function of both space and time.

We will assume the fields are continuous enough to interchange spatial and time differentiations. Substituting Equation (10–10) into Maxwell III gives

$$\nabla \times \mathbf{E} = -\frac{\partial \mathbf{B}}{\partial t} = -\nabla \times \frac{\partial \mathbf{A}}{\partial t} \tag{10–11}$$

which can be written as

$$\nabla \times \left(\mathbf{E} + \frac{\partial \mathbf{A}}{\partial t}\right) = 0 \tag{10–12}$$

Hence for the time-dependent case it is the quantity $\mathbf{E} + \partial \mathbf{A}/\partial t$ that can be expressed as the gradient of a scalar potential; that is,

$$\mathbf{E} = -\nabla \Phi - \frac{\partial \mathbf{A}}{\partial t} \tag{10–13}$$

For time-independent **A** this expression reduces to the electrostatic result.

Suppose now that we express the potentials **A** and $\Phi$ in terms of potentials $\mathbf{A}'$ and $\Phi'$ as

**Gauge Transformation**

$$\mathbf{A} = \mathbf{A}' + \nabla \chi \qquad \Phi = \Phi' - \frac{\partial \chi}{\partial t} \tag{10–14}$$

where $\chi$ is some function of $(\mathbf{r}, t)$. Then from Equations (10–10) and (10–13) we find that

$$\mathbf{B} = \nabla \times \mathbf{A}' \qquad \mathbf{E} = -\nabla \Phi' - \frac{\partial \mathbf{A}'}{\partial t} \tag{10–15}$$

Thus $\mathbf{A}'$ and $\Phi'$ are potentials that give the same **E** and **B** fields as the potentials **A** and $\Phi$. Since $\chi(\mathbf{r}, t)$ can be an arbitrary differentiable function, there exists a continuum of potential choices to represent given **E** and **B** fields. The nonuniqueness of potentials expressed in Equation (10–14) is known as a *gauge transformation.* Observable quantities like **E** and **B** are unchanged by a gauge transformation and are said to be *gauge-invariant*. Maxwell's equations are expressed in terms of the **E** and **B** fields and hence are invariant to a gauge transformation. The gauge freedom in the choice of **A** and $\Phi$ is often exploited in the solution of electromagnetic problems.

The Maxwell equations in Table 8–1 can be expressed in terms of the vector and scalar potential, using Equations (10–10) and (10–13). Maxwell I becomes

$$\nabla \cdot \mathbf{E} = -\nabla^2 \Phi - \frac{\partial}{\partial t}(\nabla \cdot \mathbf{A}) = \frac{\rho}{\varepsilon_0}$$

or

$$\nabla^2\Phi + \frac{\partial}{\partial t}(\nabla \cdot \mathbf{A}) = -\frac{\rho}{\varepsilon_0} \tag{10-16}$$

Maxwell II and III were already used in defining the potentials via Equations (10–10) and (10–13). Maxwell IV is expressed as

$$\nabla \times \mathbf{B} = \nabla \times (\nabla \times \mathbf{A}) = \mu_0 \mathbf{J} + \frac{1}{c^2}\frac{\partial}{\partial t}\left(-\nabla\Phi - \frac{\partial \mathbf{A}}{\partial t}\right) \tag{10-17}$$

Using the vector identity for the triple cross product, we get

$$\nabla(\nabla \cdot \mathbf{A}) - \nabla^2 \mathbf{A} = \mu_0 \mathbf{J} - \frac{1}{c^2}\frac{\partial}{\partial t}\nabla\Phi - \frac{1}{c^2}\frac{\partial^2 \mathbf{A}}{\partial t^2} \tag{10-18}$$

which can be rewritten as

$$\nabla^2 \mathbf{A} - \frac{1}{c^2}\frac{\partial^2 \mathbf{A}}{\partial t^2} = -\mu_0 \mathbf{J} + \nabla\left(\nabla \cdot \mathbf{A} + \frac{1}{c^2}\frac{\partial \Phi}{\partial t}\right) \tag{10-19}$$

Equations (10–16) and (10–19) are the potential form of Maxwell's equations. These equations are form-invariant under the gauge transformation of Equation (10–14), as can be verified by direct substitution (i.e., the equations in terms of $\mathbf{A}'$ and $\Phi'$ are the same as those in terms of $\mathbf{A}$ and $\Phi$). Inasmuch as the physical $\mathbf{E}$ and $\mathbf{B}$ fields from $\mathbf{A}$, $\Phi$ and $\mathbf{A}'$, $\Phi'$ are identical, we use the gauge freedom to simplify the potential equations. In particular, we choose the arbitrary function $\chi$ in Equation (10–14) such that we have the following:

**Lorentz Gauge**

$$\nabla \cdot \mathbf{A} + \frac{1}{c^2}\frac{\partial \Phi}{\partial t} = 0 \tag{10-20}$$

This choice is known as the Lorentz gauge. (Another gauge that is sometimes employed is $\nabla \cdot \mathbf{A} = 0$, which is called the *Coulomb gauge*.) The differential equation that $\chi$ must satisfy is obtained by substitution of Equation (10–14) into Equation (10–20), giving

$$\nabla^2\chi - \frac{1}{c^2}\frac{\partial^2 \chi}{\partial t^2} = -\left(\nabla \cdot \mathbf{A}' + \frac{1}{c^2}\frac{\partial \Phi'}{\partial t}\right) \tag{10-21}$$

where the inhomogeneous term is determined by the choice of the $\mathbf{A}'$ and $\Phi'$ potentials.

Substituting the condition of Equation (10–20) into Equations (10–16) and (10–19), we obtain

**Wave Equations for Potentials in Lorentz Gauge**

$$\nabla^2\Phi - \frac{1}{c^2}\frac{\partial^2\Phi}{\partial t^2} = -\frac{\rho}{\varepsilon_0}$$
$$\nabla^2\mathbf{A} - \frac{1}{c^2}\frac{\partial^2\mathbf{A}}{\partial t^2} = -\mu_0\mathbf{J} \qquad \textbf{(10-22)}$$

which are the wave equations for the scalar and vector potentials in the Lorentz gauge. The **A** and $\Phi$ potentials satisfying Equations (10–22) are still not unique, since further gauge transformations that satisfy

$$\nabla^2\chi - \frac{1}{c^2}\frac{\partial^2\chi}{\partial t^2} = 0$$

can be made in the Lorentz gauge. In the next section we find solutions to the wave equations for the potentials.

# 10-3 RETARDED POTENTIAL SOLUTIONS

In the static case the wave equations of Equation (10–22) reduce to

$$\nabla^2\Phi(\mathbf{r}) = -\frac{\rho(\mathbf{r})}{\varepsilon_0} \qquad \nabla^2\mathbf{A}(\mathbf{r}) = -\mu_0\mathbf{J}(\mathbf{r})$$

which are the Poisson equations. In unbounded regions of space these equations have the solutions

$$\Phi(\mathbf{r}) = \frac{1}{4\pi\varepsilon_0}\int\frac{\rho(\mathbf{r}')\,dV'}{R} \qquad \mathbf{A}(\mathbf{r}) = \frac{\mu_0}{4\pi}\int\frac{\mathbf{J}(r')\,dV'}{R}$$

where $R = |\mathbf{r} - \mathbf{r}'|$. Solutions to the wave equations of Equation (10–22) in the time-dependent case have a similar form, if we allow for the interval of time $R/c$ required for a light signal to travel from a given source point $\mathbf{r}'$ to the field point $\mathbf{r}$.

The four wave equations of Equation (10–22) are each of the form

$$\left(\nabla^2 - \frac{1}{c^2}\frac{\partial^2}{\partial t^2}\right) f(\mathbf{r}, t) = -s(\mathbf{r}, t) \qquad \textbf{(10-23a)}$$

One way to solve this equation is to work in terms of the harmonic components of $f(\mathbf{r}, t)$. One projects the frequency $\omega$ part of the two sides of the equation by multiplying by $e^{i\omega t}$ and integrating over time. The result is the equation

$$(\nabla^2 + k^2)F(\mathbf{r}, \omega) = -S(\mathbf{r}, \omega) \qquad \textbf{(10-23b)}$$

where $k = \omega/c$ and

**Fourier Transforms**

$$F(\mathbf{r}, \omega) = \int_{-\infty}^{+\infty} dt\, e^{i\omega t} f(\mathbf{r}, t) \qquad S(\mathbf{r}, \omega) = \int_{-\infty}^{+\infty} dt\, e^{i\omega t} s(\mathbf{r}, t) \tag{10–24a}$$

The functions $F(\mathbf{r}, \omega)$ and $S(\mathbf{r}, \omega)$ are time Fourier transforms of $f(\mathbf{r}, t)$ and $s(\mathbf{r}, t)$. The inverse Fourier transforms are as follows:

**Inverse Fourier Transforms**

$$f(\mathbf{r}, t) = \frac{1}{2\pi}\int_{-\infty}^{+\infty} d\omega\, e^{-i\omega t} F(\mathbf{r}, \omega)$$
$$s(\mathbf{r}, t) = \frac{1}{2\pi}\int_{-\infty}^{+\infty} d\omega\, e^{-i\omega t} S(\mathbf{r}, \omega) \tag{10–24b}$$

In obtaining Equations (10–24b) from Equations (10–24a), we have used the following representation for the delta function:

$$\delta(t - t') = \frac{1}{2\pi}\int_{-\infty}^{+\infty} d\omega\, e^{i\omega(t-t')} \tag{10–24c}$$

This representation can be established by first integrating over a finite interval $-a \leqslant \omega \leqslant a$ and then taking the limit $a \to \infty$.

Equation (10–23b) for the harmonic component $F(\mathbf{r}, \omega)$ is called the *Helmholtz equation*; it reduces to the Poisson equation when $k = 0$. Its solution, which goes to zero at large $r$, is

$$F(\mathbf{r}, \omega) = \int dV' \frac{C_{\mathrm{R}} e^{ikR} + C_{\mathrm{A}} e^{-ikR}}{4\pi R} S(\mathbf{r}', \omega) \tag{10–25}$$

where the constant coefficients must satisfy $C_{\mathrm{R}} + C_{\mathrm{A}} = 1$. To verify that this expression is a solution, one uses, for $R \neq 0$,

$$(\nabla^2 + k^2)\frac{e^{\pm ikR}}{R} = (\nabla_R^2 + k^2)\frac{e^{\pm ikR}}{R} = \frac{1}{R}\left(\frac{d^2}{dR^2} + k^2\right)e^{\pm ikR} = 0 \tag{10–26a}$$

At $R = 0$ we have the $\delta$ function term as in Equation (2–32), so

$$(\nabla^2 + k^2)\frac{e^{\pm ikR}}{R} = -4\pi\delta(\mathbf{R}) \tag{10–26b}$$

We now recombine the harmonic components $F(\mathbf{r}, \omega)$ to get $f(\mathbf{r}, t)$ [i.e., we use the inverse Fourier transform, Equation (10–24b)]:

$$f(\mathbf{r}, t) = \int_{-\infty}^{\infty} \frac{d\omega}{2\pi} e^{-i\omega t} \int dV' \frac{C_{\mathrm{R}} e^{ikR} + C_{\mathrm{A}} e^{-ikR}}{4\pi R} \int_{-\infty}^{\infty} dt'\, e^{i\omega t'} s(\mathbf{r}', t') \tag{10-27}$$

The $\omega$ integral can be done, yielding a $\delta$ function

$$\int_{-\infty}^{\infty} \frac{d\omega}{2\pi} e^{i\omega(-t \pm R/c + t')} = \delta\left(-t \pm \frac{R}{c} + t'\right) \tag{10-28}$$

and then the integral over $t'$ can be done. The result is

$$f(\mathbf{r}, t) = \int dV' \frac{1}{4\pi R} [C_{\mathrm{R}} s(\mathbf{r}', t_{\mathrm{R}}) + C_{\mathrm{A}} s(\mathbf{r}', t_{\mathrm{A}})] \tag{10-29a}$$

where

$$t_{\mathrm{R}} = t - \frac{R}{c} \qquad t_{\mathrm{A}} = t + \frac{R}{c} \tag{10-29b}$$

One sees that the solution to the wave equation, Equation (10–29a), is very similar to the solution (see Section 3–3) of the Poisson equation: For both, each volume element of the source $s(\mathbf{r}')\,dV'$ contributes a Coulomb term $s(\mathbf{r}')\,dV'/(4\pi R)$ to the potential field. The only difference is that for the wave equation the time at which the source element is to be evaluated is not $t$, the time at which the field is being evaluated, but rather $t_{\mathrm{R}} = t - R/c$ or $t_{\mathrm{A}} = t + R/c$. Since $R/c$ is the time interval needed for light to travel the distance $R$ from the source at point $\mathbf{r}'$ to the point $\mathbf{r}$ where the field is being evaluated, $t_{\mathrm{R}}$ is the time at which light would have to be emitted from the source point $\mathbf{r}'$ in order to arrive at the field point $\mathbf{r}$ at time $t$. Similarly, $t_{\mathrm{A}}$ is the time at which light emitted from the field point $\mathbf{r}$ would arrive at the source point $\mathbf{r}'$. (The abbreviations R and A stand for *retarded* and *advanced*, respectively; of course, $t_{\mathrm{A}} > t > t_{\mathrm{R}}$.) The coefficients $C_{\mathrm{R}}$ and $C_{\mathrm{A}}$ are arbitrary, except for their sum, $C_{\mathrm{R}} + C_{\mathrm{A}} = 1$, as far as the wave equation is concerned. They are fixed by the physical consideration that causes precede their effects. Thus the source time can only be $t_{\mathrm{R}}$, not $t_{\mathrm{A}}$; and so $C_{\mathrm{A}} = 0$ and $C_{\mathrm{R}} = 1$. Hence the physically acceptable solution of the wave equation is the *retarded solution,*

$$f(\mathbf{r}, t) = \int dV' \frac{s(\mathbf{r}', t_{\mathrm{R}})}{4\pi R} \qquad t_{\mathrm{R}} = t - \frac{R}{c} \tag{10-30}$$

The retarded solutions to the time-dependent wave equations of Equation (10–22) are then given by

**Retarded Potentials**

$$\Phi(\mathbf{r}, t) = \frac{1}{4\pi\varepsilon_0} \int \frac{\rho(\mathbf{r}', t_{\mathrm{R}})\, dV'}{R}$$
$$\mathbf{A}(\mathbf{r}, t) = \frac{\mu_0}{4\pi} \int \frac{\mathbf{J}(\mathbf{r}', t_{\mathrm{R}})\, dV'}{R} \tag{10-31}$$

# 10-4 RADIATION WITH HARMONIC TIME VARIATION

If we know the current and the charge density at each point in space for all past times, we can straightforwardly evaluate the retarded potentials of Equations (10–31). This evaluation is usually a difficult analytic task. As a simplification, we consider only simple harmonic time dependences of the sources,

$$\begin{aligned} \mathbf{J}(\mathbf{r}', t_R) &= \mathbf{J}(\mathbf{r}')e^{-i\omega t_R} = \mathbf{J}(\mathbf{r}')e^{ikR}e^{-i\omega t} \\ \rho(\mathbf{r}', t_R) &= \rho(\mathbf{r}')e^{ikR}e^{-i\omega t} \end{aligned} \tag{10–32}$$

where $k = \omega/c$. The time dependence of the potentials will now be also simple harmonic, so the potentials can be expressed as a product of space- and time-dependent terms of the form

$$\mathbf{A}(\mathbf{r}, t) = \mathbf{A}(\mathbf{r})e^{-i\omega t} \qquad \Phi(\mathbf{r}, t) = \Phi(\mathbf{r})e^{-i\omega t} \tag{10–33}$$

with the spatial dependence given, as in Equation (10–31), by

$$\mathbf{A}(\mathbf{r}) = \frac{\mu_0}{4\pi}\int \frac{\mathbf{J}(\mathbf{r}')e^{ikR}}{R}\,dV' \tag{10–34a}$$

$$\Phi(\mathbf{r}) = \frac{1}{4\pi\varepsilon_0}\int \frac{\rho(\mathbf{r}')e^{ikR}}{R}\,dV' \tag{10–34b}$$

The $e^{ikR}$ factor reflects the time retardation. It is understood that the physical fields and the sources are obtained by taking real parts. A general time dependence can be obtained by superposing harmonic solutions.

The magnetic field is determined by the curl of the vector potential, $\mathbf{B} = \nabla \times \mathbf{A}$. If there is no current density at the field point, the electric field can be determined by using the fourth Maxwell equation,

$$\frac{\partial \mathbf{E}}{\partial t} = c^2\, \nabla \times \mathbf{B}$$

For a harmonic time variation $e^{-i\omega t}$ the electric field can thereby be expressed as

$$\mathbf{E} = \frac{ic}{k}\nabla \times \mathbf{B} \tag{10–35}$$

Thus for simple harmonic time dependence both $\mathbf{E}$ and $\mathbf{B}$ can be calculated from the vector potential alone at field points where $\mathbf{J}$ vanishes.

If the factor $e^{ikR}$ in Equations (10–34) is close to unity, then the potentials are close to their static values. This situation occurs if the following limit holds:

**Quasi-static Limit**

$$kR \ll 1 \tag{10-36}$$

This requirement can be expressed in terms of wavelength $\lambda = 2\pi/k$ as follows:

**Quasi-static Limit**

$$R \ll \frac{\lambda}{2\pi} \tag{10-37}$$

This condition requires distances between source and field points to be small compared with the wavelength of the radiation. In this low-frequency (long-wavelength) regime, called the *quasi-static limit*, the displacement current and the radiated power are negligible. The quasi-static approximation applies to ordinary lumped-circuit electronics employing resistive, inductive, and capacitive elements and various nonlinear devices such as transistors. When the size of the circuit becomes comparable to wavelengths $\lambda = 2\pi c/\omega$ characteristic of the frequencies of oscillations in the circuit, the quasi-static approximation breaks down. In the TV/FM frequency range ($\approx 10^8$ Hz), $\lambda/(2\pi) \approx 0.5$ m, so care must be taken in circuit design. For microwave (e.g., radar) circuits where $\lambda$ is much smaller, wires are replaced by waveguides to suppress radiation losses.

The quasi-static regime corresponding to $kR \ll 1$ is called the *near zone*. Another regime of particular interest is the radiation zone:

**Radiation Zone**

$$r \gg \lambda \qquad (\text{or } kr \gg 1) \qquad r \gg r' \tag{10-38}$$

That is, the distance to the field point is large compared with either the wavelength or the source dimension. The intermediate zone between these regimes is difficult to deal with because fewer simplifying approximations can be made; we omit discussion of the intermediate zone and concentrate our attention on the radiation zone.

In the radiation zone where $r \gg r'$, we can approximate the expression for $R$ in Equation (4–2) by

$$R \approx r - \hat{\mathbf{r}} \cdot \mathbf{r}' \tag{10-39}$$

Then the space-dependent part of the vector potential in Equation (10–34a) for simple harmonic waves becomes

**Radiation Zone Approximation**

$$A(\mathbf{r}) \approx \frac{\mu_0}{4\pi} \frac{e^{ikr}}{r} \int \mathbf{J}(\mathbf{r}') e^{-ik\hat{\mathbf{r}} \cdot \mathbf{r}'} \, dV' \tag{10-40}$$

We can go quite far toward computing the magnetic and electric fields radiated by a general current distribution. A convenient way of expressing Equation (10–40) for this purpose is

$$\mathbf{A}(\mathbf{r}) = \frac{\mu_0}{4\pi}\frac{e^{ikr}}{r}\,\mathbf{F}(\hat{\mathbf{r}}) \tag{10-41a}$$

$$\mathbf{F}(\hat{\mathbf{r}}) = \int \mathbf{J}(\mathbf{r}')e^{-ik\hat{\mathbf{r}}\cdot\mathbf{r}'}\,dV' \tag{10-41b}$$

The factor $e^{ikr}/r$ in $\mathbf{A}(\mathbf{r})$ is an outgoing spherical wave. To compute the magnetic field, we take the curl of $\mathbf{A}(\mathbf{r})$:

$$\begin{aligned}\mathbf{B} = \nabla \times \mathbf{A} &= \frac{\mu_0}{4\pi}\nabla \times \left[\frac{e^{ikr}}{r}\,\mathbf{F}(\hat{\mathbf{r}})\right] \\ &= \frac{\mu_0}{4\pi}\left[\nabla\left(\frac{e^{ikr}}{r}\right) \times \mathbf{F}(\hat{\mathbf{r}}) + \frac{e^{ikr}}{r}\,\nabla \times \mathbf{F}(\hat{\mathbf{r}})\right]\end{aligned} \tag{10-42}$$

The essential simplification in the radiation zone is to retain only the leading term in $1/r$. Since $\hat{\mathbf{r}} = \mathbf{r}/r$, the second term in Equation (10–42) falls off as $1/r^2$, as can be quickly verified. The leading contribution to the gradient term in Equation (10–42) is $ik\hat{\mathbf{r}}e^{ikr}/r$, so the magnetic field is

$$\mathbf{B} = \frac{\mu_0}{4\pi}\,ik\,\frac{e^{ikr}}{r}\,\hat{\mathbf{r}} \times \mathbf{F} \tag{10-43}$$

By a similar argument the electric field is, with Equation (10–35),

$$\mathbf{E} = -c\hat{\mathbf{r}} \times \mathbf{B} \tag{10-44}$$

From Equations (10–43) and (10–44) we conclude that in the radiation zone the magnetic and electric fields are mutually perpendicular and both are perpendicular to the direction of propagation $\hat{\mathbf{r}}$. This result is exactly what one would expect, since far from the source the radiation wave should locally look like a plane wave. The calculation of the radiation fields reduces to the evaluation of the $\mathbf{F}$ integral of Equation (10–41b).

The radiated power can also be expressed in terms of $\mathbf{F}$. The Poynting vector is, from Equation (10–44),

$$\mathbf{Y} = \frac{1}{\mu_0}\mathbf{E} \times \mathbf{B} = -\frac{c}{\mu_0}(\hat{\mathbf{r}} \times \mathbf{B}) \times \mathbf{B}$$

or

$$\mathbf{Y} = \frac{c}{\mu_0}B^2\hat{\mathbf{r}} \tag{10-45}$$

The power is propagated outward as we would expect.

Since the Poynting vector **Y** is bilinear in the fields, we cannot simply multiply the complex fields and then later take the real part; we must first find the physical fields. The physical magnetic field is found by first including the harmonic time dependence in Equation (10–43) and then taking the real part:

$$\mathbf{B}(\mathbf{r}, t) = \frac{\mu_0 k}{4\pi r}\hat{\mathbf{r}} \times \mathrm{Re}[ie^{i(kr-\omega t)}\mathbf{F}(\hat{\mathbf{r}})]$$
$$= \frac{\mu_0 k}{4\pi r}\hat{\mathbf{r}} \times [-\mathrm{Re}\,\mathbf{F}\sin(kr-\omega t) + \mathrm{Im}\,\mathbf{F}\cos(kr-\omega t)] \tag{10-46}$$

Squaring $B(\mathbf{r}, \mathbf{t})$ then gives

$$B^2 = \left(\frac{\mu_0}{4\pi}\right)^2 \frac{k^2}{r^2}|\hat{\mathbf{r}} \times [-\mathrm{Re}\,\mathbf{F}\sin(kr-\omega t) + \mathrm{Im}\,\mathbf{F}\cos(kr-\omega t)]|^2 \tag{10-47}$$

We are usually most interested in the time average of **Y**. The time average of $\sin^2(kr - \omega t)$, for example, is found by averaging over the period $\tau = 2\pi/\omega$:

$$\langle \sin^2(kr-\omega t)\rangle_t \equiv \frac{1}{\tau}\int_0^\tau \sin^2(kr-\omega t)\,dt = \frac{1}{2} \tag{10-48}$$

Similarly, we find $\langle\cos^2(kr-\omega t)\rangle_t = \frac{1}{2}$ and $\langle\sin(kr-\omega t)\cos(kr-\omega t)\rangle_t = 0$, so the time-averaged power flux is

$$\langle \mathbf{Y}\rangle_t = \frac{\mu_0}{(4\pi)^2}\frac{k^2 c}{2r^2}[(\hat{\mathbf{r}} \times \mathrm{Re}\,\mathbf{F})^2 + (\hat{\mathbf{r}} \times \mathrm{Im}\,\mathbf{F})^2]\hat{\mathbf{r}} \tag{10-49}$$

The power radiated into a solid-angle element $d\Omega$ is then

$$\langle dP\rangle_t = r^2\, d\Omega\, \langle \mathbf{Y}\rangle_t \cdot \hat{\mathbf{r}}$$
$$= \frac{\mu_0}{(4\pi)^2}\frac{k^2 c}{2}[(\hat{\mathbf{r}} \times \mathrm{Re}\,\mathbf{F})^2 + (\hat{\mathbf{r}} \times \mathrm{Im}\,\mathbf{F})^2]\, d\Omega \tag{10-50}$$

which is independent of radius, as expected.

## 10-5 MULTIPOLE RADIATION

If the source dimensions are small compared with the radiated wavelength,

$$r' \ll \lambda \qquad (\text{or } kr' \ll 1) \tag{10-51}$$

where $\lambda = 2\pi c/\omega$, the radiation integral $\mathbf{F}$ can be evaluated term by term in an expansion. This expansion is known as the *multipole expansion* and consists of replacing the exponential in Equation (10–41b) by its Taylor series:

$$e^{-ik\hat{\mathbf{r}}\cdot\mathbf{r}'} = \sum_{n=0}^{\infty} \frac{1}{n!}(-ik\hat{\mathbf{r}}\cdot\mathbf{r}')^n \tag{10-52}$$

If the inequality of Equation (10–51) is valid, the series is rapidly convergent, and we can often approximate the vector potential by a single term of the series

$$\mathbf{F}(\hat{\mathbf{r}}) = \sum_{n=0}^{\infty} \frac{1}{n!} \int \mathbf{J}(\mathbf{r}')(-ik\hat{\mathbf{r}}\cdot\mathbf{r}')^n \, dV' \tag{10-53}$$

The lowest term in this series ($n = 0$) gives what is known as electric dipole radiation. An example of a system with this moment is charge flowing back and forth along a straight wire (a linear antenna). The $n = 1$ term is partly a magnetic dipole and partly an electric quadrupole. An example of a system that emits magnetic dipole radiation is a current loop with a time-varying current. The rotation of a static magnetic dipole moment is also a source of magnetic dipole radiation. Because of the great importance of these lowest multipoles in many applications, we will examine their field values and power distributions in some detail.

## 10-5A Electric Dipole

Electric dipole radiation arises from the $n = 0$ term in the multipole expansion of Equation (10–52) with

$$\mathbf{F}_{\mathrm{ED}} = \int \mathbf{J}(\mathbf{r}')\, dV' \tag{10-54}$$

To relate the integral to the electric dipole moment, we generalize the identity of Equation (7–3) to nonstatic currents.

We consider the product $f\mathbf{J}$, where $f(\mathbf{r}')$ is an arbitrary scalar function. Applying the divergence theorem for a bounding surface that completely encloses the current distribution, we find

$$\int \nabla\cdot(f\mathbf{J})\, dV' = \int f\mathbf{J}\cdot d\mathbf{S}' = 0 \tag{10-55}$$

where $\nabla$ acts on the source variables. If we expand $\nabla\cdot(f\mathbf{J}) = (\nabla f)\cdot\mathbf{J} + f\,\nabla\cdot\mathbf{J}$ and use the continuity equation,

$$\nabla\cdot\mathbf{J} = -\frac{\partial\rho}{\partial t} = i\omega\rho,$$

Equation (10–55) leads to the identity

$$\int (\nabla f)\cdot \mathbf{J}\, dV' = \int f \frac{\partial \rho}{\partial t}\, dV' = -i\omega \int f\rho\, dV' \qquad \textbf{(10–56)}$$

This equation, with appropriate choices of $f(\mathbf{r}')$, is useful in evaluating the various electric and magnetic multipole moments.

For the electric dipole case the appropriate choice is $f = x_i'$, the $i$th Cartesian component of $\mathbf{r}'$. The identity of Equation (10–56) gives

$$\int J_i\, dV' = -i\omega \int x_i' \rho\, dV' \qquad \textbf{(10–57)}$$

This relation can be expressed in vector form as

$$\mathbf{F}_{\text{ED}} = \int \mathbf{J}\, dV' = -i\omega \mathbf{p} \qquad \textbf{(10–58)}$$

where $\mathbf{p} = \int \mathbf{r}'\rho(\mathbf{r}')\, dV'$ is the electric dipole moment. In terms of the vector potential of Equations (10–41) we have

$$\mathbf{A}_{\text{ED}}(\mathbf{r}) = -\frac{i\mu_0\omega}{4\pi}\frac{e^{ikr}}{r}\mathbf{p} \qquad \textbf{(10–59)}$$

Once the $\mathbf{F}$ vector is known, our general results of Equations (10–43) and (10–44) almost immediately give the electric and magnetic field expressions:

$$\mathbf{B}_{\text{ED}} = \frac{\mu_0}{4\pi} k\omega \frac{e^{ikr}}{r}\, \hat{\mathbf{r}} \times \mathbf{p} \qquad \textbf{(10–60a)}$$

$$\mathbf{E}_{\text{ED}} = -\frac{\mu_0}{4\pi}\omega^2 \frac{e^{ikr}}{r}\, \hat{\mathbf{r}} \times (\hat{\mathbf{r}} \times \mathbf{p}) \qquad \textbf{(10–60b)}$$

To make these results more explicit, we can choose the electric dipole moment to lie along the $z$ axis so that

$$\hat{\mathbf{r}} \times \mathbf{p} = p\hat{\mathbf{r}} \times \hat{\mathbf{z}} = -p \sin\theta \hat{\boldsymbol{\phi}} \qquad \textbf{(10–61)}$$

We obtain

$$\mathbf{B}_{\text{ED}} = -\frac{\mu_0}{4\pi}\frac{\omega^2 p}{c}\frac{e^{ikr}}{r}\sin\theta\hat{\boldsymbol{\phi}} \qquad \mathbf{E}_{\text{ED}} = -\frac{\mu_0}{4\pi}\omega^2 p \frac{e^{ikr}}{r}\sin\theta\hat{\boldsymbol{\theta}} \qquad \textbf{(10–62)}$$

The directions of the **E** and **B** fields for electric dipole radiation are illustrated in Figure 10–4.

The time-dependent fields are obtained from Equations (10–62) by reintroducing the harmonic factor $e^{-i\omega t}$. In terms of the retarded time

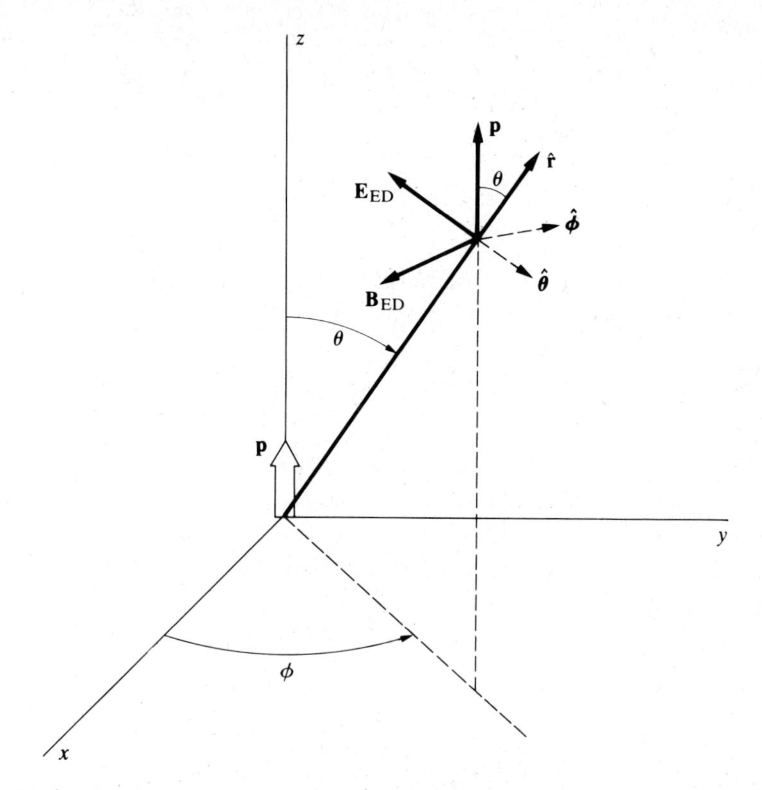


**FIGURE 10-4** Electric and magnetic field directions for electric dipole radiation

$t_R = t - (r/c)$ from the origin, the fields are as follows:

**Electric Dipole Radiation Fields**

$$\mathbf{E}_{ED}(\mathbf{r}, t) = -\frac{\mu_0}{4\pi}\omega^2 p \frac{e^{-i\omega t_R}}{r} \sin\theta\hat{\boldsymbol{\theta}}$$

$$\mathbf{B}_{ED}(\mathbf{r}, t) = -\frac{\mu_0}{4\pi}\frac{\omega^2 p}{c}\frac{e^{-i\omega t_R}}{r} \sin\theta\hat{\boldsymbol{\phi}} \quad \textbf{(10-63)}$$

These results could have been inferred from the radiation fields of an accelerated charge found in Section 10–1, as follows: Suppose that the electric dipole moment is due to harmonic oscillations of a charge $q$ with amplitude $d$. The acceleration of the charge is

$$a(t) = -\omega^2 d e^{-i\omega t} \quad \textbf{(10-64)}$$

and the electric dipole moment is

$$p(t) = qde^{-i\omega t} \quad \textbf{(10-65)}$$

giving

$$a(t) = -\frac{\omega^2}{q}p(t) \tag{10-66}$$

If we use the retarded time and this acceleration in Equations (10–2) and (10–3), the results reproduce the radiation fields derived in Equations (10–63).

The time-averaged power radiated by an oscillating electric dipole can be deduced from Equation (10–50) with $\mathbf{F} = -i\omega p\hat{\mathbf{z}}$,

$$\langle dP_{\mathrm{ED}}\rangle_t = \frac{1}{4\pi\varepsilon_0}\left(\frac{\omega^4 p^2}{8\pi c^3}\right)\sin^2\theta\, d\Omega \tag{10-67}$$

No power is radiated along the direction of $\mathbf{p}$ (i.e., $\theta = 0$ or $\pi$), in accord with our conclusion in Section 10–1. The angular distribution of the radiated power from Equation (10–67) is illustrated in Figure 10–5. Integrating Equation (10–67) over $d\Omega = \sin\theta\, d\theta\, d\phi$, we find that the time-averaged total radiated power is

$$\begin{aligned}\langle P_{\mathrm{ED}}\rangle_t &= \frac{1}{4\pi\varepsilon_0}\left(\frac{p^2\omega^4}{8\pi c^3}\right)(2\pi)\int_{-1}^{+1}(1-\cos^2\theta)\, d(\cos\theta)\\ &= \frac{1}{4\pi\varepsilon_0}\left(\frac{p^2\omega^4}{8c^3}\right)\end{aligned} \tag{10-68}$$

A charge $q$ moving in a circular orbit about an opposite charge $-q$ at the center is a rotating electric dipole, as are, equivalently, two perpendicular linear electric dipoles that oscillate harmonically 90° out of phase. Two identical charges $q$ that are always symmetrically placed with respect to the origin (i.e., the mean of their positions is fixed at the origin) give electric

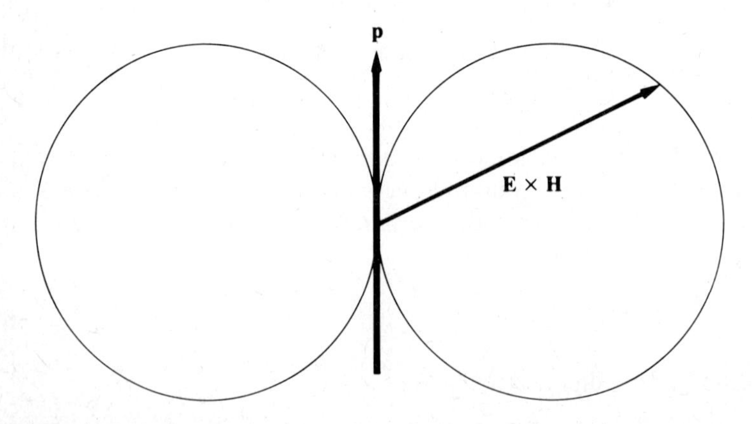


**FIGURE 10–5** Power radiated from an oscillating electric dipole

dipole moments that cancel, so only radiation from higher multipoles is emitted.

## 10-5B Magnetic Dipole and Electric Quadrupole

The $n = 1$ term in the multipole expansion of Equation (10–53) is

$$\mathbf{F}(\hat{\mathbf{r}}) = -ik \int (\hat{\mathbf{r}} \cdot \mathbf{r}')\mathbf{J}(r')\, dV' \tag{10–69}$$

The Cartesian component of this quantity is

$$F_i = -\frac{ik}{r} \sum_j \int x_j x_j' J_i \, dV' \tag{10–70}$$

We express the integral as a sum and difference:

$$\sum_j x_j \int x_j' J_i \, dV' = \sum_j x_j \int \frac{1}{2}(x_j' J_i - x_i' J_j)\, dV' + \sum_j x_j \int \frac{1}{2}(x_j' J_i + x_i' J_j)\, dV' \tag{10–71}$$

Following Equations (7–9) through (7–11), the first term on the right-hand side can be expressed in terms of the magnetic moment,

$$\mathbf{m} = \frac{1}{2} \int \mathbf{r}' \times \mathbf{J}\, dV' \tag{10–72}$$

as

$$\sum_j x_j \int \frac{1}{2}(x_j' J_i - x_i' J_j)\, dV' = (\mathbf{m} \times \mathbf{r})_i \tag{10–73}$$

The second term on the right of Equation (10–71) can be simplified by noting that for $f = x_i' x_j'$, $\nabla f = x_j' \hat{\mathbf{x}}_i + x_i' \hat{\mathbf{x}}_j$; so

$$\int (\nabla f) \cdot \mathbf{J}\, dV' = \int (x_j' J_i + x_i' J_j)\, dV' \tag{10–74}$$

With the identity in Equation (10–56) the second term in Equation (10–71), which vanishes in the quasi-static limit, becomes in the case of radiation

$$\sum_j x_j \int \frac{1}{2}(x_j' J_i + x_i' J_j)\, dV' = -i\omega \sum_j x_j \int x_i' x_j' \rho \, dV' \tag{10–75}$$

In terms of the electric quadrupole moment defined in Equation (4–9),

$$Q_{ij} \equiv \int \rho(\mathbf{r}')(3x_i' x_j' - \delta_{ij} r'^2)\, dV' \tag{10–76}$$

we can write Equation (10–75) as

$$\sum_j x_j \int \frac{1}{2}(x_j' J_i + x_i' J_j)\, dV' = -i\omega \sum_j x_j \left(\frac{1}{3} Q_{ij}\right) - i\omega x_i \int \rho(\mathbf{r}') r'^2\, dV' \qquad \textbf{(10–77)}$$

The last term does not contribute to **B** or **E** because $\nabla \times \mathbf{r} = 0$, so it can be dropped. Our final expression for **F** of Equation (10–69) is

$$\mathbf{F} = \mathbf{F}_{\mathrm{MD}} + \mathbf{F}_{\mathrm{EQ}} \qquad \textbf{(10–78)}$$

where

$$\mathbf{F}_{\mathrm{MD}} = -ik\mathbf{m} \times \hat{\mathbf{r}} \qquad \textbf{(10–79a)}$$

$$(\mathbf{F}_{\mathrm{EQ}})_i = -\frac{\omega k}{3r} \sum_j Q_{ij} x_j \qquad \textbf{(10–79b)}$$

with **m** and $Q_{ij}$ given by Equations (10–72) and (10–76), respectively. The magnetic and electric fields can now be evaluated by Equations (10–43) and (10–44). We will explicitly evaluate only the magnetic dipole field.

If we use the $\mathbf{F}_{\mathrm{MD}}$ of Equation (10–79a) in Equation (10–43), the magnetic field is

$$\mathbf{B}_{\mathrm{MD}} = \frac{\mu_0}{4\pi} k^2 \frac{e^{ikr}}{r} \hat{\mathbf{r}} \times (\mathbf{m} \times \hat{\mathbf{r}}) \qquad \textbf{(10–80a)}$$

and the electric field, by Equation (10–44), is

$$\mathbf{E}_{\mathrm{MD}} = -\frac{\mu_0}{4\pi} \omega k \frac{e^{ikr}}{r} \hat{\mathbf{r}} \times [\hat{\mathbf{r}} \times (\mathbf{m} \times \hat{\mathbf{r}})]$$

or

$$\mathbf{E}_{\mathrm{MD}} = \frac{\mu_0}{4\pi} \omega k \frac{e^{ikr}}{r} \mathbf{m} \times \hat{\mathbf{r}} \qquad \textbf{(10–80b)}$$

By comparing these magnetic dipole fields with the electric dipole fields of Equations (10–60), we see that upon the replacement

$$\mathbf{m} \to \mathbf{p}c \qquad \textbf{(10–81)}$$

the magnetic dipole fields transform to the electric dipole fields, as follows:

$$\mathbf{E}_{\mathrm{MD}} \to -c\mathbf{B}_{\mathrm{ED}} \qquad c\mathbf{B}_{\mathrm{MD}} \to \mathbf{E}_{\mathrm{ED}} \qquad \textbf{(10–82)}$$

Thus the two types of dipole radiation have the same angular distributions.

These radiations can be distinguished by the electric field polarization relative to the dipole axis.

If we take $\mathbf{m} = m\hat{\mathbf{z}}$, the magnetic dipole radiation fields in spherical coordinates are

$$\mathbf{B}_{\text{MD}} = -\frac{\mu_0}{4\pi} k^2 m \frac{e^{ikr}}{r} \sin\theta \hat{\boldsymbol{\theta}} \qquad \mathbf{E}_{\text{MD}} = \frac{\mu_0}{4\pi} ck^2 m \frac{e^{ikr}}{r} \sin\theta \hat{\boldsymbol{\phi}} \tag{10-83}$$

The directions of the **E** and **B** fields for magnetic dipole radiation are illustrated in Figure 10–6.

Finally, we observe from Equations (10–82) that $\mathbf{Y}_{\text{MD}} = \mathbf{Y}_{\text{ED}}$. Hence by the electric dipole result of Equation (10–67), replacing **p** by $\mathbf{m}/c$ and using $\mu_0 \varepsilon_0 c^2 = 1$, we deduce that the time-averaged differential power is

$$\langle dP_{\text{MD}} \rangle_t = \frac{\mu_0}{4\pi} \left( \frac{m^2 \omega^4}{8\pi c^3} \right) \sin^2\theta \, d\Omega \tag{10-84}$$

The corresponding time-averaged total radiated power is

$$\langle P_{\text{MD}} \rangle_t = \frac{\mu_0}{4\pi} \left( \frac{m^2 \omega^4}{3c^3} \right) \tag{10-85}$$

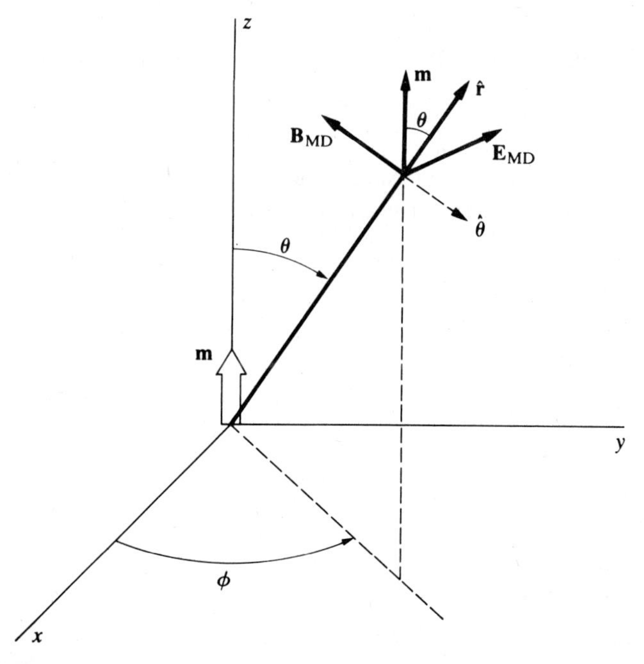


**FIGURE 10–6** Electric and magnetic field directions from an oscillating magnetic dipole

# 10-6 ANTENNA DESIGN

As most people realize, there are two types of antennas, those for transmitting and those for receiving a signal. A remarkable fact is that those qualities that make a superior design of one type also hold for the other. We begin this section by proving this fact. A discussion of some realistic antenna designs follows the proof.

## 10-6A Reciprocity Theorem

For fields with harmonic time variation Maxwell's equations for the curl of $\mathbf{E}$ and $\mathbf{B}$ are

$$\nabla \times \mathbf{E} = i\omega \mathbf{B} \qquad \nabla \times \mathbf{B} = \mu_0 \left( \mathbf{J} - \frac{i\omega}{c^2} \mathbf{E} \right) \tag{10-86}$$

If there are two sets of fields $(\mathbf{E}_a, \mathbf{B}_a)$ and $(\mathbf{E}_b, \mathbf{B}_b)$ that satisfy these equations, with current distributions $\mathbf{J}_a$ and $\mathbf{J}_b$ but the same frequency $\omega$, then

$$\begin{aligned} \nabla \cdot (\mathbf{E}_a \times \mathbf{B}_b - \mathbf{E}_b \times \mathbf{B}_a) &= \mathbf{B}_b \cdot \nabla \times \mathbf{E}_a - \mathbf{E}_a \cdot \nabla \times \mathbf{B}_b \\ &\quad - \mathbf{B}_a \cdot \nabla \times \mathbf{E}_b + \mathbf{E}_b \cdot \nabla \times \mathbf{B}_a \\ &= -\mu_0 (\mathbf{E}_a \cdot \mathbf{J}_b - \mathbf{E}_b \cdot \mathbf{J}_a) \end{aligned} \tag{10-87}$$

Integrating over all space and using the divergence theorem, we find

$$\oint_S (\mathbf{E}_a \times \mathbf{B}_b - \mathbf{E}_b \times \mathbf{B}_a) \cdot d\mathbf{S} = -\mu_0 \int_V (\mathbf{E}_a \cdot \mathbf{J}_b - \mathbf{E}_b \cdot \mathbf{J}_a) \, dV \tag{10-88}$$

The surface integral at infinity exactly vanishes. To show that it does, we take the integration surface to be a sphere, so that

$$\mathbf{E}_a \times \mathbf{B}_b \cdot d\mathbf{S} = \mathbf{E}_a \cdot (\mathbf{B}_b \times \hat{\mathbf{r}}) r^2 \, d\Omega = \frac{1}{c} \mathbf{E}_a \cdot \mathbf{E}_b r^2 \, d\Omega \tag{10-89}$$

In the last step we have used the radiation field relation of Equation (10–44), which is valid since the surface is far from the current distribution and hence $\hat{\mathbf{r}}_b \approx \hat{\mathbf{r}}$. Only the leading terms of the radiation fields, which fall off as $1/r$, will contribute to the two terms on the left side of Equation (10–88) as the integration radius becomes infinitely large. Because of Equation (10–89), these two terms cancel, and we infer that

$$\int_{\text{all space}} \mathbf{E}_a \cdot \mathbf{J}_b \, dV = \int_{\text{all space}} \mathbf{E}_b \cdot \mathbf{J}_a \, dV \tag{10-90}$$

This relation is the *reciprocity theorem.* (Remember that $a$ and $b$ fields have the same frequency.)

The reciprocity theorem can be used to prove an important antenna property. We assume that $\mathbf{J}_a$ is the current density induced in antenna $a$ due to the action of the field $\mathbf{E}_b$ generated by antenna $b$. And $\mathbf{J}_b$ is the current resulting in antenna $b$ due to $\mathbf{E}_a$ coming from antenna $a$. If antennas $a$ and $b$ are broadcasting the same power, then $\mathbf{E}_a$ at $a$ is the same as $\mathbf{E}_b$ at antenna $b$. The reciprocity theorem then says that the power absorbed by $a$ is identical to that absorbed by $b$. The conclusion is that antenna power patterns are the same for transmitting or receiving. We customarily concentrate on the theory of transmitting radiation, with the understanding that reception is reciprocal. Despite this reciprocity, antennas used for transmitting usually differ considerably from those designed for receiving because of the different power levels at which they operate.

Antennas for very low frequencies can be many kilometers in length, while microwave antennas are centimeters in size. A basic unit for measuring the size of an antenna is the wavelength of the radiated waves.

## 10-6B Long-Wire Antenna

A simple antenna that has practical application is the center-fed-wire antenna illustrated in Figure 10–7. The current in this case is symmetric and has a node at each end. A possible current density for such an antenna would be

$$\mathbf{J}(\mathbf{r}) = I_0\delta(x)\delta(y)\sin[k(\ell - |z|)]\hat{\mathbf{z}} \tag{10–91}$$

This current would be exact for a very thin wire with no resistance.

Just as for the multipole fields, the first step is to work out the $\mathbf{F}$ integral of Equation (10–41b). Substituting the current density of Equation (10–91) into the integral, we obtain

$$\mathbf{F} = I_0\hat{\mathbf{z}}\int_{-\ell}^{+\ell} dz' \sin[k(\ell - |z'|)]e^{-ikz'\cos\theta} \tag{10–92}$$

The integration can be done either by writing the sine function in terms of complex exponentials or by looking up the result in a table. Of course, we should break up the integration into two parts to evaluate the absolute value $|z'|$ as $\pm z'$. The result is then

$$\mathbf{F} = \frac{2I_0\hat{\mathbf{z}}}{k}\left[\frac{\cos(k\ell\cos\theta) - \cos k\ell}{\sin^2\theta}\right] \tag{10–93}$$

With the use of Equations (10–43) and (10–44) the $\mathbf{B}$ and $\mathbf{E}$ radiation fields follow directly. We will be mostly interested in the time-averaged power spectrum, which with the aid of Equation (10–50) is

$$\langle dP\rangle_t = \frac{\mu_0 c I_0^2}{8\pi^2}\left[\frac{\cos(k\ell\cos\theta) - \cos(k\ell)}{\sin\theta}\right]^2 d\Omega \tag{10–94}$$

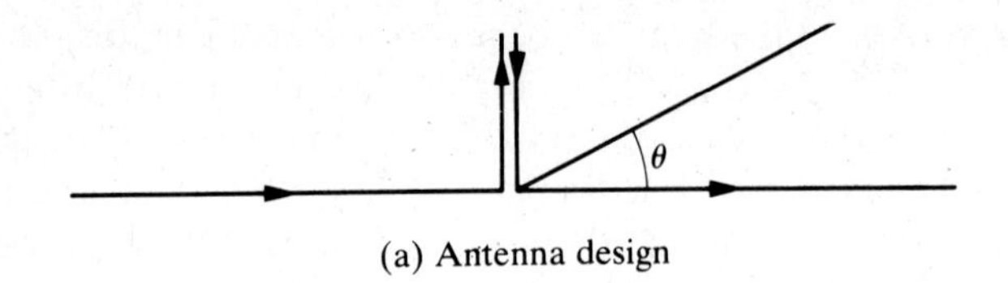


(a) Antenna design

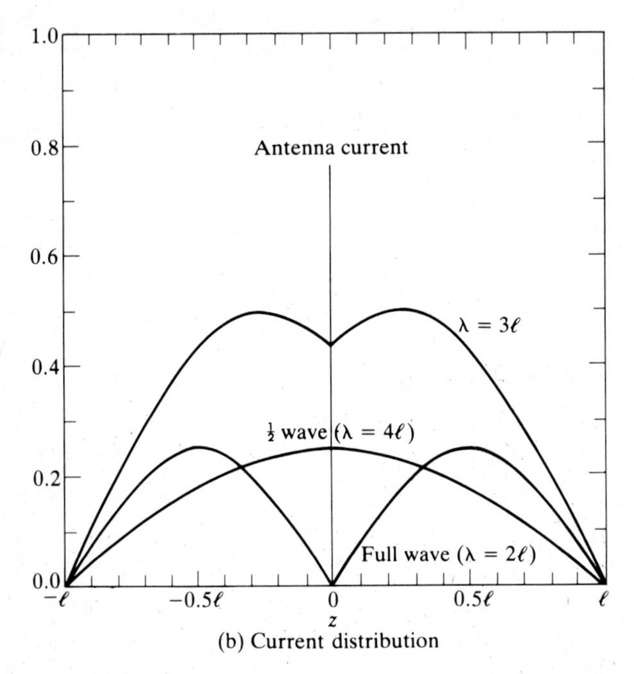


(b) Current distribution

**FIGURE 10–7** Center-fed antenna. (a) Antenna geometry. (b) Current distribution on the antenna for several ratios of antenna length to wavelength.

The corresponding time-averaged total power is then

$$\langle P \rangle_t = \frac{\mu_0 c I_0^2}{4\pi} \int_0^\pi \frac{d\theta}{\sin\theta} [\cos(k\ell \cos\theta) - \cos(k\ell)]^2 \qquad \textbf{(10–95)}$$

This integral cannot be carried out in terms of elementary functions, but it can be expressed in terms of the Si and Ci functions. With access to electronic calculators, direct integration of Equation (10–95) by numerical methods is probably easier.

The power carried by the antenna can be expressed in terms of an effective *radiation resistance*:

$$\text{power radiated} \equiv \frac{1}{2} I_0^2 R_{\text{rad}} \qquad \textbf{(10–96)}$$

where $\frac{1}{2}I_0^2$ is the time-averaged square of the current if a multiple of one-half wavelength of current is on the antenna. From Equations (10–95) and (10–96) the antenna radiation resistance is

$$R_{\text{rad}} = \frac{\mu_0 c}{2\pi} \int_0^{\pi} \frac{d\theta}{\sin\theta} [\cos(k\ell \cos\theta) - \cos(k\ell)]^2 \tag{10–97}$$

The radiation wavelength is $\lambda = 2\pi/k$, so $k\ell = 2\pi(\ell/\lambda)$. Since the coefficient $\mu_0 c/2\pi$ approximately equals $60\,\Omega$, Equation (10–97) can be expressed as

$$R_{\text{rad}} = (60\,\Omega) \int_0^{\pi} \frac{d\theta}{\sin\theta} \left[\cos\left(\frac{2\pi\ell}{\lambda}\cos\theta\right) - \cos\left(\frac{2\pi\ell}{\lambda}\right)\right]^2 \tag{10–98}$$

In Figure 10–8 we plot $R_{\text{rad}}$ as a function of $\ell/\lambda$. We see that the antenna radiation resistance is maximum when there are multiples of half wavelengths of current on the antenna. Another interesting result from Equation (10–98) is that a typical antenna has a resistance of roughly $60\,\Omega$.

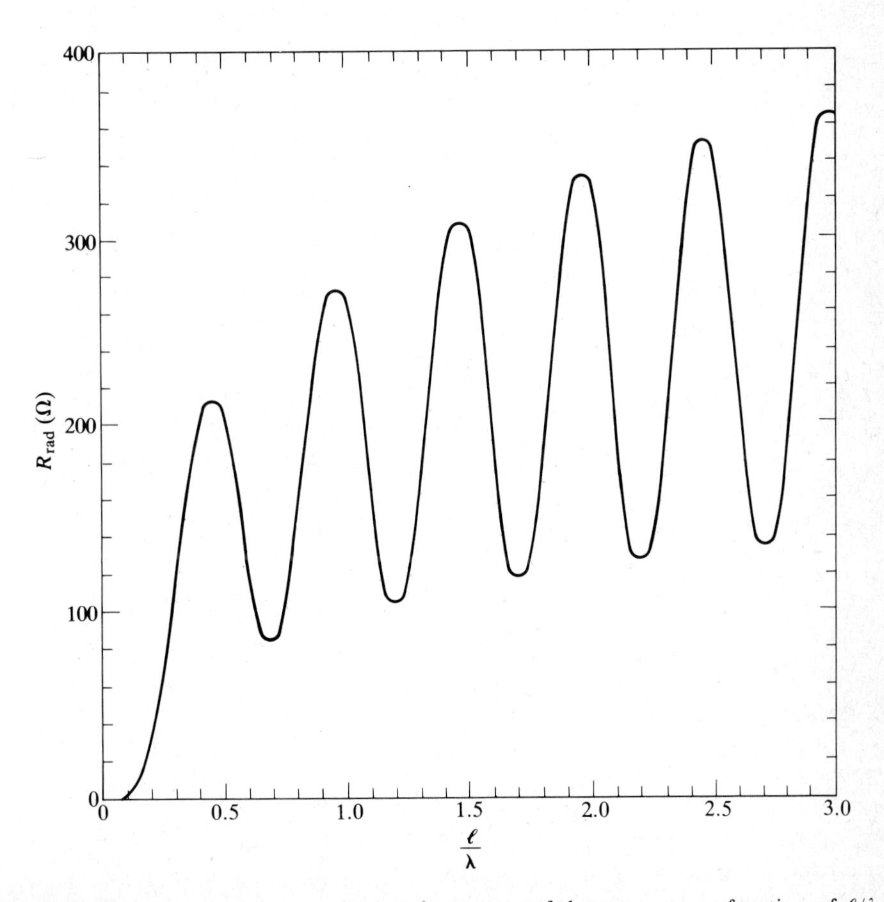


**FIGURE 10–8** Radiation resistance for a center-fed antenna as a function of $\ell/\lambda$

For this reason coaxial TV lead-in wire has an impedance of 75 Ω, giving an efficient transfer of power from the antenna to the TV set.

The power radiation pattern for the straight antenna also depends on the length in wavelengths. In Figure 10–9 the power distribution $d\langle P\rangle_t/d\Omega$ is plotted for $\ell/\lambda = \frac{1}{2}$, 1, $\frac{3}{2}$, and 2.

An important measure of the performance of an antenna is the *gain*, defined by

$$\text{gain} \equiv \frac{\langle dP\rangle_t/d\Omega|_{\max}}{\langle P\rangle_t/(4\pi)} = \frac{\text{maximum power/solid angle}}{\text{average power}/(4\pi)} \tag{10-99}$$

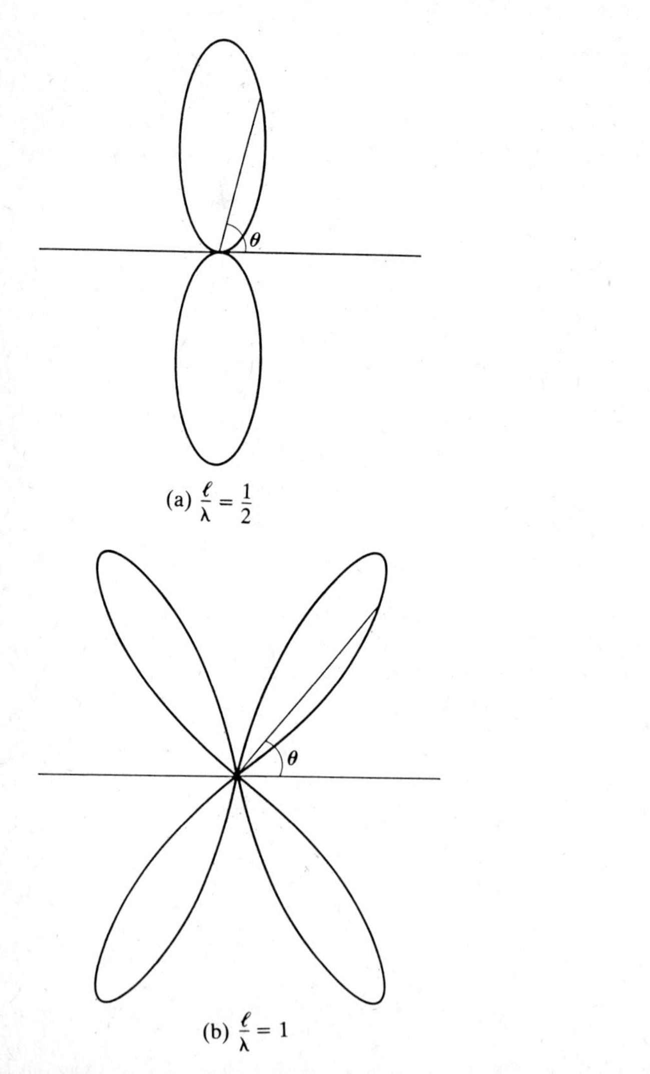


**FIGURE 10–9** Power distribution from a center-fed antenna, for (a) $\ell/\lambda = \frac{1}{2}$, (b) $\ell/\lambda = 1$, (c) $\ell/\lambda = \frac{3}{2}$, and (d) $\ell/\lambda = 2$

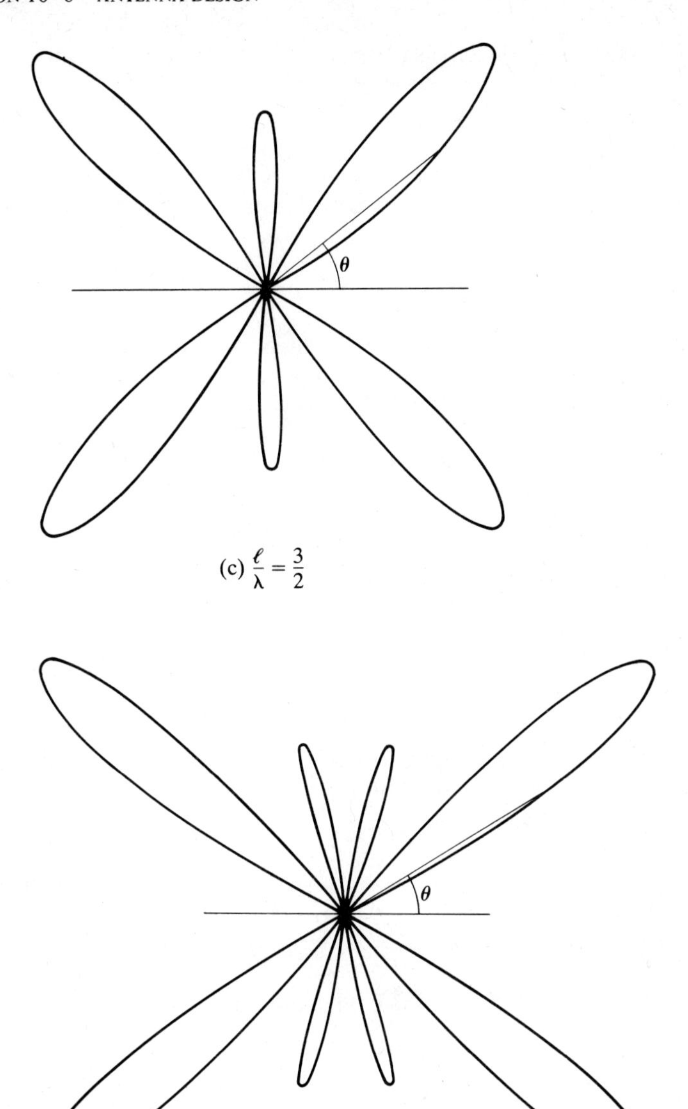


**FIGURE 10–9** *Continued*

As a simple example, we note that for a dipole antenna the gain function is, by Equations (10–67) and (10–68),

$$\text{gain}(\theta) = \frac{3}{2}\sin^2\theta \tag{10-100}$$

So a maximum gain of $\frac{3}{2}$ is achieved in the equatorial plane. The conventional way to express gain is in decibels (dB):

$$\text{gain}_{\text{dB}} = 10\log_{10}(\text{gain}) \qquad \textbf{(10-101)}$$

For the dipole the gain is 1.76 dB. A good rooftop TV antenna will have a gain of 6 dB. A large gain is associated with a high degree of directionality.

## 10-6C Antenna Arrays

Through the linking of a group of simple antenna elements, an array of high gain can be achieved. To see how this linking is achieved, we consider a group of electric dipoles each having a current density

$$\mathbf{J}_i(r') = -i\omega\mathbf{p}_i\delta(\mathbf{r}' - \mathbf{d}_i)e^{-i\beta_i}$$

and located at position $\mathbf{d}_i$. By Equation (10–41b) the $\mathbf{F}$ vector is

$$\mathbf{F} = -i\omega\sum_i \mathbf{p}_i e^{-i(k\hat{\mathbf{r}}\cdot\mathbf{d}_i + \beta_i)} \qquad \textbf{(10-102)}$$

The phase angle $\beta_i$ of each dipole can, in principle, be arbitrarily adjusted through electronic means or through the use of delay lines between the elements.

To give a simple example, we consider two dipoles of equal strength $\mathbf{p} = p\hat{\mathbf{z}}$ separated by a distance $d$ along the $x$ axis. We choose the phase angles $\beta_1 = \beta_2 = 0$. Then

$$\hat{\mathbf{r}}\cdot\mathbf{d} = d\sin\theta\cos\phi \qquad \textbf{(10-103)}$$

and the $\mathbf{F}$ vector becomes

$$\mathbf{F} = -i\omega p\hat{\mathbf{z}}(1 + e^{-ikd\sin\theta\cos\phi}) = -\omega p\hat{\mathbf{z}}[\sin\lambda + i(1 + \cos\lambda)] \qquad \textbf{(10-104)}$$

where

$$\lambda \equiv kd\sin\theta\cos\phi \qquad \textbf{(10-105)}$$

By Equation (10–50) the power distribution is

$$\frac{\langle dP\rangle_t}{d\Omega} = \frac{\mu_0}{(4\pi)^2}\frac{2\omega^4}{c}\sin^2\theta\cos^2\left(\frac{kd\sin\theta\cos\phi}{2}\right) \qquad \textbf{(10-106)}$$

The angular distribution of the power radiated by this array of two dipoles is more strongly peaked than that of the individual dipoles. The gain of the array is consequently larger than the gain of a single dipole.

If we take $kd/2 = \pi/2$, we see that in the equatorial plane ($\theta = \pi/2$)

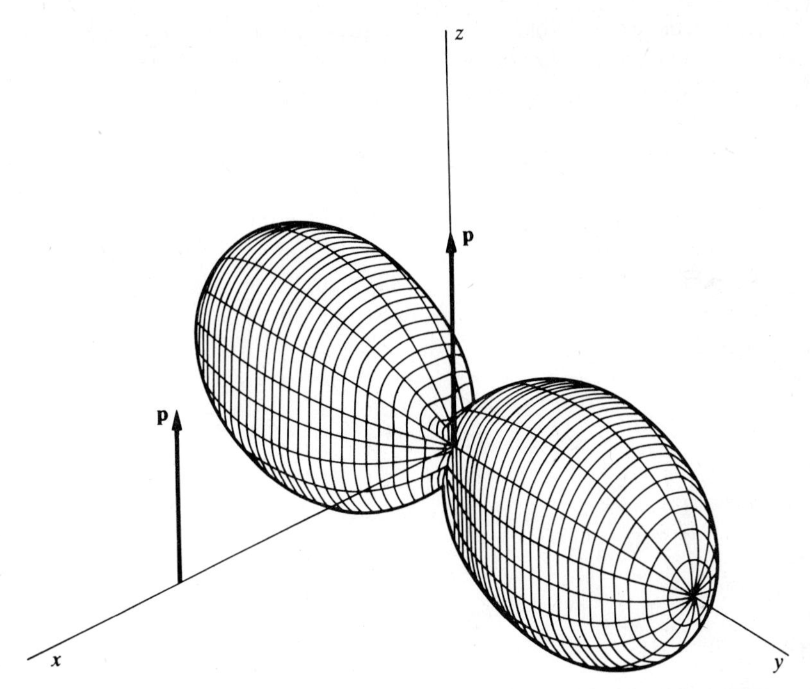


**FIGURE 10-10** Power distribution at distances far from an array of two electric dipoles separated by $d = \frac{1}{2}\lambda$

the power vanishes along the $\phi = 0$ and $\phi = \pi$ axes; that is, no power is emitted along the line connecting the dipoles. This result is quite reasonable, since $kd/2 = \pi/2$ means that the dipoles are separated by one-half wavelength and thus radiation along the $x$ axis destructively interferes. A plot of this radiation pattern is given in Figure 10–10.

Arrays of antennas are useful for many purposes. For example, a radio station on the outskirts of a large city does not wish to waste money sending energy away from the city. So it uses an antenna array. Another example of an array of antennas is the radio telescope—or even arrays of radio telescopes—which has astoundingly large gains. For example, the Very Large Array in New Mexico has a gain of about 150 dB.

## 10-7 APPLICATION: PULSARS–ROTATING MAGNETIC DIPOLES

In 1968 Jocelyn Bell and Antony Hewish of Cambridge University discovered a radio source in space that emitted pulses with a period of about 1 s. Since then, more than one hundred such *pulsars*, some of which also emit detectable visible light and X rays, have been identified; their periods range from several seconds down to 0.088 s for the Vela pulsar and 0.033 s for the Crab pulsar, a "millisec" pulsar. Figure 10–11 shows the Crab nebula and its associated pulsar.

The extreme constancy of the pulse period of each pulsar, aside from a gradual lengthening, suggests that the period is the rotational period of a compact object, a "star." To see just how compact such an object must be, we present the following argument, which gives an inequality involving the period $T$ and the density $\rho$. The condition that the gravitational force on a particle on the equator of a pulsar of radius $a$ is stronger than centrifugal force is

$$\omega^2 a < \frac{GM}{a^2} \qquad \frac{4\pi^2 a}{T^2} < \frac{G}{a^2}\frac{4}{3}\pi a^3 \rho \tag{10-107}$$

or

$$\rho > \frac{3}{GT^2} \tag{10-108}$$

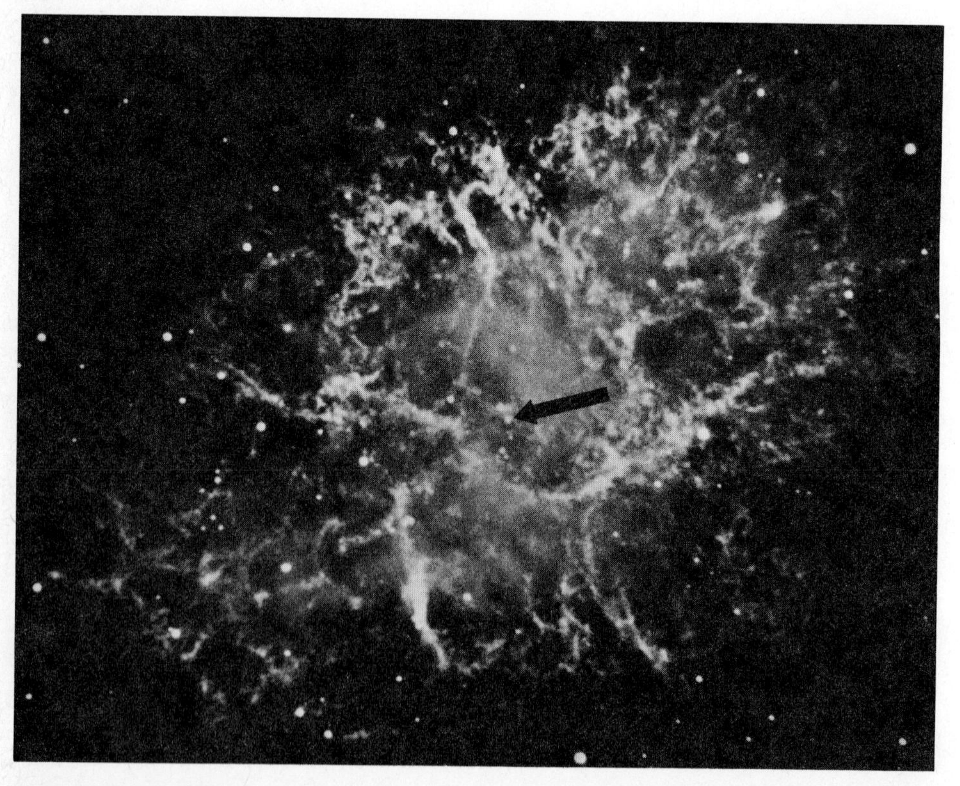

**FIGURE 10–11** Crab nebula. The pulsar that illuminates the nebula is indicated.

For $T \approx 0.1$ s, this expression gives $\rho > 1.4 \times 10^{10}$ g/cm$^3$, which is far more dense than ordinary matter. The only starlike object with this high a density is a *neutron star*, whose density is of the order of that of nuclear matter, $10^{15}$ g/cm$^3$.

A neutron star is thought to be a remnant from a supernova explosion, which results from the gravitational collapse of a star that has consumed its nuclear energy fuel. Neutron stars are stable for masses roughly in the range 0.2 to 2.0 solar masses. A larger mass would collapse, becoming a black hole; a smaller mass would expand, becoming a white dwarf. For typical neutron stars with masses between 0.5 and 1.5 solar masses, the radius is about 10 km and the density increases inward to the order of $10^{15}$ g/cm$^3$. A surface "crust" consisting of a crystal lattice of heavy-metal nuclei is believed to be about 1-km thick. For $\rho > 4.3 \times 10^{11}$ g/cm$^3$ nuclei are no longer stable, and the bulk of the neutron star consists of a degenerate neutron fluid with a minor admixture of protons and electrons. This type of matter is superconducting. Even if the neutron star retains only a small fraction of the parent star's angular momentum, it will initially rotate very rapidly, many times faster than the Crab pulsar. With age the pulsar's rotational kinetic energy is dissipated through electromagnetic radiation owing to a rotating magnetic dipole, as will be described.

Near the center of the Crab nebula, which is the remnant of the supernova event, there are two faint stars of the fifteenth and sixteenth magnitude. In 1942 Walter Baade suggested that one of these with an unusual spectrum, pure continuum, was the origin of and energy source for the Crab nebula. It is Baade's star that turned out to be the Crab pulsar. By integrating the brightness of the Crab nebula, researchers determined that it radiates a total power of

$$P_{\text{Crab}} \approx 10^{31} \text{ W} \tag{10-109}$$

For many years, where this power required to light up the Crab nebula came from was a mystery. It is now known to come from the rotational kinetic energy of the Crab neutron star.

The Crab pulsar spins with a period $T = 0.033$ s and it is slowing down at a rate of

$$\dot{T} = 36.526 \text{ ns/day} \approx 4.23 \times 10^{-13} \text{ s/s} \tag{10-110}$$

The kinetic energy of rotation is

$$K = \frac{1}{2} I\omega^2 \tag{10-111}$$

and the fractional time rate of change of rotational kinetic energy is

$$\frac{\dot{K}}{K} = \frac{I\omega\dot{\omega}}{\frac{1}{2}I\omega^2} = \frac{2\dot{\omega}}{\omega} = -\frac{2\dot{T}}{T} \tag{10-112}$$

In the last equality we have used $\omega = 2\pi/T$. Numerically, this value is

$$\frac{\dot{K}}{K} = (-2)\frac{(4.23 \times 10^{-13})}{(0.033 \text{ s})} = -2.56 \times 10^{-11} \text{ s}^{-1} \tag{10-113}$$

For a neutron star of one solar mass ($M_{\odot} \approx 2 \times 10^{30}$ kg) uniformly distributed over a radius of $10^4$ m, we have

$$\begin{aligned} K &= \frac{1}{2}\left(\frac{2}{5} M_{\odot} a^2\right)\left(\frac{2\pi}{T}\right)^2 \\ &= \frac{1}{5}(2 \times 10^{30})(10^8)\left(\frac{2\pi}{0.033}\right)^2 = 1.45 \times 10^{42} \text{ J} \end{aligned} \tag{10-114}$$

From Equations (10–113) and (10–114), we obtain

$$\dot{K} = (1.45 \times 10^{42})(2.56 \times 10^{-11}) = 3.7 \times 10^{31} \text{ W} \tag{10-115}$$

which nicely matches the observed Crab output of Equation (10–109).

A physical mechanism must be found for the conversion of the rotational kinetic energy to electromagnetic power. A logical possibility is that the magnetic flux trapped in the superconducting neutron star gives a magnetic moment that rotates with the star and thereby radiates electromagnetic power. Let us suppose that the magnetic moment of the pulsar is aligned at an angle relative to its rotation axis, as illustrated in Figure 10–12. The component parallel to the rotation axis is static in direction and magnitude and does not radiate. The radiation arises from the rotating component of the magnetic moment, $\mathbf{m}_T$, perpendicular to the rotation axis.

Taking the $z$ axis along the direction of rotation, we can write the magnetic moment $\mathbf{m}_T(t)$ as

$$\mathbf{m}_T = m_T(\hat{\mathbf{x}} \cos \omega t + \hat{\mathbf{y}} \sin \omega t) \tag{10-116}$$

According to Equation (10–116), a rotating dipole is equivalent to two linear magnetic dipoles oscillating 90° out of phase. Thus the total radiated power will be just twice that calculated in Equation (10–85),

$$\langle P \rangle_t = \frac{\mu_0}{4\pi}\left(\frac{2m^2\omega^4}{3c^3}\right) \tag{10-117}$$

By requiring that this power match the observed rate of decrease of rotational kinetic energy in Equation (10–115), we can obtain a rough estimate of the required rotating magnetic moment. We find a value

$$m_T = 3.4 \times 10^{27} \text{ A·m}^2 \tag{10-118}$$

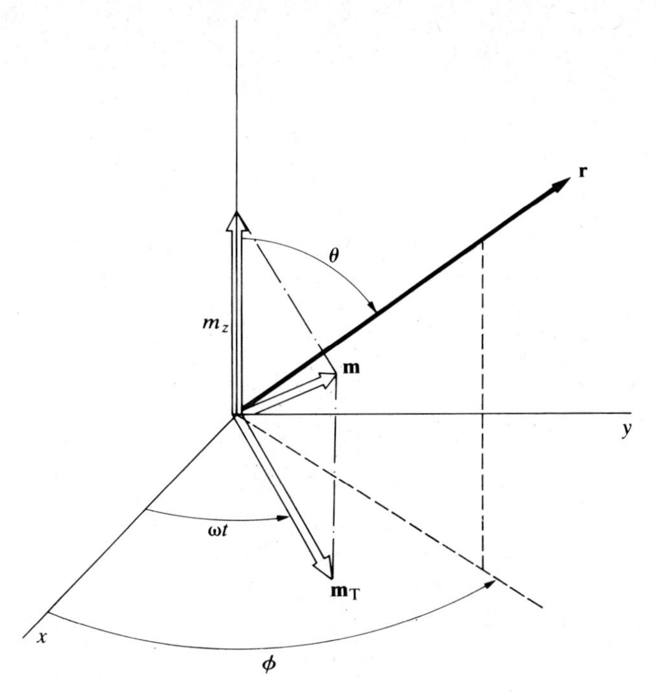


**FIGURE 10–12** Rotating magnetic dipole. The $z$ component is static and does not radiate.

Since the wavelength of this radiation, $\lambda = 10^7$ m, corresponding to the 30-Hz rotation frequency, is much longer than the pulsar radius of $10^4$ m, the magnetic field is quasi-static in the region of the pulsar surface, and **B** is given by the field of a rotating static dipole, Equation (7–13). The field at the pulsar surface is of order

$$B_P = \left(\frac{\mu_0}{4\pi}\right)\frac{m_T}{a^3} \approx 10^8 \text{ T} \qquad \textbf{(10–119)}$$

Compared with very strong laboratory fields of 10 T, this field is enormous. The plausibility that such fields exist can be qualitatively argued from the weak magnetic fields of the parent star. The sun has a surface magnetic field of $\approx 10^{-2}$ T, and many stars have been observed with considerably higher fields. Since a stellar interior is a highly conducting plasma, one might reasonably expect the magnetic field to be "frozen" to the matter (since by Faraday's law a changing magnetic field gives an emf that would induce very high currents). Hence the flux may remain roughly constant after the transformation to a neutron star. The pulsar field would then be (starting with the sun's field and radius)

$$B_P \approx B_\odot \left(\frac{a_\odot}{a_P}\right)^2 \approx (10^{-2}\ \mathrm{T}) \left(\frac{0.7 \times 10^9\ \mathrm{m}}{10^4\ \mathrm{m}}\right)^2 \approx 0.5 \times 10^8\ \mathrm{T} \tag{10–120}$$

which is comparable to the value associated with the radiation.

To a fixed observer on the pulsar, the net electromagnetic force on any charge inside the pulsar must vanish; for if it did not, the superconducting material comprising the pulsar would conduct infinite currents. For a zero Lorentz force,

$$\mathbf{F} = 0 = q(\mathbf{E} + \mathbf{v} \times \mathbf{B})$$

the $\mathbf{E}$ field must be

$$\mathbf{E} = -\mathbf{v} \times \mathbf{B} \tag{10–121}$$

From the equatorial radius and angular velocity of Equation (10–113) and the magnetic field of Equation (10–119), we estimate

$$E \approx (1.8 \times 10^6)(10^8) \approx 10^{14}\ \mathrm{V/m} \tag{10–122}$$

which is a very high field.

If there were no charged particles outside the pulsar, we would observe a flux of 30-Hz radiation and nothing else. Since a broad radiation spectrum is observed, there must be an ionized atmosphere surrounding the pulsar. The atmosphere presumably rotates along with the pulsar out to a radius where the rotational velocity approaches the speed of light. The detailed dynamics of the pulsar atmosphere are at present not well understood. Electrons must be accelerated by large electric fields to spiral along the $B$ field lines and, as they spiral, radiate the electromagnetic waves that light up the nebula surrounding the pulsar. A small fraction of this radiated energy emerges, having the periodicity of the pulsar rotation. This X-ray, optical, and radio radiation arrives as if it originated from a great rotating beacon.

## 10-8 RADIATION REACTION

The central theme of this chapter is that an accelerating charge emits radiation that propagates outward at the speed of light. For a slowly moving charge $q$ the total radiated power, from Equation (10–6), is

$$P = \left(\frac{1}{4\pi\varepsilon_0}\right)\left(\frac{2}{3}\right)\frac{q^2a^2}{c^3} \tag{10–123}$$

Conservation of energy dictates that this radiated energy must be associated with a decrease of the charge's kinetic energy and/or with work done by external forces on the charge. Hence the motion of a charged particle is not entirely determined by the action of external forces alone, since it is influenced by the radiation it emits. The fields that enter the Lorentz force are the external field and the field due to the charge itself. In this section we will consider the reaction on a charge due to its radiation.

Fortunately, the radiation reaction force $\mathbf{F}_{\text{rad}}$ usually gives only a small correction to the motion, determined by

$$\dot{\mathbf{p}} = \mathbf{F}_{\text{ext}} + \mathbf{F}_{\text{rad}} \tag{10–124}$$

In solving for $\mathbf{F}_{\text{rad}}$, we make two simplifying assumptions. First, we consider only nonrelativistic motion so that Equation (10–123) gives the total radiated power. We also assume that the external forces are such that the actual motion is periodic, so that at two times $t_1$ and $t_2$ the state of motion and the field energy are the same. Then the work done against the radiation reaction force must exactly equal the energy radiated during this time period:

$$-\int_{t_1}^{t_2} \mathbf{F}_{\text{rad}} \cdot \mathbf{v}\, dt = \left(\frac{1}{4\pi\varepsilon_0}\right)\left(\frac{2}{3}\right)\frac{q^2}{c^3}\int_{t_1}^{t_2} \dot{v}^2\, dt \tag{10–125}$$

The right side can be integrated by parts, using

$$\int_{t_1}^{t_2} \dot{v}^2\, dt = \int_{t_1}^{t_2} \dot{\mathbf{v}} \cdot (\dot{\mathbf{v}}\, dt) = \mathbf{v} \cdot \dot{\mathbf{v}}\Big|_{t_1}^{t_2} - \int_{t_1}^{t_2} \ddot{\mathbf{v}} \cdot \mathbf{v}\, dt \tag{10–126}$$

With the assumed periodicity of motion the first term on the right side vanishes, and we obtain

$$\int_{t_1}^{t_2} \mathbf{F}_{\text{rad}} \cdot \mathbf{v}\, dt = \left(\frac{1}{4\pi\varepsilon_0}\right)\frac{2q^2}{3c^3}\int_{t_1}^{t_2} \dot{\mathbf{a}} \cdot \mathbf{v}\, dt \tag{10–127}$$

This equation is satisfied if the radiation reaction force is

$$\mathbf{F}_{\text{rad}} = \left(\frac{1}{4\pi\varepsilon_0}\right)\left(\frac{2}{3}\right)\frac{q^2}{c^3}\dot{\mathbf{a}} \tag{10–128}$$

A direct treatment that calculates the effect of the field of a charge on itself gives the same result.

To obtain a feeling for the practical importance of the radiation reaction term, we consider the specific example of a charge of mass $m$ driven by a spring force $F_{\text{ext}} = -m\omega_0^2 x$. If we do not include the radiation reaction force, the equation of motion is

$$m\ddot{x} + m\omega_0^2 x = 0 \tag{10–129}$$

and has the solution

$$x(t) = A \cos \omega_0 t \tag{10-130}$$

We use this approximate solution to Equation (10–124) to estimate the magnitude:

$$\left|\frac{F_{\text{rad}}}{F_{\text{ext}}}\right| \approx \frac{(1/4\pi\varepsilon_0)(\frac{2}{3})(q^2/c^3)A\omega_0^3}{m\omega_0^2 A} = \left(\frac{1}{4\pi\varepsilon_0}\right)\left(\frac{2}{3}\right)\frac{q^2\omega_0}{mc^3} \tag{10-131}$$

The reaction force on an electron becomes comparable to the harmonic force at a frequency

$$\omega_0 \approx \left(\frac{1}{9 \times 10^9}\right)\left(\frac{3}{2}\right)\frac{(9.1 \times 10^{-31})(3 \times 10^8)^3}{(1.6 \times 10^{-19})^2}$$
$$= 1.6 \times 10^{23} \text{ radians/s} \tag{10-132}$$

which is in the very hard, gamma ray frequency range. In most cases the effect of the reaction force on the motion of a charge particle is negligible. (It is certainly not negligible in a good antenna.) In the extreme cases where this force plays a significant role in the dynamics of motion, quantum corrections are more important.

# 10-9 RADIATION FROM A POINT CHARGE

At the beginning of this chapter we gave an elementary discussion of radiation from an accelerated charge. We subsequently considered the solution of Maxwell's equations for given charge and current distributions. This discussion led us to the theory of antennas and the multipole expansion. Now we return to the subject of charges and derive the electromagnetic fields of an electric charge in arbitrary motion. Many important examples of radiation are naturally treated in terms of point electric charges. Among these examples is a charge moving in a circle (synchrotron radiation).

## 10-9A Lienart-Wiechert Potentials

For a continuous charge density $\rho(\mathbf{r}, t)$ the scalar potential solution of Equation (10–22) was found in Equation (10–31) to be

$$\Phi(\mathbf{r}, t) = \frac{1}{4\pi\varepsilon_0}\int \frac{\rho(\mathbf{r}', t_{\text{R}})}{R}\, dV' \tag{10-133}$$

where the retarded time $t_{\text{R}}$ is

$$t_{\text{R}} = t - \frac{R}{c} \tag{10-134}$$

with $\mathbf{R} = \mathbf{r} - \mathbf{r}'$, as usual. If the charge density is to represent a point particle, we take

$$\rho(\mathbf{r}', t_R) = a\delta[\mathbf{r}' - \mathbf{r}_0(t_R)] \tag{10–135}$$

where $\mathbf{r}_0(t)$ is the position of the particle at time $t$. We rewrite Equation (10–133) as

$$\Phi(\mathbf{r}, t) = \frac{q}{4\pi\varepsilon_0} \int dV' \int dt_R \frac{\delta[\mathbf{r}' - \mathbf{r}_0(t_R)]\delta[t_R - t + (R/c)]}{R} \tag{10–136}$$

The problem with using Equation (10–133) directly is that since the field point time $t$ is fixed, integration over $\mathbf{r}'$ means that $t_R$ also varies. By inserting the extra delta function, we can reverse the order of integration, since $t_R$ is then formally unrelated to $\mathbf{r}'$ until the $t_R$ integral is performed. Doing the $\mathbf{r}'$ volume integral first gives

$$\Phi(\mathbf{r}, t) = \frac{q}{4\pi\varepsilon_0} \int dt_R \frac{\delta[t_R - t + (R/c)]}{R} \tag{10–137a}$$

where now $\mathbf{R} = \mathbf{r} - \mathbf{r}_0(t_R)$. For the vector potential with $\mathbf{J}(\mathbf{r}', t_R) = q\mathbf{v}(t_R)\delta(\mathbf{r}' - \mathbf{r}_0)$, we obtain, by similar reasoning,

$$\mathbf{A}(\mathbf{r}, t) = \frac{\mu_0 q}{4\pi} \int dt_R \frac{\mathbf{v}(t_R)\delta[t_R - t + (R/c)]}{R} \tag{10–137b}$$

The particle velocity $\mathbf{v}(t_R)$ is defined by

$$\mathbf{v}(t_R) \equiv \frac{d\mathbf{r}_0(t_R)}{dt_R} \tag{10–138}$$

Although we will use only Equations (10–137) in our subsequent work, to directly perform the $t_R$ integrals in Equations (10–137) has some interest. This calculation provides an opportunity to introduce some mathematical techniques that will be applied later, since the $t_R$ integral must eventually be worked out. A useful delta function identity is

$$\int_{-\infty}^{+\infty} dx\, \delta[h(x)] f(x) = \sum_i \frac{f(x_i)}{|dh/dx|_{x=x_i}} \tag{10–139}$$

where $\{x_i\}$ is the set of simple zeros of $h(x)$. This identity is readily established by changing the independent variable to

$$y = h(x), \qquad dy = \left|\frac{dh}{dx}\right| dx \tag{10–140}$$

We have put in an absolute value sign so that $dy > 0$ if $dx$ is positive. In our case

$$h = t_{\mathbf{R}} - t + \frac{R}{c} \tag{10-141}$$

so that

$$\frac{dh}{dt_{\mathbf{R}}} = 1 + \frac{\dot{R}}{c} \tag{10-142}$$

where $\dot{R}$ is the derivative with respect to retarded time. From the definition $R^2 = r^2 + r_0^2(t_{\mathbf{R}}) - 2\mathbf{r} \cdot \mathbf{r}_0(t_{\mathbf{R}})$, we obtain, by implicit differentiation,

$$\dot{R} = \frac{dR}{dt_{\mathbf{R}}} = -\hat{\mathbf{R}} \cdot \dot{\mathbf{r}}_0 = -\hat{\mathbf{R}} \cdot \mathbf{v} \tag{10-143}$$

Putting these expressions together yields $dh/dt_{\mathbf{R}} = 1 - (\hat{\mathbf{R}} \cdot \mathbf{v}/c) \equiv b$ and

**Lienart-Wiechert Potentials**

$$\Phi(\mathbf{r}, t) = \frac{q}{4\pi\varepsilon_0 R} \frac{1}{b}\bigg|_{t_{\mathbf{R}}} \tag{10-144a}$$

$$\mathbf{A}(\mathbf{r}, t) = \frac{\mu_0 q \mathbf{v}}{4\pi R} \frac{1}{b}\bigg|_{t_{\mathbf{R}}} \tag{10-144b}$$

The notation $|_{t_{\mathbf{R}}}$ means that $\mathbf{v}$ and $R$ are to be evaluated at the retarded time.

## 10-9B Fields of a Moving Charge

We could work directly with the potentials of Equations (10–144) to compute the **E** and **B** fields, but to start with Equations (10–137) is easier. The electric field is found from

$$\mathbf{E} = -\nabla\Phi - \frac{\partial \mathbf{A}}{\partial t} \tag{10-145}$$

With aid of the potentials of Equations (10–137), the **E** field then becomes

$$\mathbf{E}(\mathbf{r}, t) = -\frac{q}{4\pi\varepsilon_0} \hat{\mathbf{R}} \frac{\partial}{\partial R} \int dt_{\mathbf{R}} \frac{\delta(t_{\mathbf{R}} - t + R/c)}{R} - \frac{\mu_0 q}{4\pi} \frac{\partial}{\partial t} \int dt_{\mathbf{R}} \frac{\mathbf{v}(t_{\mathbf{R}})\delta(t_{\mathbf{R}} - t + R/c)}{R} \tag{10-146}$$

After evaluating the space and time derivatives, we obtain

$$\mathbf{E}(\mathbf{r}, t) = \frac{q}{4\pi\varepsilon_0} \int dt_{\mathrm{R}} \left[ \frac{\hat{\mathbf{R}}}{R^2} \delta\left(t_{\mathrm{R}} - t + \frac{R}{c}\right) - \frac{\hat{\mathbf{R}} - \mathbf{v}/c}{cR} \delta'\left(t_{\mathrm{R}} - t + \frac{R}{c}\right)\right] \tag{10–147}$$

where $\delta'$ means differentiation of the delta function with respect to its argument. Such a derivative is defined by integration by parts as

$$\int_{-\infty}^{+\infty} dx\ f(x)\,\delta'[h(x)] = \int dy \frac{f}{|dh/dx|}\,\delta'(y) = \int dy \left(\frac{f}{b}\right)\delta'(y)$$
$$= -\int dy \frac{d}{dy}\left(\frac{f}{b}\right)\delta(y) = -\frac{d}{dy}\left(\frac{f}{b}\right)\Bigg|_{h=0} \tag{10–148}$$

From Equation (10–147) with $f = -(\hat{\mathbf{R}} - \mathbf{v}/c)/(cR)$, we then find the electric field to be

$$\mathbf{E}(\mathbf{r}, t) = \frac{q}{4\pi\varepsilon_0}\left[\frac{\hat{\mathbf{R}}}{R^2 b} + \frac{1}{bc}\frac{d}{dt'}\left(\frac{\hat{\mathbf{R}} - \mathbf{v}/c}{Rb}\right)\right]_{tR} \tag{10–149}$$

where again it is understood that $\mathbf{v}$ and $\mathbf{R}$ are to be evaluated at the retarded time $t_{\mathrm{R}} = t - R/c$.

When the differentiation indicated in Equation (10–149) is evaluated, we find that

$$\mathbf{E}(\mathbf{r}, t) = \frac{q}{4\pi\varepsilon_0 Rcb^2}\left[\frac{bc\hat{\mathbf{R}}}{R} + \dot{\hat{\mathbf{R}}} - \frac{\mathbf{a}}{c} - \frac{(\hat{\mathbf{R}} - \mathbf{v}/c)}{Rb}(\dot{R}b + r\dot{b})\right]_{t\mathrm{R}} \tag{10–150}$$

From Equation (10–143) with $\dot{\mathbf{R}} = -\mathbf{v}$ we observe that

$$\mathbf{a} = \dot{\mathbf{v}} \qquad \dot{\hat{\mathbf{R}}} = -\frac{1}{R}(\mathbf{v} - \hat{\mathbf{R}}\hat{\mathbf{R}}\cdot\mathbf{v}) \tag{10–151}$$

After these derivatives are substituted into Equation (10–150), we observe that the electric field can be conveniently separated into two terms,

$$\mathbf{E} = \mathbf{E}_v + \mathbf{E}_a \tag{10–152}$$

where $\mathbf{E}_v$ is independent of the particle's acceleration $\mathbf{a}$ and $\mathbf{E}_a$ is directly proportional to the acceleration. With a bit of algebra these parts turn out to be

$$\mathbf{E}_v = \frac{q}{4\pi\varepsilon_0} \frac{1 - v^2/c^2}{R^2 b^3} \left( \hat{\mathbf{R}} - \frac{\mathbf{v}}{c} \right) \Bigg|_{t_R} \tag{10-153a}$$

$$\mathbf{E}_a = \frac{q}{4\pi\varepsilon_0} \frac{\hat{\mathbf{R}} \times [(\hat{\mathbf{R}} - \mathbf{v}/c) \times \mathbf{a}/c]}{Rcb^3} \Bigg|_{t_R} \tag{10-153b}$$

That $\mathbf{E}_a$ is a radiation field is apparent, since its magnitude falls off as $1/R$.

The calculation of the magnetic field proceeds in a similar manner. From $\mathbf{B} = \nabla \times \mathbf{A}$ in Equation (10–137b)

$$\begin{aligned} \mathbf{B}(\mathbf{r}, t) &= \frac{\mu_0 q}{4\pi} \int dt_R \nabla \times \left[ \frac{\mathbf{v}(t_R)\delta(t_R - t + R/c)}{R} \right] \\ &= -\frac{q}{4\pi\varepsilon_0 c^2} \int dt_R \mathbf{v} \\ &\quad \times \left[ -\frac{\hat{\mathbf{R}}}{R^2} \delta\left( t_R - t + \frac{R}{c} \right) + \frac{\hat{\mathbf{R}}}{Rc} \delta'\left( t_R - t + \frac{R}{c} \right) \right] \end{aligned} \tag{10-154}$$

Using the identity of Equation (10–148), we obtain

$$\mathbf{B}(\mathbf{r}, t) = \frac{q}{4\pi\varepsilon_0 c^2} \left[ \frac{\mathbf{v} \times \hat{\mathbf{R}}}{R^2 b} + \frac{1}{bc} \frac{d}{dt'} \left( \frac{\mathbf{v} \times \hat{\mathbf{R}}}{Rb} \right) \right]_{t_R} \tag{10-155}$$

By direct comparison of this result to Equation (10–149), we verify that

$$\mathbf{B} = \frac{1}{c} \hat{\mathbf{R}} \times \mathbf{E} \tag{10-156}$$

so that the magnetic field can be computed from our final result in Equations (10–153) for $\mathbf{E} = \mathbf{E}_v + \mathbf{E}_a$. We should emphasize that our results, Equations (10–153) and (10–156), are exact. No small-velocity approximation has been made.

From the electric and magnetic fields of a moving charge, we can compute the power radiated. The Poynting vector is

$$\mathbf{Y} = \frac{1}{\mu_0} \mathbf{E} \times \mathbf{B} = \frac{1}{\mu_0 c} \mathbf{E} \times (\hat{\mathbf{R}} \times \mathbf{E}) = \frac{E^2}{\mu_0 c} \hat{\mathbf{R}} \tag{10-157}$$

since $\hat{\mathbf{R}} \cdot \mathbf{E} = 0$. The energy radiated during a field time interval $dt$ into a solid-angle element $d\Omega$ is

$$dU = (\mathbf{Y} \cdot \hat{\mathbf{R}}) R^2 \, d\Omega \, dt = \frac{E^2 R^2}{\mu_0 c} \, d\Omega \, dt \tag{10-158}$$

Since the velocity and acceleration are measured in terms of the particle's (retarded) time $t_{\mathrm{R}}$, computing radiation in terms of an integral over retarded time is natural. So the power radiated in this case is

$$\frac{dP(t_{\mathrm{R}})}{d\Omega} = \frac{E^2R^2}{\mu_0 c}\left(\frac{dt}{dt_{\mathrm{R}}}\right) \tag{10–159}$$

Since $t = t_{\mathrm{R}} + R/c$, by Equation (10–134) $dt/dt_{\mathrm{R}} = b$, where $b = 1 - \hat{\mathbf{R}}\cdot\mathbf{v}/c$, and the angular distribution of the power is

$$\frac{dP(t_{\mathrm{R}})}{d\Omega} = \frac{E^2R^2b}{\mu_0 c} \tag{10–160}$$

Only the $\mathbf{E}_a$ term of Equations (10–153) contributes to radiation, since at large distances $E_v^2R^2 \to 0$. For rapidly moving particles the angular dependence is primarily determined by $b$. A large power of $b$ occurs in the denominator of Equation (10–160), as can be seen from Equation (10–153b). If $\theta$ is a (small) angle between $\hat{\mathbf{R}}$ and $\mathbf{v}$, we have

$$b = 1 - \frac{v\cos\theta}{c} \approx 1 - \frac{v}{c} + \frac{v\theta^2}{2c} \tag{10–161}$$

If $v \approx c$, $b^{-N}$ is sharply peaked in the forward direction. The half-width at half-maximum occurs at an angle given by

$$\theta^2 \approx 2\left(1 - \frac{v}{c}\right) \approx 1 - \frac{v^2}{c^2}$$

or

$$\theta \approx \frac{1}{\gamma} \tag{10–162}$$

where $\gamma = (1 - v^2/c^2)^{-1/2}$. This effect can be seen in Figure 10–2.

We explicitly work out here one simple case of radiated power. If the velocity is small, the electric field from Equation (10–153b) is (with $\mathbf{v} = 0$)

$$\mathbf{E}_a = \frac{q}{4\pi\varepsilon_0}\frac{\hat{\mathbf{R}}\times(\hat{\mathbf{R}}\times\mathbf{a})}{R^2c^2} \tag{10–163}$$

And if $\theta$ is the angle between $\mathbf{R}$ and $\mathbf{a}$,

$$R^2E_a^2 = \frac{q^2a^2\sin^2\theta}{16\pi^2\varepsilon_0^2c^4} \tag{10–164}$$

$$\frac{dP}{d\Omega} = \frac{q^2a^2\sin^2\theta}{16\pi^2\varepsilon_0 c^3} \tag{10–165}$$

the power radiated into all directions is

$$P = 2\pi \int_{-1}^{+1} d(\cos\theta)\, \frac{q^2 a^2 \sin^2\theta}{16\pi^2 \varepsilon_0 c^3} = \frac{1}{4\pi\varepsilon_0}\left(\frac{2q^2a^2}{3c^3}\right) \tag{10-166}$$

which is the same result that we found from our simple physical-kink picture in Equation (10–6). This low-velocity radiation formula is known as *Larmor's radiation formula.*

## 10-9C Uniform Motion

If the charge $q$ is moving uniformly, $\mathbf{a} = 0$ and $\mathbf{E} = \mathbf{E}_v$ in Equations (10–153). The result is interesting and points the way toward the theory of relativity.

For simplicity we take the direction of motion along the $z$ axis so that

$$\mathbf{r}_0(t_R) \equiv \mathbf{v}t_R = vt_R\hat{\mathbf{z}} \tag{10-167}$$

where $v$ is a constant. From the geometry of Figure 10–13 we have

$$R^2 = x^2 + y^2 + (z - vt_R)^2 \tag{10-168a}$$

$$R = c(t - t_R) \tag{10-168b}$$

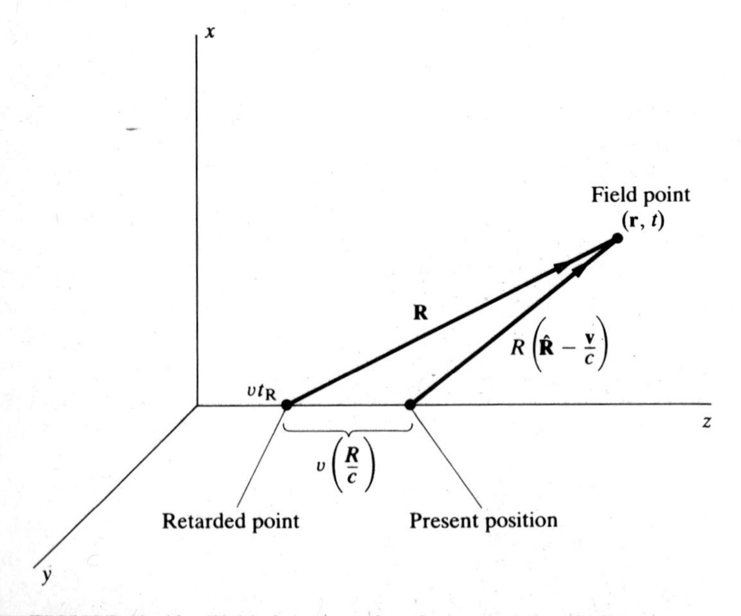


**FIGURE 10-13** Fields from a uniformly moving electric charge

By eliminating $R$ in Equations (10–168), we can solve for $t_{\mathbf{R}}$:

$$t_{\mathbf{R}} = \frac{c^2 t - vz \pm \sqrt{(c^2 t - vz)^2 + (c^2 - v^2)(x^2 + y^2 + z^2 - c^2 t^2)}}{c^2 - v^2} \tag{10–169}$$

For all $v$, and in particular $v = 0$, we must have $t_{\mathbf{R}} = t - R/c$, so the negative sign is the proper choice in Equation (10–169). With a bit of algebraic simplification Equation (10–169) can be used to evaluate

$$\begin{aligned} c\left(R - \frac{\mathbf{R} \cdot \mathbf{v}}{c}\right) &= ct - vz - t_{\mathbf{R}}(c^2 - v^2) \\ &= \sqrt{c^2 - v^2}\sqrt{x^2 + y^2 + \left[\frac{(z - vt)}{\sqrt{1 - (v^2/c^2)}}\right]^2} \end{aligned} \tag{10–170}$$

And the electric potential of Equation (10–144a) is then

$$\Phi(\mathbf{r}, t) = \frac{q}{4\pi\varepsilon_0} \frac{\gamma}{\sqrt{x^2 + y^2 + [\gamma(z - vt)]^2}} \tag{10–171}$$

where $\gamma = 1/\sqrt{1 - (v^2/c^2)}$. Thus the potential of a uniformly moving charge, which is

$$\Phi' = \frac{q}{4\pi\varepsilon_0} \frac{1}{\sqrt{x'^2 + y'^2 + z'^2}} \tag{10–172}$$

in the rest frame of the charge, takes the form of Equation (10–171) when the charge is in motion. The configuration of the lines of force from the moving charge is squashed along the direction of motion. The two results are related by the substitutions

$$x' = x \qquad y' = y \qquad z' = \gamma(z - vt) \tag{10–173}$$

These equations represent a part of the relativistic Lorentz transformation. A more complete analysis, as given in the next chapter, shows that the time coordinate also depends on the reference frame and that a clock moving with the charge seems to run slow compared with a clock in the field point rest frame.

The electric field of the moving charge is given by Equation (10–153a),

$$\mathbf{E} = \frac{q}{4\pi\varepsilon_0} \frac{[1 - (v^2/c^2)][\hat{\mathbf{R}} - (\mathbf{v}/c)]}{R^2[1 - (\hat{\mathbf{R}} \cdot \mathbf{v}/c)]^3} \tag{10–174}$$

The directions of the lines of force converge to a point along the vector $\hat{\mathbf{R}} - (\mathbf{v}/c)$, which by Figure 10–13 is the position of the charge at the present time $t$, not the retarded time $t_{\mathbf{R}}$. From Equation (10–174) we again observe that the field lines are no longer isotropic for a moving charge but are concentrated toward the plane perpendicular to the motion of the charge. This field is discussed in further detail in the next chapter.

## 10-9D Acceleration Along the Motion

When the acceleration $\mathbf{a}$ is parallel to the velocity $\mathbf{v}$ of the charge, the electric field of Equation (10–153b) is

$$\mathbf{E}_a = \frac{q}{4\pi\varepsilon_0} \frac{\hat{\mathbf{R}} \times (\hat{\mathbf{R}} \times \mathbf{a})}{Rc^2 b^3} \tag{10–175}$$

Again, if we choose the polar axis along the acceleration direction, the power angular distribution is, from Equation (10–160),

$$\frac{dP(t_{\mathbf{R}})}{d\Omega} = \frac{q^2}{16\pi^2\varepsilon_0 c^3} \frac{\sin^2\theta}{[1 - (v/c)\cos\theta]^5} \tag{10–176}$$

The effect of the denominator factor is to direct the radiation into two forward lobes when $v$ is nearly equal to $c$. An example of this effect when $v/c = 0.5$ is shown in Figure 10–2.

## 10-9E Synchrotron Radiation

When an electric charge revolves in a circular orbit, it will radiate owing to the centripetal acceleration. Radiation produced in this way is called *synchrotron radiation.* A magnetic field is the most common agent causing circular motion of a charge. Synchrotron radiation is of great importance in astrophysical processes, and it provides a limitation on the construction of electron accelerators.

Referring to the general expression of Equation (10–153b) for the radiated electric field,

$$\frac{4\pi\varepsilon_0}{q}\mathbf{E}_a = \frac{\hat{\mathbf{R}} \times [(\hat{\mathbf{R}} - \mathbf{v}/c) \times \mathbf{a}]}{Rc^2 b^3} \tag{10–177}$$

we observe that $\mathbf{E}_a$ is largest when $\hat{\mathbf{R}}$ is along the direction of motion, in which case

$$\mathbf{E}_a \propto \frac{-\mathbf{a}}{(1 - \mathbf{v}/c)^2} \tag{10–178}$$

Almost all the power radiated by a highly relativistic particle will be plane-polarized along the acceleration direction. This behavior is a characteristic signature for synchrotron radiation. An angular integration of the radiated power for a relativistic particle gives

$$\text{power} = \frac{1}{4\pi\varepsilon_0}\left(\frac{2q^2c}{3R^2}\right)\left(\frac{1}{1 - v^2/c^2}\right)^2 \tag{10–179}$$

Long before the existence of the pulsar in the Crab nebula was found, we knew that the visible light reaching us from this nebula was synchrotron radiation, because it was polarized. The only light that we see from relativistic electrons moving helically in a magnetic field is from the instant on the orbit that the electron was heading in our direction. We have seen that such light is polarized perpendicular to $\mathbf{v}$ and the magnetic field. An analysis of the observed light gives us information about the magnetic field throughout the nebula.

Another situation where synchrotron light is of great importance is when electrons are confined to circular orbits by a magnetic field, as in a particle accelerator. As we see from Equation (10–179), highly relativistic particles will radiate strongly, and synchrotron energy loss can be a real problem when electrons (the lightest charged particle) are accelerated to very high energies. The new LEP electron-positron collider in Geneva, Switzerland, has a radius of 4 km to minimize this loss, but even so a fully loaded ring ($3 \times 10^{12}$ particles) each moving with energy 60 GeV radiates $1.5 \times 10^6$ W.

Conversely, the synchrotron radiation from electrons moving in a ring can be exceedingly useful. Synchrotron sources provide the most intense ultraviolet and X-ray beams available and are of great utility to scientists in the study of solid-state and atomic physics.

For an electron (rest energy 0.5 MeV) circulating with an energy of 1 GeV $= 10^3$ MeV, the angular size of the radiation pattern is, from Equation (10–162),

$$\theta \approx \gamma^{-1} = \left(\frac{0.5}{10^3}\right) = \frac{1}{2} \times 10^{-3} \text{ rad}$$

At a distance of 10 m this angular size corresponds to a beam width of about 1 cm.

## 10-10 PROPAGATION AND EXTINCTION OF WAVES IN MATTER

In Chapters 8 and 9 we discussed in some detail waves propagating in matter. We found that when an incident wave encounters a dielectric, part of the wave is reflected and part is transmitted. The transmitted wave travels more slowly, its phase velocity given by $c/n$, where $n$ is the index of refraction. Now that we have an introduction to radiation from charges, we can view the propagation problem from a new perspective.

We are now able to trace in detail what happens in a dielectric. The incident wave will induce oscillating electric dipoles in the dielectric molecules. These molecules will in turn radiate, and hence the total field is the sum of the incident and radiated fields. As we will demonstrate, the effect of the coherent radiation from all the molecules generates both a wave propagating with velocity $c/n$ and one that cancels the incident wave inside the dielectric. This latter process is an example of the extinction theorem of Ewald and Oseen (circa 1910). In addition, the amplitude of the propagating wave is exactly as predicted by the Fresnel formula. In our example here we will consider only the simplest case of a plane wave incident normally on a dielectric.

To begin, a plane wave polarized along the $\hat{x}$ axis moves parallel to the $z$ axis and encounters a thin dielectric sheet at $z = z'$, as shown in Figure 10–14. A dipole moment,

$$d\mathbf{p} = \chi E(z')\hat{\mathbf{x}}\, dz'\, \imath\, d\imath\, d\phi \tag{10–180}$$

is induced in an element of the sheet by the electric field $\mathbf{E} = E(z')\hat{\mathbf{x}}$ of the incident wave. As usual, the physical wave is found by including the time variation factor $e^{-i\omega t}$ and taking the real part. This oscillating dipole element will radiate in all directions, and by Equation (10–59) the vector potential

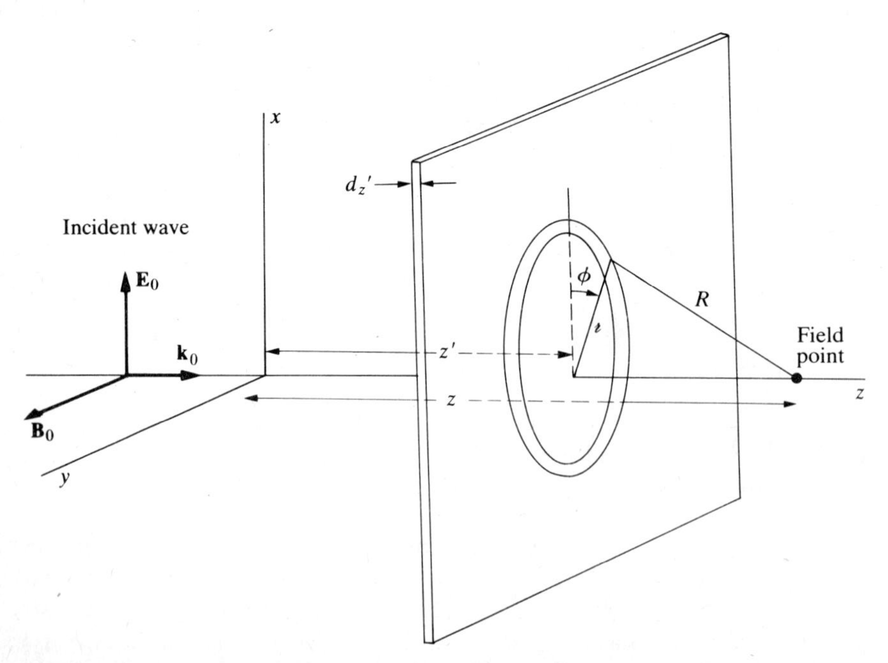


**FIGURE 10–14** Plane wave incident on a dielectric sheet. The radiation from each point on the sheet is coherently added at the field point.

from the element is

$$d\mathbf{A} = \frac{\mu_0}{4\pi}(-i\omega\, d\mathbf{p})\frac{e^{ik_0 R}}{R} = -\frac{i\omega\mu_0}{4\pi}\chi\, dz'\,\hat{\mathbf{x}}\imath\, d\imath\, d\phi\,\frac{e^{ik_0 R}}{R} \tag{10–181}$$

where the speed of light in empty space is $c = \omega/k_0$. The vector potential for the sheet is then

$$d\mathbf{A} = -\frac{i\omega\mu_0}{2}\chi\, dz'\,\hat{\mathbf{x}}\int_0^\infty \frac{\imath\, d\imath}{R}\, e^{ik_0 R} \tag{10–182}$$

By the geometry of Figure 10–14 we have

$$R^2 = \imath^2 + (z - z')^2 \tag{10–183}$$

and for fixed $z'$ we find

$$R\, dR = \imath\, d\imath \tag{10–184}$$

By changing the variable to $R$, we can perform the integral of Equation (10–182):

$$\begin{aligned} d\mathbf{A} &= -\frac{i\omega\mu_0}{2}\chi\, dz'\,\hat{\mathbf{x}}\int_{|z-z'|}^{R_m} e^{ik_0 R}\, dR \\ &= -\frac{c\mu_0\chi}{2}\, dz'\,\hat{\mathbf{x}}(e^{ik_0 R_m} - e^{ik_0|z-z'|}) \end{aligned} \tag{10–185}$$

where $R_m$ is a large, but not infinite, value of $R$. For large $R_m$ the first term, $e^{ik_0 R_m}$, oscillates rapidly but averages to zero. The induced vector potential from the sheet is

$$d\mathbf{A} = -\frac{c\mu_0\chi}{2}\hat{\mathbf{x}}E(z')\, dz' \begin{cases} e^{ik_0(z-z')}, & \text{for } z > z' \\ e^{-ik_0(z-z')}, & \text{for } z < z' \end{cases} \tag{10–186}$$

For a fixed $z'$ the upper possibility represents a transmitted wave propagating to the right, and the lower possibility is a reflected wave traveling to the left.

The corresponding electric field is

$$d\mathbf{E} = \frac{ik_0\chi}{2\varepsilon_0}E(z')\hat{\mathbf{x}}\, dz' \begin{cases} e^{ik_0(z-z')}, & \text{for } z > z' \\ e^{-ik_0(z-z')}, & \text{for } z < z' \end{cases} \tag{10–187}$$

for the transmitted and reflected waves. In obtaining Equation (10–187), we first computed $\mathbf{B} = \nabla \times \mathbf{A}$ and then used $\mathbf{E} = -c(\pm\hat{\mathbf{z}} \times \mathbf{B})$.

With the result of Equation (10–187) for the electric field of a dielectric sheet, we now compute the electric field in a solid dielectric by adding sheet contributions. We want the dielectric to fill the right-hand half-space ($z > 0$). At a given position in the dielectric there will be three contributions to the electric field: (1) the incident wave, (2) waves transmitted from sheets to the left ($z' < z$), and (3) waves reflected from sheets to the right ($z' > z$). The total electric field $x$ component $E(z)$ is then given by

$$E(z) = E_0 e^{ik_0 z} + \frac{ik_0\chi}{2\varepsilon_0}\int_0^z dz'\, E(z')e^{ik_0(z-z')} + \frac{ik_0\chi}{2\varepsilon_0}\int_z^\infty dz'\, E(z')e^{-ik_0(z-z')} \quad \textbf{(10–188)}$$

This equation is an *integral equation*, albeit a simple one, since the unknown $E(z)$ appears in the integrand. It also does not obviously converge unless we assume $E(z)$ damps fast enough so that the second integral in Equation (10–188) converges. All real dielectrics are at least slightly absorbing, so we expect $E(z)$ to drop off exponentially at large $z$. We will assume convergence and then check the resulting solution for consistency.

Since all of the functions appearing in Equation (10–188) are exponential, a good guess for the solution is

$$E(z) = Ae^{ikz} \quad \textbf{(10–189)}$$

When $\chi \to 0$, we would find that $A = E_0$ and $k = k_0$.

Substituting Equation (10–189) into Equation (10–188) and performing the integrals gives

$$Ae^{ikz} = E_0 e^{ik_0 z} + \frac{Ak_0\chi}{2\varepsilon_0}\left(\frac{e^{ikz}}{k-k_0} - \frac{e^{ik_0 z}}{k-k_0} - \frac{e^{ikz}}{k+k_0}\right) \quad \textbf{(10–190)}$$

Since $k = k_0$ cannot be a solution (except when $\chi = 0$), the coefficients of $e^{ikz}$ and $e^{ik_0 z}$ must be separately equal, giving

$$E_0 = \frac{k_0\chi}{2\varepsilon_0}\frac{A}{k-k_0} \quad \textbf{(10–191a)}$$

$$1 = \frac{k_0^2\chi/\varepsilon_0}{k^2 - k_0^2} \quad \textbf{(10–191b)}$$

The second of the above relations can be written as

$$n^2 = 1 + \frac{\chi}{\varepsilon_0} \quad \textbf{(10–192)}$$

where, as usual, $n = k/k_0$. The first relation, Equation (10–191a), then becomes

$$A = \frac{2}{n+1}E_0 \quad \textbf{(10–193)}$$

and the transmitted field is

$$E(z) = \frac{2E_0}{n+1} e^{ik_0 nz} \tag{10–194}$$

Let us go over the results of this calculation. First, we found that one part of the radiated field just canceled the incident field, so in the dielectric there is no longer a wave propagating at $c$; this phenomenon is called extinction. Second, a wave is generated by the coherent radiation of all the dipoles, which has a phase velocity of $c/n$. Third, the ratio of transmitted to incident amplitudes is $2/(n + 1)$, in agreement with the Fresnel equation result for normal incidence. Finally, as seen from Chapter 7, the susceptibility

$$\chi(\omega) = \frac{e^2 N/m}{\omega_0^2 - \omega^2 - i\omega\gamma} \tag{10–195}$$

for a simple molecular oscillator always has a positive imaginary part. The transmitted wave is then damped as expected, and the integral equation, (10–188), is well defined.

That similar results will also follow for any angle of incidence or for more general media is not difficult to imagine. The general theorem is proved and other references are given by J. De Goede and P. Mazur [*Physica*, 58: 568 (1972)].

A similar calculation sheds light on the corresponding case of light hitting a conductor. A thin sheet of conductor will have an induced current density $g\mathbf{E}$. This oscillating current will radiate, and the transmitted radiated electric field will be just 180° out of phase with the incoming wave. The result is that the field goes to zero in the conductor, and all of the energy is reflected.

# 10-11 WAVE EQUATION FOR **E** AND **B** WITH SOURCES

We end this chapter by establishing that both **E** and **B** satisfy inhomogeneous wave equations. This result will play an important role in our discussion of relativity in Chapter 11.

The Maxwell equations are coupled, first-order differential equations for the **E** and **B** fields. Uncoupled, second-order differential equations for **E** and **B** can be derived from Maxwell's equations. These second-order equations are useful because they make evident the existence of wave solutions. However, the second-order equations do not have the full content of Maxwell's equations and must be supplemented by other conditions.

Taking the curl of Maxwell III,

$$\nabla \times (\nabla \times \mathbf{E}) = -\frac{\partial}{\partial t}(\nabla \times \mathbf{B})$$

and substituting Maxwell IV, we obtain

$$\nabla \times (\nabla \times \mathbf{E}) = -\frac{1}{c^2}\frac{\partial^2 \mathbf{E}}{\partial t^2} - \mu_0 \frac{\partial \mathbf{J}}{\partial t}$$

where the relation $c = 1/\sqrt{\mu_0 \varepsilon_0}$ has been used. Using the vector identity for the triple cross product, we get

$$\nabla(\nabla \cdot \mathbf{E}) - \nabla^2 \mathbf{E} = -\frac{1}{c^2}\frac{\partial^2 \mathbf{E}}{\partial t^2} - \mu_0 \frac{\partial \mathbf{J}}{\partial t} \qquad (10\text{–}196)$$

Substituting Maxwell I, we arrive at the wave equation for **E**:

**Wave Equation for E**

$$\nabla^2 \mathbf{E} - \frac{1}{c^2}\frac{\partial^2 \mathbf{E}}{\partial t^2} = \mu_0 \left( c^2 \, \nabla \rho + \frac{\partial \mathbf{J}}{\partial t} \right) \qquad (10\text{–}197)$$

To derive the wave equation for **B**, we take the curl of Maxwell IV and substitute Maxwell III, to obtain

$$\nabla \times (\nabla \times \mathbf{B}) = \mu_0 \nabla \times \mathbf{J} - \frac{1}{c^2}\frac{\partial^2 \mathbf{B}}{\partial t^2} \qquad (10\text{–}198)$$

Again using the vector identity for the triple cross product and Maxwell II, we have the wave equation for **B**:

**Wave Equation for B**

$$\nabla^2 \mathbf{B} - \frac{1}{c^2}\frac{\partial^2 \mathbf{B}}{\partial t^2} = -\mu_0 \nabla \times \mathbf{J} \qquad (10\text{–}199)$$

## SUMMARY

### Important Concepts and Equations

$$P = \frac{\Delta U}{\Delta t} = \frac{q^2}{4\pi\varepsilon_0}\left(\frac{2a^2}{3c^3}\right)$$

*Power radiated by a charge $q$ with acceleration $a$.*

$$\mathbf{B} = \nabla \times \mathbf{A} \qquad \mathbf{E} = -\nabla \Phi - \frac{\partial \mathbf{A}}{\partial t}$$

*Expression for the magnetic and electric fields in terms of the vector and scalar potentials.*

$$\mathbf{A} = \mathbf{A}' + \nabla\chi \qquad \Phi = \Phi' - \frac{\partial\chi}{\partial t}$$

*Gauge transformation. For any scalar field $\chi$ the* **E** *and* **B** *fields are the same for the potentials* $\mathbf{A}'$ *and* $\Phi'$ *or* **A** *and* $\Phi$.

$$\nabla\cdot\mathbf{A} + \frac{1}{c^2}\frac{\partial\Phi}{\partial t} = 0$$

*Lorentz gauge condition.*

$$\nabla^2\Phi - \frac{1}{c^2}\frac{\partial^2\Phi}{\partial t^2} = -\frac{\rho}{\varepsilon_0} \qquad \nabla^2\mathbf{A} - \frac{1}{c^2}\frac{\partial^2\mathbf{A}}{\partial t^2} = -\mu_0\mathbf{J}$$

*Wave equations for potentials in the Lorentz gauge.*

$$t_\mathrm{R} = t - \frac{R}{c}$$

*Retarded time $t_\mathrm{R}$; where $R/c$ is the time required for a light signal to travel from the source to the field point.*

$$\Phi(\mathbf{r}, t) = \frac{1}{4\pi\varepsilon_0}\int\frac{\rho(\mathbf{r}', t_\mathrm{R})\,dV'}{R} \qquad \mathbf{A}(\mathbf{r}, t) = \frac{\mu_0}{4\pi}\int\frac{\mathbf{J}(\mathbf{r}', t_\mathrm{R})\,dV'}{R}$$

*Retarded potentials. Solutions to the wave equation for the potentials in the Lorentz gauge.*

$$\mathbf{A}(\mathbf{r}) = \frac{\mu_0}{4\pi}\frac{e^{ikr}}{r}\mathbf{F}(\hat{\mathbf{r}}) \qquad \mathbf{F}(\hat{\mathbf{r}}) = \int\mathbf{J}(\mathbf{r}')e^{-ik\hat{\mathbf{r}}\cdot\mathbf{r}'}\,dV'$$

*Space-dependent part of the vector potential in the radiation zone ($r \gg \lambda, r \gg r'$).*

$$\mathbf{A}(\mathbf{r}, t) = \mathbf{A}(\mathbf{r})e^{-i\omega t}$$

*Vector potential for a given angular frequency $\omega = ck$.*

$$\mathbf{B} = \frac{\mu_0}{4\pi}ik\frac{e^{ikr}}{r}\hat{\mathbf{r}}\times\mathbf{F} \qquad \mathbf{E} = -c\hat{\mathbf{r}}\times\mathbf{B}$$

*Magnetic and electric fields in the radiation zone.*

$$\langle dP\rangle_t = \frac{\mu_0}{(4\pi)^2}\frac{k^2c}{2}[(\hat{\mathbf{r}}\times\mathrm{Re}\,\mathbf{F})^2 + (\hat{\mathbf{r}}\times\mathrm{Im}\,\mathbf{F})^2]\,d\Omega$$

*Time-averaged power radiated into a solid angle element $d\Omega$.*

$$\mathbf{F}(\hat{\mathbf{r}}) = \sum_{n=0}^{\infty}\frac{1}{n!}\int\mathbf{J}(\mathbf{r}')(-ik\hat{\mathbf{r}}\cdot\mathbf{r}')^n\,dV$$

*Radiation multipole expansion.*

$$\mathbf{A}_\mathrm{ED}(\mathbf{r}) = -\frac{i\mu_0\omega}{4\pi}\frac{e^{ikr}}{r}\mathbf{p}$$

*Electric dipole vector potential.*

$$\mathbf{E}_{\mathrm{ED}}(\mathbf{r}, t) = -\frac{\mu_0 \omega^2 p}{4\pi} \frac{e^{-i\omega t_{\mathrm{R}}}}{r} \sin\theta \hat{\boldsymbol{\theta}} \qquad \mathbf{B}_{\mathrm{ED}}(\mathbf{r}, t) = -\frac{\mu_0}{4\pi} \frac{\omega^2 p}{c} \frac{e^{-i\omega t_{\mathrm{R}}}}{r} \sin\theta \hat{\boldsymbol{\phi}}$$

*Electric dipole radiation fields for an electric dipole along the z axis.*

$$\langle P_{\mathrm{ED}} \rangle_t = \frac{1}{4\pi\varepsilon_0}\left(\frac{p^2\omega^4}{8c^3}\right)$$

*Time-averaged power radiated by electric dipoles.*

$$\mathbf{A}_{\mathrm{MD}}(\mathbf{r}) = -\frac{i\mu_0 k}{4\pi} \frac{e^{ikr}}{r} (\mathbf{m} \times \hat{\mathbf{r}})$$

*Magnetic dipole vector potential.*

$$\mathbf{E}_{\mathrm{MD}} = -\frac{m}{p} \mathbf{B}_{\mathrm{ED}} \qquad \mathbf{B}_{\mathrm{MD}} = \frac{m}{pc^2} \mathbf{E}_{\mathrm{ED}}$$

*Magnetic dipole radiation fields.*

$$\langle P_{\mathrm{MD}} \rangle_t = \frac{\mu_0}{4\pi}\left(\frac{m^2\omega^4}{3c^3}\right)$$

*Time-averaged power radiated by magnetic dipoles.*

$$\int_{\text{all space}} \mathbf{E}_a \cdot \mathbf{J}_b \, dV = \int_{\text{all space}} \mathbf{E}_b \cdot \mathbf{J}_a \, dV$$

*Reciprocity theorem.*

$$\mathbf{F}_{\mathrm{rad}} = \frac{1}{4\pi\varepsilon_0}\left(\frac{2}{3}\right)\frac{q^2}{c^3}\dot{\mathbf{a}}$$

*Radiation reaction force on a charge.*

$$\Phi(\mathbf{r}, t) = \frac{q}{4\pi\varepsilon_0 R} \frac{1}{b}\bigg|_{t_{\mathrm{R}}} \qquad \mathbf{A}(\mathbf{r}, t) = \frac{\mu_0 q \mathbf{v}}{4\pi R} \frac{1}{b}\bigg|_{t_{\mathrm{R}}}$$

*Lienart-Wiechert potentials evaluated at the charge position at the retarded time $t_{\mathrm{R}} = t - R/c$.*

$$\mathbf{E} = \mathbf{E}_v + \mathbf{E}_a$$

*The electric field of a moving charged particle can be separated into a term independent of the particle's acceleration and a term proportional to* **a**. *The acceleration-dependent field at large distance corresponds to radiation from the charge.*

$$\mathbf{B} = \frac{1}{c} \hat{\mathbf{R}} \times \mathbf{E}$$

*The magnetic field of a moving charge is perpendicular to the electric field and the direction of the retarded point.*

$$\left(\nabla^2 - \frac{1}{c^2}\frac{\partial^2}{\partial t^2}\right)\mathbf{E} = \mu_0\left(c^2\,\nabla\rho + \frac{\partial \mathbf{J}}{\partial t}\right) \qquad \left(\nabla^2 - \frac{1}{c^2}\frac{\partial^2}{\partial t^2}\right)\mathbf{B} = -\mu_0\,\nabla\times\mathbf{J}$$

*Inhomogeneous wave equations satisfied by the electric and magnetic fields.*

## PROBLEMS

### Section 10-1: Wave Generation

**10-1** An electron ($q = -1.6 \times 10^{-19}$ C, $m = 9.1 \times 10^{-31}$ kg) is accelerated from rest through a distance $d$ in a uniform electric field $E$. What fraction of the energy gained is lost by radiation? Evaluate numerically for $E = 3 \times 10^6$ V/m and $d = 1$ m.

**10-2** An electron initially moves in a circular orbit of radius $r_0 = 0.5 \times 10^{-10}$ m about a proton. If classical physics applies at atomic dimensions, we expect radiation due to the centripetal acceleration. This loss of energy causes the electron to spiral toward the proton. Assume that for each revolution the orbit is nearly circular. How long does it take for a hydrogen atom of initial radius $r_0$ to collapse? Is the assumption that the radiation only slightly perturbs the orbital mechanics a good approximation? This predicted collapse of matter was one of the reasons for the formulation of quantum mechanics in the opening decades of the twentieth century.

### Section 10-2: Wave Equations for Potentials

**10-3** A large conducting sheet in the $xy$ plane carries a uniform surface current density $\mathbf{j}$ in the $x$ direction, $\mathbf{j} = j\hat{\mathbf{x}}\cos\omega t$. The size of the sheet is much greater than the wavelength $\lambda = 2\pi c/\omega$, and its thickness is much less than the skin depth. How many watts per square meter are radiated if $j = 100$ A/m?

### Section 10-3: Retarded Potential Solutions

**10-4** Show directly by substitution that the retarded solution in Equation (10–30) satisfies the wave equation (10–23a). *Hint:* Verify the identity

$$\nabla_{\mathbf{r}}^2\left(\frac{s}{R}\right) = \nabla_{\mathbf{R}}^2\left(\frac{s}{R}\right) = -4\pi s\delta^{(3)}(\mathbf{R}) + \frac{1}{R}\frac{\partial^2 s}{\partial R^2}$$

and that $\partial s/\partial t = -c(\partial s/\partial R)$.

**10-5** A long straight neutral conducting wire lies along the $z$ axis.

(i) If the wire carries a constant current $I$, what are the static electric and magnetic fields?

(ii) Suppose that instead of being constant in time, the current density is given by

$$\mathbf{J} = \begin{cases} 0, & \text{for } t < 0 \\ I\delta(x)\delta(y)\hat{\mathbf{z}}, & \text{for } t > 0 \end{cases}$$

Calculate the retarded vector potential and the electric and magnetic fields.

## Section 10-5: Multipole Radiation

**10–6** Show that the only nonzero multipole moment of the charge distribution

$$\rho_{\text{ED}} = -\mathbf{p} \cdot \nabla\delta(\mathbf{r})$$

is an electric dipole moment $\mathbf{p}$, and the only moment of the current distribution

$$\mathbf{J}_{\text{MD}} = -\mathbf{m} \times \nabla\delta(\mathbf{r})$$

is a magnetic dipole moment $\mathbf{m}$. Use the equation of continuity to find the electric dipole current distribution.

**10–7** Show that the exact electric and magnetic fields of an oscillating electric dipole of moment $\mathbf{p} = p\hat{\mathbf{z}}$ are

$$\mathbf{B} = \frac{\mu_0}{4\pi}(i\omega p)\frac{e^{ikr}}{r}\left(ik - \frac{1}{r}\right)\sin\theta\hat{\boldsymbol{\phi}}$$

$$\mathbf{E} = \frac{\mu_0}{4\pi}c^2 p\frac{e^{ikr}}{r}\left[-\frac{2}{r}\left(ik - \frac{1}{r}\right)\cos\theta\hat{\mathbf{r}} + \left(-k^2 - \frac{ik}{r} + \frac{1}{r^2}\right)\sin\theta\hat{\boldsymbol{\theta}}\right]$$

in spherical coordinates. Identify the radiation, near-zone quasi-static, and intermediate fields.

**10–8** A charge $+q$ moves in a circular orbit of radius $a$ in the $xy$ plane at a constant angular velocity $\omega_0$. Assume $\omega_0 a/c \ll 1$.

(i) What are the electric and magnetic fields at a point on the $z$ axis in the radiation zone?

(ii) What are the dominant frequency and polarization of the radiation on the $z$ axis?

(iii) What is the polarization of the radiation in the $xy$ plane?

(iv) Calculate the total radiated power $P$, and find the torque that must be supplied to the charge to keep it moving at a constant angular velocity.

(v) If instead of a single charge, suppose that there are two equal charges at the ends of a rod of length $2a$ rotating about its center. What is the dominant frequency of the radiation?

**10–9** Consider a system of $N$ interacting charged particles all having the same charge-to-mass-ratio $q/m$. The electric dipole moment of this system is

$$\mathbf{p} = \sum_{i=1}^{N} q_i\mathbf{r}_i = \frac{q}{m}\sum_{i=1}^{N} m_i\mathbf{r}_i$$

Show that this system cannot emit electric dipole radiation unless it is acted upon by an external force. (*Hint:* Recall the conditions necessary for an electric dipole to radiate and also the relation between external force and the velocity of the center of mass.)

**10–10** In Problem 10–9 you showed that a system of interacting charged particles having the same charge-to-mass ratio cannot emit electric dipole radiation unless it is acted upon by an external force. Show that this system cannot emit magnetic dipole radiation either. (*Hint:* Consider the relation between the magnetic dipole moment and the total angular momentum derived in Section 7–4.)

**10–11** A small circular loop of wire of radius $a$ carries a current $I = I_0 \cos\omega t$. The loop is located in the $xy$ plane.

(i) Calculate the first nonzero multipole moment of the system.
(ii) Give the form of the vector potential $\mathbf{A}(\mathbf{r}, t)$ for this system for $r \to \infty$, calculate the asymptotic electric and magnetic fields, and determine the angular distribution of the outgoing radiation.
(iii) Describe the main features of the radiation pattern.
(iv) Calculate the average power radiated.

## Section 10-6: Antenna Design

**10-12** The problem of a radiating electric dipole above a conducting plane can be solved by the method of images. Consider an electric dipole oriented parallel to a flat conducting surface and a distance $d$ above it. Show that the field boundary conditions are satisfied by using an image dipole $-\mathbf{p}$. Use the exact dipole field from Problem 10–7.

**10-13** The antenna gain defined in Equations (10–99) and (10–101) is a directionality property of an antenna. For an antenna with a single dominant power lobe of angular width $\Delta\theta$, estimate the decibel gain.

**10-14** As in Section 10–6C, two radiating electric dipoles are directed along the $z$ axis. One is at the origin, and the other is located at $x = d$ and has a phase $\beta$ relative to the first dipole. We observed that when $\beta = 0$ and $d = \lambda/2$, the main power lobes were directed along the $y$ axis. Compute the power distribution for $kd = \pi$ but with an arbitrary phase $\beta$. What is the effect of $\beta$? Repeat the calculation when the dipoles are separated by a full wavelength. Sketch some representative power patterns in the $xy$ plane.

**10-15** A horizontal dipole lies a distance $d$ above a conducting plane (this plane might be the earth), as shown in Figure 10–15.
(i) Are the directional properties of this antenna an improvement over the dipole alone? (Use the image property discussed in Problem 10–12.)
(ii) What is the optimum distance $d$?
(iii) For this optimum distance, sketch the antenna radiation pattern.

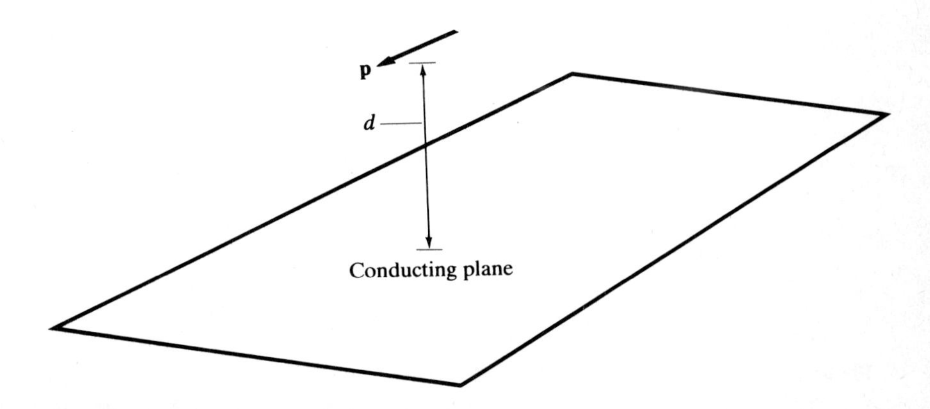


**FIGURE 10-15** Dipole above a conducting plane

## Section 10-7: Pulsars

**10-16** Compute the power radiated by the rotating dipole moments of the earth and of Jupiter at the frequency of their rotations. (The dipole moments for earth and Jupiter are given in Section 7–1; the angle between the rotation axis and the dipole moment is about 11° for both planets.)

## Section 10-8: Radiation from Charges

**10-17** For a charged particle accelerated along its direction of motion, show that the radiated power is

$$P = \frac{1}{4\pi\varepsilon_0}\left(\frac{2q^2}{3c^3}\right) a^2\gamma^6$$

where $\gamma = (1 - v^2/c^2)^{-1/2}$.

**10-18** If a plane electromagnetic wave is incident on a free particle of charge $e$ and mass $m$, it will cause the particle to oscillate and hence to radiate. The whole process can be described as scattering of radiation, with a cross section defined as

$$\sigma_\text{T} \equiv \frac{\text{radiated power}}{\text{incident power flux}}$$

Show that the *Thomson scattering cross section* $\sigma_\text{T}$ is

$$\sigma_\text{T} = \left(\frac{1}{4\pi\varepsilon_0}\right)^2 \left(\frac{8\pi}{3}\right)\left(\frac{e^2}{mc^2}\right)^2$$

For an electron show that, numerically, $\sigma_\text{T} = 0.66 \times 10^{-24}\ \text{cm}^2$.

**10-19** Using the synchrotron radiation power formula, (10–179), show that for a circulating electron the power radiated is

$$P\,(\text{watts}) = (6.8 \times 10^{-7})\frac{E^4\,(\text{GeV})}{R^2\,(\text{meters})}$$

and the energy lost per revolution is

$$\delta E\,(\text{MeV}) = 0.089\frac{E^4\,(\text{GeV})}{R\,(\text{meters})}$$

For the LEP accelerator at CERN in Geneva, $E = 60$ GeV and $R = 4243$ m, and there are a total of $3 \times 10^{12}$ electrons and positrons circulating. Evaluate the loss per revolution for each particle and the total power required to maintain the beam energy. (The rest energy of the electron is $mc^2 = 0.51$ MeV.)

## Section 10-10: Extinction of Waves in Matter

**10-20** Show that for a plane wave incident normally on a dielectric the radiation produced by the induced dielectric dipoles gives a reflected wave whose amplitude agrees with Fresnel's equations.

**10-21** A plane wave is incident normally on a good conductor of conductivity $g$. Work out the effects of the induced radiating currents on the initial wave. Show that there is no transmitted wave, define the corresponding skin depth, and find the reflectivity. (Neglect power loss by ohmic heating.)

## Section 10-11: Wave Equation for **E** and **B** with Sources

**10-22** Using the wave equations for the field strengths **E** and **B**, show that from the retarded solution, Equation (10–30), of the wave equation, one gets the results of Equations (10–43) and (10–44) for the radiation fields without the introduction of potentials.

# 11

# Relativity

## CHAPTER CONTENTS

The equations of Newtonian mechanics contain a number of relativity principles; for instance, the acceleration of a body that is a member of an isolated system depends only on its position relative to the other members of the system, rather than on its absolute position in space. Similarly, its acceleration is independent of absolute direction, velocity, or time. These relativity principles can all be illustrated by saying that by observing the motions of the members of the solar system, we cannot determine an absolute time nor an absolute ("relative to space") position, orientation, or velocity of the solar system.

The Maxwell equations satisfy these Newtonian relativity principles—except, apparently, for the relativity of velocity. The precise Newtonian statement of the relativity of velocity is that the Newtonian equations of motion have the same form using either coordinates $\mathbf{r}$ or $\mathbf{r}'$ where $\mathbf{r}$ and $\mathbf{r}'$ are related by $\mathbf{r} = \mathbf{r}' + \mathbf{v}t$; this relation is described by saying that the two coordinate frames $S$ and $S'$ are moving with respect to each other by the constant velocity $\mathbf{v}$. It is called a Galilean transformation. This form invariance of the Newtonian equations means that any internal motion of a system obeying Newton's equations, such as the solar system, is also a possible motion if coordinates $\mathbf{r}'$ are substituted for $\mathbf{r}$. That is, if all velocities of the bodies in the solar system were changed by $\mathbf{v}$, the internal motions would be unaffected.

But Maxwell's equations are not form-invariant to a Galilean transformation. A simple illustration of this fact is that according to Maxwell's equations, light (electromagnetic radiation) propagates with a speed $c$, where $c$ is a certain combination of constants, namely, $1/\sqrt{\varepsilon_0\mu_0}$. Thus if Maxwell's equations hold for an observer in one coordinate system, they will not hold for another observer in a different coordinate system moving with a velocity $\mathbf{v}$ relative to the first and related to the first by a Galilean transformation. In this second coordinate system light will propagate with a speed that depends on its direction of propagation, with a value in the range $c - v$ to $c + v$. All experiments—in particular, the measurement of the velocity of light propagating in various directions—agree with Maxwell's equations. According to the Newtonian relativity of velocity, as described above, this experimental agreement means that the earth just happens to be at rest, to the accuracy of experiments, in the frame in which Maxwell's equations hold. Before the time of Copernicus, this assumption would not have been thought absurd. In any case, there is an experimental test. Because of the acceleration of the earth by the sun, the velocity of an earth-based observer varies during the year, and a precise enough test of Maxwell's equations would find them not to hold throughout the year. The measurement was first done by Michelson and Morley (1891), who did not find the predicted failure of Maxwell's equations.

Now, in fact, Maxwell's equations have a form invariance to a transformation between coordinate frames moving relative to one another called the Lorentz transformation. Thus Maxwell's equations obey a velocity relativity principle, which is, however, different from the Newtonian velocity relativity, because the Lorentz and Galilean coordinate transformations are different.

The Michelson-Morley experiment and many others have failed to measure absolute velocity, which suggests that all the equations of physics obey a single velocity relativity principle. Einstein proposed (1905) that there is a single principle and that this principle is the electromagnetic one. Thus Newton's equations must be modified so as to have the same form invariance as Maxwell's equations. The alteration is of the order $v^2/c^2$, and thus at small $v/c$ the predictions of Newton's equations remain the same. But experiments that are done in situations where the alteration is very large establish that Einstein was correct. Today, Einstein's version of the relation between coordinates and clocks in different reference frames (the Lorentz transformation obeyed by Maxwell's equations) is simply called *relativity*.

Although two observers in relative motion agree that Maxwell's equations are satisfied, they do not see the same values of the electric and magnetic fields at a given point. The relation between the fields is given by transformation equations that are determined by the principles of relativity. The sources of the fields obey similar transformations, which are derived in this chapter.

## 11-1 LORENTZ TRANSFORMATION

To find the coordinate transformations under which Maxwell's equations are form-invariant, we use a strategy originally devised by Einstein. He considered an electromagnetic pulse emitted from the origin of a coordinate frame at $t = 0$. According to Maxwell's equations, the pulse becomes a spherical wave packet propagating outward from the origin at the speed of light. That is, the fields of the pulse are localized on a shell whose radius $r$ is given by

$$r = ct \tag{11-1}$$

Since $r^2 = x^2 + y^2 + z^2$, the square of Equation (11–1) can be written as

$$x^2 + y^2 + z^2 - c^2t^2 = 0 \tag{11-2}$$

We now describe this physical phenomenon by using a different set of coordinates $\mathbf{r}'$ and $\mathbf{t}'$ (frame of reference $S'$), which are related to the original $\mathbf{r}$ and $t$ (frame of reference $S$) in such a way that Maxwell's equations still hold; that is, Maxwell's equations are form-invariant to the transformation from frame $S$ to frame $S'$, and the speed of light $c$ is the same in both frames. See Figure 11–1. By this requirement the pulse of radiation described in terms of $\mathbf{r}'$ and $t'$ will behave just as before, and in particular, the radius of the shell will be

$$r' = ct' \tag{11-3}$$

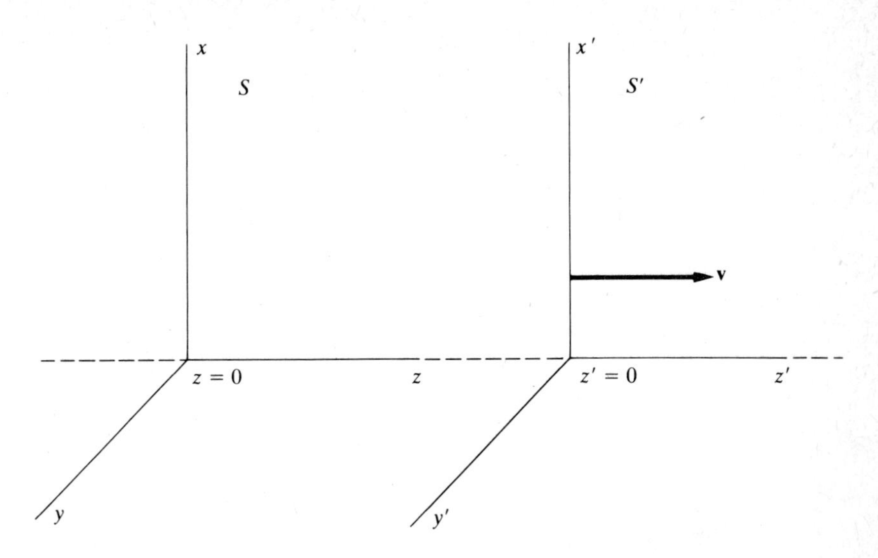


**FIGURE 11-1** Two reference frames $S$ and $S'$. The frame $S'$ is moving uniformly with velocity $\mathbf{v} = v\hat{\mathbf{z}}$ relative to frame $S$.

As before, this relation implies

$$x'^2 + y'^2 + z'^2 - c^2t'^2 = 0 \tag{11-4}$$

We now wish to determine the relationship between coordinates and times in $S$ and $S'$ frames such that Equations (11–2) and (11–4) both hold. We first try the Galilean transformation

$$\mathbf{r} = \mathbf{r}' + \mathbf{v}t' \qquad t = t' \tag{11-5}$$

under which Newton's law $\mathbf{F} = m\mathbf{a}$ has the same form in the $S$ and $S'$ frames (i.e., is form-invariant). In Equation (11–5), $\mathbf{v}$ is the relative velocity of the origins of $S$ and $S'$; see Figure 11–2. Here we assume that a universal time $t = t'$ describes events in both frames. From Equation (11–5) we obtain

$$r^2 - c^2t^2 = r'^2 - c^2t'^2 + 2\mathbf{r}' \cdot \mathbf{v}t' + v^2t'^2 \tag{11-6}$$

Hence spherical light wave fronts are not preserved by the Galilean transformation.

## 11-1A Rotations in Four Dimensions

The following analogy helps give us insight about the correct coordinate and time transformations from $S$ to $S'$. At equilibrium a droplet of water in a gravity-free region will be spherical, with a radius $R$ that depends only

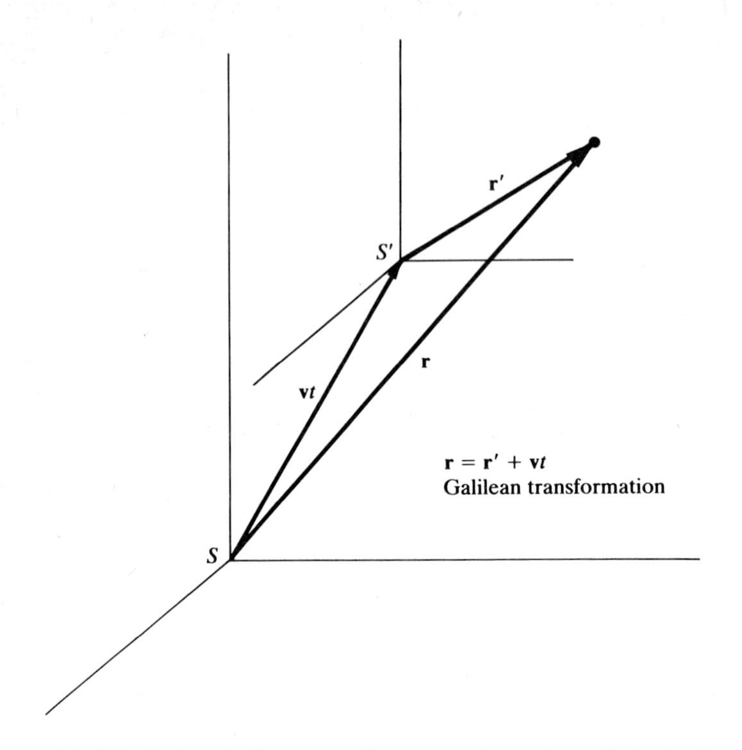


**FIGURE 11-2** Galilean coordinate transformation, which implies $\mathbf{u} = \mathbf{u}' + \mathbf{v}$ for the velocities of a point measured in $S$ and $S'$

on the number of water molecules in the drop and the ambient temperature and pressure. The surface of the drop is described by the equation

$$|\mathbf{r}|^2 = R^2 \tag{11-7}$$

that is, a point $\mathbf{r}$ is on the surface of the drop if and only if it satisfies this equation. (We take the origin of the coordinate system at the center of the drop.) Since this equation represents, in effect, a law of nature, it must have the same form in a rotated frame $S'$; see Figure 11–3. If the rotation is taken to be in the $xy$ plane, the $S$ and $S'$ coordinates are related by

$$x = x' \cos\theta - y' \sin\theta \qquad y = x' \sin\theta + y' \cos\theta \qquad z = z' \tag{11-8}$$

Since

$$r^2 = x^2 + y^2 + z^2 = x'^2 + y'^2 + z'^2 = r'^2 \tag{11-9}$$

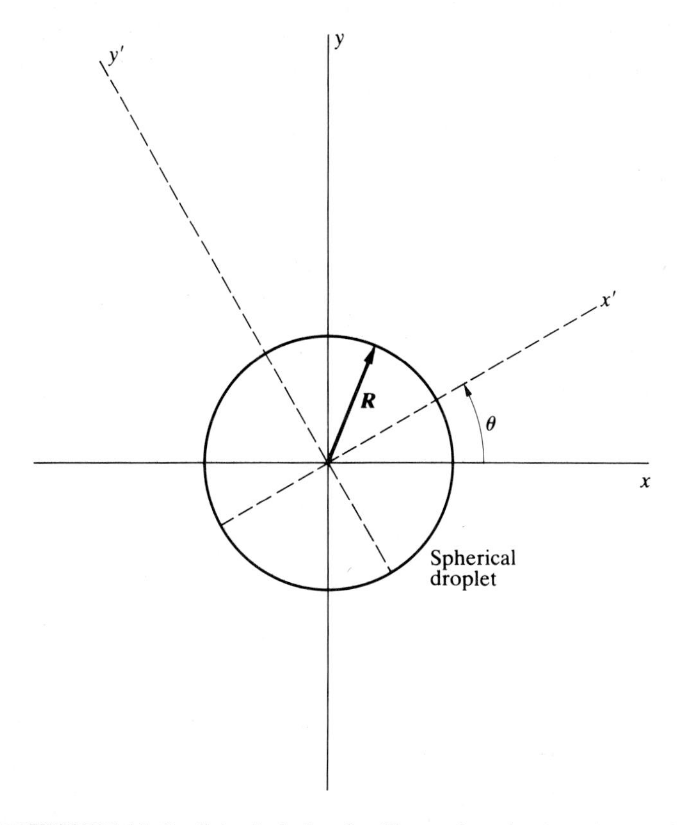


**FIGURE 11-3** Spherical droplet illustrating the invariance of the droplet radius under the rotation of the coordinate reference frame

we have $|r'^2| = R^2$, and the droplet shape is indeed invariant to the coordinate transformation.

We now return to the spherical pulse of light. We can write Equation (11–2), which describes the spherical pulse of radiation, as

$$x^2 + y^2 + z^2 + (ict)^2 = 0 \tag{11-10}$$

where $i = \sqrt{-1}$. Equation (11–10) describes a sphere of zero radius in a four-dimensional space with coordinates $(x, y, z, ict)$, known as *Minkowski space*. Just as Equation (11–7) was invariant to rotations of the three-dimensional coordinate system, Equation (11–10) is invariant to rotations of the four-dimensional coordinate system. We will show shortly that Maxwell's equations themselves are form-invariant under these transformations.

Some of the Minkowski space rotations, namely, those not involving $t$, are spatial rotations. For instance, a rotation in the $xy$ plane ($z$ and $t$ unchanged) is the spatial rotation "around the $z$ axis" of Equation (11–8).

However, a rotation involving the $ict$ coordinate is a velocity transformation. For instance, consider a rotation in the $(z, ict)$ plane ($x$ and $y$ unchanged) of the same form as Equation (11–8), with ($x$ and $z$) and ($y$ and $ict$) exchanged:

$$x = x' \qquad y = y' \qquad z = z' \cos\theta - (ict')\sin\theta$$
$$ict = z'\sin\theta + (ict')\cos\theta \tag{11-11}$$

If $z$, $z'$, $t$, and $t'$ are all to be real, $\sin\theta$ must be imaginary and $\cos\theta$ real. This result is accomplished by taking $\theta$ to be imaginary: Put

$$\theta = i\eta \tag{11-12}$$

then

$$\cos\theta = \frac{e^{i\theta} + e^{-i\theta}}{2} = \frac{e^{-\eta} + e^{\eta}}{2} = \cosh\eta \tag{11-13}$$

$$\sin\theta = \frac{e^{i\theta} - e^{-i\theta}}{2i} = -i\frac{(e^{-\eta} - e^{\eta})}{2} = i\sinh\eta \tag{11-14}$$

where sinh and cosh are the hyperbolic trigonometric functions. Equation (11–11) becomes

$$\begin{aligned} &x = x' \qquad y = y' \qquad z = z'\cosh\eta + ct'\sinh\eta \\ &ct = ct'\cosh\eta + z'\sinh\eta \end{aligned} \tag{11-15}$$

We now can verify that the $S'$ and $S$ frames are moving uniformly with respect to one another. For instance, consider a point at rest in the $S'$ frame; its motion is given by $z'(t') = z'_0$, a constant. Hence

$$dz = c\,dt'\sinh\eta \qquad dt = dt'\cosh\eta \tag{11-16}$$

and

$$\frac{dz}{dt} = \frac{c\,dt'\sinh\eta}{dt'\cosh\eta} = c\,\frac{\sinh\eta}{\cosh\eta} = c\tanh\eta \tag{11-17}$$

That is, in the $S$ frame this point is moving with velocity

$$v = c\tanh\eta \tag{11-18}$$

and similarly for any other point at rest in the $S'$ frame. We can write the transformation in terms of $v$, using

$$\cosh\eta = \frac{1}{\sqrt{1 - \tanh^2\eta}} = \frac{1}{\sqrt{1 - (v^2/c^2)}} = \gamma \tag{11-19}$$

$$\sinh\eta = \tanh\eta\cosh\eta = \frac{v/c}{\sqrt{1 - (v^2/c^2)}} = \frac{v}{c}\gamma \tag{11-20}$$

where we have introduced the customary notation

$$\gamma \equiv \frac{1}{\sqrt{1 - v^2/c^2}} \tag{11-21}$$

In this notation the general $(z, t)$ transformation of Equation (11–15) reads as follows:

**Lorentz Transformation**

$$x = x' \qquad y = y'$$
$$z = \gamma(z' + vt') \qquad t = \gamma\left(t' + \frac{v}{c^2} z'\right) \tag{11-22}$$

This transformation is for the $S'$ frame moving in the $z$ direction with velocity $v$ relative to the $S$ frame; see Figure 11–1. Since the space-time coordinates $(z', t')$ and $(z, t)$ are real numbers, the factor $\gamma$ in Equation (11–22) must be real. Because $\gamma = (1 - v^2/c^2)^{-1/2}$, the relative velocity $v$ of the two frames cannot be greater than $c$. The velocity transformation in Equation (11–22) is known as the Lorentz transformation, introduced in 1904 by the Dutch physicist H. A. Lorentz. Sometimes, the term *Lorentz transformation* is used for any rotation in the four-dimensional $(x, y, z, ict)$ space, and a velocity transformation such as the rotation in the $(z, ict)$ plane, Equation (11–15) or (11–22), is termed a *Lorentz boost*, or simply a boost.

The Lorentz transformation, Equation (11–22), can be inverted to give

$$x' = x \qquad y' = y$$
$$z' = \gamma(z - vt) \qquad t' = \gamma\left(t - \frac{v}{c^2} z\right) \tag{11-23}$$

for the $S'$ frame moving with velocity $v$ along the $z$ axis with respect to the $S$ frame. This inverse transformation is seen to follow from Equations (11–22) by the substitution of $-v$ for $v$. It is a consequence of the fact that the $S$ frame moves with velocity $-v$ with respect to the $S'$ frame; see Figure 11–4.

The quantity $\eta$ of the imaginary angle $i\eta$ of the rotation in the $(z, ict)$ plane is called the *rapidity*, corresponding to the velocity $v$. It is a useful concept because of its geometrical meaning. For instance, consider two successive rotations in the $(z, ict)$ plane, that is, two successive boosts along the $z$ direction. Frame $S'$ is rotated by the angle $i\eta$ relative to $S$, that is, $S'$ moves at velocity $v$ in $S$, where $\eta$ and $v$ are related by $v = c \tanh \eta$. Frame $S''$ is rotated by the angle $i\eta'$ relative to $S'$, that is, $S''$ moves at velocity $u'$ in $S'$, where $\eta'$ and $u'$ are again related by $u' = c \tanh \eta'$. Then just as for two successive rotations in the same plane in space, frame $S''$ is rotated by the angle $i(\eta' + \eta)$ relative to $S$, that is, frame $S''$ moves at velocity $u$ in $S$, given by

**Addition of Velocities**

$$u = c \tanh(\eta + \eta') = c\,\frac{\tanh \eta + \tanh \eta'}{1 + \tanh \eta \tanh \eta'} = \frac{v + u'}{1 + vu'/c^2} \tag{11-24}$$

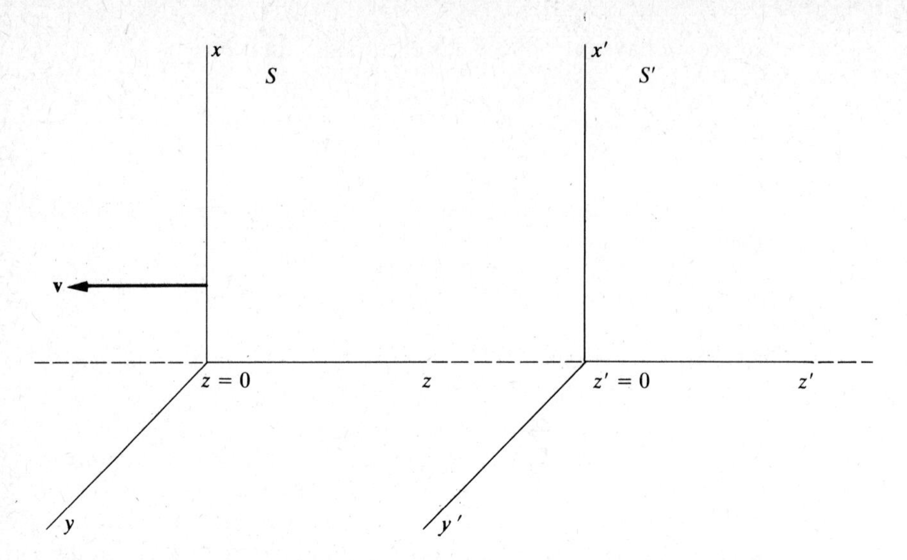


**FIGURE 11-4** Two reference frames $S$ and $S'$ as in Figure 11–1 but from the point of view of an observer at rest on $S'$

Another approach to get the velocity addition law is to take the differential of the Lorentz transformation in Equation (11–22), giving

$$dt = \gamma\left(dt' + \frac{v}{c^2}\,dz'\right) \qquad dz = \gamma(dz' + v\,dt')$$
$$dy = dy' \qquad dx = dx' \tag{11-25}$$

We divide each of these equations for the coordinate differentials in $S$ by the equation for the time differential in $S$:

$$\frac{dz}{dt} = \frac{dz' + v\,dt'}{dt' + v/c^2\,dz'} = \frac{dz'/dt' + v}{1 + v/c^2(dz'/dt')}$$
$$\frac{dy}{dt} = \frac{dy'}{\gamma[dt' + (v/c^2)\,dz']} = \frac{dx'/dt'}{\gamma[1 + (v/c^2)(dz'/dt')]} \tag{11-26}$$
$$\frac{dx}{dt} = \frac{dx}{\gamma[dt' + (v/c^2)\,dz']} = \frac{dx'/dt'}{\gamma[1 + (v/c^2)(dz'/dt')]}$$

As before, we denote the velocities in the $S$ and $S'$ frames by

$$\mathbf{u} = \frac{d\mathbf{r}}{dt} \qquad \mathbf{u}' = \frac{d\mathbf{r}'}{dt'} \tag{11-27}$$

Then Equation (11–26) becomes

$$u_z = \frac{u_z' + v}{1 + (v/c^2)u_z'} \qquad \mathbf{u}_\mathrm{T} = \frac{\mathbf{u}_\mathrm{T}'}{\gamma[1 + (v/c^2)u_z']} \tag{11–28}$$

where $\mathbf{u}_\mathrm{T}$ is the velocity transverse to the direction of the boost,

$$\mathbf{u}_\mathrm{T} = u_x\hat{\mathbf{x}} + u_y\hat{\mathbf{y}}$$

The $u_z$ addition law is, of course, the same as Equation (11–24). The inverse transformation is

$$u_z' = \frac{u_z - v}{1 - (v/c^2)u_z} \qquad \mathbf{u}_\mathrm{T}' = \frac{\mathbf{u}_\mathrm{T}}{\gamma[1 - (v/c^2)u_z]} \tag{11–29}$$

The Lorentz transformation of a velocity less than $c$ never leads to a velocity greater than $c$. For example, if $u_z' = \frac{9}{10}c$ and $v = \frac{9}{10}c$, we find, from Equation (11–28),

$$u_z = \frac{\frac{9}{10}c + \frac{9}{10}c}{1 + \frac{81}{100}} = \frac{180}{181}c$$

Whereas the Galilean transformation would have given

$$u_z = u_z' + v = \tfrac{9}{5}c$$

## 11-1B Nonrelativistic Limit of the Lorentz Transformation

The Lorentz transformation reduces to the Galilean transformation when the transformation velocity is much less than the velocity of light. For $v \ll c$ we can make a binomial series expansion of $\gamma$ in powers of $(v/c)^2$:

$$\gamma = \left[1 - \left(\frac{v}{c}\right)^2\right]^{-1/2} = 1 + \frac{1}{2}\left(\frac{v}{c}\right)^2 + \cdots \tag{11–30}$$

The Lorentz transformation of Equation (11–22) becomes

$$z = z' + vt' + O(v^2) \qquad t = t' + \frac{v}{c^2}z' + O(v^2) \tag{11–31}$$

Through terms of first order in $v$, the first equations of Equations (11–31) and (11–5) are equivalent. However, the latter of the equations are different: $t = t' + (v/c^2)z'$ compared with $t = t'$. The term $(v/c^2)z'$ represents a position-dependent difference in the origin of the times of the two frames $S$ and $S'$ (synchronization). It makes a difference only for phenomena that link different positions. An example is light sent between points $A$ and $B$ on the

$z$ axis. It is only because of the term $(v/c^2)z'$ relating $t$ and $t'$ that the velocity of light going from $A$ to $B$ and from $B$ to $A$ is the same through first order in $v/c$ in either the frame $S$ or $S'$. Otherwise, with the Galilean $t = t'$, these velocities would differ by $2v$. But for phenomena involving only things moving slowly compared with the velocity of light, the term $(v/c^2)z'$ is negligible:

$$z \approx z' + vt' \qquad t \approx t' \tag{11-32}$$

In the small-velocity limit $u_z'/c \ll 1$, $v/c \ll 1$, Equation (11–28) becomes

$$u_z \approx u_z' + v \qquad \mathbf{u}_\mathrm{T} \approx \mathbf{u}_\mathrm{T}' \tag{11-33}$$

which is the Galilean result obtained by time differentiation of Equation (11–5).

## 11-1C Simultaneity

A remarkable consequence of the Lorentz transformation is that events simultaneous to an observer using frame $S$ are generally not simultaneous to an observer using frame $S'$, which is moving with velocity $v$ with respect to $S$.

We consider two events $A$ and $B$ at locations $(x_A, y_A, z_A)$ and $(x_B, y_B, z_B)$ that appear to occur at the same instant of time $t_A = t_B$ to an observer at rest in the $S$ frame. From the inverse Lorentz transformation of Equation (11–23), an observer at rest in $S'$ sees the events occurring at the following space-time coordinates:

$$\begin{aligned} ct_A' &= \gamma\left(ct_A - \frac{v}{c} z_A\right) & ct_B' &= \gamma\left(ct_B - \frac{v}{c} z_B\right) \\ z_A' &= \gamma(z_A - vt_A) & z_B' &= \gamma(z_B - vt_B) \\ y_A' &= y_A & y_B' &= y_B \\ x_A' &= x_A & x_B' &= x_B \end{aligned} \tag{11-34}$$

The time interval between the events as seen by the observer in $S'$ is

$$c(t_A' - t_B') = \gamma c(t_A - t_B) - \gamma \frac{v}{c}(z_A - z_B) \tag{11-35}$$

Setting $t_A = t_B$, we obtain

$$c(t_A' - t_B') = -\gamma \frac{v}{c}(z_A - z_B) \tag{11-36}$$

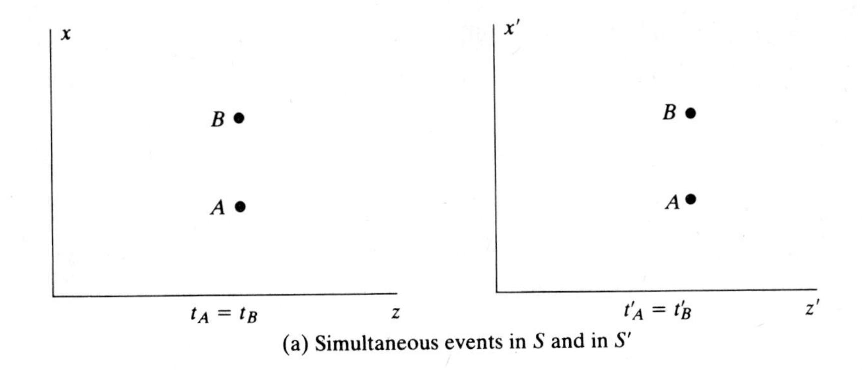


(a) Simultaneous events in $S$ and in $S'$

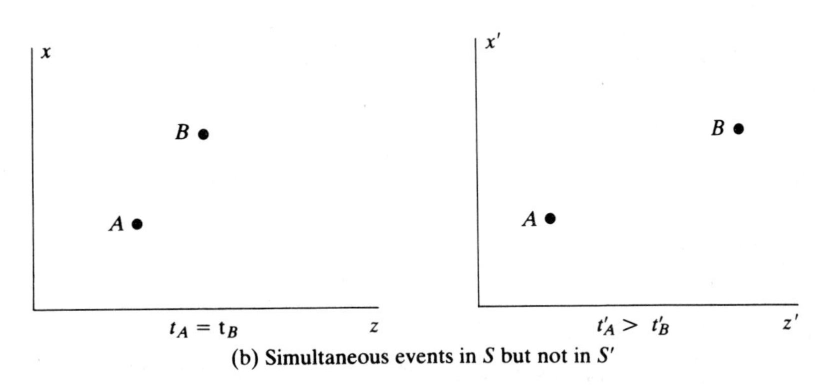


(b) Simultaneous events in $S$ but not in $S'$

**FIGURE 11–5** Simultaneity. Events that are simultaneous in one coordinate frame (a) may or (b) may not be simultaneous when viewed from a moving frame.

Only if the two events occur at the same $z$ location in $S$, $z_A = z_B$, will the events also appear simultaneous to the observer in $S'$. In general, simultaneous events for an observer in $S$ will *not* be simultaneous for an observer in $S'$. The time interval $t'_A - t'_B$ between events $A$ and $B$ as seen in the $S'$ frame is positive if $z_B > z_{A'}$ as illustrated in Figure 11–5, and negative if $z_A > z_B$.

## 11-1D Time Dilation

Next, consider two events that occur at the same $z$ location in the $S$ frame, $z_A = z_B$, but at different times, $t_A \neq t_B$. For example, the two events could be two successive ticks of a clock at rest in $S$. According to the results of the Lorentz transformation, as given in Equation (11–35), the time interval between the events in $S'$ is

$$t'_A - t'_B = \gamma(t_A - t_B) \tag{11-37}$$

which we can write more compactly as

$$\Delta t' = \gamma\, \Delta t \tag{11-38}$$

The time interval $\Delta t'$ in $S'$ is longer by a factor of $\gamma$ (i.e., dilated) than the time interval $\Delta t$ in the system in which the clock is at rest. This situation is illustrated in Figure 11–6. The time interval in the system in which the clock is at rest is called the "proper time" of the clock. (*Proper* in this expression is not being used in the sense of "correct" or "fitting" but, rather, in the sense of "belonging to," as in the word *property*.)

If, instead, the clock were at rest in $S'$, $z'_A = z'_B$, then from the Lorentz transformation of Equation (11–22) we have

$$ct_A = \gamma\left(t'_A + \frac{v}{c} z'_A\right) \qquad ct_B = \gamma\left(t'_B + \frac{v}{c} z'_B\right) \tag{11-39}$$

These relations yield

$$t_A - t_B = \gamma(t'_A - t'_B) \tag{11-40}$$

or more compactly,

$$\Delta t = \gamma\, \Delta t' \tag{11-41}$$

This result is not unexpected since interchanging the roles of the frames is equivalent to changing the transformation to the inverse transformation, which means changing $v$ to $-v$ and, in Equation (11–38), $\gamma(-v) = \gamma(v)$. The clock that is moving always runs slower.

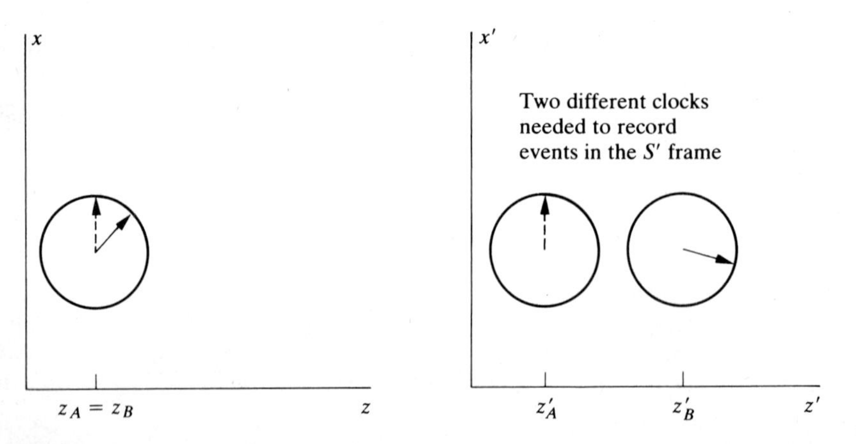


**FIGURE 11–6** Time dilation. Time intervals are longer if viewed from a moving reference frame.

## 11-1E Experimental Verification of Time Dilation

The positively charged pi meson ($\pi^+$) decays into a positively charged muon ($\mu^+$) and a neutrino ($\nu$), $\pi^+ \rightarrow \mu^+ + \nu$. A collection of $\pi^+$ mesons at rest obeys, on the average, an exponential decay law,

$$N(t') = N(0)e^{-t'/\tau_{\pi^+}} \qquad \textbf{(11-42)}$$

where $N(t')$ is the number of $\pi^+$ mesons still existing at time $t'$, out of a number $N(0)$ present at time $t' = 0$. This exponential decay law can be derived from the theory of quantum mechanics. The decay law can be expressed in differential form as

$$\frac{dN}{N} = -\frac{dt'}{\tau_{\pi^+}} \qquad \textbf{(11-43)}$$

The constant $\tau_{\pi^+}$ is called the *mean lifetime* of the $\pi^+$ meson. The experimental value of the mean lifetime is

$$\tau_{\pi^+} \approx 2.6 \times 10^{-8}\text{ s} \qquad \textbf{(11-44)}$$

Suppose now we consider the decay of $\pi^+$ mesons that are moving with velocity $v$ with respect to the observer's rest frame $S$. The number of $\pi^+$ mesons present is independent of the frame in which the mesons are observed. In the $S'$ frame for which the mesons are at rest, the decay law is Equation (11–42). To an observer at rest in $S$ the time interval in which a number $dN$ are observed to decay is

$$dt = \gamma\, dt' \qquad \textbf{(11-45)}$$

according to time dilation. From Equations (11–43) and (11–45) we have

$$\frac{dN}{N} = -\frac{dt/\gamma}{\tau_{\pi^+}} \qquad \textbf{(11-46)}$$

In integrated form the decay law in the $S$ frame is

$$N(t) = N(0)e^{-t/(\gamma\tau_{\pi^+})} \qquad \textbf{(11-47)}$$

Hence the observed lifetime in the $S'$ frame of $\pi^+$ mesons moving with velocity $v$ in the $S$ frame is

$$\tau_{\pi^+}(v) = \gamma\tau_{\pi^+} \qquad \textbf{(11-48)}$$

The $\pi^+$ mesons in motion live, on the average, a factor of $\gamma$ times longer than $\pi^+$ mesons at rest, as illustrated in Figure 11–7. At the Fermi National Accelerator Laboratory in Batavia, Illinois, $\pi^+$ mesons can be

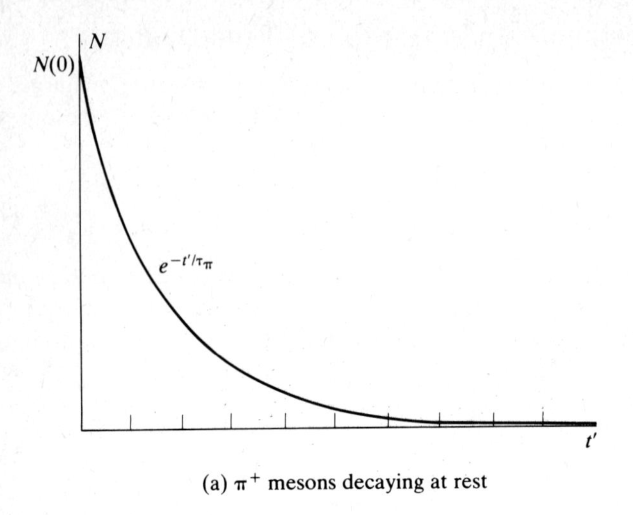


(a) $\pi^+$ mesons decaying at rest

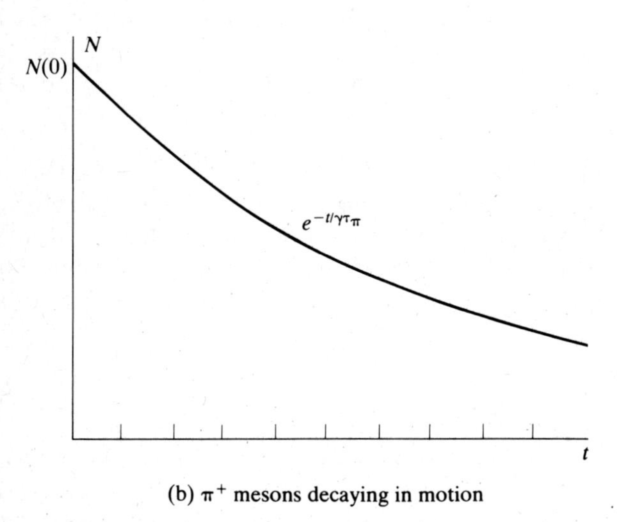


(b) $\pi^+$ mesons decaying in motion

**FIGURE 11-7** Pi mesons remaining when viewed (a) at rest and (b) in motion. The mesons appear to live longer if they are moving.

produced with

$$\gamma = 2000 \tag{11-49}$$

corresponding to a velocity

$$v = 0.999999875c \tag{11-50}$$

The mean lifetime of the $\pi^+$ mesons in the laboratory is then

$$\tau_{\pi^+}(v) = (2000)(2.6 \times 10^{-8}\text{ s}) = 5.2 \times 10^{-5}\text{ s} \tag{11-51}$$

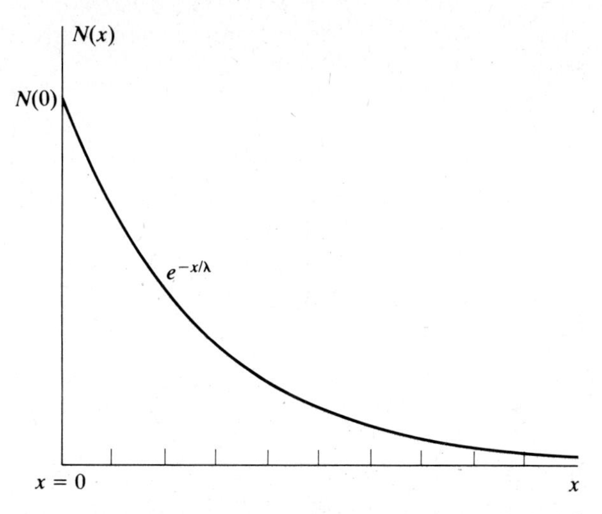


**FIGURE 11–8** Pi mesons that have not decayed after they have moved a distance $x$

The distance $x$ that a $\pi^+$ meson with velocity $v$ in the $S$ frame moves in time $t$ is $x = vt$. The decay law of Equation (11–47) can be written as a function of distance as

$$N(x) = N(0)e^{-x/v\gamma\tau} \tag{11-52}$$

where $N(x)$ is the number of mesons decaying at distance $x$, as illustrated in Figure 11–8. The quantity

$$\lambda = v\gamma\tau_{\pi^+} = vt_{\pi^+}(v) \tag{11-53}$$

is called the mean decay length. Without time dilation the mean decay length is 7.8 m. The mean decay length of $\pi^+$ mesons with $\gamma = 2000$ is

$$\lambda \approx ct_{\pi^+}(v) = (3 \times 10^8 \text{ m/s})(5.2 \times 10^{-5} \text{ s}) = 15.6 \text{ km} \tag{11-54}$$

Even though the decay lifetime in their rest frame is very short, the $\pi^+$ mesons go a long way before decaying. The fact that $\pi^+$ mesons at Fermilab travel several kilometers before decaying is a dramatic confirmation of the time dilation. Any clock, including biological clocks, will show time dilation when in motion compared with at rest.

## 11-1F Length Contraction

Suppose a rod of rest length $L$ is moving lengthwise with velocity $v$ relative to an observer at rest in the $S$ frame. The rod is at rest with respect to the

$S'$ frame, as illustrated in Figure 11–9. An observer at rest in $S'$ sees the two ends of the rod at $z$ locations $z'_A$ and $z'_{B'}$ with

$$z'_A - z'_B = L \tag{11-55}$$

An observer in the $S$ frame makes a simultaneous measurement (i.e., $t_A = t_B$) of the two ends of the rod. According to the inverse Lorentz transformation,

$$z'_A = \gamma(z_A - vt_A) \qquad z'_B = \gamma(z_B - vt_B) \tag{11-56}$$

Hence we find

$$z'_A - z'_B = \gamma(z_A - z_B) \tag{11-57}$$

Since $z_A - z_B = L(v)$ is the length of the rod moving with speed $v$ relative to the observer, we conclude that the lengths of the rod in the two frames are related as

$$L(v) = \frac{1}{\gamma} L \tag{11-58}$$

Thus the length of a rod that is in motion relative to the observer appears to be contracted by a factor $1/\gamma$ from its rest length.

If, instead, the rod was at rest in the $S$ frame, we would have

$$z_A - z_B = L \tag{11-59}$$

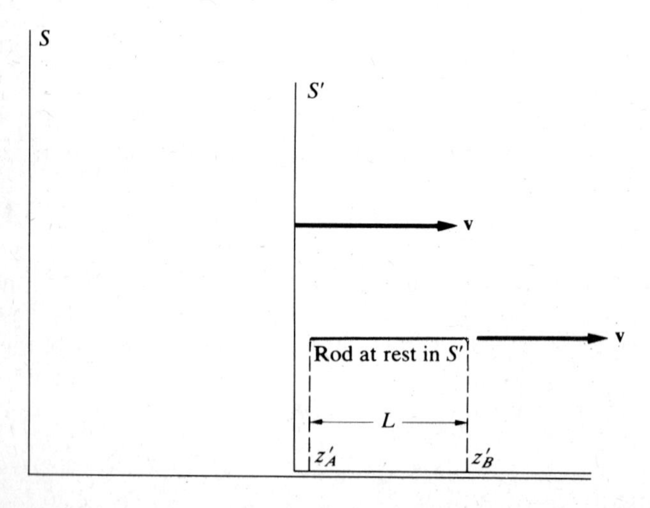


**FIGURE 11-9** Rod of length $L$ at rest in $S'$. From the reference frame $S$ it appears shorter by $\gamma^{-1}$

To determine the length of the rod in the $S'$ frame, an observer in that frame could photograph the ends of the rod simultaneously at time $t'_A = t'_B$. Taking the difference of the Lorentz transformations,

$$z_A = \gamma(z'_A + vt'_A) \qquad z_B = \gamma(z'_B + vt'_B) \tag{11-60}$$

we find

$$L = z_A - z_B = \gamma(z'_A - z'_B) \tag{11-61}$$

Since the rod moves with velocity $-v$ relative to the $S'$ frame, as shown in Figure 11–10, we denote its length in the $S'$ frame by $L(-v)$:

$$z'_A - z'_B = L(-v) \tag{11-62}$$

Thus Equation (11–61) gives

$$L(-v) = \frac{1}{\gamma} L \tag{11-63}$$

As with time dilation, we could have used the argument that interchanging frames is equivalent to $v \to -v$, and hence Equation (11–63) could have been deduced from Equation (11–58).

All observers would agree that the length of an object, as measured from a frame in motion relative to the object, appears to be shorter than the

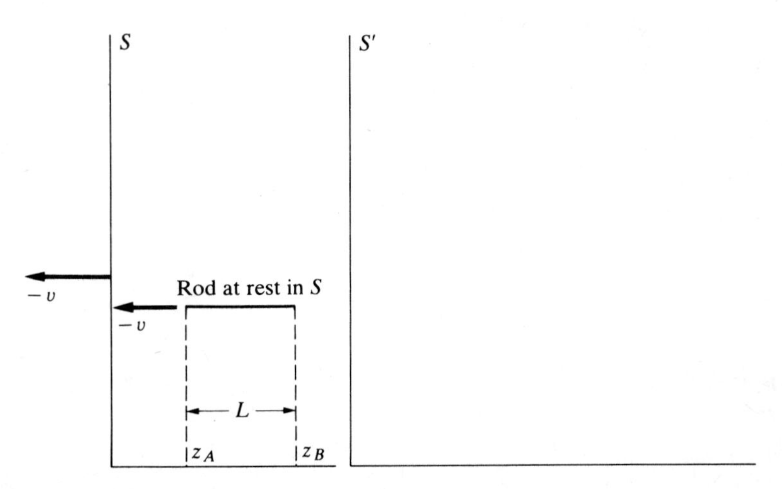


**FIGURE 11-10** Rod of length $L$ at rest in $S$. When it is viewed from frame $S'$, the rod has a length $L/\gamma$.

length in the rest frame. The rest length is called the "proper length" of the object. This conclusion about the length contraction is not contradictory since a simultaneous measurement to one observer does not appear to have been made simultaneously to an observer in a different frame. That is, for $z_A \neq z_B$ and $z'_A \neq z'_B$, $t_A = t_B$ implies $t'_A \neq t'_B$ in the Lorentz transformation; similarly, $t'_A = t'_B$ implies that $t_A \neq t_B$.

## 11-2 LORENTZ TRANSFORMATION OF FIELDS AND SOURCES

In ordinary three-dimensional coordinate space, dot products of vectors are invariant under rotations of coordinate axes. For example, $(\mathbf{r}_1 - \mathbf{r}_2)\cdot(\mathbf{r}_1 - \mathbf{r}_2)$ has the same value in any rotated coordinate frame. A similar invariance holds for other quantities that transform like vectors, such as the operator $\nabla$; the Laplacian operator $\nabla^2 = \nabla \cdot \nabla$ has the same form in any Cartesian frame; that is, in the $S$ frame

$$\nabla^2 = \frac{\partial^2}{\partial x^2} + \frac{\partial^2}{\partial y^2} + \frac{\partial^2}{\partial z^2} \tag{11-64a}$$

and in the rotated frame $S'$

$$\nabla'^2 = \frac{\partial^2}{\partial x'^2} + \frac{\partial^2}{\partial y'^2} + \frac{\partial^2}{\partial z'^2} \tag{11-64b}$$

### 11-2A Covariance

In four-dimensional Minkowski space the dot product of two four-dimensional vectors is invariant under rotations. We introduce the notation

$$x \equiv (x_1, x_2, x_3, x_4) \equiv (x, y, z, ict) \tag{11-65}$$

for the space-time coordinate of a point. Then, for example, the interval between two space-time points,

$$(x - x')\cdot(x - x') = (x_1 - x'_1)^2 + (x_2 - x'_2)^2 + (x_3 - x'_3)^2 + (x_4 - x'_4)^2 \tag{11-66}$$

is invariant under Lorentz transformations (i.e., rotations in Minkowski space). By definition, the components of a general four-vector,

$$a \equiv (a_1, a_2, a_3, a_4) \tag{11-67}$$

transform under four-dimensional rotations in the same way as the components of the four-vector $x$. The dot product of 2 four-vectors $a$ and $b$ is

**Einstein Summation Convention**

$$a \cdot b = a_1 b_1 + a_2 b_2 + a_3 b_3 + a_4 b_4$$
$$= \sum_{\mu=1}^{4} a_\mu b_\mu$$
$$= a_\mu b_\mu \tag{11-68}$$

where a summation over the repeated vector index $\mu$ is implied in $a_\mu b_\mu$. This notation is the Einstein summation convention.

The four-dimensional generalization of the del operator is

$$\square \equiv \left(\frac{\partial}{\partial x_1}, \frac{\partial}{\partial x_2}, \frac{\partial}{\partial x_3}, \frac{\partial}{\partial x_4}\right) = \left(\frac{\partial}{\partial x}, \frac{\partial}{\partial y}, \frac{\partial}{\partial z}, \frac{1}{ic}\frac{\partial}{\partial t}\right) \tag{11-69}$$

The four-dimensional analog of the Laplacian operator is the *D'Alembertian operator*,

$$\square^2 \equiv \square_\mu \square_\mu = \frac{\partial^2}{\partial x_1^2} + \frac{\partial^2}{\partial x_2^2} + \frac{\partial^2}{\partial x_3^2} + \frac{\partial^2}{\partial x_4^2}$$
$$= \frac{\partial^2}{\partial x^2} + \frac{\partial^2}{\partial y^2} + \frac{\partial^2}{\partial z^2} - \frac{1}{c^2}\frac{\partial^2}{\partial t^2} \tag{11-70a}$$

Since the D'Alembertian is the dot product of 2 four-vector operators, it has the same form in a frame $S'$ related to $S$ by a four-dimensional rotation (Lorentz transformation):

$$\square'^2 = \frac{\partial^2}{\partial x'^2} + \frac{\partial^2}{\partial y'^2} + \frac{\partial^2}{\partial z'^2} - \frac{1}{c^2}\frac{\partial^2}{\partial t'^2} \tag{11-70b}$$

In ordinary three-dimensional coordinate space, the laws of physics are invariant to spatial rotations. A necessary condition for this invariance to be true is that the laws of physics be expressed as vector (or tensor) equations. For example, if $S$ and $S'$ are inertial frames related by a rotation of coordinate space axes, Newton's law of motion for a particle of mass $m$ expressed as the vector equation

$$\mathbf{F} = m\ddot{\mathbf{r}} \tag{11-71}$$

holds equally for $S$ and $S'$ frames. For example, the components of Newton's law in terms of $S$ frame quantities read

$$F_x = m\ddot{x} \qquad F_y = m\ddot{y} \qquad F_z = m\ddot{z} \tag{11-72}$$

In terms of $S'$ frame quantities, where $S'$ is a frame rotated by an angle $\theta$ about the $z$ axis [see Equation (11–8)], these equations read

$$F_{x'} \cos\theta - F_{y'} \sin\theta = m(\ddot{x}' \cos\theta - \ddot{y}' \sin\theta)$$
$$F_{x'} \sin\theta + F_{y'} \cos\theta = m(\ddot{x}' \sin\theta + \ddot{y}' \cos\theta)$$
$$F_{z'} = F_z \qquad \textbf{(11–73)}$$

These equations have different forms from the original equations in frame $S$. However, because the vectors $\mathbf{F}$ and $\ddot{\mathbf{r}}$ transform the same way, linear combinations of these equations can be taken, to arrive at

$$F_{x'} = m\ddot{x}' \qquad F_{y'} = m\ddot{y}' \qquad F_{z'} = m\ddot{z}' \qquad \textbf{(11–74)}$$

which have the same forms as the original equations in the $S$ frame.

Similar considerations apply in four-dimensional Minkowski space. In order that the laws of physics have the same form in two frames related by a Lorentz transformation of the space-time coordinates, the equations must be expressed as scalars, four-vectors, or tensor products of four-vectors. We now proceed to show that Maxwell's electromagnetic theory can be expressed in such a covariant form.

## 11-2B Current and Charge Density Transformations

We begin with the equation of continuity, Equation (5–14),

$$\nabla \cdot \mathbf{J} + \frac{\partial \rho}{\partial t} = 0$$

which we rewrite as

$$\nabla \cdot \mathbf{J} + \frac{\partial}{\partial(ict)}(ic\rho) = 0 \qquad \textbf{(11–75)}$$

Recognizing that $\nabla$ and $\partial/\partial(ict)$ are components of the $\square_\mu$ operator, we see that the covariant form of Equation (11–75) must be

**Equation of Continuity**

$$\square_\mu J_\mu = 0 \qquad \textbf{(11–76)}$$

with components of the four-vector current density $J_\mu$ identified as

$$J = (J_x, J_y, J_z, ic\rho) \qquad \textbf{(11–77)}$$

Therefore the components of $J$ must transform under rotations in four-space in the same way as the components of $x$. This result could be directly verified, using the definitions of $\mathbf{J}$ and $\rho$ in the equation of continuity and the Lorentz transformations of $\mathbf{r}$ and $t$. For a Lorentz boost in the $z$ direction the sources will transform in correspondence with Equation (11–22) as

$$J_x = J'_x \qquad J_y = J'_y \qquad J_z = \gamma(J'_z + v\rho') \qquad \rho = \gamma\left(\rho' + \frac{v}{c^2} J'_z\right) \tag{11-78}$$

## 11-2C Covariant Form of Maxwell's Equations

Turning next to the wave equations, we rewrite Equations (10–197) and (10–199) as

$$\Box^2 \mathbf{E} = \mu_0 \left( c^2 \, \nabla\rho + \frac{\partial \mathbf{J}}{\partial t} \right) \tag{11-79a}$$

$$\Box^2 \mathbf{B} = -\mu_0 \nabla \times \mathbf{J} \tag{11-79b}$$

First, consider the $z$ component of the wave equation for $\mathbf{B}$:

$$\Box^2 B_3 = -\mu_0(\nabla_1 J_2 - \nabla_2 J_1) = -\mu_0(\Box_1 J_2 - \Box_2 J_1) \tag{11-80}$$

Anticipating that this expression will be a tensor equation in its covariant form, we introduce a tensor $F_{\mu\nu}$ with component $F_{12}$, defined as

$$F_{12} \equiv B_3 \tag{11-81}$$

Then Equation (11–80) is the $\mu = 1$, $\nu = 2$ component of

$$\Box^2 F_{\mu\nu} = -\mu_0(\Box_\mu J_\nu - \Box_\nu J_\mu) \tag{11-82}$$

Since the right-hand side of Equation (11–82) is antisymmetric under interchange of $\mu$ and $\nu$, we must also have

$$F_{\mu\nu} = -F_{\nu\mu} \tag{11-83}$$

As a consequence, all diagonal components of the $F_{\mu\nu}$ tensor vanish:

$$F_{11} = F_{22} = F_{33} = F_{44} = 0 \tag{11-84}$$

We now show that all components of the wave equations, Equations (11–79), are components of Equation (11–82). The $x$ and $y$ components of Equation (11–79b) are similarly given by Equation (11–82), if we define

$$F_{23} = B_1 \tag{11-85a}$$

$$F_{31} = B_2 \tag{11-85b}$$

The $\mu = 1$, $\nu = 4$ component of Equation (11–82) is

$$\Box^2 F_{14} = -\mu_0(\Box_1 J_4 - \Box_4 J_1) = -\mu_0 \left( ic\,\nabla_1 \rho - \frac{1}{ic}\frac{\partial}{\partial t} J_1 \right)$$

$$= \frac{\mu_0}{ic}\left( c_2\,\nabla_1 \rho + \frac{\partial J_1}{\partial t} \right) \tag{11-86}$$

The right-hand side of Equation (11–86) is the same as that of Equation (11–79a), so we make the identification

$$F_{14} = \frac{E_1}{ic} \tag{11-87a}$$

and similarly,

$$F_{24} = \frac{E_2}{ic} \tag{11-87b}$$

$$F_{34} = \frac{E_3}{ic} \tag{11-87c}$$

Then Equation (11–82) is equivalent to Equations (11–79) for **E** and **B**, but written in covariant form. In matrix representation the electromagnetic field tensor is

**Electromagnetic Field Tensor**

$$F_{\mu\nu} = \begin{bmatrix} 0 & B_3 & -B_2 & \dfrac{E_1}{ic} \\ -B_3 & 0 & B_1 & \dfrac{E_2}{ic} \\ B_2 & -B_1 & 0 & \dfrac{E_3}{ic} \\ -\dfrac{E_1}{ic} & -\dfrac{E_2}{ic} & -\dfrac{E_3}{ic} & 0 \end{bmatrix} \tag{11-88}$$

Maxwell's equations themselves can likewise be expressed in covariant form, using the tensor $F_{\mu\nu}$. The result is as follows:

**Covariant Form of Maxwell's Equations**

$$\Box_\nu F_{\mu\nu} = \mu_0 J_\mu \tag{11-89a}$$

$$\frac{1}{2}\varepsilon_{\alpha\beta\mu\nu}\Box_\beta F_{\mu\nu} = 0 \tag{11-89b}$$

where $\varepsilon_{\alpha\beta\mu\nu}$ is an antisymmetric unit tensor ($\varepsilon = +1$ for cyclic indices, e.g., 1234; $\varepsilon = -1$ for anticyclic indices, e.g., 2134; and $\varepsilon = 0$ otherwise).

We now verify that these tensor equations are the Maxwell equations. Taking $\mu = 3$ in Equation (11–89a), we obtain

$$\sum_{i=1}^{3} \frac{\partial}{\partial x_i} F_{3i} + \frac{\partial}{\partial(ict)} F_{34} = \mu_0 J_3 \tag{11-90}$$

From Equation (11–88) this expression becomes

$$\frac{\partial B_2}{\partial x_1} - \frac{\partial B_1}{\partial x_2} - \frac{1}{c^2}\frac{\partial E_3}{\partial t} = \mu_0 J_3 \tag{11-91}$$

When we use $c^2 = (\mu_0\varepsilon_0)^{-1}$, this equation is identical to the three-component of Maxwell IV, in Table 8–1. A similar correspondence holds for the $\mu = 1$ and $\mu = 2$ components. With $\mu = 4$ in Equation (11–89a) we find

$$\sum_{i=1}^{3} \frac{\partial}{\partial x_i} F_{4i} = \mu_0 ic\rho \tag{11-92}$$

which is equivalent to Maxwell I.

Turning to Equation (11–89b), we first consider $\alpha = 3$. Using the antisymmetric property of Equation (11–83) to combine terms, we obtain

$$\Box_1 F_{24} + \Box_2 F_{41} + \Box_4 F_{12} = 0 \tag{11-93}$$

or by Equation (11–88)

$$\frac{1}{ic}\left(\frac{\partial E_2}{\partial x_1} - \frac{\partial E_1}{\partial x_2}\right) + \frac{1}{ic}\frac{\partial}{\partial t} B_3 = 0 \tag{11-94}$$

This equation is the third component of Maxwell III; $\alpha = 1$ and $\alpha = 2$ components also correspond to Maxwell III. With $\alpha = 4$ in Equation (11–89b), we obtain

$$\Box_1 F_{23} + \Box_2 F_{31} + \Box_3 F_{12} = 0 \tag{11-95}$$

which reduces to

$$\frac{\partial B_1}{\partial x_1} + \frac{\partial B_2}{\partial x_2} + \frac{\partial B_3}{\partial x_3} = 0 \tag{11-96}$$

which is Maxwell II. This step completes the verification of the equivalence of Equations (11–89) to Maxwell's equations.

## 11-2D Field Transformations

Since $E_i$ and $B_i$ are components of the same tensor $F_{\mu\nu}$ of Equation (11–88), we might expect that a Lorentz transformation will involve both **E** and **B** fields. The Lorentz transformation in the $z$ direction in Equation (11–22) can be written as a matrix equation,

$$x_\mu = a_{\mu\nu} x'_\nu \tag{11-97}$$

where the matrix $a_{\mu\nu}$ is

$$a_{\mu\nu} = \begin{bmatrix} 1 & 0 & 0 & 0 \\ 0 & 1 & 0 & 0 \\ 0 & 0 & \gamma & -i\gamma \dfrac{v}{c} \\ 0 & 0 & i\gamma \dfrac{v}{c} & \gamma \end{bmatrix} \tag{11-98}$$

Any four-vector transforms in the same way. Each index of a tensor transforms like a four-vector. Hence the Lorentz transform of $F_{\alpha\beta}$ is

$$F_{\mu\nu} = a_{\mu\alpha} a_{\nu\beta} F'_{\alpha\beta} \tag{11-99}$$

Using Equation (11–98), we find

$$F_{12} = F'_{12} \qquad F_{13} = \gamma F'_{13} - \frac{i\gamma v}{c} F'_{14} \qquad F_{14} = \frac{i\gamma v}{c} F'_{13} + \gamma F'_{14}$$

$$F_{23} = \gamma F'_{23} - \frac{i\gamma v}{c} F'_{24} \qquad F_{24} = \frac{i\gamma v}{c} F'_{23} + \gamma F'_{24} \tag{11-100}$$

$$F_{34} = \gamma^2 F'_{34} + \frac{\gamma^2 v^2}{c^2} F'_{43} = F'_{34}$$

Expressing this result in **E** and **B** components from Equation (11–88), we obtain the field transformations

$$E_x = \gamma(E'_x + vB'_y) \qquad B_x = \gamma\left(B'_x - \frac{v}{c^2} E'_y\right)$$

$$E_y = \gamma(E'_y - vB'_x) \qquad B_y = \gamma\left(B'_y + \frac{v}{c^2} E'_x\right) \tag{11-101}$$

$$E_z = E'_z \qquad B_z = B'_z$$

where the $S'$ frame moves along the $z$ axis with velocity $v$ relative to the $S$ frame.

The inverse field transformations can be found from Equation (11–101) either by directly solving or by just interchanging primed and unprimed fields and letting $v \to -v$. Since the transformation velocity is in the $z$ direction $\mathbf{v} = v\hat{\mathbf{z}}$, we have

$$(\mathbf{v} \times \mathbf{B}')_x = -vB'_y \qquad (\mathbf{v} \times \mathbf{B}')_y = vB'_x \tag{11-102}$$

Hence we can write the field transformations in the form

$$\mathbf{E}_\mathrm{T} = \gamma\left(\mathbf{E}'_\mathrm{T} - \frac{\mathbf{v}}{c} \times c\mathbf{B}'\right) \qquad \mathbf{E}_{/\!/} = \mathbf{E}'_{/\!/}$$

$$c\mathbf{B}_T = \gamma\left(c\mathbf{B}'_T + \frac{\mathbf{v}}{c} \times \mathbf{E}'\right) \qquad c\mathbf{B}_{/\!/} = c\mathbf{B}'_{/\!/} \tag{11-103}$$

where $_T$ and $_{/\!/}$ denote directions transverse and parallel to **v**, respectively.

## 11-2E Invariance of Electric Charge

Charge is a physical attribute of elementary particles, such as the electron and proton, which is experimentally found to be independent of the velocity of the particle. In atoms the electrons and the protons move at different velocities; for example, in lead the innermost electrons travel with velocities that are a significant fraction of the speed of light. Since atoms are electrically neutral, this behavior also establishes that charge is independent of velocity.

We will now verify theoretically this experimental fact by showing that charge is invariant under a Lorentz transformation. We consider charges at rest in the $S'$ frame:

$$J'_z = 0 \tag{11-104}$$

The total charge in the rest frame of the charges is

$$Q' = \int \rho'\, dx'\, dy'\, dz' \tag{11-105}$$

In the $S$ frame the charge density is given by Equation (11–78) as

$$\rho = \gamma\rho' \tag{11-106}$$

The differential coordinate elements in the $S$ frame, as determined at the same instant of time in the $S$ frame, are

$$dx = dx' \qquad dy = dy' \qquad dz = \frac{1}{\gamma}\, dz' \tag{11-107}$$

The length of the $z$ differential is contracted by a factor $1/\gamma$. Thus the total charge in the $S$ frame is

$$Q = \int \rho\, dx\, dy\, dz = \int (\gamma\rho')\, dx'\, dy' \left(\frac{dz'}{\gamma}\right) = \int \rho'\, dx'\, dy'\, dz' = Q' \tag{11-108}$$

Thereby we have established the invariance of the charge to a Lorentz boost.

# 11-3 APPLICATIONS OF LORENTZ TRANSFORMATION FOR FIELDS AND SOURCES

When electric charges are at rest, the fields are electrostatic. For an observer in motion relative to these charges, the charges are moving. For this observer there are magnetic as well as electric fields. Hence Coulomb's law and the Biot-Savart law are intimately connected. The electromagnetic field of a moving charge can be calculated by Lorentz-transforming the electrostatic field of a charge at rest. In this section the Lorentz transformation of fields and sources is illustrated in some cases of physical interest.

## 11-3A Fields of a Point Charge Moving with Constant Velocity

We consider the problem of finding the electric and magnetic fields of a uniformly moving charge. If a charge $q$ is at rest at the origin of the $S'$ frame, its electric and magnetic fields are

$$\mathbf{E}' = \frac{q\mathbf{r}'}{4\pi\varepsilon_0 r'^3} \qquad \mathbf{B}' = 0 \tag{11-109}$$

where $\mathbf{r}' = \boldsymbol{\imath}' + z'\hat{\mathbf{z}}$. In frame $S$ the charge moves with constant velocity $\mathbf{v} = v\hat{\mathbf{z}}$ along the $z$ axis. The field transformations of Equations (11–103) provide the electric field in the $S$ frame:

$$E_z = E_z' = \frac{qz'}{4\pi\varepsilon_0 r'^3} \qquad \mathbf{E}_\mathrm{T} = \gamma\mathbf{E}_\mathrm{T}' = \frac{\gamma q\boldsymbol{\imath}'}{4\pi\varepsilon_0 r'^3} \tag{11-110}$$

where

$$\boldsymbol{\imath}' = x'\hat{\mathbf{x}} + y'\hat{\mathbf{y}}$$

From the coordinate transformation of Equation (11–23),

$$z' = \gamma(z - vt) \qquad \boldsymbol{\imath}' = \boldsymbol{\imath} \tag{11-111}$$

and hence $\mathbf{r}' = \imath\hat{\boldsymbol{\imath}} + \gamma(z - vt)\hat{\mathbf{z}}$. The electric field in $S$ is then

$$\mathbf{E} = \frac{q}{4\pi\varepsilon_0 r'^3}\gamma[\boldsymbol{\imath} + \hat{\mathbf{z}}(z - vt)] \tag{11-112}$$

For convenience we introduce the relative coordinate of the field point and the charge point,

$$\mathbf{r}_\mathrm{p} = \boldsymbol{\imath} + (z - vt)\hat{\mathbf{z}} \tag{11-113}$$

which is a vector to the field point measured in frame $S$ from the origin of $S'$ at time $t$; see Figure 11–11. The electric field of Equation (11–112) can be expressed in terms of $\mathbf{r}_\mathrm{p}$ as

$$\mathbf{E} = \frac{\gamma q\mathbf{r}_\mathrm{p}}{4\pi\varepsilon_0 r'^3} \tag{11-114}$$

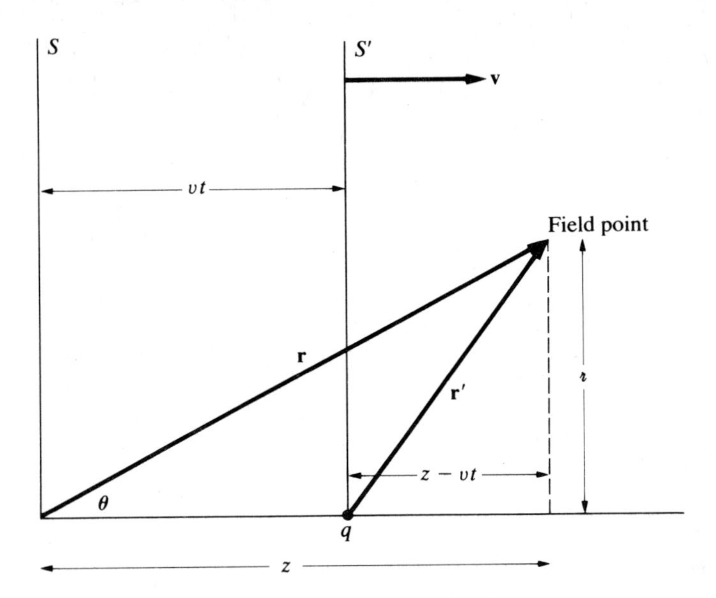


FIGURE 11–11 Charge at rest in $S'$. In this frame it has a coulomb electric field. The field transformations then give the field of a moving charge.

where $r'^2 = \imath^2 + \gamma^2(z - vt)^2$. Note that the electric field is along $\mathbf{r}_p$, a vector from the instantaneous location of the charge in $S$ to the point where $\mathbf{E}$ is measured in $S$. The magnetic field in $S$ found from Equation (11–109) and the field transformations of Equation (11–103) is

$$\mathbf{B} = \frac{\gamma \mathbf{v} \times \mathbf{E}'}{c^2} = \frac{\gamma q \mathbf{v} \times \mathbf{r}'}{4\pi\varepsilon_0 r'^3 c^2} = \frac{\gamma q \mathbf{v} \times \mathbf{r}_p}{4\pi\varepsilon_0 r'^3 c^2} \tag{11–115}$$

By comparison with Equation (11–114), we have

$$\mathbf{B} = \frac{\mathbf{v} \times \mathbf{E}}{c^2} \tag{11–116}$$

The electric field lines that were isotropic in $S'$ tend to concentrate into a plane perpendicular to $\mathbf{v}$ when viewed from frame $S$. To see why, we consider, without loss of generality, time $t = 0$, that is, when $q$ is at the origin of $S$. Then as in Figure 11–11, we define

$$z = r\cos\theta \qquad \imath = r\sin\theta \tag{11–117}$$

Then from Equation (11–114) we have

$$\mathbf{E} = \frac{q\mathbf{r}_p\gamma}{4\pi\varepsilon_0 r^3(\sin^2\theta + \gamma^2\cos^2\theta)^{3/2}} \tag{11–118}$$

Since $\gamma = (1 - v^2/c^2)^{-1/2}$ and $\mathbf{r}_p = \mathbf{r}$ at $t = 0$, this expression can be further simplified to

$$E = \frac{q\hat{\mathbf{r}}}{4\pi\varepsilon_0 r^2} \frac{1 - (v^2/c^2)}{[1 - (v^2/c^2)\sin^2\theta]^{3/2}} \tag{11-119}$$

We should note that this electric field is exactly the same as that obtained from the Lienart-Wiechert potentials in Equation (10–174). This result again illustrates the consistency between relativity and Maxwell's equations. In Equation (11–119) we observe that $|\mathbf{E}|$ is maximum at fixed $|\mathbf{r}|$ when $\hat{\mathbf{r}}$ is perpendicular to $\mathbf{v}$ (i.e., $\theta = \pi/2$), and hence the lines of $\mathbf{E}$ concentrate in the $\theta = \pi/2$ plane, as shown graphically in Figure 11–12.

For low charge velocities, $v \ll c$, the $\mathbf{B}$ field in Equation (11–115)

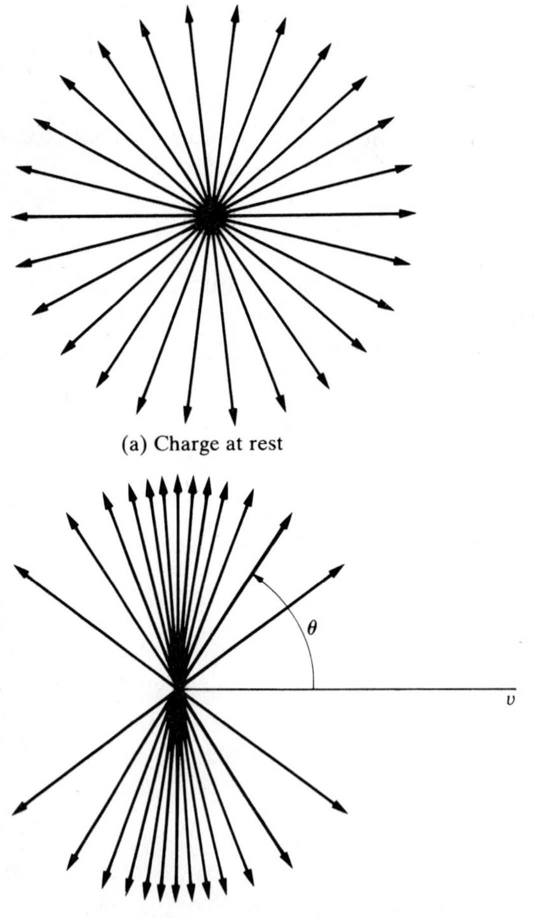


**FIGURE 11-12** Electric field of (a) a charge at rest and (b) a moving charge. The field concentrates into a plane perpendicular to the motion for the moving charge.

becomes

$$\mathbf{B} = \frac{q\mathbf{v} \times \mathbf{r}_p}{4\pi\varepsilon_0 c^2 r'^3} \tag{11-120}$$

Here $q\mathbf{v}$ is an effective current element $I\,d\mathbf{r}$, $\mathbf{r}_p$ points from the current element to the field point, and $\mathbf{r}' \approx \mathbf{r}_p$ in the nonrelativistic limit. Using $\mu_0 = (\varepsilon_0 c^2)^{-1}$, we see that Equation (11–120) reproduces the Biot-Savart law of Equation (5–25) for a current element. This result demonstrates the close relationship between the magnetic and electric fields.

## 11-3B Current Flow in a Wire

An ordinary metallic conductor consists of a fixed crystal lattice of positive ions and nearly free, negative conduction electrons. In the absence of external electric fields the random motion of the conduction electrons yields, on the average, no current flow. When an external electric field is applied, the conduction electrons move in response with an average velocity $v$ along the conductor. In the ion rest frame $S$ the number of electrons and ions per unit length is equal, so the wire is electrically neutral. If it were otherwise, compensating electrons would flow from the environment until neutral equilibrium was established.

We now analyze the fields and the currents as viewed from two frames: $S$, in which the ions are at rest [Figure 11–13(a)], and $S'$, in which the electrons are, on the average, at rest [Figure 11–13(b)]. Multiplying the transformations for charge and current density in Equation (11–78) by the cross-sectional area of the wire, we obtain the following transformation for currents and linear charge densities:

$$I_\pm = \gamma(I'_\pm + v\lambda'_\pm) \tag{11-121a}$$

$$\lambda_\pm = \gamma\left(\lambda'_\pm + \frac{v}{c^2} I'_\pm\right) \tag{11-121b}$$

where the plus and minus subscripts refer to the ions and the electrons, respectively. In $S'$ we know that $I'_- = 0$, and from Equation (11–121b)

$$\lambda_- = \gamma\lambda'_- \tag{11-122}$$

In $S$ the net charge per unit length must vanish, so from Equation (11–121b)

$$0 = \lambda_+ + \lambda_- = \lambda_+ + \gamma\lambda'_- \tag{11-123}$$

and hence

$$\lambda'_- = \frac{-\lambda_+}{\gamma} \tag{11-124}$$

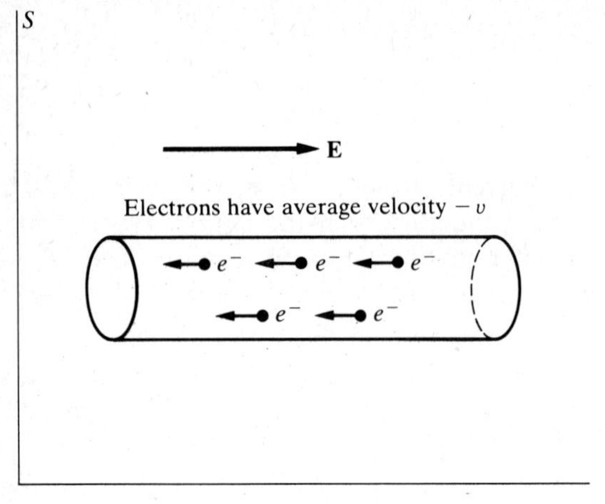


(a) Ions at rest

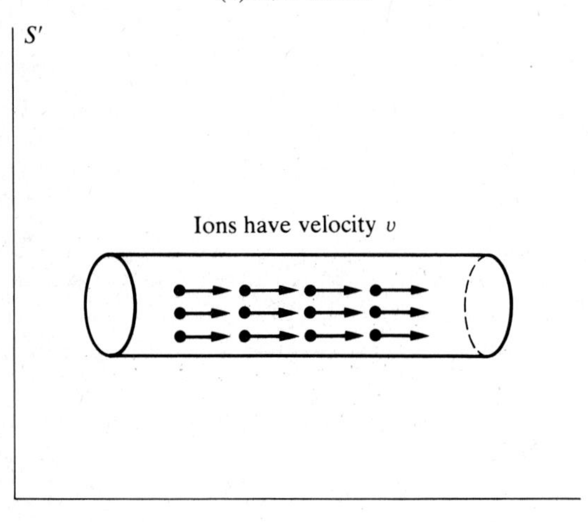


(b) Electrons have no net motion along the wire

**FIGURE 11-13** Metallic conduction as viewed from (a) the ion and (b) the electron rest frames

The fields in $S'$ due to $\lambda'_-$ are

$$\mathbf{E}'_- = \frac{\lambda'_- \hat{\boldsymbol{\imath}}}{2\pi\varepsilon_0 \imath'} \qquad \mathbf{B}'_- = 0 \tag{11-125}$$

and the fields in $S$ due to $\lambda_+$ are

$$\mathbf{E}_+ = \frac{\lambda_+ \hat{\boldsymbol{\imath}}}{2\pi\varepsilon_0 \imath} \qquad \mathbf{B}_+ = 0 \tag{11-126}$$

Using the field transformations, (11–103), and noting that $\imath' = \imath$, we find that the total fields in $S$ are

$$\mathbf{E} = \mathbf{E}_+ + \mathbf{E}_- = \frac{\hat{\boldsymbol{\imath}}}{2\pi\varepsilon_0 \imath}(\lambda_+ + \gamma\lambda'_-) = 0 \tag{11–127}$$

$$\mathbf{B} = \frac{\gamma\mathbf{v}}{c^2} \times \mathbf{E}' = \frac{\gamma\lambda'_- v}{2\pi\varepsilon_0 c^2}\frac{\hat{\mathbf{z}} \times \hat{\boldsymbol{\imath}}}{\imath} \tag{11–128}$$

Since $I'_- = 0$ by Equation (11–121a),

$$I_- = \gamma(I'_- + v\lambda'_-) = \gamma v\lambda'_- \tag{11–129}$$

Using $\hat{\mathbf{z}} \times \hat{\boldsymbol{\imath}} = \hat{\boldsymbol{\phi}}$ and $\mu_0 = (\varepsilon_0 c^2)^{-1}$, we obtain

$$\mathbf{B} = \frac{\mu_0 I_-}{2\pi\imath}\phi \qquad \mathbf{E} = 0 \tag{11–130}$$

which is exactly the Biot-Savart result for a current $I_-$ flowing in a wire.

We have seen that a magnetic field due to current flow is a relativistic effect. The remarkable fact that it is so easily observable is due to cancellation of the much larger coulomb effects by charge neutrality. There are analogous gravitational effects predicted by the theory of general relativity, but these are usually very small compared with the ordinary (Newtonian) gravitational force because negative masses do not exist, and so unlike the electrostatic force, the Newtonian force cannot be canceled by the use of opposite-sign sources.

## 11-4 DOPPLER SHIFT AND ABERRATION

A plane wave observed in frame $S$ has propagation vector $\mathbf{k}$ and frequency $\omega$, with magnitudes related by $\omega/k = c$. If this same wave is viewed from another reference frame $S'$, it will in general appear to propagate in a new direction $\mathbf{k}'$ and with a different frequency $\omega'$, with $\omega'/k' = c$. This change in direction is called *aberration*, and the frequency change is known as the *Doppler shift*. Both have played important roles in developments in physics and cosmology. In this section we derive the relevant formulas that describe these phenomena.

Suppose a plane wave has an electric field in frame $S$ given by

$$\mathbf{E}(\mathbf{r}, t) = \mathbf{E}_0 \cos(\mathbf{k} \cdot \mathbf{r} - \omega t)$$

We now consider this wave in a frame $S'$. By the Lorentz transformation the coordinates $\mathbf{r}$ and $t$ are linear combinations of $\mathbf{r}'$ and $t'$, and so the linear combination $\mathbf{k} \cdot \mathbf{r} - \omega t$ expressed in terms of $\mathbf{r}'$ and $t'$ is again a linear combination, which we can write in the form

$$\mathbf{k} \cdot \mathbf{r} - \omega t = \mathbf{k}' \cdot \mathbf{r}' - \omega' t' \tag{11–131}$$

Each side of Equation (11–131) has the form of a scalar product $\kappa \cdot x = \kappa_\mu x_\mu$

of a four-vector

$$\kappa = \left(\mathbf{k}, \frac{i\omega}{c}\right) \tag{11-132}$$

called the *wave number four-vector* with the position four-vector $x = (\mathbf{r}, ict)$. From this fact the solution of Equation (11–131) for $\mathbf{k}'$ and $\omega'$ is easy: $\mathbf{k}'$ and $\omega'$ are related to $\mathbf{k}$ and $\omega$ by the same Lorentz transformation that relates $\mathbf{r}'$ and $t'$ to $\mathbf{r}$ and $t$, because this transformation yields Equation (11–131).

We choose our coordinate system such that $\mathbf{k}$ and $\mathbf{k}'$ lie in the $xz$ plane and are given by

$$\mathbf{k} = k(\hat{\mathbf{z}} \cos\theta + \hat{\mathbf{x}} \sin\theta) \qquad \mathbf{k}' = k'(\hat{\mathbf{z}} \cos\theta' + \hat{\mathbf{x}} \sin\theta') \tag{11-133}$$

as illustrated in Figure 11–14. With the Lorentz transformation of Equations (11–22) and (11–133), Equation (11–131) becomes

$$k \sin\theta x' + k \cos\theta\gamma(z' + vt') - \omega\gamma\left(t' + \frac{v}{c^2} z'\right)$$
$$= k' \sin\theta' x' + k' \cos\theta' z' - \omega' t' \tag{11-134}$$

Equating the coefficients of $x'$, $z'$, and $t'$ separately provides

$$k \sin\theta = k' \sin\theta' \tag{11-135a}$$

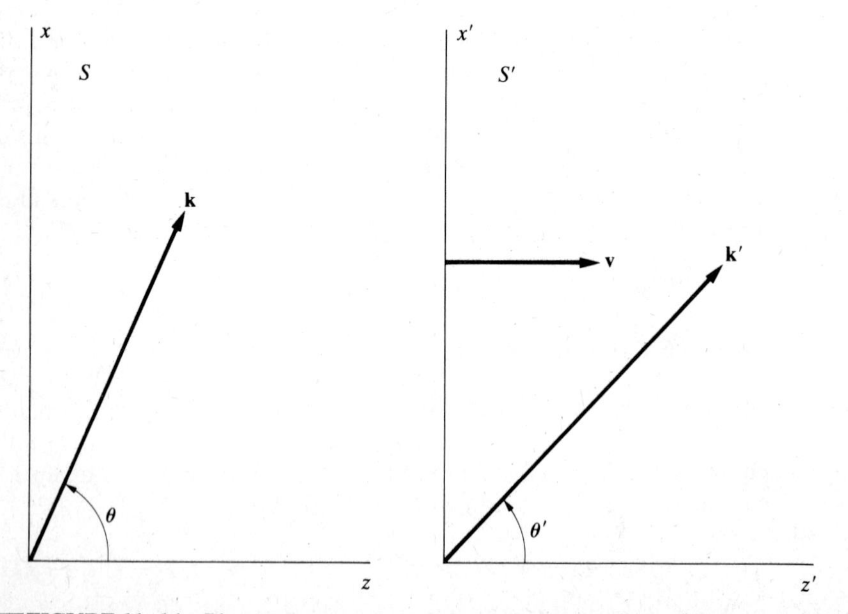


**FIGURE 11-14** Plane wave propagation as seen in two reference frames

$$k\gamma\left(\cos\theta - \frac{v}{c}\right) = k'\cos\theta' \tag{11-135b}$$

$$k\gamma(v\cos\theta - c) = -ck' \tag{11-135c}$$

Only two of these three equations are independent. Equations (11–135) provide a full description of the aberration and Doppler phenomena.

## 11-4A Doppler Shift

The relativistic Doppler wavelength shift formula follows directly from Equation (11–135c):

**Doppler Shift**

$$\lambda' = \frac{\lambda}{\gamma}\frac{1}{[1 - (v/c)\cos\theta]} \tag{11-136}$$

For a wave propagating along the positive $z$ axis, $\theta = 0$ and Equation (11–136) is expressed as follows:

**Longitudinal Doppler Shift**

$$\lambda' = \lambda\sqrt{\frac{1 + v/c}{1 - v/c}} \tag{11-137}$$

This expression is the exact longitudinal Doppler shift. We note that $\lambda' > \lambda$; that is, visible wavelengths seen by an observer moving away from the source appear to be shifted toward longer wavelengths (i.e., for visible light, toward the red end of the spectrum). This shift is known as the *red shift*. For nonrelativistic velocities we obtain, from Equation (11–137), the shift

$$\frac{\lambda' - \lambda}{\lambda} = \frac{v}{c} \qquad \left(\frac{v}{c} \ll 1\right) \tag{11-138}$$

This nonrelativistic Doppler shift holds for any wave motion, including sound waves. The longitudinal Doppler shift is a very useful way of measuring velocity. Applications range from automobile velocity measurements by reflected microwaves to astrophysical measurements of the velocities of distant galaxies by stellar spectral line shifts.

With $\theta = \pi/2$ in Equation (11–136) we obtain a transverse Doppler shift:

**Transverse Doppler Shift**

$$\frac{\lambda'}{\lambda} = \sqrt{1 - \frac{v^2}{c^2}} \tag{11-139}$$

This shift has also been experimentally observed; it is usually much smaller than the longitudinal Doppler shift, owing to the absence of a linear term in $v/c$.

## 11-4B Aberration

From the ratio of Equations (11–135b) and (11–135c) we obtain the general aberration formula:

**Aberration**

$$\cos\theta' = \frac{\cos\theta - (v/c)}{1 - (v/c)\cos\theta} \tag{11-140}$$

For low velocities we can neglect $v^2/c^2$ and higher-order terms. Setting $\theta' = \theta + \Delta\theta$ and observing from Equation (11–140) that $\Delta\theta$ must be small, we expand the left-hand side of Equation (11–140) by using the cosine expansion

$$\cos(\theta + \Delta\theta) = \cos\theta\cos\Delta\theta - \sin\theta\sin\Delta\theta \approx \cos\theta - \Delta\theta\sin\theta$$

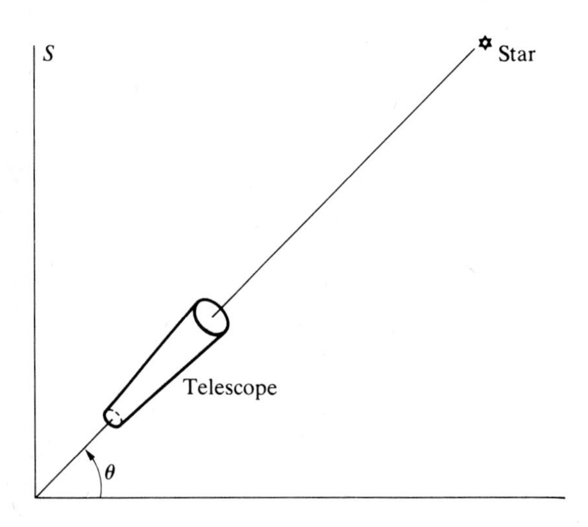


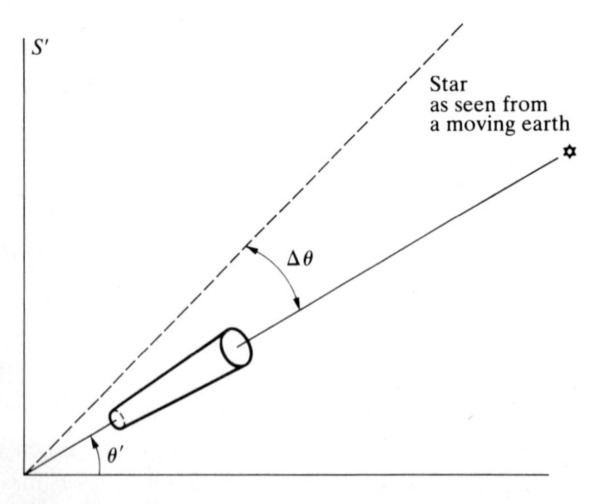


**FIGURE 11-15** Stellar aberration in star positions due to relative motion

where $\Delta\theta \ll 1$. The denominator of the right-hand side is expanded for $v/c \ll 1$ as

$$\left(1 - \frac{v}{c}\cos\theta\right)^{-1} \approx 1 + \frac{v}{c}\cos\theta + \cdots$$

Equation (11–140) is then approximated by

$$\cos\theta - (\sin\theta)\,\Delta\theta \approx \left(\cos\theta - \frac{v}{c}\right)\left(1 + \frac{v}{c}\cos\theta\right) \approx \cos\theta - \frac{v}{c}\sin^2\theta$$

$$\Delta\theta \approx \frac{v}{c}\sin\theta \tag{11–141}$$

Figure 11–15 illustrates stellar aberration, comparing observations of a star with a telescope in a rest frame $S$ and with a telescope on the moving

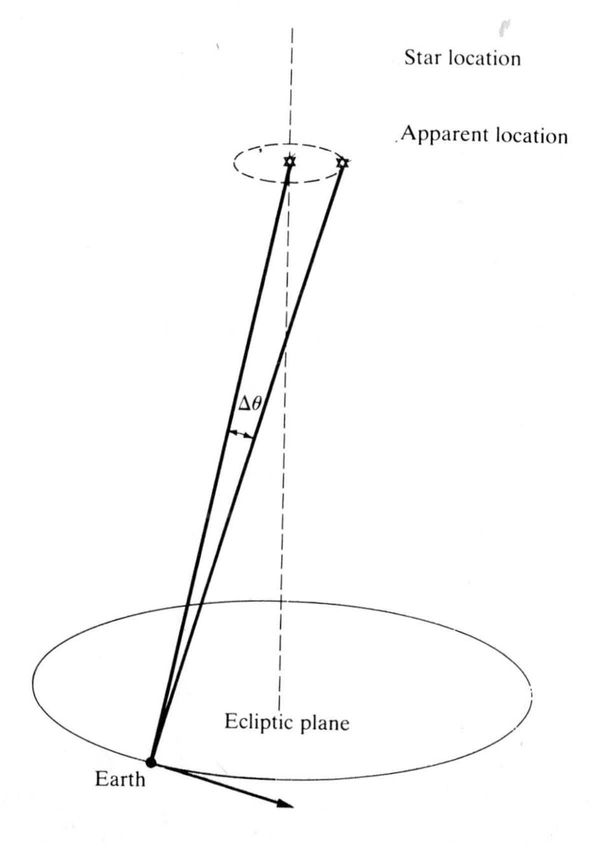


**FIGURE 11–16** Apparent star motion due to the motion of the earth about the sun

earth. Aberration is analogous to the observation that vertically falling rain appears to fall toward a moving automobile when observed from the automobile. A star directly above the ecliptic plane of the earth will appear to move in a circle as the earth moves around the sun (see Figure 11–16) with an angular displacement of

$$\Delta\theta = \frac{v}{c} = \frac{3 \times 10^4 \text{ m/s}}{3 \times 10^8 \text{ m/s}} = 10^{-4} \text{ rad} = 20.6'' \tag{11-142}$$

Stars not directly above the ecliptic plane seem to move in ellipses. This effect is known as *stellar aberration* and was first found by James Bradley in 1725. Strictly speaking, aberration is not a relativistic effect since it is proportional to $v/c$ and not $(v/c)^2$. We include it because of its historical importance and because it is a natural result of the Lorentz invariance of the phase of a plane wave.

## 11-5 VECTOR AND SCALAR POTENTIALS

In Section 10–2 we showed that Maxwell's equations could be expressed in terms of a vector and a scalar potential $\mathbf{A}$ and $\Phi$. The electric and magnetic fields could then be computed by

$$\mathbf{B} = \nabla \times \mathbf{A} \qquad \mathbf{E} = -\nabla\Phi - \frac{\partial \mathbf{A}}{\partial t} \tag{11-143}$$

The potentials are not unique since the set $(\mathbf{A}', \Phi')$ gives the same $\mathbf{E}$ and $\mathbf{B}$ fields as $(\mathbf{A}, \Phi)$ if they are related by the gauge transformation:

**Gauge Transformation**

$$\mathbf{A}' = A + \nabla\chi \qquad \Phi' = \Phi - \frac{\partial \chi}{\partial t} \tag{11-144}$$

When Equations (11–143) are substituted into the remaining Maxwell equations, a set of two coupled equations for the vector potentials are obtained. We use the choice of a specific gauge:

**Lorentz Gauge**

$$\nabla \cdot \mathbf{A} + \frac{1}{c^2}\frac{\partial \Phi}{\partial t} = 0 \tag{11-145}$$

Then the equations for $\mathbf{A}$ and $\Phi$ decouple into

**Wave Equations for Potentials in Lorentz Gauge**

$$\begin{aligned} \nabla^2\Phi - \frac{1}{c^2}\frac{\partial^2 \Phi}{\partial t^2} &= -\frac{\rho}{\varepsilon_0} \\ \nabla^2\mathbf{A} - \frac{1}{c^2}\frac{\partial^2 \mathbf{A}}{\partial t^2} &= -\mu_0 \mathbf{J} \end{aligned} \tag{11-146}$$

These expressions are the wave equations for the scalar and vector potentials in the Lorentz gauge.

Noting that the differential operator on the left-hand sides of Equations (11–146) is $\Box^2$ and that the $(J, ic\rho)$ on the right-hand sides are the components of a four-vector $J$, we conclude that the potentials form a four-vector

$$A = \left(\mathbf{A}, \frac{i\Phi}{c}\right) \tag{11-147}$$

The potential wave equation in covariant form is then

**Covariant Wave Equation**

$$\Box^2 A_\mu = -\mu_0 J_\mu \tag{11-148}$$

The Lorentz gauge of (11–145), in which Equation (11–148) holds, can be written in covariant form as

**Lorentz Gauge Condition**

$$\Box_\mu A_\mu = 0 \tag{11-149}$$

In terms of the four-vector potential, the electromagnetic tensor of Equation (11–88) can be written as

**Field Tensor from Vector Potential**

$$F_{\mu\nu} = \Box_\mu A_\nu - \Box_\nu A_\mu \tag{11-150}$$

To verify this result, we compare the components of Equation (11–150) and Equation (11–88). For example,

$$F_{12} = \nabla_1 A_2 - \nabla_2 A_1 = B_3.$$

$$F_{14} = \nabla_1\left(\frac{i\Phi}{c}\right) - \frac{\partial A_1}{\partial(ict)} = \frac{1}{ic}\left(-\nabla_1\Phi - \frac{\partial \mathbf{A}}{\partial t}\right) = \frac{E_1}{ic} \tag{11-151}$$

In covariant form the gauge transformation of Equations (11–144) is

$$A_\mu = A'_\mu + \Box_\mu \chi \tag{11-152}$$

The electromagnetic tensor $F_{\mu\nu}$ of Equation (11–150) is clearly form-invariant under the gauge transformation. The Lorentz condition, Equation (11–149), applied to Equation (11–152) shows that the gauge function $\chi(x_\mu)$ is not completely arbitrary but must satisfy $\Box^2\chi = 0$.

## 11-6 RELATIVISTIC MECHANICS

We have seen that the principle of relativity forces a modification of the concepts of time intervals and velocity addition. The equations of mechanics must also be form-invariant under the Lorentz transformation. We will now generalize the definitions of momentum and energy so that in the absence of external forces the momentum and the energy of a system of particles are conserved. Then we generalize Newton's law to describe the relativistic motion of charged particles in electric and magnetic fields.

### 11-6A Momentum and Energy

The requirement that physical laws must be covariant under a Lorentz transformation means that momentum and energy must be defined somewhat differently from the usual Newtonian form. The resulting relativistic momentum and energy are perhaps the most accurately tested aspect of the theory of special relativity, through particle physics experiments where speeds approaching that of light are routinely achieved.

The proper time interval $d\tau$ is defined as follows:

**Proper Time Interval**

$$d\tau \equiv \sqrt{-\frac{1}{c^2}\,dx_\mu\,dx_\mu} \tag{11-153}$$

where the infinitesimal invariant interval $dx_\mu\,dx_\mu$ is given by

$$-dx_\mu\,dx_\mu = c^2(dt)^2 - (d\mathbf{r})^2 \tag{11-154}$$

A proper time interval at the same point in space ($d\mathbf{r} = 0$) is just the ordinary time interval $dt$. Proper time is the invariant generalization of the usual concept of time. Dividing Equation (11–153) by the time interval $dt$, we obtain

$$\frac{d\tau}{dt} = \sqrt{1 - \frac{u^2}{c^2}} = \frac{1}{\gamma} \tag{11-155}$$

where $u \equiv d|\mathbf{r}|/dt$ is the particle speed.

The momentum four-vector is defined to be

$$p_\mu \equiv m\,\frac{dx_\mu}{d\tau} \tag{11-156}$$

where the constant $m$ is the inertial mass. The space part (three-vector) of $p_\mu$ is

$$p_i = m\,\frac{dx_i}{d\tau} = mu_i\,\frac{dt}{d\tau} = mu_i\gamma$$

$$\mathbf{p} = m\mathbf{u}\gamma \tag{11-157}$$

For low velocities this relativistic momentum reduces to the Newtonian momentum $m\mathbf{u}$. The time component of $p_\mu$ is

$$p_4 = mic\frac{dt}{d\tau} = icm\gamma \equiv \frac{iE}{c} \tag{11-158}$$

As we will see later, the quantity $E$ is the relativistic energy.

The relativistic momentum $p_\mu$ is clearly a four-vector by its definition in Equation (11–156), since $dx_\mu$ is a Lorentz four-vector and $m$ and $d\tau$ are Lorentz scalars. The four-vector $(\mathbf{p}, iE/c)$ will Lorentz-transform similarly to the coordinate four-vector $(\mathbf{r}, ict)$, so

$$\begin{aligned} &p_x = p'_x \qquad p_y = p'_y \qquad p_z = \gamma\left(p'_z + \frac{v}{c}\frac{E'}{c}\right) \\ &\frac{E}{c} = \gamma\left(\frac{E'}{c} + \frac{v}{c}p'_z\right) \end{aligned} \tag{11-159}$$

The inverse transformation is obtained from these equations by $v \to -v$, yielding

$$\begin{aligned} &p'_x = p_x \qquad p'_y = p_y \qquad p'_z = \gamma\left(p_z - v\frac{E}{c^2}\right) \\ &E' = \gamma(E - vp_z) \end{aligned} \tag{11-160}$$

The most fundamental property of momentum is that in the absence of outside forces the momentum of a system remains constant no matter how the components of the system interact.

We assume that in the $S'$ reference frame the particle motions are nonrelativistic. From Equations (11–157) and (11–158) the momentum and the energy of a given particle are

$$\mathbf{p}' \approx m\mathbf{u}' \tag{11-161a}$$

$$E' \approx mc^2\gamma = mc^2\left(1 - \frac{u'^2}{c^2}\right)^{-1/2} \approx mc^2 + \frac{1}{2}mu'^2 \tag{11-161b}$$

For a system of interacting particles in the $S'$ frame, the total momentum is a constant vector. The sum

$$\sum_i E'_i = \sum_i m_i c^2 + \sum_i \frac{1}{2} m_i u_i'^2$$

is also expected to be constant since mass conservation and, in the absence of friction and neglecting potential energy, the system kinetic energy should be separately conserved.

In the $S'$ frame we therefore expect from Newtonian mechanics that

$$\sum_i \mathbf{p}_i' = \mathbf{P}' = \text{constant vector} \qquad \sum_i E_i' = E' = \text{constant} \tag{11-162a}$$

The first of these equations is valid whatever the nature of internal forces among the particles $\{i\}$. When this system is viewed from an arbitrary second system $S$, from Equations (11–160) we have the following relations:

**Relativistic Momentum-Energy Conservation**

$$\sum_i \mathbf{p}_i = \mathbf{P} = \text{constant vector} \qquad \sum_i E_i = E = \text{constant} \tag{11-162b}$$

These quantities are also constants of the motion. We can finally observe that if the net relativistic momentum $\mathbf{P}$ is conserved for any internal forces, then the total relativistic energy will also be conserved, even if frictional forces are present.

From Equations (11–157) and (11–158) we find that the energy and the momentum of a particle of mass $m$ are related as

$$E = [|\mathbf{p}c|^2 + (mc^2)^2]^{1/2} \tag{11-163}$$

## 11-6B Newton's Law

The motion of a charged particle in an electric or magnetic field requires knowledge of how the momentum and the energy change under the influence of the Lorentz force. The nonrelativistic treatment of this problem is to solve Newton's second law. Newton's law must be generalized to apply to the motion of particles moving with velocities near the speed of light.

Consider a particle of charge $q$ moving with velocity $\mathbf{v}$ with respect to reference frame $S$ in a region with electric and magnetic fields $\mathbf{E}$ and $\mathbf{B}$. The Lorentz force on this charge is then

$$\mathbf{F} = q(\mathbf{E} + \mathbf{v} \times \mathbf{B}) \tag{11-164}$$

In the charge rest frame $S'$ the Lorentz force is

$$\mathbf{F}' = q\mathbf{E}' \tag{11-165}$$

The fields in $S$ and $S'$ are related by the inverse field transformations from Equations (11–103),

$$\mathbf{E}_T' = \gamma(\mathbf{E}_T + \mathbf{v} \times \mathbf{B}) \qquad \mathbf{E}_{/\!/}' = \mathbf{E}_{/\!/} \tag{11-166}$$

and hence the forces in $S$ and $S'$ are related, since the charge $q$ is a relativistic invariant, by

$$\mathbf{F}'_{\mathrm{T}} = \gamma \mathbf{F}_{\mathrm{T}} \qquad \mathbf{F}'_{/\!/} = \mathbf{F}_{/\!/} \tag{11–167}$$

In the particle rest frame the nonrelativistic Newton's law is correct, and

$$\mathbf{F}' = \frac{d\mathbf{p}'}{dt'} \tag{11–168}$$

In the $S'$ reference frame we have, by Equation (11–155), where $d\tau \equiv dt'$ in the particle rest frame $S'$,

$$dt' = \frac{1}{\gamma}\,dt \tag{11–169}$$

For the component transverse to the boost, $\mathbf{p}'_{\mathrm{T}} = \mathbf{p}_{\mathrm{T}}$, and so by Equation (11–169)

$$\frac{d\mathbf{p}'_{\mathrm{T}}}{dt'} = \gamma\,\frac{d\mathbf{p}_{\mathrm{T}}}{dt} \tag{11–170}$$

For the parallel component we have, from Equations (11–159),

$$\frac{d\mathbf{p}_{/\!/}}{dt} = \gamma\left(\frac{d\mathbf{p}_{/\!/}}{dt'} + \frac{v}{c^2}\frac{dE'}{dt'}\right)\frac{dt'}{dt} \tag{11–171}$$

Since $E' = mc^2[1 - (u'/c)^2]^{-1/2}$ is quadratic in $u'$, $dE'/dt' = m\gamma^3 u'(du'/dt')$ vanishes at $u' = 0$ (i.e., the rest frame), and thus Equations (11–169) and (11–171) give

$$\frac{d\mathbf{p}_{/\!/}}{dt} = \frac{d\mathbf{p}'_{/\!/}}{dt'} \tag{11–172}$$

Comparing the Lorentz force transformation of Equations (11–167) with the time rate of change of momentum transformations of Equations (11–170) and (11–172), we observe that

$$\mathbf{F} = \frac{d\mathbf{p}}{dt} \tag{11–173}$$

holds in an arbitrary frame $S$. This expression is the correct relativistic form for Newton's second law. Interestingly, it was in this form, and not as $\mathbf{F} = m\mathbf{a}$, that Newton originally formulated his law. Of course, Newton knew nothing of relativity, but for this and other purposes the momentum form is more fundamental. We encountered one of these situations earlier in Section 8–7 when we considered systems whose momentum was partly mechanical and partly carried in the electromagnetic field.

# SUMMARY

## Important Concepts and Equations

$$x = x' \qquad y = y'$$
$$z = \gamma(z' + vt') \qquad t = \gamma\left(t' + \frac{v}{c^2} z'\right)$$

*Lorentz transformation relating coordinates and times in the two coordinate frames S and S'. The S' frame moves with velocity v along the positive z direction relative to S, and*

$$\gamma = \left(1 - \frac{v^2}{c^2}\right)^{-1/2}$$

$$u_z = \frac{u'_z + v}{1 + (v/c^2)u'_z} \qquad \mathbf{u}_\mathrm{T} = \frac{\mathbf{u}'_\mathrm{T}}{\gamma[1 + (v/c^2)u'_z]}$$

*Velocity addition formula. The S' frame moves with velocity v along the positive z axis. The velocity of a point, as measured in the S' frame, is* $(\mathbf{u}'_\mathrm{T}, u'_z)$*, and the velocity relative to the S frame is* $(\mathbf{u}_\mathrm{T}, u_z)$*, where* $\mathbf{u}_\mathrm{T} = u_x\hat{\mathbf{x}} + u_y\hat{\mathbf{y}}$.

$$\Delta t = \gamma\,\Delta t'$$

*Time dilation. A clock at rest in S' appears to run slower when viewed from a moving frame S. A time interval* $\Delta t$ *in S is longer by a factor* $\gamma$ *than the corresponding time interval* $\Delta t'$ *as observed from S'.*

$$L(v) = \frac{1}{\gamma} L$$

*Length contraction. The length of a rod that is in motion with speed v relative to the observer appears to be shorter than its rest length L by a factor* $1/\gamma$.

$$a_\mu b_\mu = \sum_{\mu=1}^{4} a_\mu b_\mu$$

*Einstein summation convention. An implied summation over a repeated tensor index is understood.*

$$\Box = \left(\frac{\partial}{\partial x_1}, \frac{\partial}{\partial x_2}, \frac{\partial}{\partial x_3}, \frac{\partial}{\partial x_4}\right) = \left(\frac{\partial}{\partial x}, \frac{\partial}{\partial y}, \frac{\partial}{\partial z}, \frac{1}{ic}\frac{\partial}{\partial t}\right)$$

*Four-dimensional del operator.*

$$\Box^2 = \Box_\mu \Box_\mu = \frac{\partial^2}{\partial x^2} + \frac{\partial^2}{\partial y^2} + \frac{\partial^2}{\partial z^2} - \frac{1}{c^2}\frac{\partial^2}{\partial t^2}$$

*D'Alembertian operator.*

$$\Box_\mu J_\mu = 0$$

*Equation of continuity in covariant notation.*

$J_\mu = (J_x, J_y, J_z, ic\rho)$

*Four-vector current density.*

$$F_{\mu\nu} = \begin{bmatrix} 0 & B_3 & -B_2 & \dfrac{E_1}{ic} \\ -B_3 & 0 & B_1 & \dfrac{E_2}{ic} \\ B_2 & -B_1 & 0 & \dfrac{E_3}{ic} \\ -\dfrac{E_1}{ic} & -\dfrac{E_2}{ic} & -\dfrac{E_3}{ic} & 0 \end{bmatrix}$$

*Electromagnetic field tensor.*

$\Box_\nu F_{\mu\nu} = \mu_0 J_\mu \qquad \frac{1}{2}\varepsilon_{\alpha\beta\mu\nu}\Box_\beta F_{\mu\nu} = 0$

*Covariant form of Maxwell's equations, where $\varepsilon_{\alpha\beta\mu\nu} = +1$ for cyclic indices, $\varepsilon_{\alpha\beta\mu\nu} = -1$ for anticyclic indices, and $\varepsilon_{\alpha\beta\mu\nu} = 0$ otherwise.*

$$a_{\mu\nu} = \begin{bmatrix} 1 & 0 & 0 & 0 \\ 0 & 1 & 0 & 0 \\ 0 & 0 & \gamma & -i\gamma\dfrac{v}{c} \\ 0 & 0 & i\gamma\dfrac{v}{c} & \gamma \end{bmatrix}$$

*Lorentz transformation matrix for a boost along the positive z axis with velocity v.*

$x_\mu = a_{\mu\nu} x'_\nu$

*Lorentz transformation in matrix notation.*

$F_{\mu\nu} = a_{\mu\alpha} a_{\nu\beta} F'_{\alpha\beta}$

*Field tensor transformation.*

$$\mathbf{E}_\mathrm{T} = \gamma\left(\mathbf{E}'_\mathrm{T} - \frac{\mathbf{v}}{c} \times c\mathbf{B}'\right) \qquad \mathbf{E}_{/\!/} = \mathbf{E}'_{/\!/}$$

$$c\mathbf{B}_\mathrm{T} = \gamma\left(c\mathbf{B}'_\mathrm{T} + \frac{\mathbf{v}}{c} \times \mathbf{E}'\right) \qquad c\mathbf{B}_{/\!/} = c\mathbf{B}'_{/\!/}$$

*Electric and magnetic field transformations between frames S and S′, where S′ is in motion with velocity **v** relative to S.*

$$A_\mu = \left(A_x, A_y, A_z, \frac{i\Phi}{c}\right)$$

*Four-vector potential. The space components are the vector potential components, and the time component is proportional to the scalar potential.*

$$\Box^2 A_\mu = -\mu_0 J_\mu$$

*Wave equation for the four-vector potential with the source term in the Lorentz gauge.*

$$\Box_\mu A_\mu = 0$$

*Lorentz gauge condition in covariant notation.*

$$F_{\mu\nu} = \Box_\mu A_\nu - \Box_\nu A_\mu$$

*Field tensor written in terms of the four-vector potential.*

$$A_\mu = A'_\mu + \Box_\mu \chi$$

*Gauge transformation in covariant notation, where $\chi$ is a scalar field.*

$$c\, d\tau = \sqrt{-dx_\mu\, dx_\mu}$$

*Proper time interval $d\tau$.*

$$dt = \gamma\, d\tau \qquad \text{(time dilation)}$$

*Relation of time interval $dt$ in a moving frame to the proper time interval $d\tau$.*

$$p_\mu = m \frac{dx_\mu}{d\tau}$$

*Four-vector momentum.*

$$\mathbf{p} = m\mathbf{u}\gamma$$

*Relation of three-momentum to the velocity* $\mathbf{u}$.

$$p_4 = \frac{iE}{c}$$

*Relation of time component of $p_\mu$ to the energy $E$.*

$$p_x = p'_x \qquad p_y = p'_y \qquad p_z = \gamma\left(p'_z + \frac{v}{c}\frac{E'}{c}\right) \qquad \frac{E}{c} = \gamma\left(\frac{E'}{c} + \frac{v}{c}p'_z\right)$$

*Lorentz transformation of energy and momentum from a frame $S$ to a frame $S'$ that moves with velocity $v$ along the positive $z$ direction relative to $S$.*

$$E = [|\mathbf{p}c|^2 + (mc^2)^2]^{1/2}$$

*Relation of energy and momentum for a particle of rest mass $m$.*

$$\mathbf{F} = \frac{d\mathbf{p}}{dt}$$

*Relativistic form of Newton's second law.*

# PROBLEMS

## Section 11-1: Lorentz Transformation

**11-1** A "clock" is constructed so that the interval is fixed by the time taken by a light pulse to travel a distance $L$ from a source to a mirror and back (i.e., $\Delta t_0 = 2L/c$).

(i) In a reference frame in which the mirror and the source are moving with velocity $v$ tangent to the mirror surface, show that this time interval is $\Delta t = \Delta t_0/\sqrt{1 - (v^2/c^2)}$. Use the assumption that light travels at the same speed in all frames.

(ii) Assuming that the clock interval is independent of orientation, show that in a frame where the mirror and the source are moving at velocity $v$ perpendicular to the mirror, the distance between the source and mirror is $L_0\sqrt{1 - v^2/c^2}$.

**11-2** A Lorentz transformation rotates one four-vector $A'_\nu$ into a second four-vector $A_\mu$ by $A_\mu = a_{\mu\nu}A'_\nu$. Show that the Cartesian component matrix $a_{\mu\nu}$ corresponding to a boost along the $z$ axis is

$$a_{\mu\nu} = \begin{bmatrix} 1 & 0 & 0 & 0 \\ 0 & 1 & 0 & 0 \\ 0 & 0 & \gamma & \dfrac{-iv}{c}\gamma \\ 0 & 0 & \dfrac{iv}{c}\gamma & \gamma \end{bmatrix}$$

Find components of the inverse matrix $a^{-1}$. Two successive boosts are represented by the product of the individual boosts. Show that the velocity addition formula results from two boosts along the $z$ axis.

**11-3** A plane electromagnetic wave propagates through a medium of index of refraction $n_0$ so that its phase velocity is given by $u_0 = c/n_0$. If the medium moves with velocity $v$ in the same direction, show that the apparent index of refraction and the phase velocity of the wave relative to the laboratory are

$$n = \frac{n_0 + v/c}{1 + n_0 v/c} \qquad u = \frac{c/n_0 + v}{1 + v/(n_0 c)}$$

and if $v \ll c$

$$u \approx \frac{c}{n_0} + v\left(1 - \frac{1}{n_0^2}\right)$$

The quantity $1 - 1/n_0^2$ is called the *Fresnel dragging coefficient.*

## Section 11-2: Lorentz Transformation of Fields and Sources

**11-4** Consider a long cylindrical column of electrons of uniform charge density $\rho_0$ and radius $a$.

(i) Derive the repulsive force on an electron at a distance $r < a$ from the axis.

(ii) An observer in the laboratory sees a beam of circular cross section and density $\rho$

moving at velocity $v$. What force does the observer see on an electron of the beam at distance $r < a$ from the axis?

(iii) If $v$ is near the velocity of light, what is the force of part (ii) as seen by an observer moving with the beam? Compare this force with the answer to part (ii), and comment.

(iv) If $n = 2 \times 10^{10}\ \text{cm}^{-3}$ and $v = 0.99c$ ($c$ = light velocity), what gradient of a transverse magnetic field (in teslas per meter) would just hold this beam from spreading in one of its dimensions?

**11-5** Using the field transformation equations, show that $E^2 - c^2B^2$ and $\mathbf{E} \cdot \mathbf{B}$ are Lorentz invariants.

**11-6** For perpendicular **E** and **B** fields (i.e., $\mathbf{E} \cdot \mathbf{B} = 0$), show that you can transform to a moving system having either a pure electric or magnetic field except when $E = cB$.

**11-7** The quantity $I \equiv F_{\mu\nu}F_{\mu\nu}$ is invariant under a Lorentz transformation; that is, it has the same numerical value for fields evaluated in any reference frame. What is this invariant in terms of electric and magnetic fields?

**11-8** Two electrons with equal velocities $v$ are moving side by side a distance $a$ apart. Midway between them is an infinite sheet of fixed positive charges with a surface charge density $\sigma$ in its rest frame. In the frame $S$ $\sigma$ is at rest, and in the frame $S'$ the electrons are at rest.

(i) Find the charge density $\sigma'$ of the sheet in $S'$.

(ii) Use Gauss's law to determine the electric field $E'$ in $S'$ due to the charge sheet.

(iii) How large must $v$ be in order that the electrons maintain the separation $a$?

## Section 11-3: Fields of a Moving Point Charge

**11-9** What do the electric and magnetic fields of a point charge moving at a constant velocity $v$ look like in the limit $v \to c$?

**11-10** Calculate the electric flux of a moving charge from Equation (11–119). What is the significance of the result?

## Section 11-5: Vector and Scalar Potentials

**11-11** A point charge $e$ moves with constant velocity $v$ in the $z$ direction so that at time $t$ it is at the point $Q$ with coordinates $x = 0$, $y = 0$, and $z = vt$. Now consider the time $t$ and the point $P$ with coordinates $x = b$, $y = 0$, and $z = 0$.

(i) Find the scalar potential $\Phi$.

(ii) Find the vector potential **A**.

(iii) Find the electric field in the $x$ direction, $E_x$.

## Section 11-6: Relativistic Mechanics

**11-12** Using the Lorentz transformation for momentum and energy, show explicitly that $E^2 - p^2c^2$ is an invariant. Evaluate this invariant with the single-particle values for $E$ and $p$ to show

$$E^2 = (mc^2)^2 + (pc)^2$$

Show that the velocity of the particle is $v = pc^2/E$.

**11-13** Quantum theory states that when a wave is emitted or absorbed, it is quantized into packets of energy $E = h\nu$ and momentum $p = h\nu/c$. If the classical and quantum pictures are to be consistent, show that the light quantum's mass must be zero to obtain the correct ratio of energy to momentum. Compare the result of Problem 11–13 with the energy and momentum fluxes obtained in Sections 8–6 and 8–7.

**11-14** Compton effect: The basic assumption in quantum theory is that electromagnetic radiation interacts with matter only in energy multiples of $h\nu$, where $\nu$ is the frequency (in hertz) of the radiation and $h$ is Planck's constant. The collision between a photon and an electron initially at rest is treated as a collision between a massless particle, having energy $h\nu$ and momentum $h\nu/c$, an an electron with rest energy $mc^2$. The collision process is shown in Figure 11–17. Apply the conservation laws of momentum and energy to obtain the relation

$$\lambda' - \lambda = \frac{h}{mc}(1 - \cos\phi)$$

where $\lambda$ and $\lambda'$ are the photon's wavelengths before and after the collision. This equation is Compton's formula. In Maxwell's classical theory $\lambda = \lambda'$ always. For what wavelengths is the classical theory seriously wrong?

**11-15** Show that the kinetic energy, $T \equiv E - mc^2$, of the recoiling electron in Compton scattering $\gamma e \to \gamma e$ (see Problem 11–14) is

$$T = h\nu \frac{2h/(\lambda mc)(\sin^2\theta/2)}{1 + [2h/(\lambda mc)(\sin^2\theta/2)]}$$

where $\theta$ is the angle of the electron measured from the incident direction.

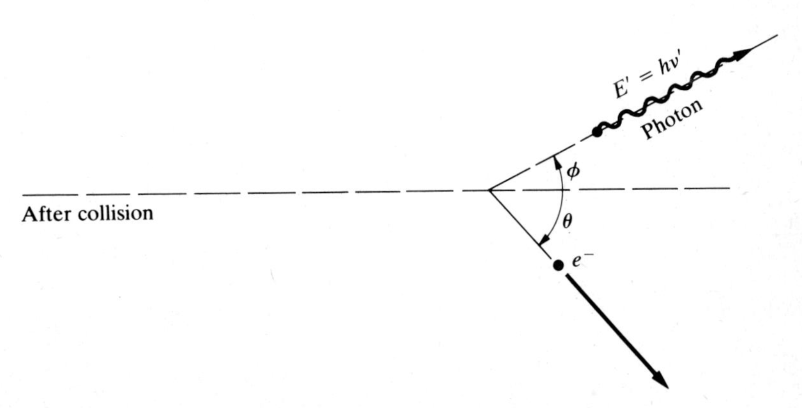


**FIGURE 11-17** Compton collision of a photon and an electron

**11-16** A positron $e^+$ with mass $m_e$ and momentum $p$ strikes an electron $e^-$ also with mass $m_e$ initially at rest. Show that the velocity of the center-of-momentum frame is $v = pc^2/(m_e c^2 + E)$, where $E^2 = (pc)^2 + (m_e c^2)^2$. Relate the momenta in the center-of-momentum system to $p$. The result of the collision is a $\pi^+\pi^-$ pair, each having mass $m_\pi > m_e$. What is the smallest laboratory momentum $p$ that will allow this reaction to proceed?

**11-17** Show that an electron $e^-$ and a positron $e^+$ (identical except opposite charge) cannot annihilate into a single photon. (Annihilation into two or more photons is allowed.)

**11-18** Our galaxy is about $10^5$ light years across, and the most energetic particles known have an energy of about $10^{19}$ eV. How long would it take a proton with this energy to traverse the galaxy as measured in the rest frame of (i) the galaxy and (ii) the particle?

**11-19** Show that the covariant equation of motion of a particle of charge $q$ and mass $m$ moving in an electromagnetic field is

$$\frac{dp_\mu}{d\tau} = \frac{q}{c} F_{\mu\nu} u_\nu$$

where $u_\nu$ is the four-vector velocity defined by $u_\nu \equiv dx_\nu/d\tau$. What is the interpretation of the time component of the equation of motion?

**11-20** A charged particle moves at right angles to a uniform magnetic field. Show that the orbit is circular, and determine the relativistically correct frequency and radius for a given momentum. Find an expression for $rB$ in terms of the energy of the particle.

# Appendix A

## DIFFERENTIAL VECTOR OPERATIONS IN CARTESIAN, SPHERICAL, AND CYLINDRICAL COORDINATE SYSTEMS

In the Cartesian system the operations are as follows:

**Gradient**

$$\nabla s = \hat{\mathbf{x}}\frac{\partial s}{\partial x} + \hat{\mathbf{y}}\frac{\partial s}{\partial y} + \hat{\mathbf{z}}\frac{\partial s}{\partial z}$$

**Divergence**

$$\nabla \cdot \mathbf{V} = \frac{\partial V_x}{\partial x} + \frac{\partial V_y}{\partial y} + \frac{\partial V_z}{\partial z}$$

**Curl**

$$\nabla \times \mathbf{V} = \hat{\mathbf{x}}\left(\frac{\partial V_z}{\partial y} - \frac{\partial V_y}{\partial z}\right) + \hat{\mathbf{y}}\left(\frac{\partial V_x}{\partial z} - \frac{\partial V_z}{\partial x}\right) + \hat{\mathbf{z}}\left(\frac{\partial V_y}{\partial x} - \frac{\partial V_x}{\partial y}\right)$$

**Laplacian**

$$\nabla^2 s = \frac{\partial^2 s}{\partial x^2} + \frac{\partial^2 s}{\partial y^2} + \frac{\partial^2 s}{\partial z^2}$$

In the spherical system the operations are as follows:

**Gradient**

$$\nabla s = \hat{\mathbf{r}}\frac{\partial s}{\partial r} + \hat{\boldsymbol{\theta}}\frac{1}{r}\frac{\partial s}{\partial \theta} + \hat{\boldsymbol{\phi}}\frac{1}{r \sin\theta}\frac{\partial s}{\partial \phi}$$

**Divergence**

$$\nabla \cdot \mathbf{V} = \frac{1}{r^2}\frac{\partial}{\partial r}(r^2 V_r) + \frac{1}{r \sin\theta}\frac{\partial}{\partial \theta}(\sin\theta V_\theta) + \frac{1}{r\sin\theta}\frac{\partial V_\phi}{\partial \phi}$$

**Curl**

$$\nabla \times \mathbf{V} = \hat{\mathbf{r}}\frac{1}{r\sin\theta}\left[\frac{\partial}{\partial \theta}(\sin\theta V_\phi) - \frac{\partial V_\theta}{\partial \phi}\right]$$
$$+\hat{\boldsymbol{\theta}}\frac{1}{r}\left[\frac{1}{\sin\theta}\frac{\partial V_r}{\partial \phi} - \frac{\partial}{\partial r}(rV_\phi)\right] + \hat{\boldsymbol{\phi}}\frac{1}{r}\left[\frac{\partial}{\partial r}(rV_\theta) - \frac{\partial V_r}{\partial \theta}\right]$$

**Laplacian**

$$\nabla^2 s = \frac{1}{r^2}\frac{\partial}{\partial r}\left(r^2\frac{\partial s}{\partial r}\right) + \frac{1}{r^2\sin\theta}\frac{\partial}{\partial\theta}\left(\sin\theta\frac{\partial s}{\partial \theta}\right) + \frac{1}{r^2\sin^2\theta}\frac{\partial^2 s}{\partial \phi^2}$$

Note that

$$\frac{1}{r^2}\frac{\partial}{\partial r}\left(r^2\frac{\partial s}{\partial r}\right) = \frac{1}{r}\frac{\partial^2}{\partial r^2}(rs)$$

In the cylindrical system the operations are as follows:

**Gradient**

$$\nabla s = \hat{\mathbf{r}}\frac{\partial s}{\partial r} + \hat{\boldsymbol{\phi}}\frac{1}{r}\frac{\partial s}{\partial \phi} + \hat{\mathbf{z}}\frac{\partial s}{\partial z}$$

**Divergence**

$$\nabla \cdot \mathbf{V} = \frac{1}{r}\frac{\partial}{\partial r}(r V_r) + \frac{1}{r}\frac{\partial V_\phi}{\partial \phi} + \frac{\partial V_z}{\partial z}$$

**Curl**

$$\nabla \times \mathbf{V} = \hat{\mathbf{r}}\left(\frac{1}{r}\frac{\partial V_z}{\partial \phi} - \frac{\partial V_\phi}{\partial z}\right) + \hat{\boldsymbol{\phi}}\left(\frac{\partial V_r}{\partial z} - \frac{\partial V_z}{\partial r}\right) + \hat{\mathbf{z}}\frac{1}{r}\left[\frac{\partial}{\partial r}(r V_\phi) - \frac{\partial V_r}{\partial \phi}\right]$$

**Laplacian**

$$\nabla^2 s = \frac{1}{r}\frac{\partial}{\partial r}\left(r\frac{\partial s}{\partial r}\right) + \frac{1}{r^2}\frac{\partial^2 s}{\partial \phi^2} + \frac{\partial^2 s}{\partial z^2}$$

# Appendix B

## SYSTEMS OF UNITS

In this appendix we discuss the two major systems of electromagnetic units in use today, the SI (or mks) and Gaussian systems. When one is dealing with macroscopic quantities that will be measured, the SI system is preferable since measuring instruments give readings in these units. For purely theoretical work other units may be desirable for calculational simplicity. In any system the defined units must be consistent with all equations governing electromagnetic phenomena. This requirement leads to a natural way of introducing different systems of units. If the basic defining equations for electromagnetic behavior are written by using arbitrary constants of proportionality, various relations that these constants must satisfy can be derived. The choice of the constants determines the nature of the system of units.

We begin by writing Maxwell's equations with arbitrary proportionality constants:

$$\nabla \cdot \mathbf{E} = k_1 \rho \qquad \text{(I)} \qquad \nabla \times \mathbf{E} = -k_2 \frac{\partial \mathbf{B}}{\partial t} \qquad \text{(II)}$$

$$\nabla \cdot \mathbf{B} = 0 \qquad \text{(III)} \qquad \nabla \times \mathbf{B} = k_3 \frac{\partial \mathbf{E}}{\partial t} + k_4 \mathbf{J} \qquad \text{(IV)}$$

These constants are not independent. Taking the divergence of Maxwell IV, using I, gives

$$k_4 \nabla \cdot \mathbf{J} + k_3 k_1 \frac{\partial \rho}{\partial t} = 0$$

which compared with the continuity equation, implies

$$k_4 = k_3 k_1 \qquad \textbf{(B-1)}$$

Solving Maxwell's equations for waves in empty space gives the wave equation for either **E** or **B**:

$$\left(\nabla^2 - k_2 k_3 \frac{\partial^2}{\partial t^2}\right)\begin{bmatrix}\mathbf{E}\\ \mathbf{B}\end{bmatrix} = 0$$

which implies

$$k_2 k_3 = \frac{1}{c^2} \qquad \textbf{(B-2)}$$

If the Lorentz force law, the Biot-Savart law, and Coulomb's law are written in terms of proportionality constants, a set of relations can be derived that defines the various systems of units. We write the Lorentz force law as

$$\mathbf{F} = q(\mathbf{E} + k_2 \mathbf{v} \times \mathbf{B}) \tag{B-3}$$

By an argument similar to that presented in Section 6–2, the constant in Equation (B–3) is the same $k_2$ as in Maxwell equation II. (A given flux change yields the same emf whether it is due to a time-varying magnetic field or a motion of the circuit.) The force per unit length between two infinitely long, current-carrying wires can be determined from the Lorentz force law and Maxwell IV,

$$d\mathbf{F} = k_2 I \, d\boldsymbol{\ell} \times \mathbf{B} \qquad \oint \mathbf{B} \cdot d\mathbf{r} = 2\pi \imath B = k_4 I'$$
$$\frac{F}{\ell} = \frac{k_4 k_2 I I'}{2\pi \imath} \tag{B-4}$$

Finally, Maxwell I and Equation (B–3) can be used to obtain an expression for Coulomb's law:

$$\oint \mathbf{E} \cdot d\mathbf{S} = 4\pi r^2 E = k_1 q' \qquad \mathbf{F} = q\mathbf{E} = \frac{k_1 q q'}{4\pi r^2} \hat{\mathbf{r}} \tag{B-5}$$

Since **E** and **B** are theoretical constructs (i.e., the units are not directly expressed in terms of mechanical quantities), we are free to relate them at our convenience by fixing the constant $k_2$ or, alternatively, constant $k_3$ from Equation (B–2). A unit of charge or current must be chosen that specifies either $k_1$ or $k_4$. The only experimental parameter introduced by Maxwell's equations is the speed of light.

## B-1 SI (OR mks) SYSTEM

In the SI system of units the constant $k_2$ is chosen to be unity. The other arbitrary constant is found from the definition of the fundamental unit of electromagnetism in the SI system, the ampere. As was stated in Section 5–4, a current of 1 ampere is that current, carried in two infinitely long parallel wires separated by 1 meter in empty space, that produces a force per unit length of $2 \times 10^{-7}$ newtons per meter between the wires. Equation (B–4) then defines $k_4$, known as the permeability of free space:

$$k_4 = \frac{F}{\ell} \times \frac{2\pi \imath}{I^2}$$
$$\mu_0 = k_4 \equiv 4\pi \times 10^{-7} \text{ N/A}^2 \tag{B-6}$$

Equation (B–2) defines $k_3$ to be

$$k_3 = \frac{1}{c^2} \tag{B-7}$$

while Equation (B–1) gives $1/k_1$, known as the permittivity of free space, as

$$\frac{1}{k_1} = \frac{1}{\mu_0 c^2} \equiv \varepsilon_0 \approx (9 \times 4\pi \times 10^9 \text{ N·m}^2/\text{A}^2)^{-1} \tag{B-8}$$

These choices define the rationalized SI system, which, in contrast to the Gaussian system to be discussed next, puts the geometric factors of

**TABLE B-1** Electromagnetic relations

| Quantity | SI or mks (Coulomb, Ampere, Ohm) | cgs (Statcoulomb, Statampere, Second-Centimeter$^{-1}$) |
|---|---|---|
| Potentials | $\Phi = \frac{1}{4\pi\varepsilon_0} \sum_i \frac{q_i}{r}$ | $\Phi = \sum_i \frac{q_i}{r}$ |
| | $\mathbf{A} = \frac{\mu_0}{4\pi} \sum_i \frac{\mathbf{I}_i}{r}$ | $\mathbf{A} = \frac{1}{c} \sum_i \frac{\mathbf{I}_i}{r}$ |
| | $\varepsilon_0 \approx \frac{1}{36\pi} \times 10^{-9}$ | $c$ = speed of light in vacuum $\approx 3 \times 10^{10}$ cm/s |
| | $\mu_0 = 4\pi \times 10^{-7}$ | |
| Fields | $\mathbf{E} = -\nabla\Phi - \frac{1}{c}\frac{\partial \mathbf{A}}{\partial t}$ | $\mathbf{E} = -\nabla\Phi - \frac{1}{c}\frac{\partial \mathbf{A}}{\partial t}$ |
| | $\mathbf{B} = \nabla \times \mathbf{A}$ | $\mathbf{B} = \nabla \times \mathbf{A}$ |
| Matter | $\mathbf{D} = \varepsilon\mathbf{E}, \mathbf{B} = \mu\mathbf{H}$ | $\mathbf{D} = \varepsilon\mathbf{E}, \mathbf{B} = \mu\mathbf{H}$ |
| Force | $\mathbf{F} = q(\mathbf{E} + \mathbf{v} \times \mathbf{B})$ | $\mathbf{F} = q\left(\mathbf{E} + \frac{\mathbf{v}}{c} \times \mathbf{B}\right)$ |
| Maxwell's equations | $\nabla \cdot \mathbf{E} = \frac{\rho}{\varepsilon_0}$ | $\nabla \cdot \mathbf{E} = 4\pi\rho$ |
| | $\nabla \times \mathbf{E} = -\frac{\partial \mathbf{B}}{\partial t}$ | $\nabla \times \mathbf{E} = -\frac{1}{c}\frac{\partial \mathbf{B}}{\partial t}$ |
| | $\nabla \times \mathbf{B} = \mu_0\varepsilon_0 \frac{\partial \mathbf{E}}{\partial t} + \mu_0\mathbf{J}$ | $\nabla \times \mathbf{B} = \frac{1}{c}\frac{\partial \mathbf{E}}{\partial t} + \frac{4\pi}{c}\mathbf{J}$ |
| | $\nabla \cdot \mathbf{B} = 0$ | $\nabla \cdot \mathbf{B} = 0$ |

$4\pi$ in Coulomb's law and the Biot-Savart law, rather than in Maxwell's equations.

## B-2 GAUSSIAN SYSTEM

The Gaussian system is based on the cgs system of mechanical units. The unit of charge, the statcoulomb, is defined by writing Coulomb's law as

$$F = \frac{qq'}{r^2} \tag{B-9}$$

That is, the force between two charges, each of magnitude 1 statcoulomb separated by 1 centimeter, is 1 dyne. Comparing this expression with Equation (B–2), we see that

$$k_1 = 4\pi \tag{B-10}$$

In the Gaussian system all six fields (**E**, **D**, **P**, **B**, **H**, **M**) have the same units, so

$$k_2 = \frac{1}{c} \tag{B-11}$$

From Equation (B–2)

$$k_3 = \frac{1}{c} \tag{B-12}$$

The equation-of-continuity requirement (B–1) then yields

$$k_4 = \frac{4\pi}{c} \tag{B-13}$$

We conclude by summarizing the major equations in the two systems (Table B–1) and by providing a table to convert from Gaussian to SI units (Table B–2).

**TABLE B-2** Conversion factors

| Gaussian Unit | = | Conversion Factor | × | SI Unit |
|---|---|---|---|---|
| Dyne | | $10^{-5}$ | | Newton |
| Erg | | $10^{-7}$ | | Joule |
| Gauss | | $10^{-4}$ | | Tesla |
| Maxwell | | $10^{-4}$ | | Weber |
| Oersted | | $\frac{1}{4\pi} \times 10^3$ | | Ampere per meter |
| Statampere | | $3.33 \times 10^{-10}$ | | Ampere |
| Statcoulomb | | $3.33 \times 10^{-10}$ | | Coulomb |
| Statvolt | | 300 | | Volt |

# Appendix C

## NUMERICAL METHODS

There are many uses for computers in solving electricity and magnetism problems. Computer-generated graphics were used extensively in the preparation of this text; most diagrams in the book were initially drawn by a computer-driven plotter. As mentioned in Section 3–8, boundary value problems involving Laplace's equation can be solved to any desired accuracy by employing computers to obtain numeric solutions. In this appendix two examples of programs used to calculate equipotentials and field lines, and a third routine solving Laplace's equation, are given.

The field lines and equipotentials for a given field are governed by first-order differential equations (see Problem 1–8). In most cases no closed-form solution exists, and the differential equations must be solved numerically. Many different numerical methods have been derived for solving first-order differential equations. All have the same general form. Starting with a differential equation

$$\frac{dy}{dx} = f(x, y)$$

In integrated form we have

$$y_{n+1} \equiv y(x_{n+1}) = y_n + \int_{x_n}^{x_{n+1}} f(x', y)\, dx'$$

Given an initial point $(x_n, y_n)$, we determine a new point $(x_{n+1}, y_{n+1})$ by

$$x_{n+1} = x_n + h \qquad y_{n+1} = y_n + hg$$

where $h$ is the step size for the independent variable and $g$ depends on $f(x, y)$. The new point is used as the initial point for the next calculation, and the process is continued as needed. A simple example of the $g$ function is

$$g = f(x_n, y_n)$$

With this choice of $g$, the approximation is known as Euler's method. More sophisticated, and more complicated, $g$ functions will give more accurate results. These are based on knowing not only $y_n$ but some previous values, such as $y_{n-1}$ and $y_{n-2}$. We provide two examples to demonstrate some considerations that enter in choosing a particular numerical method.

We first consider an uncharged conducting sphere in a uniform external electric field, as discussed in Section 3–5C. The field lines and the equipotentials for this problem, shown in Figure 3–9, were generated by Computer Program 1 to be presented shortly. The points calculated by the program are shown in Figure C–1. The program uses the Adams extrapolation-interpolation formulas:

**Adams Extrapolation Formula**

$$y_{n+1} = y_n + \frac{h}{24}(55f_n - 59f_{n-1} + 37f_{n-2} - 9f_{n-3}) + \text{error term}$$

**Adams Interpolation Formula**

$$y_{n+1} = y_n + \frac{h}{24}(9f_{n+1} + 19f_n - 5f_{n-1} + f_{n-2}) + \text{error term}$$

where $f_n \equiv f(x_n, y_n)$. The Adams method is relatively fast compared with other routines with the same level of accuracy, because the derivatives at previous points are reused in calculating the new point, which decreases the number of computations needed for each new coordinate. For the initial four points of the solution the coordinates must be evaluated by another method, which also must be done if the step size is changed part way along. These limitations presented no difficulty for this problem. For other problems alternative methods may be easier to apply.

In Computer Program 2 we calculate the field lines and equipotentials, as shown in Figure 1–10, for two point charges $Q_1 = +3Q$ and $Q_2 = -2Q$ separated by a distance $D$. The points calculated by this program are shown in Figure C–2. For this example the step size remains constant but the step variable, defined to be the variable changing the most, and the step direction change. This plot was generated by using the Runge-Kutta method:

$$y_{n+1} = y_n + \tfrac{1}{6}k_1 + \tfrac{1}{3}k_2 + \tfrac{1}{3}k_3 + \tfrac{1}{6}k_4 + \text{error term}$$

$$k_1 = hf(x_n, y_n) \qquad k_2 = hf(x_n + \tfrac{1}{2}h,\ y_n + \tfrac{1}{2}k_1)$$

$$k_3 = hf(x_n + \tfrac{1}{2}h,\ y_n + \tfrac{1}{2}k_2) \qquad k_4 = hf(x_n + h,\ y_n + k_3)$$

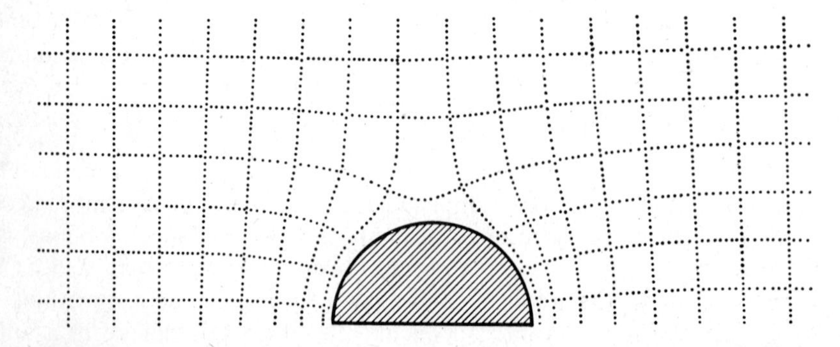

**FIGURE C–1**

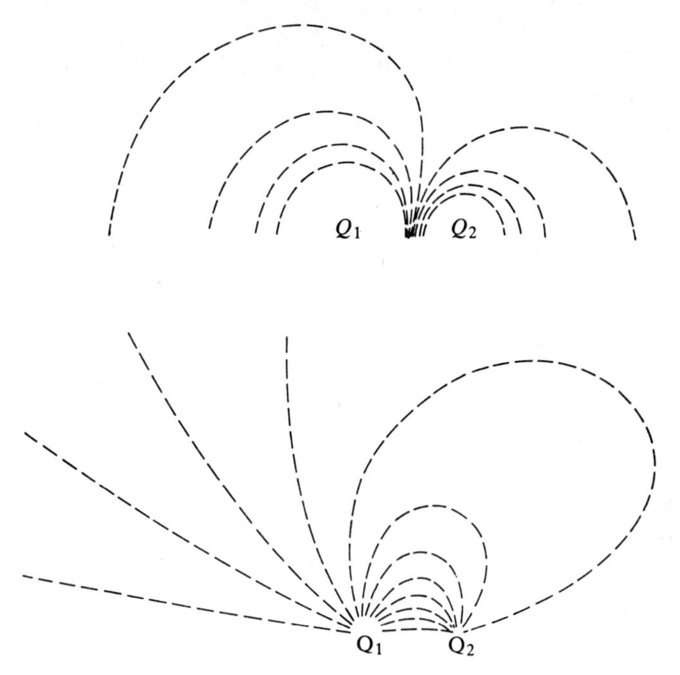


**FIGURE C–2**

Each new coordinate value requires four evaluations of the function $f(x, y)$, thus making this routine considerably slower than the Adams method. However, changes in the step variable, step direction, or step size cause no problems, and this consideration outweighs the longer calculation time.

Computer Program 3 uses the finite-difference routine to solve Laplace equations for the example given in Section 3–8, where the method is discussed extensively.

All of the programs are written in FORTRAN.

Computer Program 1
Adams Method

```
C
C             This program calculates the field lines (FP='FIELD) and
C             equipotentials (FP='POTENTIAL') for an uncharged conducting
C             sphere with radius "RAD" in an initially uniform electric field
C             directed along the X-axis.
C
      COMMON /LINE TYPE/FP  /PARAMS/I,RAD
      CHARACTER*9 FP
      RAD=1.0
C
C             Output the field line coordinates.  Only the coordinates in the
C             first quadrant need be calculated since the problem has four-
C             fold symmetry about the X and Y-axes.
C
```

```
      FP='FIELD'
      DO 100 K=0,5
         J=0
         I=0
C
C              Output the field line number.
C
         WRITE (6,200)K
C
C              Take the initial coordinate far enough away from the sphere
C              that the field lines are approximately parallel and equally
C              spaced.  Continue computing field lines as long as the line is
C              outside the sphere (DIST > RAD) and the X-coordinate is greater
C              than zero.
C
         X0=4*RAD
         Y0=RAD*(1+2*K)/4.
C
C              Output the initial coordinate.
C
         WRITE (6,220)X0,Y0
         DIST=SQRT(X0**2+Y0**2)
         DO WHILE (DIST .GT. RAD .AND. X0 .GT. 0)
            CALL ADAM(X0,Y0,X1,Y1)
            I=1
            X0=X1
            Y0=Y1
            DIST=SQRT(X0*X0+Y0*Y0)
            J=J+1
C
C              Output every twentieth coordinate.
C
            IF (J .GT. 20) THEN

               WRITE (6,220)X0,Y0
               J=0
            END IF
         END DO
C
C              Output the last point.

C
         WRITE (6,220)X0,Y0
 100  CONTINUE
C
C              Output the equipotential coordinates.
C
      FP='POTENTIAL'
      DO 120 K=0,7
         J=0
         I=0
C
C              Output the potential number.
C
         WRITE (6,210)K
C
C              Take the initial coordinate far enough away from the sphere
C              that the equipotentials are approximately equally spaced.
C              Continue computing as long as the Y-coordinate is greater than
C              zero.  Note that Y is now the incremented variable.
C
         X0=RAD*(1+2*K)/4.
         Y0=3*RAD
C
C              Output the initial point.
C
```

```
            WRITE (6,220)X0,Y0
            DO WHILE (Y0 .GT. 0)
               CALL ADAM(Y0,X0,Y1,X1)
               I=1
               X0=X1
               Y0=Y1
               J=J+1
C
C                 Output every twentieth coordinate.
C
               IF  (J .GT. 20) THEN
                  WRITE (6,220)X0,Y0
                  J=0
               END IF
            END DO
C
C                 Output the last point.
C
            WRITE (6,220)X0,Y0
120      CONTINUE
C
200      FORMAT ('0',8X,'FIELD LINE * ',I1/9X,'X',13X,'Y')
210      FORMAT ('0',7X,'EQUIPOTENTIAL * ',I1/,8X,'X',15X,'Y')
220      FORMAT (1X,2G 15.6)
         END
C
C                 Adams Subroutine
C
         SUBROUTINE ADAM(X0,Y0,X1,Y1)
         COMMON  /LINE TYPE/FP  /PARAMS/I,RAD
         REAL DY(-3:1)
         CHARACTER*9  FP
         DATA H/.01/
C
C                 Initialize the parameters for the first point.
C
         IF (I .EQ. 0) THEN
            DY(-3)=0
            DY(-2)=0
            DY(-1)=0
         END IF
         DY(0)=DIFEQ(X0,Y0)
         X1=X0-H
         Y1=Y0-H*(55*DY(0)-59*DY(-1)+37*DY(-2)-9*DY(-3))/24.
         DY(1)=DIFEQ (X1,Y1)
         Y1=Y0-H*(9*DY(1)+19*DY(0)-5*DY(-1)+DY(-2))/24
         DY(-3)=DY(-2)
         DY(-2)=DY(-1)
         DY(-1)=DY(0)
         RETURN
         END
C
C                 This function defines the differential equations which governs
C                 the field lines (FP='FIELD') and the equipotentials
C                 (FP='POTENTIAL').  Note that when the equipotentials are being
C                 calculated, X and Y are interchanged.
C
         FUNCTION DIFEQ (X,Y)
         COMMON/LINE TYPE/FP  /PARAMS/I,RAD
         CHARACTER*9 FP
C
C                 EX and EY are the X and Y components of the electric field.
C
         EY=3.0*(RAD**3)*X*Y
         IF (FP .EQ. 'FIELD') THEN
            EX=(Y*Y+X*X)**2.5 + (RAD**3)*(2.0*X*X-Y*Y)
```

```
         DIFEQ=EY/EX
      ELSE
         EX=(Y*Y+X*X)**2.5 + (RAD**3)*(2.0*Y*Y-X*X)
         DIFEQ=-EY/EX
      END IF
      END
```

Computer Program - 2
Runge-Kutta Method

```
C
C           This program calculates the field lines (FP='FIELD') and
C           equipotentials (FP='POTENTIAL') for two charges of magnitudes
C           Q1 and Q2 located on the X-axis at -D and +D, respectively.  It
C           is assumed that Q1 > 0 and Q2 < 0.
C
      COMMON /LINE TYPE/FP  /PARAMS/I,Q1,Q2,D
      CHARACTER*9 FP
      DATA PI/3.141592654/
C
C           Select charges and the position.
C
      Q1=3.0
      Q2=-2.0
      D=1.5
C
C           Calculate the field line coordinates.  Only the coordinates for
C           Y positive need be calculated since the problem has symmetry
C           about the X-axis.
C
      FP='FIELD'
      DO 100 K=0,11
         J=0
         I=0
C
C           Output the field line number.
C
         WRITE (6,200)K
C
C           Take the initial coordinate near Q1 so that the field lines are
C           approximately equiangularly spaced.  Continue computing field
C           lines as long as the line is on the page and the line does not
C           terminate on Q2.  The page size is defined to be 16*D by 12*D.
C
         X0=.1*COS (5*K*PI/60+7.5*PI/180)-D
         Y0=.1*SIN (5*K*PI/60+7.5*PI/180)
         D1=SQRT((X0-D)**2+Y0**2)
C
C           Output the initial point.
C
         WRITE (6,220)X0,Y0
C
         DO WHILE (D1 .GT. .1 .AND. ABS(X0) .LT. 8*D .AND. ABS(Y0) .LT. 6*D)
            CALL RUNKUT(X0,Y0,X1,Y1)
            I=1
            X0=X1
            Y0=Y1
            D1=SQRT((X0-D)**2+Y0**2)
            J=J+1
C
C           Output every twentieth coordinate.
C
            IF (J .GT. 20) THEN
               WRITE (6,220) X0,Y0
               J=0
```

```
              END IF
           END DO
C
C                 Output the last point.
C
           WRITE (6,220)X0,Y0
C
100     CONTINUE
C
C                 Calculate the equipotential coordinate.  V0 is the potential
C                 difference between the lines.
C
        FP='POTENTIAL'
        V0=0.2
        DO 120 K=-4,4
           J=0
           I=0
C
C                 Output the equipotential number.
C
           WRITE (6,210)K+5
C
C                 Determine the initial X-coordinate by solving for X in terms of
C                 the potential, which is then incremented in steps of V0.
C

           IF (K .EQ. 0)THEN
              X0=D*(Q2+Q1)/(Q1-Q2)
           ELSE IF (K .LT. 0) THEN
              X0=(Q1+Q2+SQRT ((Q1+Q2)**2+4*D*K*V0*(Q2-Q1+K*V0*D)))/(2*K*V0)
           ELSE
              X0=(-(Q1+Q2)+SQRT((Q1+Q2)**2-4*D*K*V0*(Q2-Q1-K*V0*D)))/(2*K*V0)
           END IF
           Y0=.1
C
C                 Output the initial point.
C
           WRITE (6,220)X0,Y0
C
C                 Compute equipotential coordinates as long as Y > 0.
C
           DO WHILE (Y0 .GT. 0)
              CALL RUNKUT(X0,Y0,X1,Y1)
              I=1
              X0=X1
              Y0=Y1
              J=J+1
C
C                 Output every twentieth coordinate.
C

                 IF (J .GT. 20) THEN
                    J=0
                    WRITE (6,220)X0,Y0
                 END IF
           END DO
C
C                 Output the last point.
C
           WRITE (6,220)X0,Y0
120     CONTINUE
C
200     FORMAT('0',8X,'FIELD LINE # ',I2/,9X,'X',13X,'Y')
210     FORMAT('0',7X,'EQUIPOTENTIAL # ',I1/,8X,'X',15X,'Y')
220     FORMAT(1X,2G15.6)
```

```
C
      END
C
C
C             Runge-Kutta Subroutine
C
      SUBROUTINE RUNKUT(X0,Y0,X1,Y1)
      COMMON /LINE TYPE/FP  /PARAMS/I,Q1,Q2,D  /STEP VAR/STEP
      CHARACTER *9 FP
      CHARACTER *1 STEP
      REAL K1,K2,K3,K4
C
C             The beginning step variable and direction are determined by the
C             slope at the initial point.
C
      IF (I .EQ. 0) THEN
         STEP='X'
         IF (ABS(DIFEQ(X0,Y0)) .LT. 1.0) THEN
            STEP='X'
            H=.02*ABS(DIFEQ(X0,Y0))/DIFEQ(X0,Y0)
         ELSE
            STEP='Y'
            H=.02
         END IF
      END IF
C
C             Compute the Runge-Kutta parameters.
C
30    K1=H*DIFEQ (X0, Y0)
      K2=H*DIFEQ (X0+H/2, Y0+K1/2)
      K3=H*DIFEQ (X0+H/2, Y0+K2/2)
      K4=H*DIFEQ (X0+H, Y0+K3)
      DELTA=(K1+2*K2+2*K3+K4)/6
      IF (STEP .EQ. 'X') THEN
         X1=X0+H
         Y1=Y0+DELTA
      ELSE
         X1=X0+DELTA
         Y1=Y0+H
      END IF
C
C             Step in the variable which is changing the most.  This is
C             determined by comparing the magnitudes of the changes in X and
C             Y.  If the step variable changes, the sign of the step is
C             determined by the sign of the slope multiplied by the sign of
C             the previous step direction.
C
      IF (STEP .EQ. 'X') THEN
         IF (ABS(X0-X1) .LT. ABS(Y0-Y1)) THEN
            STEP='Y'
            H=.02*(Y1-Y0)/ABS(Y1-Y0)
         END IF
      ELSE
         IF (ABS(Y0-Y1) .LT. ABS(X0-X1)) THEN
            STEP='X'
            H=.02*(X1-X0)/ABS(X1-X0)
         END IF
      END IF
      RETURN
      END
C
C
C             This function defines the differential equation which governs
C             the field lines (FP='FIELD') and the equipotentials
C             (FP='POTENTIAL').
```

```
C
      FUNCTION DIFEQ(X,Y)
      COMMON /LINE TYPE/FP  /PARAMS/I,Q1,Q2,D  /STEP VAR/STEP
      CHARACTER*9 FP
      CHARACTER*1 STEP
C
C             EX and EY are the X and Y-components of the electric field.
C
      EX=Q1*(X+D)/((X+D)**2+Y*Y)**1.5 + Q2*(X-D)/((X-D)**2+Y*Y)**1.5
      EY=Q1*Y/((X+D)**2+Y*Y)**1.5  +  Q2*Y/(X-D)**2+Y*Y)**1.5
      IF (FP .EQ. 'FIELD') THEN
         IF (STEP .EQ. 'X') DIFEQ=EY/EX
         IF (STEP .EQ. 'Y') DIFEQ=EX/EY
      ELSE
         IF (STEP .EQ. 'X') DIFEQ=-EX/EY
         IF (STEP .EQ. 'Y') DIFEQ=-EY/EX
      END IF
      END
```

Computer program 3
Finite-Difference Method for Solution of Laplace Equation

```
C
C             This program works the example given in Section 3.8.
C
C             Initialize the lattice.
C
      REAL PHI (0:5,0:5)/6*3.,0.,1.5,4*2.,0.,4*1.,2.,0.,4*1.,2.,0.,4*1.,
     1                   2.,6*1./
C
C             AVE(I,J) is the average of the values adjacent to the
C             site (i,j).
C
      AVE(I,J)=.25*(PHI(I-1,J)+PHI(I+1,J)+PHI(I,J-1)+PHI(I,J+1))
C
C             This program does ten iterations.
C
      DO N=1,10
              PHI(1,1)=AVE(1,1)
              PHI(2,1)=AVE(2,1)
              PHI(3,1)=AVE(3,1)
              PHI(4,1)=AVE(4,1)
              PHI(4,2)=AVE(4,2)
              PHI(4,3)=AVE(4,3)
              PHI(4,4)=AVE(4,4)
              PHI(3,4)=AVE(3,4)
              PHI(2,4)=AVE(2,4)
              PHI(1,4)=AVE(1,4)
              PHI(1,3)=AVE(1,3)
              PHI(1,2)=AVE(1,2)
              PHI(2,2)=AVE(2,2)
              PHI(3,2)=AVE(3,2)
              PHI(3,3)=AVE(3,3)
              PHI(2,3)=AVE(2,3)
C
C             Output the lattice after each iteration.
C
              PRINT100, N, ((PHI(I,J),J=0,5),I=5,0,-1)
C
      END DO
      STOP
100   FORMAT (/'0End of iteration #',I3/('0',6F8.4/))
      END
```

# Index

# VECTOR OPERATORS

Cartesian Coordinates:

$$\nabla\Phi = \hat{\mathbf{x}}\frac{\partial \Phi}{\partial x} + \hat{\mathbf{y}}\frac{\partial \Phi}{\partial y} + \hat{\mathbf{z}}\frac{\partial \Phi}{\partial z}$$

$$\nabla\cdot\mathbf{E} = \frac{\partial E_x}{\partial x} + \frac{\partial E_y}{\partial y} + \frac{\partial E_z}{\partial z}$$

$$\nabla\times\mathbf{E} = \hat{\mathbf{x}}\left(\frac{\partial E_z}{\partial y} - \frac{\partial E_y}{\partial z}\right) + \hat{\mathbf{y}}\left(\frac{\partial E_x}{\partial z} - \frac{\partial E_z}{\partial x}\right) + \hat{\mathbf{z}}\left(\frac{\partial E_y}{\partial x} - \frac{\partial E_x}{\partial y}\right)$$

$$\nabla^2\Phi = \frac{\partial^2 \Phi}{\partial x^2} + \frac{\partial^2 \Phi}{\partial y^2} + \frac{\partial^2 \Phi}{\partial z^2}$$

Cylindrical Coordinates:

$$\nabla\Phi = \hat{\boldsymbol{\varrho}}\frac{\partial \Phi}{\partial \varrho} + \hat{\boldsymbol{\phi}}\frac{1}{\varrho}\frac{\partial \Phi}{\partial \phi} + \hat{\mathbf{z}}\frac{\partial \Phi}{\partial z}$$

$$\nabla\cdot\mathbf{E} = \frac{1}{\varrho}\frac{\partial}{\partial \varrho}(\varrho E_\varrho) + \frac{1}{\varrho}\frac{\partial E_\phi}{\partial \phi} + \frac{\partial E_z}{\partial z}$$

$$\nabla\times\mathbf{E} = \hat{\boldsymbol{\varrho}}\left(\frac{1}{\varrho}\frac{\partial E_z}{\partial \phi} - \frac{\partial E_\phi}{\partial z}\right) + \hat{\boldsymbol{\phi}}\left(\frac{\partial E_\varrho}{\partial z} - \frac{\partial E_z}{\partial \varrho}\right) + \hat{\mathbf{z}}\frac{1}{\varrho}\left[\frac{\partial}{\partial \varrho}(\varrho E_\phi) - \frac{\partial E_\varrho}{\partial \phi}\right]$$

$$\nabla^2\Phi = \frac{1}{\varrho}\frac{\partial}{\partial \varrho}\left(\varrho\frac{\partial \Phi}{\partial \varrho}\right) + \frac{1}{\varrho^2}\frac{\partial^2 \Phi}{\partial \phi^2} + \frac{\partial^2 \Phi}{\partial \phi^2}$$

Spherical Coordinates:

$$\nabla\Phi = \hat{\mathbf{r}}\frac{\partial \Phi}{\partial r} + \hat{\boldsymbol{\theta}}\frac{1}{r}\frac{\partial \Phi}{\partial \theta} + \hat{\boldsymbol{\phi}}\frac{1}{r\sin\theta}\frac{\partial \Phi}{\partial \phi}$$

$$\nabla\cdot\mathbf{E} = \frac{1}{r^2}\frac{\partial}{\partial r}(r^2 E_r) + \frac{1}{r\sin\theta}\frac{\partial}{\partial \theta}(\sin\theta E_\theta) + \frac{1}{r\sin\theta}\frac{\partial E_\phi}{\partial \phi}$$

$$\nabla\times\mathbf{E} = \hat{\mathbf{r}}\frac{1}{r\sin\theta}\left[\frac{\partial}{\partial \theta}(\sin\theta E_\phi) - \frac{\partial E_\theta}{\partial \phi}\right] + \hat{\boldsymbol{\theta}}\frac{1}{r}\left[\frac{1}{\sin\theta}\frac{\partial F_r}{\partial \phi} - \frac{\partial (rF_\phi)}{\partial r}\right]$$
$$+ \hat{\boldsymbol{\phi}}\frac{1}{r}\left[\frac{\partial (rE_\theta)}{\partial r} - \frac{\partial E_r}{\partial \theta}\right]$$

$$\nabla^2\Phi = \frac{1}{r^2}\frac{\partial}{\partial r}\left(r^2\frac{\partial \Phi}{\partial r}\right) + \frac{1}{r^2\sin\theta}\frac{\partial}{\partial \theta}\left(\sin\theta\frac{\partial \Phi}{\partial \theta}\right) + \frac{1}{r^2\sin^2\theta}\frac{\partial^2 \Phi}{\partial \phi^2}$$